INTRODUCTION TO
SYSTEM DYNAMICS MODELING
with DYNAMO

Jay W. Forrester, editor

Industrial Dynamics, Jay W. Forrester, 1961

Growth of a New Product: Effects of Capacity-Acquisition Policies, Ole C. Nord, 1963

Resource Acquisition in Corporate Growth, David W. Packer, 1964

Principles of Systems,* Jay W. Forrester, 1968

Urban Dynamics, Jay W. Forrester, 1968

Dynamics of Commodity Production Cycles,* Dennis L. Meadows, 1970

World Dynamics,* Jay W. Forrester, 1971

The Life Cycle of Economic Development,* Nathan B. Forrester, 1973

Toward Global Equilibrium: Collected Papers,* Dennis L. Meadows and Donella H. Meadows, eds.

Study Notes in System Dynamics,* Michael R. Goodman, 1974

Readings in Urban Dynamics (vol. 1),* Nathaniel J. Mass, 1974

Dynamics of Growth in a Finite World,* Dennis L. Meadows, William W. Behrens III, Donella H. Meadows, Roger F. Naill, Jorgen Randers, and Erich K. O. Zahn, 1974

Collected Papers of Jay W. Forrester,* Jay W. Forrester, 1975

Economic Cycles: An Analysis of Underlying Causes,* Nathaniel J. Mass, 1975

Readings in Urban Dynamics (vol. 2),* Walter W. Schroeder III, Robert E. Sweeney, and Louis Edward Alfeld, eds. 1975

Introduction to Urban Dynamics,* Louis Alfeld and Alan K. Graham, 1976

DYNAMO User's Manual (5th ed.), Alexander L. Pugh III, 1976

Managerial Applications of System Dynamics, Edward B. Roberts, ed., 1978

Elements of the System Dynamics Method, Jorgen Randers, ed., 1980

Ecosystem Succession: A General Hypothesis and a Test Model of a Grassland, Luis T. Gutierrez and William R. Fey, 1980

Corporate Planning and Policy Design: A System Dynamics Approach, James M. Lyneis, 1980

Introduction to System Dynamics Modeling with DYNAMO, George P. Richardson and Alexander L. Pugh III, 1981

* Originally published by Wright-Allen Press and now distributed by the MIT Press

INTRODUCTION TO
SYSTEM DYNAMICS MODELING
with DYNAMO

George P. Richardson
Alexander L. Pugh III

The MIT Press
Cambridge, Massachusetts, and
London, England

© 1981 by The Massachusetts Institute of Technology

All rights Reserved. No part of this book may be reproduced in any form or by any means, electronic or mechanical, including photocopying, recording, or by any information storage and retrieval system, without permission in writing from the publisher.

Library of Congress Cataloging in Publication Data

Richardson, George P.
    Introduction to system dynamics modeling with DYNAMO.

    (MIT Press/Wright-Allen series in system dynamics)
    Includes index.
    1. System analysis - Data processing. 2. DYNAMO (Computer program language) 3. Digital computer simulation. I. Pugh, Alexander L. II. Title. III. Series.
QA402.R48   003   81-12371
ISBN 0-262-18102-9   AACR2

Publisher's Note

This format is intended to reduce the cost of publishing certain works in book form and to shorten the gap between editorial preparation and final publication. Detailed editing and composition have been avoided by photographing the text of this book directly from the authors' camera-ready copy.

To our parents
    Nathalie and Paul
    Anne and Al

CONTENTS

1 The System Dynamics Approach   1
   1.1 Problems and Models 1
   1.2 Feedback Systems 3
   1.3 Overview of the Systems Dynamics Approach 14

2 Problem Identification and System Conceptualization   18
   2.1 Defining Problems Dynamically 19
   2.2 Representing Feedback Structure 25
   2.3 Model Purpose 38
   2.4 An Example of Problem Conceptualization 45
   2.5 Summary of the Conceptual Phases of Model
        Development 61

3 Introduction to DYNAMO   67
   3.1 An Overview of DYNAMO 69
   3.2 A Computed Example 69
   3.3 Types of Equations 75
   3.4 Obtaining Output 88
   3.5 Delays, Smooths, and Averages 103
   3.6 Functions 115

4 Model Formulation   133
   4.1 Formulating Rate Equations 134
   4.2 Auxiliary Equations 159
   4.3 What is a Level? 176
   4.4 Formulating a Project Model 190
   4.5 Documentation 213
   4.6 Parameters and Initial Values 230
   4.7 Debugging 247
   4.8 Principles of Model Formulation 261

5 Model Testing and Further Development   267
    5.1 Understanding Model Behavior 268
    5.2 Sensitivity 277
    5.3 Refinement and Reformulation 293
    5.4 The Question of Validity 310

6 Policy Analysis   321
    6.1 Parameter Changes as Policy Alternatives 322
    6.2 Structural Changes as Policy Alternatives 332
    6.3 Policy Recommendations and the Question of
         Validity 349

7 Advanced Topics in DYNAMO Use   360
    7.1 User Defined Macros 360
    7.2 Examples of Macros 368
    7.3 Arrays in DYNAMO III 374
    7.4 Functions on Arrays 389
    7.5 Printing and Plotting Arrays 393
    7.6 Debugging DYNAMO III Models 396

Scaling Letters for DYNAMO Output   399

Acknowledgements for Reprinted Material   400

References   402

Index   407

PREFACE

Since the publication of *Industrial Dynamics* in 1961, the field of system dynamics has grown to encompass scholarly and applied activity in more than thirty countries around the world. Computer simulation models developed using the tools and perspectives of the field range from the intuitive to the highly technical, from ten equations to two thousand equations and more, and from the micro concerns of cellular biology to macro issues of national and global economics. Practitioners include corporate researchers, business consultants, government staffers, and academics, and their work ranges from philosophy and methodology to applied, problem-solving, policy-oriented investigations.

This book describes the system dynamics approach, from the initial stages of problem identification to the final recommendations of policy analyses. The primary purpose of the book is to provide an introduction to the field for corporate managers, government policy makers, university researchers, their staffs, and students who have an interest in using system dynamics in their work. To the extent we were able to make it so, it is a how-to-do-it book. Some people, however, may read it not to become modelers but rather to gain a perspective on the relationships between system dynamics and other quantitative modeling methodologies. Thus a secondary purpose of the book is to contribute to understandings of the potential roles of quantitative modeling in the social and policy sciences.

We assume that different readers will procede to different levels of detail. A nonquantitative introduction to the basic concepts of the system dynamics approach can be obtained from chapters 1 and 2. Adding the first three sections of chapter 3, section 4.4, and a judicious selection from chapters 5 and 6 produces an overview of how a system dynamics model is conceived, formulated, tested, and used for policy analysis. Readers who wish to try to build models of their own should find the remainder of chapters 3, 4, 5, and 6 essential. Chapter 7 contains material particularly useful after the earlier chapters have been thoroughly exercised.

Although it might be useful to read the book with pencil in hand, occasionally doing an exercise or verifying an equation, the book is designed to read more like a story than a standard text or reference manual. Nonetheless, we are convinced that no one can learn to approach problems from the system dynamics modeling perspective by merely reading about it. Once in a while the narative stops and suggests a fruitful task for the reader: sketch a graph over time to help define a certain problem, develop a causal-loop diagram or write equations for a particular structure, perform some computer experiments with a given model, or refine a model in some suggested way. Although such exercises are not required for understanding later material, those who desire more than an overview will profit from trying them. More exercise material can be found in the references at the end of the book, notably in Forrester (1968b), Goodman (1974), Alfeld and Graham (1976), and Lyneis (1980).

The computer simulation language DYNAMO has been associated with system dynamics from the beginning. Its immediate ancestor was a 1958 computer program called SIMPLE -- Simulation of Industrial Management Problems with Lots of Equations. The current implementations include DYNAMO II and II/F, DYNAMO III (with the capability to handle subscripted variables), and Mini-DYNAMO (a slightly scaled-down version of DYNAMO II which runs on mini-computers with as little as 20K of core memory). Currently in the testing stage are Micro-DYNAMO (a version capable of running on tiny micro-computers with disk storage) and DYNAMO IV, the most powerful implementation yet, including the capabilities of DYNAMO III plus higher-order, variable step-size integration methods. All of these are developed, distributed, and maintained by Pugh-Roberts Associates, 5 Lee St., Cambrige, Massachusetts, 02139.

The introduction to DYNAMO in chapter 3 applies to the core of all of these versions of the language. With a few modifications it applies as well to NDTRAN and DYSMAP, simulation languages fashioned after DYNAMO.[1] The introduction to simulation contained here requires no previous experience

---

[1] NDTRAN was developed under the direction of William Davisson and John Uhran at Notre Dame. DYSMAP was produced by the System Dynamic Research Group at the University of Bradford, U.K., under the direction of R. Geoffrey Coyle.

with computer languages. The experienced programmer, however, will readily see how system dynamics models can be formulated in a range of computer languages. Although DYNAMO and system dynamics have become almost synonymous to some, and it is indeed a good match, it should be noted that the system dynamics approach is language-free.

This book brings together insights and contributions of many people. Foremost, of course, are those of Jay Forrester. It is remarkable how often one finds that some guideline for model conceptualization or formulation traces its origins to a remark or an appendix in *Industrial Dynamics*. Without a direct hand in this book, his influence on it is enormous. In addition, we have benefitted from a long history of communications with people asociated at one time or another with the System Dynamics Group at the Massachusetts Institute of Technology and Pugh-Roberts Associates. We wish to acknowlege in particular the contributions Ed Roberts and David Andersen.

We also owe a great debt to the many people who supplied energy and talent to the mechanics of producing the book. Bill Shaffer had a hand in initiating the project, and he commented on early drafts. Cherie Wallett helped us through several deadlines, and performed with grace the onerous task of formatting the book for an IBM 6670 laser printer. Final proof-reading and polishing were handled very competently by Pilar Carrasco. Aided in the final days by Nicole Harris, Diane Leonard-Senge produced all of the artwork with great skill and care. Finally, a special note of thanks goes to Karl Clauset, who read every page and most of the revisions, made detailed comments and criticisms, and provided encouragement throughout.

At its core the book tries to tell how to practice an art. There are undoubtedly other ways to practice this art than the ones on which we focus. But rather than leave a budding artist alone with his model, wondering about which brush or color to use first, we have chosen to be direct about guidelines that we think are most helpful. As readers pursue the subject, they will develop their own styles. We hope that the introduction this book provides speeds that development.

George P. Richardson
Alexander L. Pugh III

Cambridge, Massachusetts
June, 1981.

INTRODUCTION TO
SYSTEM DYNAMICS MODELING
with DYNAMO

1 THE SYSTEM DYNAMICS APPROACH

1.1 PROBLEMS AND MODELS

System dynamics is a methodology for understanding certain kinds of complex problems. It began some thirty years ago as industrial dynamics, focusing on problems arising in the corporate setting. The early work concerned itself with management problems such as instabilities in production and employment, slack or inconsistent corporate growth, and declining market share. Within a few years of its beginnings, the methods of industrial dynamics were being applied to a far wider range of problems, from managing a research and development project to combating urban stagnation and decay, understanding the implications of exponential growth in a world of finite and declining natural resources, and even testing theories relating to diabetes.

The name "industrial dynamics" soon gave way to the more general term "system dynamics." We shall not define the word "system," noting merely that its use here should connote some generality of purview, some complexity in the problems one might address, and a wholeness of perspective--a systems approach--which one attempts to achieve for a given problem. The focus of a system dynamics study is not a system, whatever that is, but a problem--a warning we shall have occasion to repeat.

Dynamic Problems

The problems that one addresses from the perspective of system dynamics have at least two features in common. First, they are dynamic: they involve quantities which change over time. They can be expressed in terms of graphs of variables over time. Oscillating levels of employment in an industry, a decline in a city's tax base and quality of life, and the dramatically rising pattern of health care costs are dynamic problems. So, too, are cost overruns of a building project, growth of government (a problem to some people), cancer, patterns of use and abuse of a state or national park, and psychological depression. Skill in defining problems dynamically is a first step toward learning the system dynamics approach.

A second feature of problems to which the system dynamics perspective applies involves the notion of feedback. The concept will be defined and discussed in 1.2. We merely note here the broad generality of the idea: it has appeared to engineers in servo-mechanisms and closed-loop control systems, to physiologists as homeostasis, and to social scientists as the notions of the vicious circle and the self-fulfilling prophecy. To those who design high-fidelity amplifiers or suffer the screeches of public address systems, it is known simply as feedback.

The system dynamics approach applies to dynamic problems arising in feedback systems. We assert that organizations, economies, societies -- in fact, all human systems -- are feedback systems. Viewing them as such provides great leverage for understanding societal problems.

Formal Models

Perhaps the most visible feature of the system dynamics approach is its use of formal, quantitative computer models. As used here, the term "model" stands for a representation, essentially a simplification, of some slice of reality. A system dynamics model is a laboratory tool. It allows repeated experimentation with the system, testing assumptions or altering management policies. The purpose is to gain understanding, so that the problem to which the model is addressed may be solved or minimized.

A formal model has two advantages over the informal, so-called mental models on which most human decisions are based. First, formal models are more explicit and communica-

---

[1] At a much deeper level, however, the general assumptions of a modeling methodology are seldom explicit. System dynamics embodies a number of philosophical assumptions about the structure of human systems, the nature of research, what is knowable, and so on, which this book may or may not communicate. Other methodologies have their own priors, usually unstated. Let the buyer beware. See Meadows (1980).

ble. A system dynamics model exposes its assumptions about a problem for criticism, experimentation, and reformulation.[1] A mental model, on the other hand, is fuzzy and implicit. Its fuzziness is a result of its rich intuitive detail and is a source of its range and adaptability. The implicit nature of mental models, however, is the cause of occasional misunderstanding, miscommunication, and misapplication.

Second, a formal model handles complexity more easily. Unlike a mental model, a system dynamics computer model can reliably trace through time the implications of any messy maze of assumptions and interactions, without stumbling over phraseology, emotional bias, or gaps in intuition. Computer models have these two advantages not because computers are so smart but, in a sense, because they are so dumb: they love the boring, repetitive computations involved in tracing a model through time, and they require absolutely every model assumption to be spelled out in computer code, explicitly.

It is attractive, indeed, to think that experimentation with appropriate computer models might lead to the understandings we require to solve or minimize the host of complex problems we face. But the history of the application of formal models to policy problems does not produce great confidence.[2] Computer models, it seems, are like children: most people tend to like their own better than others', some look a lot better than they behave, and, we all must admit, when they are bad, they are horrid. It is the perspective of this book that computer modeling, simulation, and policy analysis promise to realize their greatest potential when they are combined with understandings and applications of the concept of feedback.

1.2 FEEDBACK SYSTEMS

Most succinctly, *feedback is the transmission and return of information.*[3] The emphasis, inherent in the word feedback itself, is on the return. A heating system produces heat to warm a room. A thermostat in the room, connected to the heating system, returns information about the room's temperature back to the heating system, turning it on or off and thereby controlling the room's temperature. A thermostat is a feedback device. Together with the furnace, pumps, and radiators or vents, it forms a feedback system.

An inventory control system is also a feedback system. Shipments deplete inventory, so, as it drops below some desired level, someone in the inventory department places

---

[2] See Greenberger, et al. (1976) for an overview.
[3] Wiener (1961), p. 96

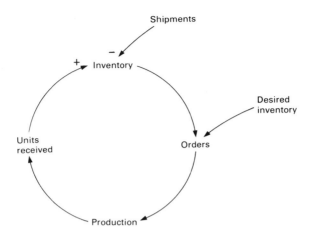

Figure 1.1: A simple inventory feedback loop

orders with producers to bring inventory back up. Information (the size of current inventory) is transmitted (to the ordering department and then to producers) and eventually returns (in the form of widgets received into inventory from the producer). Figure 1.1 captures the essence of this simple inventory-ordering system.

When diagrammed, as in figure 1.1, feedback systems characteristically form loops of interconnections--loops of causes and effects. Without resorting to the broad interpretation of the word information implicit in our above definition of feedback, we may define informally: *A feedback loop is a closed sequence of causes and effects, a closed path of action and information.* An interconnected set of feedback loops is a *feedback system.* There is a growing awareness that biological, environmental, industrial, economic, and societal systems are feedback systems. Understanding the dynamic behavior of such systems requires acknowledging the role of feedback.

Open-loop Thinking

Approaching a problem without applying the concept of feedback is sometimes referred to as open-loop thinking.[4]

---

[4] The term open-loop is really a misnomer--a loop that is open is not really a loop--but it connotes the right notion: something is missing.

Feedback Systems 5

Figure 1.2: A comparison of open-loop and feedback approaches to problem solving

Traditionally, when we discover a problem, we cogitate about it, devise a plan to deal with it, and then act on our plan, thinking we are done with the problem-solving process. The pattern is much like the problem-plan-action sequence shown in the solid arrows in figure 1.2. Usually forgotten is the fact that our action then alters the state of the system, as suggested by the dotted arrow in figure 1.2, resulting in a new understanding of the problem, perhaps a new problem definition, or a new set of problems that must be addressed.

The feedback view tries to close such open loops. Consider, as an example, the problem of managing a public recreational area such as a state park, a lake area, or a mountain people enjoy climbing. As more and more people discover the delights of camping and hiking or just sightseeing in such areas, park management is faced with a dilemma: how to protect and preserve the natural character and beauty of an area while at the same time making it available to the public to enjoy. Some may view the situation as in figure 1.3.

The figure shows that crowding is a function of visitors and area: the more visitors per season in a given contact area the more crowded the area. Crowding diminishes the quality of the wilderness experience while at the same time increases environmental damage, such as litter, cutting trees, soil compaction, and erosion. Park management can provide various services that attempt to offset these negative effects of visitors: hauling away litter, maintaining trails, managing crowds, and so on.

The view in figure 1.3, lacking the loops characteristic of the feedback perspective, suggests that a reasonable management policy to minimize environmental damage and preserve the quality of the experience for visitors is to try to increase the contact area and the services provided by the park. Encourage the use of less-traveled trails, build more trails, build more camping areas with proper toilet and litter facilities, provide educational facilities such as park centers staffed by knowledgable rangers.

While some such policies may be necessary and desirable, the open-loop perspective that spawned them here is inadequate. Feedback effects have been ignored. For example,

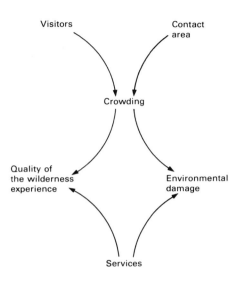

Figure 1.3: Open-loop view of problems in a popular recreational area

information (feedback) about the value of the wilderness experience will have an obvious effect on the number of visitors, as shown in figure 1.4.

The additional arrow connecting value and visitors in figure 1.4 is intended to represent an assumption about the attractivness of the area: the greater value visitors receive from visiting this public recreational area, the more visitors will come each year. Other such links operate in this system, but this one addition serves to show the inadequacy of the open-loop view. The conclusion to increase services is now not so clear. An increase in services increases the value of the experience, which in creases the number of visitors per year, increasing crowding, which increases damage to the environment and *decreases* the value of the experience. The long-term implications of the policy of increasing services are not as apparent as they were in figure 1.3. They are even less so in the more complete feedback view shown in figure 1.5.

The Behavior of Feedback Systems

Understanding the behavior of feedback systems is a goal of the system dynamics approach. The behavior of systems

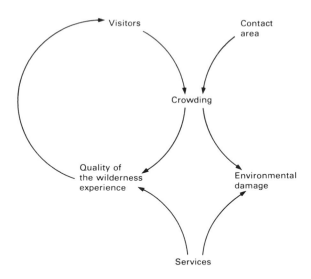

Figure 1.4: A feedback loop in a public recreational area system

of interconnected feedback loops often confounds common intuition and analysis, even though the dynamic implications of isolated loops may be reasonably obvious. The feedback structures of real problems are often so complex that the behavior they generate over time can usually be traced only by simulation.

Insight into the behavior of feedback systems begins with the observation that feedback processes divide naturally into two categories, which are labeled *positive* and *negative*. Recall the inventory control loop in figure 1.1. The loop is clearly intended to keep inventory at its desired level; if shipments cause it to wander, orders are placed for increased production to bring inventory back to its desired level. If shipments slack off because of a lack of demand, the firm will lower production in an effort to reduce inventory to its desired level. Such a loop is a *goal-seeking* loop. Desired inventory is the goal for actual inventory, and the function of the loop is to try continually to keep the actual level close to the desired.

A thermostat-heating system is a goal-seeking system, as well, as are the simple feedback systems shown in figure 1.6. These loops are called *negative feedback loops* because each attempts to negate any deviation from some equilibrium or

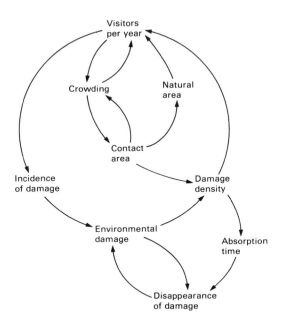

Figure 1.5: A public recreational area viewed as a feedback system

goal state. The negating property can be demonstrated by tracing a change around a loop. In the pendulum loop, for example, if the pendulum is displaced from its vertical equilibrium position, gravity introduces a restoring force that tries to push the pendulum back to vertical. The push of the restoring force increases the speed of the pendulum, and, in swinging back, the pendulum overshoots its equilibrium position. Now displaced in the opposite direction, it feels a restoring force pulling it back toward vertical. Eventually, it slows down and reverses direction, heading back toward its goal, the vertical equilibrium position. The swings are caused by the tendency of the restoring force to oppose any perturbation from equilibrium.

The other loops in figure 1.6 can be described in somewhat similar terms, although they do not necessarily oscillate like a pendulum. Each acts in some way to oppose or negate a disturbance from equilibrium, as the reader might wish to check. In various contexts such loops are described as stabilizing, equilibrating, or self-correcting. Precise dynamic implications are not possible, even for single loops: we know

Feedback Systems

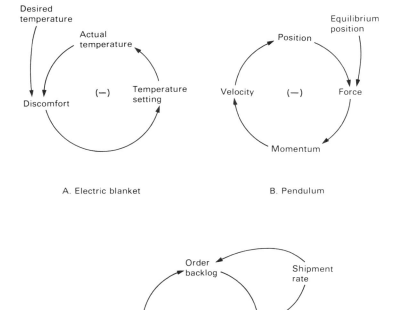

Figure 1.6: Examples of negative feedback loops: electric blanket, mass on a spring, delivery delay

a pendulum oscillates, while we are often able to adjust the temperature of an electric blanket smoothly without hunting, without oscillations zeroing in on the desired setting. Both are negative feedback systems, both are goal-seeking, but these terms only give a general idea of behavior.

In contrast, *positive feedback loops* amplify deviations or disturbances around the loop. Positive loops are variously characterized as destabilizing, disequilibrating, growth pro-

10   The System Dynamics Approach

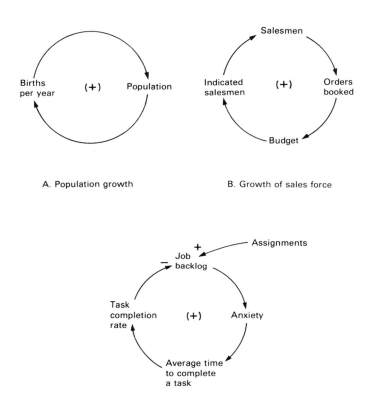

Figure 1.7: Examples of positive feedback loops: population growth, salesman loop, job backlog and anxiety

ducing, or self-reinforcing. Figure 1.7 shows several elementary examples of positive feedback loops. Their common characteristic is a tendency to reinforce, rather than negate, a change in an element in the loop.

The job backlog-anxiety loop illustrates it well. The loop purports to describe a situation of overwork. Heavy assignments increase the backlog of things to do, while the task completion rate dispenses with them. But in a situation of very high backlog, a massive number of tasks to complete causes anxiety to rise, making it more difficult to concentrate and complete any given task. The time it takes to complete a

Feedback Systems

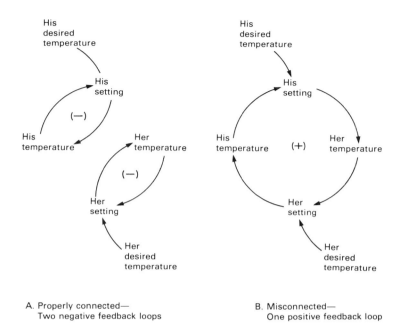

A. Properly connected—
Two negative feedback loops

B. Misconnected—
One positive feedback loop

Figure 1.8: The old double-bed electric blanket snafu. Properly connected: two negative feedback loops; mis-connected: one positive feedback loop

task rises, so the rate of completion of tasks slows down. Thus, the job backlog is depleted less rapidly. If assignments exceed completions, the job backlog rises still further, and so will anxiety and the inability to cope with tasks. Things get worse and worse until some external agent not represented in this loop acts to break the vicious cycle.

The other loops in figure 1.7 have the same reinforcing, destabilizing character. A whole class of additional examples appears with the observation that any vicious cycle is a positive feedback loop.

The distinction between the stabilizing and destabilizing character of negative and positive feedback loops is neatly captured in the story of the misconnected electric blanket. The newlyweds were given an elecric blanket for their queen-size double bed. The blanket had separate temperature settings for the two sides of the bed, one for him and one for her. Properly connected, there should have been two separate negative feedback systems, each attempting to control the temperature of the blanket for the comfort of each individual, as in figure 1.8A. The story goes that the newlyweds misconnected the blanket so that his setting controlled her blanket temperature and hers controlled his. The result is the nasty positive feedback shown in figure 1.8B. She felt cold, turned up her setting, making his side too warm for him so he turned down his setting, making her even colder, so she raised her setting even further, and so on. How such a scenario would end is left up to the fertile imagination of the reader.

The point of the story here is the vivid contrast it provides between the stabilizing or controlling character of negative feedback loops and the destabilizing character of positive feedback loops. It should not be interpreted, however, to suggest that positive feedback loops are simply ill-formed or ill-connected systems. Self-reinforcing processes, such as the positive feedback loop involving births and population, appear naturally in almost all complex, dynamic systems.

Problems in Feedback Systems

If, as we assume, human systems are feedback systems, then solving complex problems in such systems is likely to require understandings of the relationships between feedback structure and the problematic behavior observed. It has been suggested that humans typically respond to complex problems as if they arise in oversimplified negative feedback systems. Some examples follow.[5]

Problem	Response
insect damage to crops	spray with pesticide
traffic congestion	build more highways
crime	hire policemen
urban slums	build low-cost housing
automobile pollution	install catalytic converters
increasing energy costs	fix energy prices
tooth decay	fill cavities

As an exercise, the reader may wish to draw or describe in words the negative feedback loop each response assumes. A simple negative feedback loop centering on traffic congestion and highway building could include such variables as the number of cars or amount of driving, extent of the high-

---

[5] The stimulus for the table is Hardin (1972), p. 54, with additional examples contributed by Donella Meadows.

Feedback Systems

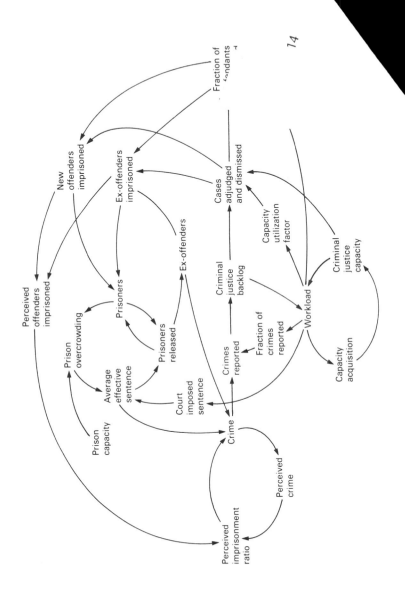

Figure 1.9: Feedback structure of a basic criminal justice system. (Source: Shaffer (1976), p. 56)

The System Dynamics Approach

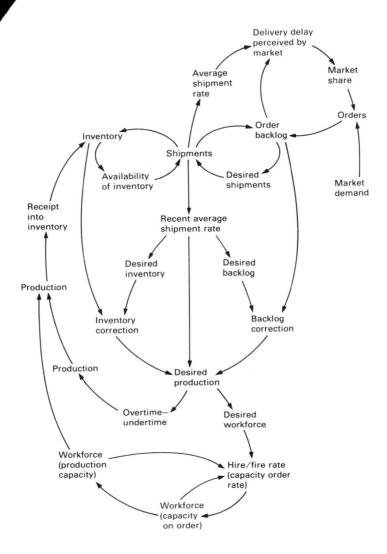

Figure 1.10: Feedback structure of a basic production sector

way system, traffic congestion, driving safety and comfort, and highway construction. As the number of cars or the amount of driving increases, traffic congestion increases, reducing driving comfort and safety. That sets up pressures that lead to more highway construction to lessen traffic congestion. The loop is negative because an increase in traffic congestion eventually results in pressures to decrease (negate) it.

After working out several of these simple negative loops, the reader should try is to discover and represent other feedback loops in these problems that conspire to defeat these simple responses. It is this second part of the exercise that reveals the purpose of looking at problems from the feedback perspective.

Real problems often become so complex that understanding behavior and predicting responses to policies are impossible without a formal model. Figure 1.9 shows the feedback structure of a formal model of a basic criminal justice system. It is far from easy to anticipate how such a model would behave when, for example, criminal justice capacity is increased, and yet such understanding is what we expect of those charged with designing and implementing public policy for real systems.

Figure 1.10 shows a similarly complex set of feedback loops in an industrial setting, focusing on problems of inventory, production, employment, and capacity acquisition. The system dynamics approach attempts to provide the understanding of complex feedback systems that we need in order to design workable policies for improvement of system behavior.

1.3 OVERVIEW OF THE SYSTEMS DYNAMICS APPROACH

The system dynamics approach to complex problems focuses on feedback processes. It takes the philosophical position that feedback structures are responsible for the changes we experience over time. The premise is that *dynamic behavior is a consequence of system structure.* Somewhat later in this book that premise will become meaningful and powerful for the reader. At this point, it may be treated as a postulate, or perhaps as a conjecture yet to be demonstrated.

As both a cause and a consequence of the feedback perspective, the system dynamics approach tends to look *within* a system for the sources of its problem behavior. Problems are not seen as being caused by external agents outside the system. Inventories are not assumed to oscillate merely because consumers periodically vary their orders. A ball does not bounce merely because someone drops it. A pendulum does not oscillate merely because it was displaced from the vertical. The system dynamicist prefers to take the point of view that these systems behave as they do for reasons *internal* to each system. A ball bounces and a pendulum oscillates because there is something about their internal structure that gives them the tendency to bounce or oscillate.

In practice, this internal point of view results in models of feedback systems that bring external agents inside the system. Customer orders become endogenous to a production system, part of the feedback structure of the system. Orders affect production; production affects orders. Part and parcel with the notion of feedback, the endogenous point of view helps to characterize the system dynamics approach.

There are roughly seven stages in approaching a problem from the system dynamics perspective:

1. problem identification and definition,

2. system conceptualization,

3. model formulation,

4. analysis of model behavior,

5. model evaluation,

6. policy analysis,

7. model use or implementation.

The process begins and ends with understandings of a system and its problems, so it forms a loop, not a linear progression. Figure 1.11 shows these stages and the likely progression through them, together with some arrows that represent the cycling, iterative nature of the process. At a number of stages along the way one's understanding of the system and the problem are enhanced by the modeling process, and that increased understanding further aids the modelng effort.

Figure 1.11 shows that final policy recommendations from a system dynamics study come not merely from manipulations with the formal model but also from the additional understandings one gains about the real system by iterations at a number of stages in the modeling process. Done properly, a system dynamics study should produce policy recommendations that can be presented, explained, and defended without resorting to the formal model. The model is a means to an end, and that end is understanding.

Overview of the Systems Dynamics Approach

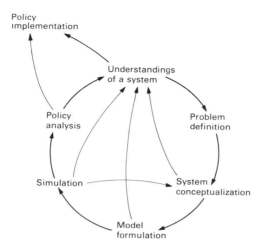

Figure 1.11: Overview of the system dynamics modeling approach

The chapters of this book follow an organization similar to the stages described by the dark path in figure 1.11. The necessary linear organization of the chapters will unfortunately tend to obscure the iterative, cycling nature that characterizes the system dynamics modeling approach.

2 PROBLEM IDENTIFICATION AND SYSTEM CONCEPTUALIZATION

Identifying and defining a problem and seeing it as the product of a system of interacting influences are related undertakings. The problem identifier will have some idea of the root causes of his problem. They will lead to a view of aspects of reality--a system--that conspire to create the problem. The system hypothesized might then suggest new understandings of the problem, a refined definition, or even additional problems that constitute the problem as originally understood.

The conceptual phases of the system dynamics approach are among the most difficult, for both beginners and experienced modelers. System conceptualization is an art. This chapter tries to communicate something of the arts involved, first, by introducing some perspectives that are helpful and second, by providing some diagramming tools. The causal-loop and flow diagrams introduced here help not only to communicate a view of a problem to others, but also to provide a framework for thinking about a problem. They are the first steps in constructing and understanding a feedback view of a problem.

Two very broad guidelines are helpful in reducing the complexity of the problem definition and conceptualization phases: (1) have a clear purpose for the modeling effort, and (2) focus on a problem, not on a system. Too general to be of much help, you might say, and yet these two prerequisites are essential to a successful modeling effort. A problem focus and a clear modeling purpose act as critically important filters, screening out unnecessary details and centering the modeler's attention on the significant aspects of a feedback system. They save a person who is applying a systems approach from having to think about *everything* in order to think about *something*.

For more detailed guidance in conceptualization, we remember that system dynamics applies to problems that are dynamic (they involve change over time) and arise in feedback systems. These two characteristics provide the most specific help in the problem definition and model conceptualization phases. One defines problems dynamically, that is, in terms of graphs of variables over time, and for problems so defined, one searches for feedback structure. In the analogy of filters, these two characteristics are the lenses of the system dynamicist. They help one see certain things clearly, and, to be fair, they undoubtedly blur other things.

2.1 DEFINING PROBLEMS DYNAMICALLY

Any problem viewed from the system dynamics perspective is likely to be first seen in terms of a graph of one or more variables over time. The process one uses to develop such graphs could be termed dynamic thinking. It does not require, as some might expect, that the modeler have access to explicit numerical data or well-defined functions to graph. While data is very helpful, one is often faced with a dynamic problem in which a key variable is not traditionally quantified or tabulated. It is even more likely, however, that the modeler or the client knows the dynamic behavior of interest without referring to data. What is required is the tendency to focus on patterns over time: periods of increase and decrease, phase relationships among variables, peaks and valleys, and so on.

The Dynamic Behavior of Important Variables

Graphing variables over time forces us, rather obviously, to identify those variables in a system that are the symptoms of the problems we wish to study. The phrase urban problems, for example, covers a multitude of troubles and concerns, but, if we say what concerns us at the moment is the pattern of demographic behavior represented by the graphs in figure 2.1, we have a vivid problem focus. (The pattern of growth-to-equilibrium-to-decline is characteristic of the populations of many of the mature, major cities of the United States.) Alternatively, we might say that our concerns center mainly around financial variables such as expenditures, income, cash flow, and accumulated debt.

Such graphs focus our attention. They define, or help us to define, our problem. While we know that urban financial difficulties undoubtedly interact with demographic patterns,

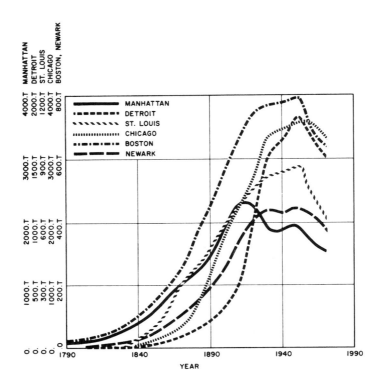

Figure 2.1: Population growth curves of large U.S. cities (Source: Schroeder and Strongman (1974), p. 201).

we also know that if the problem is pictured initially as in figure 2.1 we will attend to a somewhat different slice of urban reality than we would had we sketched graphs for an urban financial crisis.

In a process important to defining problems dynamically, one graph leads to a number of others. An urban population graph leads us to graph the pattern of births, deaths, and migration that directly produce the aggregate population pattern. The problem initially described in a single declining population curve probably becomes focused on migration patterns and the factors affecting the relative attractiveness of the city over time. Jobs, or opportunities per capita, housing availability, and the cultural benefits of the city suggest themselves, and we are led to draw a number of other graphs over time showing the pattern of business growth or decline,

Defining Problems Dynamically

housing construction, and so on. Similarly additional graphs would be drawn for a financial focus, showing variables directly connected to the acquisition of long-term debt such as income, expenditures, spending programs, tax base, and tax rate.

The Time Horizon

The time horizon, or time frame, is the period of time over which the problem plays itself out. It would appear as the extremes of the time scale on graphs drawn for the problem. It is the length of simulated time over which one will eventually run a dynamic model. It may be a number of hours (in a biological problem dealing with, say, insulin secretion and depletion) or hundreds of years (in a problem dealing with the lifecycle of a resource such as petroleum or the transition between primary energy sources). Most important, the time horizon helps to define the particular problem being addressed.

An example makes the point most clearly. Suppose we are concerned about problems connected with the production, consumption, and cost of natural gas. If we decide that the time horizon of our concern is, say, 1960 through 1990, we know we must consider political factors, oil embargos, imports, exports, different global sectors, and so on. But, if instead we take the time horizon to be from 1800 through 2100, we acquire a totally different focus. Over such a long time frame, embargos would be mere noise--tiny, perhaps imperceptible bumps on a long-range pattern. Consequently, the focus of the modeler's attention would be on other, more long-range components of the natural gas production and consumption system such as exploration, discovery, production and usage trends, and long-range costs. Furthermore, the graphical patterns of reserves, production, and consumption would be rather simple to draw over the longer time horizon, likely to appear roughly as shown in figure 2.3, while their counterparts over the years 1960 through 1990 would be much more difficult to divine.

Specifying the time horizon, then, is an important step in defining a dynamic problem, and it appears naturally in the process of sketching graphs of problem variables.

The Reference Mode

Graphing important variables, and inferring graphs of other significantly related variables, produces the problem focus for a system dynamics study. The graphs over time

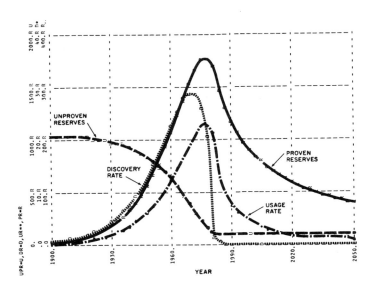

Figure 2.3: Likely patterns of natural gas production and reserves, 1800 to 2100. (Source: Naill (1973), p. 238)

are the *reference behavior modes* for the study. They are the patterns over time that will be referred to again and again in the modeling effort. One test of the validity of a system dynamics model is its ability to reproduce the reference modes identified at the outset. Thus reference modes help to define a problem, they help to focus model conceptualization, and they figure in the later validation stages of a study. It is a rare successful system dynamics study that omits them.

Thinking in Terms of Graphs over Time

We have seen that the habit of viewing problems in terms of patterns of variables over time is an integral part of the system dynamics approach. We will rely on it to help us start the process that will lead to the formulation of formal, quantitative feedback models. However, even if a formal model is not constructed, some insight can be gained from thinking in terms of graphs over time.

Figure 2.4 gives an example of dynamic thinking in the absence of a formal model. The graph represents hypotheses and inferences--thought experiments--about the long-term dynamic consequences of advertising. The figure illustrates the use of graphs over time in the conceptualization of a

Defining Problems Dynamically 23

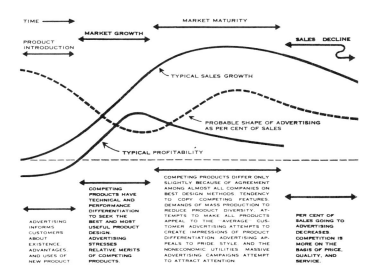

Figure 2.4: Advertising and a typical life-cycle of a product (Source: Forrester (1959))

problem. It shows a typical life-cyle of a product, focusing on the relationships between the periods of growth, peak, and decline in profitability, sales, and advertising expenditures.

Our focus here is on the use of graphs to describe, define, and analyze a problem, not on the specific problems of advertising addressed by this example. Readers interested in further development of this particular example should consult Forrester's original article, which contains several other short- and long-term graphical perspectives related to advertising.

Practice

To exercise understandings of reference modes and the processes involved in defining problems dynamically, the reader may wish to try thinking dynamically about the following problems:

1. A firm producing toothpicks has experienced considerable fluctuations in its workforce over the years. Draw a graph over time showing those (hypothetical) fluctuations. What other variables within the firm are

related to the workforce and its fluctuations? Draw comparable graphs of those other variables over the same time frame. Pay particular attention to the timing of the relationships between the curves. What influences external to the firm might be influential? Sketch their graphs as well.

2. Predator-prey ecosystems have been observed to show significant oscillations with periods that seem to depend more on the characteristics of the animals rather than external influences such as availability of water and food or severity of weather. Sketch a pair of graphs over time showing the populations of some predator and its primary prey: lynx and snowshoe rabbits or wasps and gypsie moths, for example. How are the peaks and valleys in the graphs related? If food for the prey were considered to be a factor in the instability of the system, how would its graph compare to the predator and prey graphs?

3. Sketch graphs over time that could help to define the problem of increasing use, degradation, and maintenance of a typical national park.

4. Define dynamically the problem of urban traffic congestion. How might the graphs you draw look as the time horizon is extended into the future? Sketch graphs over the same time horizon of variables that are significantly related to or responsible for the behavior you propose.

5. Think of some policy that has failed to achieve the desirable results expected of it. It could be a government policy dealing with some problem in the public sector, a policy of a corporation, some policy dealing with a personal problem, or the like. View the policy you select, and the problem to which it was addressed, from the dynamic perspective; translate it or recast it from static to dynamic terms, if necessary. Sketch three graphs over time: (a) a graph of the pattern of dynamic behavior to which the policy was directed--the problem addressed by the policy-maker(s); (b) a graph of the pattern of behavior that would have resulted had the policy been deemed successful; and (c) a graph of the behavior that actually resulted when the policy was implemented. You may have to sketch the patterns of more than one variable on each graph to capture the essence of the problem and the policy.

Defining Problems Dynamically

In the Absence of a Reference Mode

Occasionally, a problem is presented without a clear reference mode. A client might raise a question about the effects of a change of policy in advance of any understanding of a problem phrasable in terms of variables over time. Experienced system dynamicists often try to focus on the system involved by generating lists of possible policy interventions and ideas about possible effects. Such policy lists can sometimes substitute for or replace a clear dynamic view.
Important system variables are suggested by the policy levers being proposed. More variables can be inferred from considerations of possible responses of the system to the proposed interventions.

The interaction with people exceptionally familiar with the system is crucial in such an approach. Suffice to say, developing a formal system dynamics model in the absence of reference modes is difficult. The reader should try to avoid doing so, at least for a while.

2.2 REPRESENTING FEEDBACK STRUCTURE

Once a problem has been identified, significant variables located, and reference modes defined, the task of the modeler is to explore interconnections among those variables and others that are related. Searching for feedback structure, the modeler pursues chains of cause-and-effect until they reconnect with themselves, forming loops. While some circular patterns appear in verbal statements, the concept of a loop is primarily a visual one, and its simplest expression is in the form of a diagram. Two kinds of diagrams are common in the system dynamics literature. Causal-loop diagrams are most often used in early stages of model conceptualization and in later intuitive descriptions of model structure for nontechnical presentations. So-called DYNAMO flow or rate/level diagrams capture more detail and match much more closely a complete, quantitative description of a model. Model conceptualization commonly, but not necessarily, begins with causal-loops and moves to rate/level flow diagrams and finally to explicit equations capturing the diagrammed structure.

Causal-loop Diagrams

The diagrams shown in chapter 1 are causal-loop diagrams. They are sometimes referred to as influence diagrams or, in more mathematical circles, directed graphs. As shown in figure 2.5, the individual links in such diagrams can be labeled to show whether the nature of the causal-link is is "positive" or "negative." Loosely speaking, a plus sign indicates that the variables at the opposite ends of the arrow tend to move in the same direction (direct variation) while a minus sign indicates an inverse relationship. In figure 2.5a, for example, there is a direct relationship between the gap and the decision to pour: a bigger gap suggests more pouring, a smaller gap suggests less pouring. Hence, the link from gap to decision to pour is a positive link. The link from the level of wine to the gap is a negative link because the two variables are inversely related: the higher the level of wine, the less the gap, and vice versa. Similar comments hold for the links in figure 2.5b, as the reader should verify.

Some positive and negative links in causal-loop diagrams should be interpretted simply as addition or subtraction. The link from the pouring rate to the level of wine in figure 2.5a is an example. Pouring simply adds to the level of the wine. The link is labeled with a plus sign to show that addition. Unlike the other positive links in figure 2.5, this link does not represent a proportional relationship: as the pouring rate decreases, for example, the level of wine does not decrease (unless somebody's tippling)--it simply increases less rapidly. Pouring always adds to the level of the wine. Using the same symbol for proportional relationships and accumulations causes few problems as long as we are careful. The specific meaning of a sign is almost always obvious from the context, and the simplicity we gain is one of the strengths of these diagrams.

Hence, we define:

$$A \xrightarrow{+} B$$

A causal link from A to B is positive (1) if A adds to B, or (2) if a change in A produces a change in B in the *same* direction).

$$A \xrightarrow{-} B$$

A causal link from A to B is negative (1) if A subtracts from B, or (2) if a change in A produces a change in B in the *opposite* direction.

To determine the polarity (sign) of a proportional-type link, ask what happens to B when A increases or decreases. If an increase in A results in an increase in B, the link is positive; if a decrease in B results, then the link *is* negative. Be especially careful in testing link polarity using a *decrease* in A: if a decrease in A produces an *increase* in B, the link is negative, not positive (check the definition). There is a natural tendency to label increases with plus signs

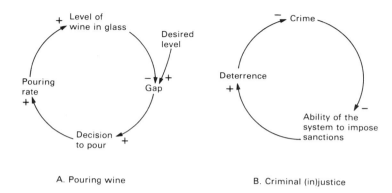

Figure 2.5: Causal-loop diagrams a. Pouring wine, b. Criminal (in)justice

and decreases with minus signs, and that tendency produces exactly the wrong labels in tracing the effect of a *decrease* in A.

Polarity of Feedback Loops in Causal-Loop Diagrams

Reading around a feedback loop the cummulative effects of its causal links gives an idea of the character of the loop. Suppose, in figure 2.5b, crime were to rise enough to cause a decrease in the ability of the system to impose sanctions. Courts become overloaded, prisons so full that imprisonment of one criminal means immediate release of another, sentences are shortened through acclerated parole or plea bargaining, and so on. The loop claims that deterrence will then decrease, and the level of crime will increase further. The feedback loop is clearly self-reinforcing: a sufficiently high level of crime leads (in the assumptions of this isolated loop) to an even higher level of crime. The loop is a *positive feedback loop*, a potential source of instability in a criminal justice system. Note that it acquires its self-reinforcing character from its two negative causal links. Each reverses the effect of the other, resulting in a loop that feeds on itself.

We can generalize to the following characterizations:

- A feedback loop is positive if it contains an even number of negative causal links.

- A feedback loop is negative if it contains an odd number of negative causal links.

Thus, the polarity (sign) of a feedback loop is the algebraic product of the signs of its links.

As described in chapter 1, and as may be evident from these characterizations, positive feedback loops amplify deviations and destabilize, while negative feedback loops strive to control and stabilize. Negative, in this context, has nothing to do with bad, and positive has no necessary connection to good.

Guidelines for Causal-loop Diagrams

The apparent simplicity of causal-loop diagrams is deceptive. It is easy for would-be modelers to go astray with them. The following suggestions may help to prevent the more common difficulties.

1. Think of variables in causal-loop diagrams as *quantities* that can rise or fall, grow or decline, or be up or down. But do not worry if you can not readily think of existing measures for them. Corollaries:

    - Use nouns or noun phrases in causal-loop diagrams, not verbs. The actions are in the arrows. (See figure 2.6.)

    - Be sure it is clear what it means to say a variable increases or decreases. (Not "attitude toward crime," but "tolerance for crime," for example.)

    - Do not use causal-links to mean "and then... ."

2. Identify the units of the variables in causal-loop diagrams, if possible. If necessary, invent some: some psychological variables might have to be thought of in "stress units" or "pressure units," for example. Units help to focus the meaning of a phrase in a diagram.

3. Phrase most variables positively ("emotional state" rather than "depression.") It is hard to understand what it means to say "depression increases" when testing link and loop polarities.

4. If a link needs explanation, disaggregate it -- make it a sequence of links. For example, a study of heroin-related crime claimed a positive link from heroin

Representing Feedback Structure 29

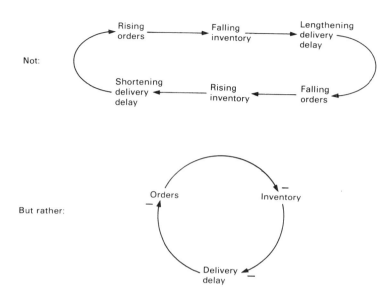

Figure 2.6: Loops illustrating that the action in causal-loop diagrams is best left to the arrows.

price to heroin-related crime. The link is clearer if disaggregated as in figure 2.7 into the sequence of positive links from heroin price to money required per addict, frequency of crimes per addict, and finally heroin-related crime. Some might feel a high price deters addicts and so lowers the number of addicts as it well might, but that is another link (see figure 2.7).

5. Beware of interpreting open loops as feedback loops. Figure 2.7, for example, does not show a feedback loop.

To exercise the understandings developed thus far about causal-loop diagrams, the reader may wish to draw some causal-loop diagrams of feedback loops for each of the following systems:

1. The arms race: find a positive loop and at least one negative loop. Is the structure symmetric for both protagonists?

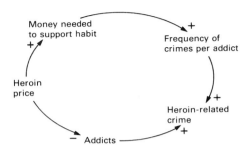

Figure 2.7: Links relating heroin price and crime; a pseudo-loop.

2. A doctor's workload: define the workload as the product of the number of patients in the practice, the average number of visits by or to each patient per year, and the average number of hours per visit. There is at least one positive loop, which leads to a growing practice. Many negative loops limit it.

3. An inventory control system: include orders, an order backlog, shipments, an inventory from which to fill orders, and production to replenish inventory. The structure is likely to contain no positive feedback loops.

4. Some recurring phenomenon: select some repetitive happening with which you are familiar (such as breathing, eating, arguments in a family, colds, business cycles, changing jobs, street cleaning, the rise and fall of civilizations, oscillations in the width of neckties or length of skirts) and sketch one or more causal-loops which seem to account for the recurring pattern over time. (Take care not to use causal links to mean "and then".) You should find at least one negative loop. (Why?)

Accumulations

Causal-loop diagrams treat only implicitly an essential aspect of feedback structure that we have largely ignored up to this point: the notion of accumulation. Wine in a wine glass accumulates as it is poured in. Production accumulates in an inventory that is depleted by shipments. We implicitly acknowledged such accumulations when we noted that some links in causal-loop diagrams should be interpreted as addition or subtraction rather than proportional change. Accumulations are conceptually quite different from the flows that affect them. We could see an accumulation, for example, and measure its amount even if we had only a still photograph of the system, stopping time. However, we could not measure from a photograph the rates of flow affecting these accumulations.

We wish explicitly to represent accumulation processes in our diagrams because they affect the dynamics of feedback systems and because the equations one uses to model them quantitatively are significantly different from ordinary algebraic relationships, as we shall see in chapter 3. Hence we introduce the notion of a system *level* and a diagrammatic convention for representing the accumulation processes levels represent.

Levels, Rates, and Flow Diagrams

Accumulations in feedback systems are variously called stocks, state variables, or levels. The term "level" is intended to invoke the image of the level of a liquid accumulating in a container. The system dynamicist takes the simplifying view that feedback systems involve continuous, fluid-like processes, and our terminology reinforces that interpretation. The flows increasing and decreasing a level are called *rates*. *Flow diagrams* picture rates and levels as stylized valves and tubs, as shown in figure 2.8, further emphasizing the analogy between accumulation processes and the flow of a liquid.

The cloud-like symbols appearing in figure 2.8 represent *sources* and *sinks* for the material (whatever it is) flowing into and out of the level. For a level of inventory affected by production and shipment rates, these cloudlike symbols represent where inventory comes from when it is produced and where it goes after it is shipped. Their presence indicates that the real-world accumulations they represent lie outside the boundary of the system being modeled. In figure 2.9, for example, showing levels and rates in a rabbit population model, sources and sinks are used to indicate that in this model we are not interested in where rabbit babies come from or where old rabbits go when they die.

Figure 2.9 also shows the disaggregation, or splitting up, of a single population into two levels. Certain problem settings require such disaggregation, as, for example, a workforce in which there is a significant training delay before workers reach full productivity. Figure 2.10 shows such a situation.

The decision of whether to represent a given accumulation as a single level or to disaggregate it into a series of levels depends on issues relating to problem definition and the

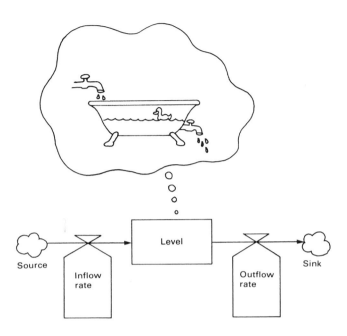

Figure 2.8: Flow diagram, showing conventional symbols for levels and rates.

dynamics of the system, which are best dealt with a bit later. Here we merely note several suggestive examples.

Figure 2.11 shows the flows and accumulations of people assumed in a large study of the dynamics of heroin addiction and traffic. The diagram omits the valve symbols for rates in order to concentrate the reader's attention on the structure of the levels. Levels such as "addicts in methadone programs" illustrate the fact that the potential policies that one would like to explore with the model influence model structure. The absence of an inflow to the "potential drug

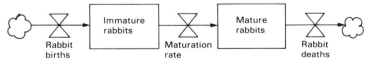

Figure 2.9: Flow diagram of a rabbit population showing immature and mature rabbits

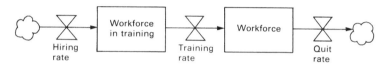

Figure 2.10: Flow diagram of a workforce disaggregated in two levels

users" level shows that this model purposely ignored births, choosing to regard the population as essentially constant except for migration.

Figure 2.12 shows the rates and levels assumed in a study of solid waste generation. Note that the presence of a level such as "processed raw materials" is necessary to allow policies associated with recycling to be explored. One could have aggregated several of the levels together, producing a model containing only "natural resources" and "solid waste," but doing so would limit the range of potential investigations with the model.

Representing Material Flows and Information Links

Material flowing out of a level diminishes the level, but information about that level passing to other parts of the system leaves the level itself unchanged. Thus information feedback links are significantly different from physical flows. In their simplicity, causal-loop diagrams ignore this difference and represent both types of links with the same sort of arrow. Flow diagrams explicitly showing rates and levels make a distinction between physical flows and information links. Physical flows are conventionally indicated by solid arrows, while information links are shown by dotted arrows. Figure 2.13 shows two examples of familiar feedback loops drawn with explicit rates, levels, physical flows, and information links.

Flow diagrams of more complex systems will undoubtedly have some variables appearing in the the information paths. There are pieces of information that we will want to name and understand in our feedback models, in addition to the rate and level variables that we have alluded to thus far. In the next chapter we will term such variables auxiliaries. Complete flow diagrams, including auxiliaries, are so close to the fully quantified form of a model that we shall postpone further development of them until chapter 3.

It is becoming increasingly common to represent system structure in a kind of hybrid diagram showing explicit rates and levels within what is otherwise a simple causal-loop dia-

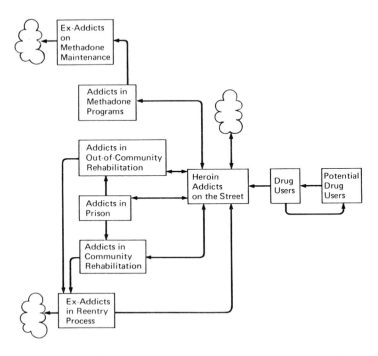

Figure 2.11: Flow of people in a study of heroin addiction in an urban community (Source: Levin, Roberts, and Hirsh (1975), p. 20)

gram. Solid lines are drawn throughout. The conserved material flows linking rates and levels are usually drawn as straight paths, while the information links in the system appear as the sinuously curved arrows of ordinary causal-loop diagrams. Figure 2.14 shows an example. The advantages of such a representation are that significant accumulations and flows are shown explicitly, while the diagram retains an appealing visual simplicity. The explicit presence of rates and levels enables a reader to infer more reliably some of the dynamic implications of the feedback structure represented. The actual dynamic consequences of the model assumptions are thus more readable in this form than in simple causal-loop form. Of course, complexity will still force us to code the structure for computer simulation to be certain of the dynamic behavior it can generate. As we have mentioned, the addition of rates and levels in our dia-

Representing Feedback Structure 35

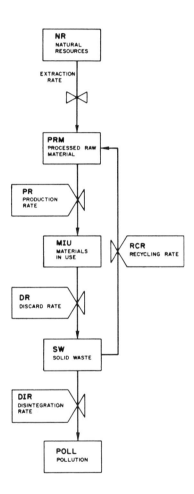

Figure 2.12: Rates and levels in the processing of natural resources into solid waste (Source: Randers and Meadows (1973), p. 191)

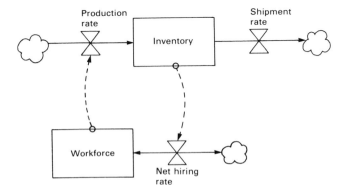

A. Inventory—workforce

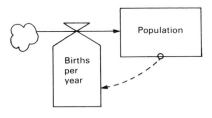

B: Population

Figure 2.13: Flow diagrams of elementary positive and negative feedback loops (see figures 1.6 and 1.7 for comparison).

Representing Feedback Structure 37

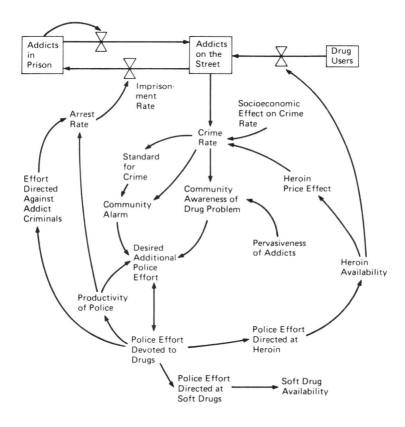

Figure 2.14: Example of a hybrid diagram, a causal-loop diagram with explicit rates and levels. The structure shows the primary variables in the police response system assumed in a study of heroin addiction, crime, and community response. (Source: Levin, Roberts, & Hirsch (1975), p. 27).

grams takes us one step closer to the equations that the computer will trace through time.

2.3 MODEL PURPOSE

> *A model without a purpose is like a ship without a sail,*
> *A boat without a rudder, a hammer without a nail...*
> (old blues variant)

From the system dynamics perspective, a model is developed to address a specific set of questions. One models problems, not systems. We emphasize the point because the purpose of a model helps guide its formulation. The problem to be addressed, the audience for the results of the study, the policies one wishes to experiment with, the implementation desired, all influence the content of the model.

The Fundamental Purpose of a System Dynamics Model

The purpose of a system dynamics model is understanding. The goal of a modeling effort is to improve understandings of the relationships between feedback structure and dynamic behavior of a system, so that policies for improving problematic behavior may be developed.

In this book we emphasize the goal of policy analysis and improvement. However, all models are theories, and sometimes a system dynamics model is developed solely for the purpose of testing a theory. Can a particular set of feedback assumptions generate a certain pattern of behavior over time? Such testing can not prove a theory. It can only disprove it by showing the desired behavior can not be produced from a given set of assumptions. The fact that a model is capable of generating certain patterns of behavior does not show that other models (theories) can not. (The process of building concensus for the theory expressed in a feedback model is addressed in section 5.4.) While theory testing is undoubtedly a valuable model purpose, we prefer here to focus on models and modeling projects whose major goal is policy design and improvement.

To prevent policy improvement from becoming an academic exercise, the modeler must be concerned from the beginning

---

[1] See Greenberger, Crensen, and Crissy (1976) for examples and discussion.

about eventual implementation. Conclusions from formal models often fail to be implemented.[1] At least one source of the failures is the lack of attention to implementation at the outset of the project, in the problem identification and model purpose stages. E. B. Roberts lists a set of strategies for effective implementation of complex corporate models[2]

1. Project Selection:
   - Solve a problem;
   - Problem must be important to client;
   - Objectives must be credible.
2. The modeling process:
   - Involve the client as much as possible;
   - Expedite initial model development;
   - Match model detail to audience;
   - Gear validity testing toward management;
   - Design measures of effectiveness into model, consistent with real-world measures.
3. Recommendations for Change:
   - Account for ability to absorb change;
   - Consider possible impact on other systems;
   - Include management re-education and/or explicit decision rules.

The general message in these guidelines is that implementation will not result merely from having built a good model and analyzed it well. Implementation must be consciously planned for from the beginning. Suffice to say, the modeler who evolves a statement of model purpose without considering these principles runs a high risk of working for no purpose at all.

Statements of Model Purpose

Some examples of statements of model purpose may be helpful to reader trying to write his own. For the production/distribution model he developed in *Industrial Dynamics*, J. W. Forrester writes:

> In constructing a useful dynamic model of corporate behavior it is essential to have clearly in mind the purposes of the model. Only by knowing the questions to be answered can we safely judge the pertinence of factors to include in or omit from the system formulation.

---

[2] Roberts (1980), pp. 77-85.

> Therefore, we shall define our immediate objective as an examination of possible fluctuating or unstable behavior arising from the principal organizational relationships and management policies at the factory, distributor, and retailer. We shall explore the way in which the simple, central structure of the system tends to accentuate or modify the external disturbances that impinge on the system.[3]

The focus of the model to be developed is instability caused by management policies, not caused by disturbances external to the production-distribution system. The statement of purpose does not mention the client or potential audience for the study, but its context implies the audience comprises students of industrial policy and behavior, presumably both academic and commercial.

The statement of purpose for the model central to the controversial *Limits to Growth* study[4] is instructive, partly because some of the controversy surrounding that project stems from differing opinions about its purpose:

> Figure [2.15] illustrates the four possible behavior modes that a growing population can exhibit over time. The mode actually observed in any specific case will depend on the characteristics of the carrying capacity -- the level of population that could be sustained indefinitely by the prevailing physical, political, and biological systems -- and on the nature of the growth process itself. One of these basic behavior modes must characterize any physically growing quantity, such as pollution, productive capital, or food output. The purpose of WORLD3 is to determine which of the behavior modes shown in figure [2.15] is most characteristic of the globe's population and material outputs under different conditions and to identify the future policies that may lead to a stable rather than an unstable behavior mode.[5]

Again, the audience for the study is only implicitly stated. The client supporting the work was the Club of Rome, which was not in a position to implement results. The level of implementation one could aim for in such a study is limited to a change in awareness of global problems and the implications of certain policies.

---

[3] Forrester (1961), p. 137.
[4] Meadows *et al.* (1972)
[5] Meadows *et al.* (1974), p. 8

Model Purpose 41

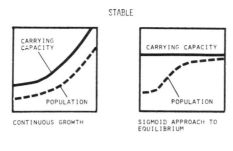
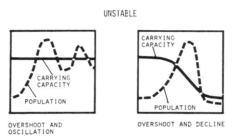

Figure 2.15: Four possible modes of population growth (Source: Meadows et al. (1974), p. 8)

The statement of purpose for the model explored in *The Persistent Poppy* is contained in extended form in the book's first chapter. The following is a heavily edited skeletal statement of purpose:

> [T]here are two heroin problems, not just one... The criminal side of the problem is the one most visible to and most impactful upon the community... The addict is also a victim -- of his drug, and of society's attitude toward it and him... The two problems of heroin interact in a perverse fashion. Effective present actions, guided by policies aimed at reduction in numbers of the addict-criminal or the addict-victim, may exert an unintended and undesired impact upon the number of addicts who will constitute the future problem... The following chapters develop a perspective on the heroin problem that is sufficiently comprehensive to encompass all of the symptoms described above...[The] computer model permits experimental manipulation of the parameters and variables of the system which...are the primary raw materials used in our search for rational heroin

policy...Many well-informed persons will disagree with our conclusions. Our hope is that they will join us in debate and dialogue. We suggest the present model, and others that might be constructed, as a vehicle for facilitating that process.[6]

The audience for the model is both experts in the field of drug-related problems and interested laypeople. The purpose is development of understandings about a heroin system encompassing the dynamics of addicts, heroin-related crime, and community attitudes and responses. The model is seen as a vehicle for focussing discussion and analysis for the design of policies having the best chance of success, however limited that chance may be in practice.

These examples of statements of model purpose contain within themselves implications for the scope of the models for which they were written. A clear understanding of the purposes of a modeling effort helps to answer questions relating to the system boundary--what should be included and what should be excluded.

System Boundary

The boundary of a system is the imaginary line separating what is considered (for modeling purposes) to be inside the system and what is considered to be outside. Concepts located outside the system boundary are excluded from a model of the system. Within the system boundary go all concepts and quantities considered by the modeler to relate significantly to the dynamics of the the problem being addressed. Such a boundary is drawn out of necessity: if one insists on the view that everything is connected to everything else, one is paralyzed, prevented from ever concluding an analysis that always stretches on to yet more variables and effects. At the end of a system dynamics study one must always ask if the policy recommendations hold up even if the boundary of the system is enlarged, but some boundary to what is being considered always exists.

What determines where the system boundary ought to be drawn? From the system dynamics perspective, one of the first criteria for a correctly drawn system boundary is the closing of feedback loops in the system. One can characterize the open-loop thinking discussed in section 1.2 as arising from an improperly drawn system boundary. Left outside are the variables which act to close the open loops. We try to

---

[6] Levin, Roberts, and Hirsch (1975), pp. 14-16.

Model Purpose

include inside the boundary all quantities that are dynamically significant for the purposes of the model.

In his classic paper "Market Growth as Influenced by Capital Investment,"[7] J. W. Forrester writes of the system boundary as follows:

> In defining a system, we start at the broadest perspective with the concept of the closed boundary. The boundary encloses the system of interest. It states that the modes of behavior under study are created by the interaction of the system components within the boundary. The boundary implies that no influences from outside of the boundary are necessary for generating the particular behavior being investigated. So saying, it follows that the behavior of interest must be identified before the boundary can be determined. From this it follows that one starts not with the construction of a model of a system but rather one starts by identifying a problem, a set of symptoms, and a behavior mode which is the subject of the study. Without a purpose, there can be no answer to the question of what system components are important. Without a purpose, it is impossible to define the system boundary.
>
> But given a purpose, one should then define the boundary which encloses the smallest possible number of components. One asks not if a component is merely present in the system. Instead, one asks if the behavior of interest will disappear or be improperly represented if the component is omitted. If the component can be omitted without defeating the purpose of the system study, the component should be excluded and the boundary thereby made smaller. An essential basis for identifying and organizing a system structure is to have a sharply and properly defined purpose.

Examples of System Boundaries

In the "Market Growth" paper, Forrester proceeds to define the purpose of the study:

> Here the objective is to identify and to explain one of the systems which can cause stagnation of sales growth even in the presence of an unlimited market. In particular, we deal here with that system which causes sales stagnation,

---
[7] Forrester (1968a)

44   Problem Identification and System Conceptualization

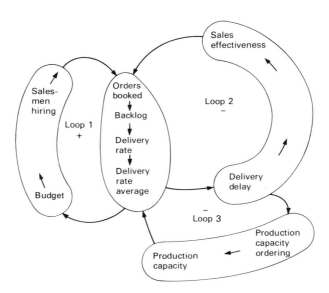

Figure 2.16: Loop structure for sales growth, delivery delay and capacity expansion, illustrating variables within the system boundary in Forrester's "Market Growth" study (Source: Forrester(1968a)).

or even sales decline, to arise out of an overly cautious capital investment policy.

The boundary follows:

Figure [2.16] illustrates the scope of the system being considered. The closed boundary surrounds the relationships shown. No other influences from the outside are necessary for creating the sales growth and stagnation patterns which will presently be developed.

Note that such potential influences as inflation, energy shortages, and recessions are not included within this system boundary. They are apparently not necessary to generate the behavior of interest in the study and are therefore excluded from the model.

Model Purpose and Problem Definition

The definition of a problem and a statement of model purpose are related but not identical. A well-defined problem can exist without a model designed to to address it. We define problems dynamically, in terms of graphs of variables over time. Statements of model purpose focus less on the nature of the problem and more on the audience for the study, potential leverage points for policies, and the kind of implementation intended.

At this point, however, these generalities are no doubt hard to assimilate. To clarify the stages, strategies, and skills involved in defining problems and conceptualizing models from the system dynamics perspective, we turn now to a specific problem, which will serve as the focus for much of the remainder of the book. Section 2.4 introduces the problem of managing a project (such as a research and development project) to a successful completion, without overruns in costs, manpower, or schedule. The general skills and concepts of sections 2.1, 2.2, and 2.3 will be placed in the context in which the reader will use them.

2.4 AN EXAMPLE OF PROBLEM CONCEPTUALIZATION

As a vehicle for illustrating the details of the system dynamics approach, we will trace the development of a model of the dynamics of a research and development project. This section begins that process by describing the problem to be addressed, developing a focused problem definition, and sketching in nonquantitative terms a first attempt at a feedback model. That model will be formulated quantitatively in DYNAMO in chapter 4, refined in chapter 5 as we test it and uncover deficiencies in its conceptualization and formulation, and used in chapter 6 as the primary medium for discussing policy analysis with a system dynamics model. The discussion in these chapters will take the form of a story of a hypothetical (one might even say model) system dynamics study. The reader should take as active a part in the proceedings as possible, repeatedly testing himself to see if he could have made a certain step without the book's direction.

In this section the concepts and skills discussed in the previous sections will be used and noted. The purpose of the section is not to increase understandings of project dynamics, although that may happen for some readers as we extend the discussion in later chapters. Our purpose is to use a particular, well-focused problem to make more vivid and real the concepts, skills, and strategies of the system dynamics approach.

The Problem

We imagine that we, or a client of ours, have had some experience with large research and development projects (R&D). Our purpose is to improve the management of such projects in such a way as to eliminate or minimize the problems that have been observed.

A common problem of large R&D projects, such as federally funded development programs, are overruns: cost overruns, the need to hire and train additional personnel midway through the project, and overruning the scheduled time allotted. No doubt, not all research and development projects have such problems, but we proceed here on the assumption that they have persisted in our experiences, or our client's, in spite of reasonable attempts to avoid them.

We are considering here a large R&D project, involving a number of people, a large number of detailed tasks, and a relatively long time frame of, say, around four years. However, the problem of overruns in personnel and schedule has analogs in smaller projects in dramatically different contexts, such as writing a paper (or a book), painting a house, or putting on a dinner party. In many ways the project model we will develop here could be interpreted, with a few changes, to represent the feedback structure inherent in any undertaking sandwiched into a specified time frame. Our intuitions about the structure and dynamics of all such projects will help us in conceptualizing our model of R&D project dynamics.

Defining the Problem Dynamically

We begin with a view of the problem in terms of graphs over time. There are two sets of graphs to consider, representing the dynamics of a well-run, problem-free project and the dynamics of a project with overruns.

What would graphs of personnel and schedule look like over time if the project proceeded flawlessly from inception to the final report on the final day? Presumably, management would have correctly assessed from the beginning the number of people and the length of time required to carry out the project. If the work is such that a rather uniform effort is required throughout, we would imagine there would be little change in the number of personnel, the intensity of effort (hours per week per person), or the scheduled finishing date. Horizontal lines over time. Alternatively, we could imagine that the project would have begun with a few workers, hired the necessary number, and then proceeded to

An Example of Problem Conceptualization

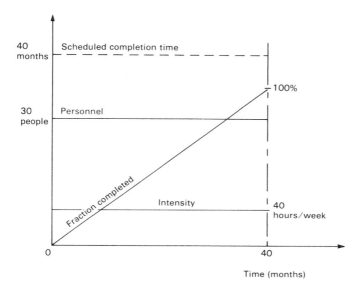

Figure 2.17: Overview of the dynamics of a hypothetical flawless R & D project.

completion with a constant workforce--producing horizontal straight lines after a brief start-up period.

Progress on the project might even be linear as well, a constantly rising fraction of the job completed, starting at zero when the project begins and ending at one (100%) at the termination of the project. Figure 2.17 shows the graphs one might draw for such a flawless project.

A project showing overruns would deviate from these ideal linear patterns. A need for additional personnel would eventually result in the hiring and training of additional people -- a rise in personnel over time. Prior to such a rise in personnel, the intensity of effort on the project would have increased. A symptom of the increase in effort would probably be a rise in the number of hours worked per week per person on the project. A schedule overrun could be represented graphically as a similar rising pattern. Suppose that we represent the schedule by a number of months of work required. If, as the project proceeds, it is perceived that more time is required to finish, that number of months initially estimated is revised upward--a rise over time in the number of months of effort scheduled for the project. Figure 2.18 shows a possible set of such patterns.

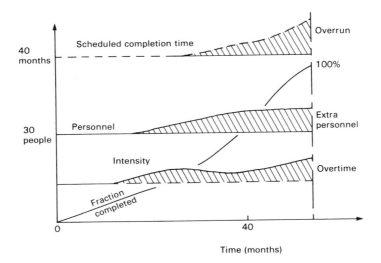

Figure 2.18: The dynamics of an R&D project showing overruns in personnel and schedule.

What would the graph of progress look like for such a project? If it were the simple linear pattern suggested for the perfect project, but with a smaller slope, it seems reasonable that people would quickly perceive that the project was not on target and would take some steps to set it right, perhaps by hiring additional personnel early in the project. It is more likely that the progress *perceived* by those involved is not linear in such a project, whether or not their actual progress is constant.

Things seem to be going well for a while, but then problems begin to appear. Parts of the project that were thought to be completed are seen to have some flaws requiring rework; it begins to appear that more man-months of work remain than are available; progress is perceived to have slowed down somehow. For a time, the fraction of the job perceived to be completed rises less rapidly than initially; it might even become horizontal for a time, showing a lack of perception of much forward motion in the project. The graph of perceived fraction completed shown in figure 2.19 reflects these observations. The graph of the actual fraction completed shows a typical pattern of real progress, based on a straight-line path from zero to 100 percent completion, but rising with increases in effort and slowing a bit as time is taken to acquaint new personnel with the project.

An Example of Problem Conceptualization

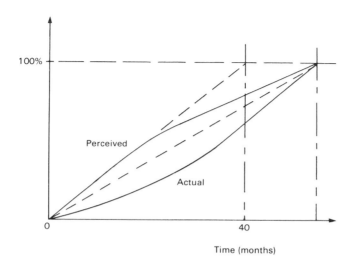

Figure 2.19: Actual and perceived progress in an R&D project with overruns.

Reference Modes

The graphs in figures 2.17, 2.18, and 2.19 are the *reference behavior modes* for the problem of overruns in an R&D project. We shall require that the models we develop for the problem be able to generate similar patterns over time. In addition to specifying clearly the problematic behavior of the system in which we are interested, the reference modes indicate the *time horizon* over which the model must run. The R&D project model we will develop must run forty to fifty months (perhaps even longer) to the completion of the project.

Usually, the reference modes for a modeling study are based on actual numerical data. Here we assume that our clients' experiences provide the data for these figures. Frequently, however, some of the graphs may have to be inferred. It might be difficult, for example, to have data over time for both perceived and actual fractions completed. In such a case the modeler would proceed as we did, trying to deduce what the shapes over time would have to be to fit with other known patterns. We might lack data about the hiring rate, yet we can infer what the graph would have to look like by noting where the workforce curve rises and falls, and how steeply.

The modeler will probably always lack hard data for subjective concepts, such as pressure that results from being behind schedule, and yet they may be critically important to the behavior of the system. The intuitions of those closest to the problem will provide reliable dynamic views of such phenomena. They will be used to incorporate endogenously within the model such notions as schedule pressure and its effects.

Model Purpose and Audience

We assume that our clients are people responsible for the management of large R & D projects and that they wish some guidance to prevent or minimize overruns. The model should be a tool that allows our clients to experiment with policies for improving project managment. (In fact, however, the real purpose of the little project model we develop here is to enable the reader to attack problems of his own from the system dynamics perspective, whether or not those problems have anything to do with project dynamics.)

How would a project model--this project model--be used? Its greatest implementation would be its adoption by the client for ongoing, future policy testing of project management. Such an implementation would require the training of someone in the client company in many of the details of system dynamics and simulation. A less complete implementation would be the one-time adoption of a set of policy recommendations stemming from the modeling study. Further along toward the weak end of the spectrum of model implementation is the use of the model as an educator or consciousness raiser. The degree of implementation desired affects not only the style of the modeler interacting with the client, but even the content of the model itself. Here, we suggest for our hypothetical client a hypothetical model use, somewhere between conscious-raising about project dynamics and the one-time adoption of a set of policy recommendations.

System Boundary

The model purpose indicates that the model should focus upon aspects of R&D project dynamics that are potentially within the control of people on the project. The following readily suggest themselves:
   -the project definition (tasks to be performed),
   -project personnel (the project workforce),
   -hiring and terminations,
   -productivity,
   -undertime and overtime,

An Example of Problem Conceptualization

- progress,
- rework,
- perception of man-months required,
- the time schedule,
- alterations in the schedule,
- costs.

One might also propose to include within the system boundary funding, changes in the project definition, existing procedures for reporting to project management, technological complexity of tasks in the project, experience of personnel, constraints on the availability of qualified personnel, time spent in organization and communication tasks, and so on. We chose not to place these influences explicitly within the system boundary.

The reasons for exclusion deserve comment. We distinguish among system components represented explicitly within the boundary, those implicitly represented within the boundary by way of aggregation or interpretation, and those outside the boundary. In the latter category, one may also usefully distinguish between environmental factors near to the system and those far from it. Those near could be defined to be components linked to variables within the system boundary by a one-way path of influence -- linked to the system but not looped into it. Those far from the system could be defined to be components so removed from the problem that they can be viewed as completely uninvolved.

Applying these distinctions, we choose to place funding and technological complexity outside the system boundary, as environmental factors near to the problem, but not endogenous. They could be assumed to be constant for the duration of the project. Reporting procedures, on the other hand, while not represented explicitly, will be implicitly included within the system boundary, aggregated in the notion of perceptions within the system. Experience of personnel will be similarly aggregated in the notion of productivity. Finally, time spent in communication tasks and changes in the project definition will be placed outside the system boundary, not because they necessarily belong there, but because on our first pass at modeling project dynamics we prefer to omit them for simplicity.

To summarize, in setting the system boundary one considers what system components are necessary to generate the behavior of interest, excluding where possible, and aggregating where useful for simplicity. It is also useful to include indicators, such as costs, by which relative improvements in system behavior can be judged. Further guiding our choices of what to include within the system boundary are three questions:

- What are the physical processes in the system that are relevant to the problem?

- What are the perceptions of those processes and how are they formed?

- How do those perceptions combine to create pressures influencing the physical processes?

Processes, perceptions, and pressures provide a starting point for system conceptualization.

A Feedback View of Project Management

With the problem defined, reference modes clear, model purpose explicit, and a preliminary system boundary drawn, we are ready to develop a view of the system in feedback terms. System conceptualization most easily begins with the physical processes. In an R&D project the physical processes include the performance of the tasks which compose the project, the movement (into and out of the system) of the people who perform those tasks, and the movement of the project along toward completion. We begin with the physical processes involving the workforce.

If it should prove necessary to change the project workforce in the course of the project, management would have some idea of the number of personnel required to complete the project in the time alloted. Hiring or firing would adjust the workforce to such a desired or indicated workforce, as shown in the figure 2.20. The simple negative feedback loop in figure 2.20 would adjust the workforce, after some hiring or firing delay, to equal the indicated workforce. Other influences not yet shown in the diagram will determine just how large a workforce is indicated.

Pursuing the physical processes in the system, we note that the function of the workforce is to make progress on the project. Let us assume that the project definition is in terms of a number of tasks. A simple view of progress suggests that an average worker completes some average number of tasks in the project every month. This average productivity and the size of the workforce together determine the rate of progress on the project. Cummulative progress would simply be the sum of the number of tasks completed every month.

Knowing that the workforce affects cummulative progress, we ask the obvious feedback question: does cummulative progress affect the workforce, and, if so, how? Cummulative progress and the initial project definition determine the number of tasks remaining. Knowing the tasks and time remain-

An Example of Problem Conceptualization

Figure 2.20: Adjusting the workforce

ing and the productivity of the workforce, project management could compute the number of people required to complete the project on schedule. These sentences complete a feedback loop moving from workforce around to indicated workforce, as shown in figure 2.21.

The goal of the negative feedback loop in figure 2.21 is to keep the project on schedule, keeping cummulative progress on target by adjusting the workforce. Note that the units of the variables around the loop help its conceptualization. "Tasks remaining" divided by productivity measured in "tasks/person/month" yields "man-months" of effort required to complete the project:

$$\frac{\text{tasks}}{(\text{tasks/person/month})} = \frac{\text{tasks}}{\text{tasks/(man-month)}}$$

$$= \frac{1}{1/(\text{man-months})}$$

$$= \text{man-months}.$$

Such a justification of units is most useful in the equation-writing stage, but here it helps to show that the links determining effort required in figure 2.21 have the correct polarity and do, indeed, produce the quantity claimed. The link from tasks remaining is positive because the variable would be in the numerator of the computation, while the link from productivity is negative because it would be in the denominator. Higher productivity would result in a *lowered* estimate of the man-months of effort required to complete the project, so the link should be negative.

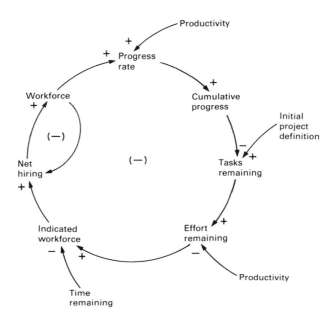

Figure 2.21: Progress

The problem definition, focusing on schedule overruns, indicates that the scheduled completion date should be regarded as a variable that can be adjusted by project management. Again, a simple negative feedback loop suggests itself, as shown in figure 2.22.

On what would management base its perception of the completion date indicated by current project conditions? Aided again by the units of the variables, we can see that if management knows the size of its workforce and how many man-months of effort remain to finish the project, it can divide to compute the months required to finish the project: "man-months" divided by "people" yield "months." Thus, the workforce helps to determine the scheduled completion date, and the scheduled completion date helps to determine the size of the workforce required. The feedback loop outlined here is shown in figure 2.23.

An Aside about Purpose and Process

The purpose of this discussion is to demonstrate how a simple system begins to be conceptualized. Ideally, the narative would stop periodically and ask the reader to try to construct the next feedback loop. However, repeated

An Example of Problem Conceptualization

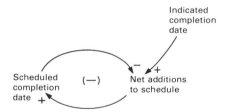

Figure 2.22: Changing the schedule

interruptions for that purpose would be tedious. Nonetheless, the reader should try to anticipate the next figure as often as suits his purposes in reading the book. A very useful exercise at this point would be to close the book and draw a composite diagram containing all the feedback loops uncovered thus far.

A Simplified Overview of Project Structure

Figure 2.24 shows the feedback structure described up to this point. There are two clear policy points in the system. Management can control hiring, and it can fix the schedule. In case of a project behind schedule, management could decide to hold the workforce constant and alter the schedule, increase the workforce and hold to the current scheduled completion date, or alter both workforce and schedule somewhat to return the project to some target path toward completion. The structure omits the most likely response to being behind schedule: overtime (and undertime) will be added to the more complete model developed in chapter 5.

Nothing in figure 2.24 gives any hint of why an R&D project would show overruns in schedule, personnel, or cost. If the project gets off-schedule for some reason not yet shown in the causal-loop diagram, it appears that the mechanisms adjusting the workforce and the schedule would become active and place the project on target again with little hiatus. We seek some hint of a structure that could, in the course of a project, produce tendencies for overruns. The concept of rework, listed among the variables that belong inside the system boundary, is a likely candidate.

The Dynamic Hypothesis

The dynamic hypothesis in a system dynamics study is a statement of system structure that appears to have the potential to generate the problem behavior. The dynamic hypothesis can be given verbally or in terms of a diagram. It might

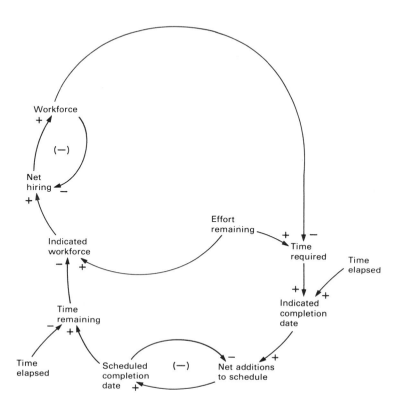

Figure 2.23: The time required and the time remaining.

contain one or a number of potential causes of the problem behavior. In complex studies a clear dynamic hypothesis, isolated from the maze of whole-system structure, may not appear until models are formulated and model behavior extensively analyzed. For the illustrative problem of overruns in an R&D project, we propose a simple dynamic hypothesis centering around the notion of rework.

Not all work done in the course of a large project is flawless. Some fraction of it is less than satisfactory and must be redone. Unsatisfactory work is not discovered right away, however. For some time it passes as real progress, until the need for reworking the tasks involved shows up. Hence we are led to suggest that the project workforce accumulates two kinds of progress: real and illusory. We shall term the lat-

An Example of Problem Conceptualization

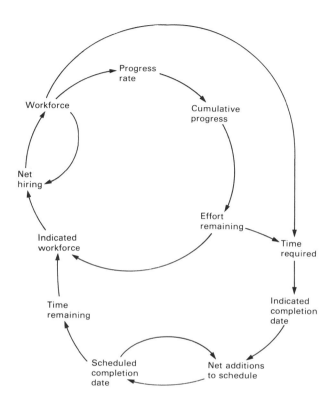

Figure 2.24: Simplified overview of project structure.

ter undiscovered rework. Together, cummulative real progress and undiscovered rework constitute what is perceived within the project as actual cummulative progress, as sketched in figure 2.25.[8]

The inclusion of undiscovered rework moves the conceptualization from strict, physical processes to perceptions. Real progress is not perceived in the system. Besides the illusory progress involving work that must be redone, there are also inadvertant miscommunications to and from project manage-

---
[8] Though simplified in this presentation, there is ample evidence for such a notion of perceived progress. See, for example, Cooper (1980).

58   Problem Identification and System Conceptualization

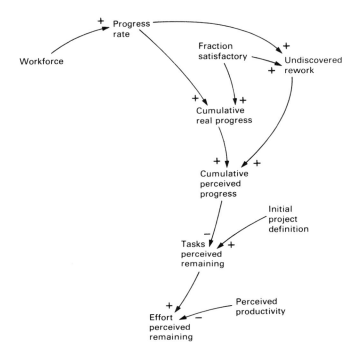

Figure 2.25: Rework and cummulative perceived progress

ment: reporting requires time, and so information about the state of the project is always delayed or projected. Here for simplicity we represent explicitly only the notion of undiscovered rework.

The distinction between reality and the perception of reality is vitally important in system dynamics models -- it might be said to be characteristic of them. The modeler must realize that not all the quantities represented in the model are knowable within the system being modeled. Some information is inaccessable to actors in the system. The result is that some model variables can not be directly linked together, even though in the model it would be trivial to do so.

Note in figure 2.25 that the man-months of effort perceived remaining are determined by *perceived* productivity, rather than actual productivity. The thinking behind the formulation is the same as for effort remaining in figure 2.24, with the realization that the true productivity of the project workforce is not knowable by actors in the system.

An Example of Problem Conceptualization

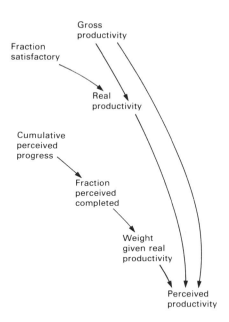

Figure 2.26: Real and perceived productivity

Having been forced to introduce the notion of perceived productivity, we must now discern how it is determined within the system. In a real study with a real client, the modeler would ask the actors in the system how management would figure out average worker productivity at various stages in the project lifecycle. Here we propose a simple, but potentially realistic formulation. Let gross productivity represent unsatisfactory as well as satisfactory work, and let real productivity represent only satisfactory work. Then set perceived productivity to be somewhere between the two -- a combination of the two that could be captured as a weighted average. Toward the beginning of the project, one would imagine that perceived productivity would be almost entirely dominated by gross productivity -- all progress being seen as real progress. Toward the end of the project, as more and more unsatisfactory work has been discovered and rework completed, the perception of productivity would have to shift and be dominated by real productivity. Figure 2.26 shows such a structure.

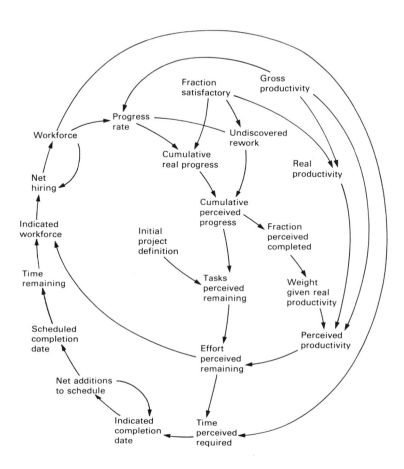

Figure 2.27: Complete structure of a simple project model.

Feedback Stucture of a Simple Project Model

Figure 2.27 gives the complete structure that results when the loops described are joined together. The structure is admittedly oversimplified. It is simple because realism at this point is less important than a clear presentation of the process of model conceptualization. Also, it is simple because every model should begin simply. The modeling process is iterative: a structure is developed, equations written (as we will soon do), behavior simulated, and the model compared

An Example of Problem Conceptualization 61

with reality, at which point refinements or complications can then be added. Complicating an essentially correct simple structure is easier than constructing and understanding a complex structure from scratch. It is adviseable always to begin simply, holding complexity in abeyance until a simple structure is well understood.

The diagram in figure 2.27 represents in causal-loop terms a model intended to show overruns in an R&D project in personnel and schedule. The intuitions used to develop the structure may convince us that the dynamic behavior of the model would indeed exhibit such overruns. However, to be sure, we must code the model for simulation by a computer. Simulation would also be necessary to ascertain whether policies intended to minimize the overruns would have their desired effects. Chapter 3 presents an introduction to the DYNAMO computer language in which the model will be coded for simulation, but we pause first to summarize a number of observations from the preceding model conceptualization.

2.5 SUMMARY OF THE CONCEPTUAL PHASES OF MODEL DEVELOPMENT

Problem identification and model conceptualization are the apparently less technical stages of a system dynamics study. Within these stages the modeler develops a statement of the context and symptoms of a problem, sketches reference behavior modes, articulates the purposes of the modeling study, settles on a system boundary, and develops a view of system structure in terms of feedback loops of action and information. Figure 2.28 summarizes how these stages and concerns fit together, and gives a preview of the later more quantitative stages in the modeling process.

The connections shown in figure 2.28 suggest the overlapping nature of the stages in the modeling process. A clear statement of model purpose, for example, contributes to to the process of model conceptualization as well as to the definition of the problem. Feedback structure is likewise a focus of two different stages of model development, nonquantitative conceptualization in terms of diagrams and formulation in DYNAMO equations.

Stripped of all explanatory detail and illustrations, the conceptual stages in the modeling process can be summarized as follows:

- *Problem identification* includes the awareness of a problem and an unambiguous definition of it. State the *context* and *symptoms* of the problem verbally. Define the problem dynamically in terms of *reference behavior modes*. There

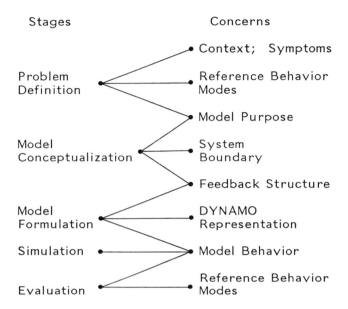

Figure 2.28: Stages and concerns in the system dynamics modeling process.

may be as many as three sets of reference modes: graphs showing the *problem behavior*, graphs showing *more desirable system behavior*, and, if policies have been tried in the past, graphs of the *observed policy behavior*.

- A statement of *model purpose* includes the *audience* for the study, an initial view of the *policies* one would like to be able to simulate with the model, and, most important, the kind or degree of *implementation* desired. The model purpose and implementation desired can range from education and consciousness raising, through actual adoption of policy recommendations, to adoption of the model by the client as an ongoing policy-testing tool (rare).

- The *system boundary* encloses the parts of system structure that are necessary to *generate the behavior of interest*. The boundary must be large enough to *close loops* of action and information within the system that are important for system behavior. Within the boundary should appear *policy levers* (points of intervention for policy testing) as well as variables (such as costs) by which policy results will be evaluated in the real system. The system boundary should *exclude components not relevant to the problem behavior of the system*. (*Focus on a problem, not a system.*) In determining what is inside and what is outside the system boundary, distinguish what is explicitly in from what is explicitly out and what is implicitly contained within by way of aggregation or interpretation of variables.

- *Model conceptualization* begins most easily with a *top-down approach*, building up the model in simple pieces, sectors, and functional areas. Develop the *physical structure* of the system first, then the clear *information flows*, followed by *perceptions*, and finally focus on *pressures* arising from perceptions that influence change in the system. Note that many decisions take a classic form: the actual state of the system, the perceived state of the system, a desired state, planning and pressures to close the gap, and resulting action changing the actual and perceived states of the system and closing a feedback loop. Availability of information is a primary concern; the modeler must represent the information on which actors in the system actually base decisions. Feedback loops result from repeated focus on the question, If A affects B, how does B affect A?

- The *dynamic hypothesis* is a statement of feedback structures that are conjectured to have the power to create or at least contribute to the problem behavior. A dynamic hypothesis should be attempted in the stage of model conceptualization, although a well-focused, consistent statement may be possible only after several iterations of conceptualization, formulation, simulation, and evaluation.

The Endogenous Point of View

One characteristic of the system dynamics approach deserves special emphasis. The system boundary is drawn so that all the components necessary to generate the problem behavior are contained within it. The approach searches for the causes and cures of a problem within some boundary. The focus is internal, not external.

The internal view creates a dramatically different problem focus. The external view places an individual, a firm, a city, or whatever, at the mercy of exogenous events. The internal view searches for structures within, which can create or exacerbate the system's problem behavior. The external view is frequently predisposed to search for blame:

"instabilities in our workforce and inventory are caused by erratic and seasonal customer orders;" "we are losing market share, so the vice president in charge of marketing should be fired;" "our city is dying because the federal government refuses to increase economic aid;" "this auto company is heading toward bankruptcy because of excessive environmental and safety regulations;" "the Great Depression began with the restrictive monetary policies of the Federal Reserve."

The external view is thus frequently event oriented: "Tommy punched Jimmy because Jimmy took Tommy's torque wrench;" "the Dow-Jones average dropped 14.25 points today in response to the President's remarks on the deteriorating situation in Outer Wistwithe;" "the Great War started when the archduke was assasinated on the beach at Pfumfnageln." Events do appear to have great explanatory power. The system dynamics approach, however, seeks to see events in a patterned context and to search for *structural* reasons for the pattern of events. The conviction is that *dynamic behavior is a consequence of structure.*

The contrast between the internal and external views of problems is often the unstated source of controversy. In studying the problem of stagnation and decay seen in some large U.S. central cities (see figure 2.1 for the population reference mode), J. W. Forrester drew the system boundary around the the city itself, leaving the suburbs in the environment of the system.[9] Migration of people and industry to and from the city was seen to be influenced by the internal dynamics of the city, affecting its *relative attractiveness* to other places. Some critics argued that the suburbs were the main cause of the cities' problems and so by not including a suburb sector the model was hopelessly flawed to begin with. Yet the internal view was able to generate the reference mode behavior of the cities! The external factors that others saw to be central to the problem, which certainly may contribute something to it, were not necessary to generate the problem behavior. To the extent that the urban model captured urban structure and dynamics accurately, therefore, it showed that some of the problems of urban decay lie within the scope of control of the city itself -- an important understanding not produced by the external view and one critically necessary for urban policy planning.

---

[9] See Forrester (1969). For an very readable and instructive development of a sequence of models leading to Forrester's, see Alfeld and Graham (1976),
[10] Lyneis (1980), chapter 2.

Summary of the Conceptual Phases of Model 65

Similar observations can be made about examples of corporate problems. Lyneis describes a system dynamics study of a large manufacturing company, a leading producer of industrial equipment.[10] The firm suffered a gradual loss of market share, declining profitability, and instability in production and employment. In the context of wage-price controls, supplier problems, and the firm's attempt to diversify, many might have attributed the problems to external factors.

The internal view, in the form of an extensive system dynamics study involving a comprehensive model of the structure of the firm, showed that the interaction among several operating policies designed to improve corporate performance was actually a significant cause of the decline. Policies recommended by the study, implemented in spite of their opposition to traditional practices of the firm, allowed the firm to improve its market share during the 1970 business downturn. One company manager estimated that the adopted policy earned in the neighborhood of $10 million in profits. Insistence on the view that problems are caused by external events would not have discovered the policy improvement.

An observation about seasonal demand completes the point. In the first book applying the feedback perspective to industrial problems, Forrester showed that a firm's reponses to random fluctuations in demand can actually create the impression of a seasonal demand. He noted:

> We find many company situations in which such an erroneous conclusion about seasonal sales had led to the establishment of employment, inventory, and advertising policies that in succeeding years caused a seasonal manufacturing pattern and thereby confirmed the original error. The likelihood of of self-generated seasonal sales patterns must be carefully considered in the design of management policies.[11]

The point here is not to denigrate a world view that holds that events cause events, because such a view is valid and has explanatory power. Rather, the purpose of the discussion is to note the distinction between the internal and external views, and to suggest that the internal view that seeks to derive patterns of events (behavior) from system structure has perhaps even more explanatory power.

The internal view is axiomatic to system dynamics. Once a boundary is drawn, the focus is inward. The emphasis on feedback loops is then no accident: without loops, all causal

---
[11] Forrester (1961), pp. 29, 443-449.

links would have to connect eventually only to influences outside the system boundary, and all causes of behavior would be traced to external influences. Feedback loops enable the inward view and give it structure.

3 INTRODUCTION TO DYNAMO

3.1 AN OVERVIEW OF DYNAMO

DYNAMO is a computer simulation language. Its name, a merger of the words "dynamic models," indicates its intended use: modeling real-world systems so that their dynamic behavior over time may be traced (imitated, simulated) by a computer.

A model written in the language DYNAMO is a view of a feedback system as if it were *continuous* over time. Orders received and items shipped, for example, are viewed as continuous flows into and out of the "tub" inventory. If the inflow and outflow of items into inventory happen to be equal, the level of inventory would not change over time; inventory would remain constant. If the inflow of items is greater than the outflow, inventory will build up, just as the level of water in a tub would build up if the faucet is open more than the drain.

To see most accurately what DYNAMO does to simulate a continuous feedback system, consider the case of constant inflow and outflow rates. Suppose we are shipping from inventory 100 units per month, and receiving only 80 units per month from our supplier. Inventory, then, is dropping constantly at the rate of 20 units per month. Its dynamic behavior is linear: its graph over time would be a declining straight line. Any high school algebra student ought to be able to compute by hand the change in inventory merely by multiplying the elapsed time and the constant rate of change of inventory:

$INV_{now} = INV_{past} +$ (intervening time)*(constant rate of change)

If inventory were 1000 units 6 months ago, and the receipt and shipment of items were 80 and 100 units per month, respectively, then

INV_{now} =
  INV_{past} + (intervening time)*(constant inflow - outflow)
   = 1000 units + (6 months)*(80-100 units/month)
   = 1000 + (6)*(-20)
   = 880 units.

The situation only becomes interesting when the inflow and outflow affecting inventory are not constant, but changing. The high school algebra student must give up in despair, or wait until he takes a calculus course, and even then obtaining the precise values of inventory over time may prove impossible. Enter the computer and DYNAMO (drum roll and trumpets). As the rate of receipts of orders and shipments vary over time, DYNAMO merely chops up continuous time into discrete bites: one tiny interval of time, then the next, and so on. *Within* each small interval of time, DYNAMO assumes the varying rates are *constant* and computes like the algebra student. The computation looks like the following:

INV_{now} = $INV_{just\ past}$ +
  (intervening time)*(assumed constant rate of change.)

The result is an approximation, to be sure. If, however, the time between computations is small enough, and the rates of change not too violent, the computed results will closely match those obtained (if it were possible to do so) in closed form by the application of sophisticated mathematics.

DYNAMO notation helps to communicate precisely how the computation described is carried out. Variables in DYNAMO have subscripts indicating their place in time. K denotes the present; J the point in time just past (just preceding K), and L the point in time in the immediate future (immediately following K). The symbol DT is used to represent the length of time elapsing between J and K, or K and L. See figure 3.1.

The inventory equation in DYNAMO notation thus becomes the following:

    INV.K = INV.J + DT*(ORDRCV.JK-SHPMTS.JK)
where
    INV.K   = value of inventory now
    INV.J   = value of inventory a time interval ago
    DT   = length of intervening time
    ORDRCV.JK   = orders received over the time interval JK

An Overview of DYNAMO

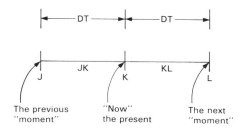

Figure 3.1: Timescripts J, K, and L in DYNAMO

SHPMTS.JK = shipments over the time interval JK

Thus, to simulate a dynamic feedback system, DYNAMO steps through time a formal, quantitative model of the system, one DT at a time. We could perform the calculations by hand (and we will in section 3.2 to insure understanding), but with any complexity at all in the system and any reasonably small computation interval DT we would be overwhelmed with the arithmetic. The task is ideal for a digital computer. In fact, there are a number of computer languages in which such computations (with different notation) could be performed, FORTRAN, BASIC, PASCAL, among others. Why DYNAMO? Simply stated, simulating such feedback systems is what DYNAMO was designed to do. One can hammer a nail with the heel of a shoe, but it's a lot easier with the tool designed for the purpose.

DYNAMO is intended to be usable by people who are not familiar with computers, as well as by experienced programmers. With a simple PLOT statement DYNAMO produces graphical output, even selecting its own scales if desired. One can ask DYNAMO to PRINT the output of a model simulation, and it will appear neatly arranged in tabular form. If programming errors occur (and they will), DYNAMO is designed to identify them for the modeler in understandable English. Finally, the language itself contains functions and features that actually aid the conceptualization and formulation of dynamic models.

3.2 A COMPUTED EXAMPLE

Coffee Cooling

To demonstrate the computation scheme used by DYNAMO to simulate a system, we consider the simple example of the cooling of a cup of coffee. Newton's law of cooling, the classic view of temperature dynamics, suffices for our purpose

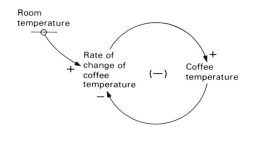

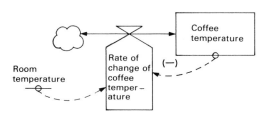

Figure 3.2: Coffee cooling loop

here. Newton postulated that coffee should cool at a rate directly proportional to the difference between the room temperature surrounding the cup and the temperature of the coffee itself. That is,

Rate of change in coffee temperature =
(constant)*(room temperature - coffee temperature)

where the constant represents such things as the circulation of air in the room, the conductivity of the cup, an adjustment for the kinds of degrees temperature is measured in, and so on. Since we will be considering hot coffee, rather than iced coffee, the change in coffee temperature will be negative; it represents the rapidity with which the temperature of the coffee is falling. (For simplicity we ignore the actual transfer of heat in the form of calories and such subtleties as the temperature of the cup and the minute warming of the room as the coffee cools.) We have a simple feedback loop, shown in figure 3.2.

Note that, as the coffee temperature falls, the cooling rate also changes: as the coffee temperature approaches the room temperature, the magnitude of the change in the coffee tem-

A Computed Example

perature will decrease. The change in coffee temperature is not constant. Thus, to compute the changes in temperature and trace the coffee temperature over time, we (and DYNAMO) resort to chopping continuous time into discrete intervals. For each small interval we assume, without too much error, that the change in coffee temperature is momentarily constant. In so doing, we can compute the coffee temperature as

$\text{Temp}_{now} = \text{Temp}_{past} +$
(intervening time)*(change in coffee temperature).

Let us assign the following mnemonic abbreviations (in anticipation of the kinds of variables one encounters in DYNAMO):

COFFEE = coffee temperature (degrees Celsius)
ROOM = room temperature (degrees Celsius)
CHNG = rate of change in coffee temperature (degrees Celsius per minute)
CONST = proportionality constant for Newton's Law (1/minutes)

Our complete quantitative model then consists of the following computations:

$\text{COFFEE}_{now} = \text{COFFEE}_{past} +$
(intervening time)*$(\text{CHNG}_{past\ interval})$
$\text{CHNG}_{next\ interval} = \text{CONST} * (\text{ROOM} - \text{COFFEE}_{now})$

For computing by hand, let us take the following values:

COFFEE = 90 (degrees Celsius, initially)
ROOM = 20 (degrees Celsius, constant)
CONST = 0.2 (dimensionless constant)

We decide (rather arbitrarily) to compute every half minute. Initially, COFFEE = 90 and ROOM = 20, so CHNG = 0.2*(20 - 90) = -14 degrees Celsius per minute. Assuming this cooling rate holds constant for the next half minute, the change in COFFEE during the next minute is -7 degrees. Therefore, after one-half minute

COFFEE = 90 + (-7) = 83 degrees Celsius.

Then the computations repeat. With COFFEE = 83 and ROOM still = 20, the cooling rate CHNG equals 0.2*(20-83) = -12.6 degrees Celsius per minute. At that rate the change in COF-

FEE during the next half minute is -6.3 degrees, making COFFEE after one full minute equal to 76.7 degrees Celcius. Table 3.1 gives these results and carries on the computation for seven full minutes. The reader may wish to check his understanding by verifying the next row or two in the table.

Table 3.1: Hand simulation of coffee cooling loop

Time (min.)	COFFEE (Coffee Temp, deg)	ROOM (Room Temp, deg)	ROOM - COFFEE (deg)	CHNG (Cooling Rate, deg/min)	Change in Coffee Temp in 0.5 min. (deg)
0	90	20	-70	-14	-7
.5	83	20	-63	-12.6	-6.3
1	76.7	20	-56.7	-11.3	-5.7(a)
1.5	71.0	20	-51	-10.0	-5.1
2	65.9	20	-45.9	-9.2	-4.6
2.5	61.3	20	-41.3	-8.2	-4.1
3	57.2	20	-37.2	-7.4	-3.7
3.5	53.5	20	-33.5	-6.7	-3.4
4	50.1	20	-30.1	-6.0	-3.0
4.5	47.1	20	-27.1	-5.4	-2.7
5	44.4	20	-24.4	-4.9	-2.4
5.5	41.9	20	-21.9	-4.4	-2.2
6	39.7	20	-19.7	-4.0	-2.0
6.5	37.8	20	-17.8	-3.6	-1.8
7	36.0	20	-16.0	-3.2	-1.6

(a) Rounding begins.

A graph of these results carried out for 20 minutes is shown in figure 3.3. The computations and the graph show the goal-seeking behavior one would expect for the cooling of a cup of coffee.

DYNAMO Equations for the Coffee-Cooling Loop

Using the subscripts described in section 3.1 to designate intervals of time, we write the DYNAMO equations for the coffee-cooling computations as follows:

COFFEE.K=COFFEE.J+(DT)*(CHNG.JK)
CHNG.KL=CONST*(ROOM-COFFEE.K),

with DT = 0.5, ROOM = 20, CONST = 0.2, and the initial value of COFFEE equal to 90.

A Computed Example

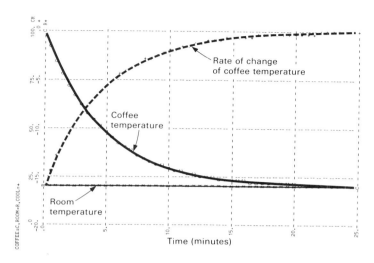

Figure 3.3: Coffee cooling over time

Note that we, and DYNAMO, do the computations "in the present," that is, at time K. The coffee temperature "now" is computed from the "past" coffee temperature and the rate of temperature change enduring from "then" to "now". Knowing the present coffee temperature, we can compute the rate of change in coffee temperature for the next time interval (KL). Once that is accomplished, we shift our attention in time to the next moment, calling it "now" (time K) and repeating the process.

No matter how complex the feedback system, how interconnected the variables, or how intriguing the dynamics, the computation scheme outlined in this simple example is precisely the one followed by DYNAMO in all applications.

A Word About Accuracy

The reader no doubt observed that the selection of the interval between computations (DT) was apparently arbitrary. We could have taken DT equal to 1 minute, or 0.0625 minutes, or whatever. The question naturally arises, what effect would changing the computation interval have on the computed results? For that matter, the selection of the computation interval was a necessary part of an apparently gross approximation, simulating a continuously changing system by irreverently chopping time into discrete intervals.

How much error, one might ask, is introduced in such a discrete approximation to a continuous process? Sweeping aside for the moment the deep issue of limited computer precision, we may assert that once the computation interval (DT) is "small enough," small changes in it have very little effect on the computations. Furthermore, in theory, it can be taken small enough so that differences between the discrete computations and the continuous solution become as small as one likes.

Table 3.2 shows a comparison of the values of the COFFEE temperature computed at different intervals. The first observation one should make from this table is that only one of the columns is out of line. Reasonably small values of the computation interval DT give approximately the same results. Second, as DT gets smaller the values get closer together and closer in their values to the continuous solution obtained through a bit of a calculus. As will be seen, the focus in system dynamics is on the *shape* of graphs over time, so we are not disturbed by these slight differences.

Table 3.2: Comparison of five different simulations of the coffee cooling loop.

Time	COFFEE Value for DT =				Continuous Solution $(20 + 70\ e^{-0.2t})$
	0.125	0.5	1	5	
0	90.0	90.0	90.0	90.0	90.0
1	77.2	76.7	76.0		77.3
2	66.7	65.9	64.8		66.9
3	58.1	57.2	55.8		58.4
4	51.1	50.1	48.7		51.5
5	45.4	44.4	42.9	20.0	45.8
6	40.8	39.8	38.4		41.1
7	37.0	36.0	34.7		37.3
8	33.8	33.0	31.7		34.1
9	31.3	30.5	29.4		31.6
10	29.2	28.7	27.5	20.0	29.5

Apparently, however, one has to be careful about taking DT too big: in the fourth column, DT = 5 clearly produces markedly different behavior. In one minute the coffee temperature jumps to equal the room temperature, and stays there. We can illustrate why this happens if we substitute the expression for CHNG into the equation for COFFEE:

$$COFFEE_{now} = COFFEE_{past} + DT*(CHNG_{past\ interval})$$

A Computed Example

$$= \text{COFFEE}_{past} + DT*CONST*(ROOM-\text{COFFEE}_{past}).$$

When DT = 5, DT*CONST = (5)*(0.2) = 1, so in the first computation interval the entire discrepancy between the ROOM temperature and the COFFEE temperature is made up:

$$\text{COFFEE}_{now} = \text{COFFEE}_{past} + (1)*(ROOM - \text{COFFEE}_{past})$$
$$= ROOM.$$

From that point on in time, the CHNG rate is zero because there is no discrepency between COFFEE and ROOM. The lesson is clear: select a computation interval small enough to prevent this sort of nonsense. Later we will give some rules of thumb for selecting DT.

3.3 TYPES OF EQUATIONS

Variables and Constants

All the quantities appearing in a DYNAMO model can be classified in two broad groups: constants, whose values cannot change at all in the course of a simulation, and variables, whose values can. The distinction in DYNAMO is easy to see at a glance. All variables are written with *time subscripts:* K, J, KL, or JK. Constants are written with no timescripts. Thus SALES.K could represent a *varying* quantity of sales of a product; SALES (without the timescript) would represent a *constant* quantity of sales.

A constant acquires its value in a DYNAMO model in a statement labeled on the left with a C. For example, in the coffee-cooling model, the constant representing the room temperature and the proportionality constant for Newton's law of cooling would appear in the following two statements:

```
C   ROOM=20
C   CONST=0.2
```

For the *variables* appearing in the hand simulation described in section 3.2, there were two rather different types of computations: the *algebraic* computation for the CHNG rate, and the *accumulating* computation for the COFFEE temperature. Since COFFEE depended for its current value on its immediately past value, it accumulated all the effects of past changes. In this particular example, COFFEE accumulated a stream of negative values for the cooling rate, and thus it declined. Other accumulations in the real-world may grow, such as global population; or grow and decline or cycle, such as a business inventory influenced by varying production and shipment rates. The identification of such accumulations in a problem is a central concern of model formulation from the system dynamics perspective. They are called levels.

Level Equations

A variable that accumulates over time an inflow and/or an outflow is a *level* variable, and in DYNAMO the equation in which the acccumulation is computed is called a *level equation*. (In other terminologies levels are known as integrations, stocks, or state variables.) The system dynamics terminology is intended to invoke an image of the level of a liquid in a tub, as noted in section 3.1.

A *level equation* is always written in the same form, preceded in a DYNAMO model by the letter L signifying the equation is a level equation:

L   LEVEL.K=LEVEL.J+DT*(INFLOW.JK-OUTFLOW.JK)

The equation for COFFEE in the coffee-cooling loop is a level equation:

L   COFFEE.K=COFFEE.J+DT*(CHNG.JK)

where the variable CHNG represents the net inflow-outflow affecting the temperature level. The subscripts, as described in section 3.2, show that the value of a level variable at time K (the present) is computed from its value at time J (the previous moment) and the change in its values resulting from the values of the flows affecting it over the time interval JK (from the previous moment to the present). DT is a constant whose value is the length of time from J to K; it is called the *computation interval*, or *solution interval*, for the model.

Other examples of likely level equations are the following:

L   INV.K=INV.J+DT*(ORDRCV.JK-SHPMNTS.JK)
where
    INV   = level of inventory (units),
    ORDRCV = rate of receiving orders into inventory
                (units/month),
    SHPMTS = rate of shipping from inventory;

L   POP.K=POP.J+DT*(BIRTHS.JK-DEATHS.JK)
where
    POP   = level of population (people),
    BIRTHS   = birth rate (people/year),

Types of Equations

DEATHS = death rate (people/year);

L   PRDCAP.K=PRDCAP.J+DT*(CNSTR.JK-DEM.JK)
where
  PRDCAP = production capacity (plants),
  CNSTR = construction rate (plants/year),
  DEM   = demolition rate (plants/year).

For some applications, a number of inflows and outflows must be accounted for; for example, the population of a city over a fifty-year time span is influenced not only by births and deaths but also by migration:

L   CITPOP.K=CITPOP.J+DT*(BRTHS.JK-DTHS.JK+INMIG.JK-
X   OUTMIG.JK)
where
  CITPOP = city population (people),
  BRTHS = birth rate (people/year),
  DTHS   = death rate (people/year),
  INMIG = in-migration rate, or immigration
           (people/year),
  OUTMIG = out-migration rate, or emigration
           (people/year).

As noted in section 2.2, the conventional symbol for a level equation in a flow diagram is a *rectangle.*

Before continuing with further discussion of DYNAMO equations, we pause briefly here to address some issues which these examples implicitly raise.

Naming Quantities

DYNAMO recognizes up to six characters for the name of a quantity. It is for this reason that the names given thus far may have occasionally looked strange. The first character in a name must be a letter, but any characters after that may be either letters or numbers. Any number of characters from one through six is acceptable. Thus inventory could be INV, I, INVTRY, or even INV1 and INV2 if one wishes to distinguish different levels of inventory. The choice of names is up to the modeler, of course, but it is wise to select names that convey the meaning of the quantity. While X1, X2, X3, and Z may be convenient shorthand for hand-written mathematics, in a DYNAMO model they would be unforgivably uncommunicative.

Arithmetic and Algebra in DYNAMO

The symbols required by DYNAMO for arithmetic computation are the common ones; we have been using them throughout without comment. The complete list is as follows:

    addition   +
    substraction   -
    multiplication *
    division   /
    exponentiation[1] **

Multiplication may also be performed by enclosing variables in parentheses and omitting the asterisk, as, for example,

    (LABOR.K)(PROD.K),

which yields the product of labor (people) and productivity (units produced per person per year).

In algebraic computations, such as the CHNG rate in the coffee-cooling model, DYNAMO performs the operations just as a (talented) algebra student would: powers and roots first, then multiplications and divisions, and finally additions and subtractions. Parentheses can be used to modify that order, as, for example, the CHNG rate

    CHNG.KL=CONST*(ROOM-COFFEE.K)

in which the subtraction is computed first, followed by the multiplication. Otherwise, arithmetic is done left to right, as the operations appear.

Format

Two subtleties arise in typing equations for the DYNAMO compiler. First, DYNAMO accepts *no spaces* in its equations. The character designating the type of equation is followed by one or more spaces, but, once the equation starts, there can be no spaces. If you find that DYNAMO is confused by an equation that looks fine to you, check to be sure that it contains no spaces.

Second, no equation may extend beyond column 72. Information in columns 73 through 80 (counting left to right) is ignored. Unfortunately, some computer systems may not warn you when you have typed beyond column 72.

---

[1] In Mini-DYNAMO and DYNAMO II/370 a quantity is raised to a power using the EXP and LOG functions, as shown in section 3.6.

Types of Equations

Consequently, if you type a line too long, and then ask the editor to print out your model, you may see a complete, apparently correct, equation where DYNAMO sees a truncated one. Worse yet, DYNAMO may find nothing wrong with what it sees, and may run your model without reporting any errors. So, if you obtain strange results from an apparently correct model, check to be sure its longest equations are not too long.

If an equation is complicated enough to go beyond column 72, it must be broken into pieces and continued on the next line (or lines) of the program. Continuations are labeled at the left in the first column with the letter X (for extension). For example, the level equation for CITPOP given earlier could be written as follows:

```
L    CITPOP.K=CITPOP.K+DT*(BRTHS.JK-DTHS.JK+
X    INMIG.JK-OUTMIG.JK)
```

The break in the equation can be made anywhere, but it is good practice to call attention to the continuation by leaving an obviously incomplete equation that breaks right after an algebraic operator.

With these understandings of arithmetic, algebra, and format in DYNAMO, we may now return to the types of equations appearing in continuous models of feedback systems. As noted, accumulations of flows are computed in level equations. The rest of the computations appearing in such models are simple algebraic computations. They are divided in DYNAMO into two classes: rate equations and auxiliary equations.

Rate Equations

The variables representing the inflows and outflows in level equations are usually computed in equations termed *rate equations* and designated in DYNAMO by the letter R. An example is the cooling rate from the coffee model:

```
R    CHNG.KL=CONST*(ROOM-COFFEE.K)
```

(In calculus terminology, a rate equation computes some or all of the derivative of a level.) As the previous discussions of DYNAMO computations imply, rate equations are computed at time K, and their values hold constant over the time interval from K to L (from the present to the next moment or next computation). The subscript for the computation of a rate is written as KL to indicate this timing.

Unlike level equations, rate equations have no standard form. The following examples, discussed in later sections of this book, indicate the wide range of commonly encountered possibilities:

        R   BIRTHS.KL=BRF*POP.K
    where
        BIRTHS   = birth rate (people per year),
        BRF   = birth rate fraction (births per year
                   per person),
        POP   = population (people);

        R   DEATHS.KL=POP.K/AVLIFE
    where
        DEATHS   = death rate (people per year),
        POP   = population (people),
        AVLIFE   = average lifespan (years);

        R   ORDRS.KL=AVSHIP.K+(DSINV.K-INV.K)/TAI
    where
        ORDRS   = orders (units per month),
        AVSHIP   = average shipments (units per month),
        DSINV   = desired inventory (units),
        INV   = actual inventory (units),
        TAI   = time to adjust inventory (months).

As these examples indicate, no one form of rate equation will suffice for all situations. Level equations, with their easily understandable and regular form, are rather trivial to write; rate equations are difficult. Each one represents a fresh understanding of some process of change in some particular system. It is reasonable to state that the majority of the effort of system conceptualization and model formulation is directed toward the proper specification of rate equations. Our focus here, however, is merely on the correct form of the subscripts and arithmetic in such equations.

The conventional symbol for a rate equation in a flow diagram is a valvelike device, as shown in section 2.2.

Auxiliary Equations

Thus far, it would appear that formulating a model in DYNAMO requires identifying the variables that will be modeled as levels, determining their rate equations, and assigning values to all the constants appearing in these equations. Usually, however, it is very difficult to write a rate equation for a level without first doing some other algebraic computations, capturing information needed in the formulation of the

Types of Equations

rate equation. These additional algebraic computations in DYNAMO are termed *auxiliary* equations; the variables computed in them are usually called simply *auxiliaries*.

An auxiliary equation is a computation representing information in a feedback system. Auxiliaries, as their name implies, aid in the formulation of rate equations. In DYNAMO an auxiliary equation is identified with the letter A in the far left-hand column. In the coffee-cooling model for example, we could have computed the difference between the ROOM temperature and the COFFEE temperature as an auxiliary:

    A   DISC.K=ROOM-COFFEE.K

DISC.K, which represents the discrepancy at time K (the present), would then be used in computing the CHNG rate:

    R   CHNG.KL=CONST*DISC.K

In this particular instance there is little to be gained by this use of an auxiliary equation, since the rate equation is not very complicated to write without it, and we are not particularly interested in the information represented by DISC. Other examples may better indicate the need and the diversity of auxiliary equations:

    A   LFO.K=BLDNGS.K*LPB/AREA
where
    LFO   = land fraction occupied (dimensionless),
    BLDNGS = buildings (buildings),
    LPB   = average land per building (acres),
    AREA   = city area (acres);

    A   DSINV.K=DIC*AVSALE.K
where
    DSINV   = desired inventory (widgets),
    DIC   = desired inventory coverage (weeks),
    AVSALE = average sales rate (widgets per week);

    A   DD.K=BKLG.K/AVSHIP.K
where
    DD   = delivery delay (weeks),
    BKLG   = order backlog (orders for thingumabobs),
    AVSHIP = average shipment rate (orders per week);

    A   EPREM.K=(CPD.K-CPPRG.K)/PPROD.K
where
    EPREM = effort perceived remaining (man-months),
    CPD   = current project definition (tasks),

CPPRG = cumulative perceived progress (tasks),
PPROD = perceived productivity (tasks/person/month),

Like rate equations and unlike levels, auxiliary equations have no single standard form. However, auxiliaries are always computed in the present from the present values of other variables, which may be levels, rates, or other auxiliaries.[2] Thus the timescript appearing on the left in an auxiliary equation is always K, and the timescripts appearing on the right are either K or, occasionally, JK. DYNAMO will remind you with an error message if you violate these timescript conventions, and then it will go ahead and compute as if you had written the timescript correctly.

An auxiliary equation is symbolized in a flow diagram by a *circle*.

Table Functions

It is often desirable to use an auxiliary that is not a simple algebraic combination of other variables in the model. Frequently, a nonlinear relationship is required. Fortunately, if a graph can be sketched, it is very simple to capture the relationship in DYNAMO in a *table function*.

Consider a simple example of salesmen generating orders to be filled by a firm.[3] The rate at which orders are booked could be written

    R   OB.KL=S.K*SE.K
where
    OB = orders booked,
    S   = salesmen (people),
    SE = sales effectiveness (orders booked/person/month).

Now suppose that sales effectiveness is believed to depend upon the *delay* the firm imposes in filling orders. SE could be formulated as a function of the delivery delay recognized in the market, DDRM. Figure 3.4 shows a hypothetical

---

[2] It is poor practice to put rate variables on the right of auxiliary (or rate) equations, unless a SMOOTH or DELAY function is employed. See section 3.5 for an introduction to these functions. See section 4.8 for a discussion of the conceptual and computational issues involved in writing the subscripts .KL or .JK on the right side of an auxiliary equation.
[3] Adapted from Forrester (1968a).

Types of Equations

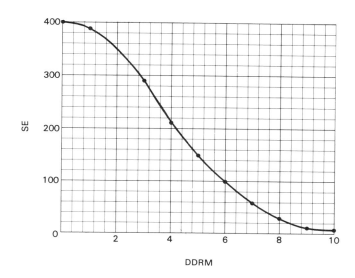

Figure 3.4: Graph of sales effectiveness, SE, as a function of the delivery delay recognized in the market, DDRM, showing points specified in a DYNAMO table function.

relationship that will serve as our example of a table function.

One could approximate such a nonlinear relationship by noting several x- and y-coordinates of points along the graph,

X	DDRM	0	1	2	3	4	5	6	7	8	9	10
Y	SE	400	388	348	292	212	152	100	60	32	12	8

and computing other values of SE by interpolation between these values. Note that the x-values are completely determined, if they are equally spaced, by specifying only the minimum value, the maximum value, and the increment between values, in this case, 0, 10, and 1, respectively.

Thus, to capture a good approximation of the relationship sketched in figure 3.4, it suffices to locate eleven points by specifying only the minimum, maximum, and increment of points on the x-axis and the successive y-values of the eleven points shown in the graph. DYNAMO accomplishes this in two lines:

    A   SE.K=TABLE(TSE,DDRM.K,0,10,1)

T   TSE=400/388/348/292/212/152/100/60/32/12/8

(In the line labeled T, which lists the successive y-values of points on the graph, DYNAMO also accepts commas instead of slashes, should commas more suit your tastes.)
    TABLE(__,__,__,__,__) in this auxiliary equation indicates that the value of SE is to be computed from a functional relationship that is specified by listing points on its graph. TSE, the first argument in the TABLE function is the (arbitrary) name of the table of y-values in the graph. In this case the name TSE was chosen as a mnemomic for "table for SE." The second argument in a TABLE function is the value of the input variables, the x-axis variable in the graph. In this case the input variable is DDRM.K, the delivery delay recognized in the market. The third, fourth, and fifth arguments in the TABLE function specify the x-axis of the graph, representing, respectively, the minimum, maximum, and increment values of x. The T equation completes the specification of the table by giving, in order, the y-values of the successive specified points on the graph, beginning with the y-value for the minimum value of x and ending with the y-value for the maximum value of x.
    As indicated above, values of SE are computed from this table by linear interpolation: if an input value DDRM.K is not equal to one of the specified points, the corresponding y-value computed by DYNAMO lies on the straight line connecting the specified points on the graph on either side of that input value. The slight inaccuracies caused by the linear interpolation -- the sudden shifts in slope between successive intervals on the interpolated graphs -- are of little computational concern and are vastly outweighed by the simplicity and flexibility of the table function formulation. (If you wish, you could specify a nonlinear relationship exactly using a polynomial passing smoothly through the points specified, but attention to such accuracy is hardly ever justifiable in system dynamics modeling.)
    Thus, to summarize, an auxiliary variable can be specified in a table function, using the following pattern:

    A   var.k = TABLE(table name, input to table,
            min-x, max-x, x-inc.)
    T   table name = y0 / y1 / y2 /.../ yn

(without spaces, of course). The lower case letters indicate phrases that must be replaced by terms acceptable to DYNAMO.
    The system dynamics modeler commonly uses two types of table functions, denoted TABLE and TABHL.[4] They differ

Types of Equations

only in the ways they treat values of the input variable that fall *outside* the specified range from minimum x to maximum x. If table is used, DYNAMO notes every time the input value leaves or returns to the specified range. The compiler prints messages to that effect before showing the results of a simulation. The messages about table overruns are for information only. DYNAMO goes ahead and computes in spite of them. If the input value is above the maximum specified, the *last* y-value listed in the table is used. If the input is less than the minimum specified, the *first* y-value is used.

TABHL performs exactly as TABLE except that sojourns outside the range of the table are not reported. DYNAMO keeps quiet about them, assuming that the modeler does not care to know. The letters HL in the name TABHL represent "high-low," as a reminder of how this function tacitly handles values of the input that are outside the specified range. In the example of sales effectiveness SE, TABHL might profitably have been used just in case DDRM goes above 10 months in a simulation. However, TABHL hides some information about model behavior. Since TABLE and TABHL compute the same way, the modeler would be well-advised to use TABLE exclusively until the more advanced stages of a project.

The reader should note that here we have been concerned only with the form of the DYNAMO equations for table functions. The art of conceptualizing such functions -- and it is a subtle art indeed -- is discussed in chapter 4.

The conventional symbol for a table function in a flow diagram is a circle with two lines drawn across it, as shown in the summary at the end of this section in figure 3.5.

---

[4] DYNAMO provides two other kinds of table functions for very special circumstances: TABXT and TABPL. TABXT is a seldom used variation in which y-values for inputs falling outside the table are computed by *extrapolation.* Rather than extending horizontally beyond the range of inputs specified, TABXT extends the table function obliquely in whatever direction the graph was heading in its last interval. TABPL is employed in those rare situations in which the corners introduced by linear interpolation in the other table functions cause problems in a simulation. TABPL passes a smooth, polynomial function through the specified points of the table, thereby eliminating corners and linear segments.

N Equations

All equations appearing in DYNAMO are labeled on the left with a letter indicating the type of equation. We have discussed C, L, R, A, and T equations, representing, respectively, constants, level equations, rates, auxiliaries, and tables. The remaining equation type is labeled N.

The primary use of the N equation is to assign an initial value to a level variable. The N equation usually makes the most sense if it is placed in the model listing right after the level equation to which it corresponds. The following is an example:

```
L    INV.K=INV.J+(DT)*(ORDRCV.JK-SHPMTS.JK)
N    INV=1000
```

These equations are intended to represent a level of inventory assigned an initial value of 1000 units. The simulation would begin with INV equal to 1000, deviating from that value in the course of the simulation if orders received and shipments become unequal.

All level variables in a model must be assigned initial values, so for each L equation there must be an N equation. Note that variables in N equations appear *without* timescripts, as only their initial values are involved.

DYNAMO has extensive capabilities with initial values and N equations. An N equation, for example, may be written with auxiliaries and rates on the right, as well as constants, even if those variables themselves do not have explicit N equations. DYNAMO will work through the computations in the model to obtain appropriate initial values, treating the A and R equations as N equations. For the inventory example above, the N equation might have been written

```
N    INV=DIC*AVSHIP,
```
where
    DIC   = Desired Inventory Coverage (weeks),
    AVSHIP = Average Shipment Rate (units/week).

DIC would probably be a constant (4 weeks?) while AVSHIP might be computed later in an auxliary equation. DYNAMO will try to produce an initial value for AVSHIP, even if the modeler does not specify one. If it eventually finds that not enough information about constants and initial values has been provided in the model, it will say so in an error message and the modeler can then add what is needed. Section 4.7 provides the details.

Types of Equations

N equations may also be used for computing a constant, which may or may not be initial value of some level. Suppose, for a simple example, one wished to compute a birth rate fraction, BRF, for a rabbit population, knowing the number of litters per fertile female rabbit per year, the number of offspring per litter, the fraction of offspring surviving, the fraction of rabbits that are female, and the fraction of females that are mature. BRF would be the product of all these, and could be computed in an N equation

```
N    BRF=LPFY*OPL*FS*FF*FM    birth rate fraction
                              (rabbits/rabbit/year)
C    LPFY=3                   litters per female per year,
C    OPL=8                    offspring per litter,
C    FS=0.6                   fraction surviving,
C    FF=0.5                   fraction of rabbits female,
N    FM=(AVLIFE-1)/AVLIFE     fraction of females mature,
C    AVLIFE=5                 average lifespan (years).
```

One could, of course, do all the computations by hand and simply write

```
C    BRF=5.76,
```

but there will be times when the computation capability of the N equation is quite necessary.

Summary of Types of Equations

We have discussed seven types of equations in a DYNAMO model. Each equation is labeled on the left with a single letter indicating its type:

- L   level equation,
- R   rate equation,
- A   auxiliary equation,
- C   assigning a value to a constant,
- T   assigning y-values in a table function,
- N   assigning or computing an initial value.

L equations are accumulations or integral equations. R and A equations are algebraic computations. C, T, and N statements assign values to parameters which remain constant throughout a simulation. An X in the left-most column indicates that an equation is being continued from the previous line.

Each of these equation types has conventions about the timescripts for the variables which appear in it. DYNAMO uses these timescripts to check that the model formulation is consistent with the simulation process, which steps the continuous model through time by computing at discrete time intervals.

Figure 3.5 shows the conventional symbols used for diagraming these equations.

3.4 OBTAINING OUTPUT

Once a model has been conceptualized and properly formulated in terms of levels, rates, and auxiliaries, only two or three additional statements in DYNAMO are required to perform a computer simulation and produce numerical and/or graphical output. Always required is the information contained in the SPEC statement, specifying several parameters required for a simulation. The user then has the option of including a PLOT statement and/or a PRINT statement, requesting output plotted graphically and/or printed numerically in tabular form.

The SPEC Statement

In one statement, labeled on the left with the letters SPEC, the modeler can specify four parameters (constants) determining details of the computer simulation. In the SPEC statement, values are assigned for DT, LENGTH, PRTPER, and PLTPER.

DT is the *computation interval* or *solution interval,* the model time elapsing between computations in the simulation. The letters DT stand for delta time, the mathematical symbol usually used to represent a change in time or a difference in time. A rule of thumb for selecting DT will be given in section 3.5. Suffice to say at this point, DT should be small enough that small changes in it have no significant effect on the results but not so small as to require inordinate amounts of computer time for the simulation. If DT is to be less than 1, it is a good idea to assign it a value that is a power of 2 (such as 0.5 or 0.125.) to minimize computer round-off error.

LENGTH is a constant indicating the *time at which the simulation should stop.* Time in DYNAMO is represented by the variable TIME.K. When TIME.K first becomes greater than or equal to LENGTH, the simulation ceases. Unless told otherwise, DYNAMO assumes that the initial value of TIME is zero. Thus the parameter LENGTH usually represents the length of time the simulation is carried out. However, if the model contains an initial-value equation for time, such as

N   TIME=1960

Obtaining Output

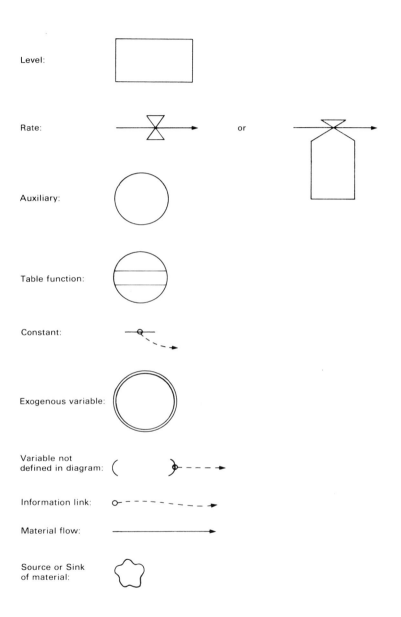

Figure 3.5: Symbols for flow diagrams

and LENGTH is set at 2000, the simulation would last for 40 units of model time, not 2000. The time coordinates printed on the graph would begin with 1960 and end with 2000.

PRTPER is a constant that stands for the *print period*, the time interval between successive prints of numerical output from the simulation. Unless told otherwise, DYNAMO assumes PRTPER is zero, meaning that no printed output is called for. If PRTPER is 5, the value of the printed variables would be printed out every five units of model time, beginning with the initial values. Whenever PRTPER is assigned a nonzero value in the SPEC statement, a PRINT statement should be included (see below), specifying which quantities are to be printed.

PLTPER is a constant representing the *plot period*, the time interval between successive points plotted on graphical output from the simulation. The statements about the print period PRTPER apply analogously to PLTPER. If plotted output is desired, PLTPER must be assigned a value greater than zero in the SPEC statement, and one or more PLOT statements must be included to specify what quantities are to be plotted.

The SPEC statement is really just a speedy way of assigning values to these four constants, e.g.:

SPEC DT=0.5/LENGTH=100/PRTPER=10/PLTPER=2

(Commas may be used here instead of slashes, if desired.) Assuming the initial value of time is zero for this example, this SPEC statement would result in 11 lines of numerical output printed in tabular form and 51 lines of graphical output (one line for TIME = 0, followed by 100/2 or 50 lines of computed output). Computing every 0.5 time units, the model would calculate each variable four times between appearances of plotted output and 20 times between successive values of printed numerical output.

The following example calls only for graphical output, since DYNAMO assumes PRTPER is zero:

SPEC DT=0.25/LENGTH=20/PLTPER=0.5

It is equivalent to

SPEC DT=0.25/LENGTH=20/PRTPER=0/PLTPER=0.5

PRINT and PLOT Statement

The PRINT statement simply lists, separated by commas, the variable names whose values are to be printed out in tab-

Obtaining Output

ular form. The following example calls for the values of inventory, orders received, shipments, and average sales to be printed every PRTPER:

PRINT INV,ORDRCV,SHPMTS,AVSALE

Note that timescripts are not used in the PRINT statement.[5]

A PLOT statement names the variables to be plotted together on a single graph. One may select a character (any single character) to be typed on the graph for each variable plotted, indicating the location of each plotted point and identifying which points belong to which variable. The following example calls for a graph from a predator/prey model in which the variables represent the rabbit population, the lynx population, and the average kills per lynx:

PLOT RAB=R/LYNX=L/AVKPL=+

Points for the rabbit population would be identified with the symbol R. The symbols L and + would locate, respectively, points for the lynx population and the average number of kills per lynx. The *slashes* between the parts of this PLOT statement signal DYNAMO to plot each of the curves on *separate scales*. In this case, DYNAMO will select its own scales and will print (possibly) three different ones for the variables plotted on a single graph.

It is often desirable to force DYNAMO to plot different variables on the *same scale*. Birth and death rates for a population, for example, might be graphed on the same scale so that it is trivial to observe from the graphs when births and deaths are equal or when one is greater than the other. To signal DYNAMO to plot several variables on the *same scale*, list them together in the PLOT statement, separating them by *commas* rather than slashes. The following example from a predator/prey model shows such a usage, requesting the rabbit birth rate, death rate, and kill rate to be plotted

---

[5] It is also possible to specify the column in which a variable will be printed. The statement

PRINT 1)INV/2)ORDRCV,SHPMTS/3)AVSALE

would produce printed output organized in three columns, with the values for orders received and shipments printed one under the other in the same column at each print period. This format is sometimes useful for making comparisons between variables.

on the same scale, while allowing different scales for the other variables:

PLOT   RAB=R/RBR=1,RDR=2,RKR=K/LYNX=L/AVKPL=+

In this case, DYNAMO would again select all the scales itself, but it would plot RBR, RDR, and RKR on the same scale.

If desirable, the modeler can specify the scale for any variables in a PLOT statement. The following PLOT statement sets the scale for the graph of the rabbit population to range from 0 to 400:

PLOT   RAB=R(0,400)/LYNX=L/AVPL=+

DYNAMO will still select the scales for the other two variables. To specify their scales as well, the modeler needs only to write in parentheses the minimum and maximum scale values immediately following the character to be plotted. Finally, a common scale for several variables can be specified by combining the comma notation with the specified scale, as in the following, in which the rabbit birth rate, death rate, and kill rate would all be plotted on a scale from 0 to 800:

PLOT   RAB=R/RBR=1,RDR=2,RKR=K(0,800)/LYNX
X   =L/AVKPL=+

The other scales would be selected by DYNAMO.

To produce more than one graph for a simulation run, the modeler simply includes more than one PLOT statement.

Documenting a Model and its Simulations

A model must communicate not only with the computer but with anyone who might be interested in its structure and applicability. Symbols like PRDCAP and INV no doubt communicate more information than variables named X or Z, also but there is still a danger that a DYNAMO model might be unreadable by anyone other than its creator. Even the modeler himself might forget what version of the model produced which simulation runs. So, to enhance the readability of a model and to identify particular simulation runs, DYNAMO provides the *, NOTE, and RUN statements.

It is useful to place an   *   statement as the first line of a DYNAMO model. Whatever is written in this statement will be printed out at the top of a model listing and will appear on all its simulation runs. The * statement should probably contain the name of the model and perhaps a date or a number or phrase identifying the particular version of the model. The

Obtaining Output

following shows a likely example, with the asterisk appearing in the first space on the line to identify the statement as the model heading:

```
*     INVENTORY CONTROL MODEL, V 3.7
```

Every simulation run of a model containing that statement would show the phrase "INVENTORY CONTROL MODEL, V 3.7" just as it appears in the * statement. The label in the statement can be at most 50 characters long, including spaces. Unlike most other DYNAMO statements, spaces are acceptable in the * statement.[6]

NOTE statements are used to document a model listing -- to identify different sectors of a model, to give meanings to possibly cryptic names for quantities, to note the units intended for each quantity, and to place in the text of the model any other comments desired by the modeler. In a NOTE statement, the word NOTE begins in the left-most space in the line and can be followed by any string of characters, including spaces. The following are possible examples:

```
NOTE
NOTE      *** PRODUCTION SECTOR ***
NOTE
```

Here asterisks were used by the modeler to attract attention to a key section of his model. They have no connection to multiplication or to the asterisk statement. The pattern they form would be printed out in every listing of the model but would not be printed on simulation runs of the model. The following example shows a NOTE statement defining a model variable from an epidemic model:

```
L     SICK.K=SICK.J+DT*(INF.JK-CURE.JK)
N     SICK=2
NOTE         INFECTED POPULATION (PEOPLE)
```

(In this example, the SICK level is increased by the infection rate INF and decreased by the cure rate CURE; its initial value in the simulation is 2.) The NOTE statement defines the quantitiy name SICK (in case it was not obvious) and gives in parentheses its units of measure -- a wise practice, as will be seen.

---

[6] Asterisk statements play special roles in DYNAMO's documentor program; see section 4.5.

There is a second way of documenting equations in DYNAMO which some users may prefer. Whenever an equation in DYNAMO is followed by one or more spaces, a comment or note may be written *on the same line* as the equation. The following, for example, is an acceptable variation of the previous example identifying the variable SICK as representing "infected people":

L SICK.K=SICK.J+DT*(INF.JK-CURE.JK) INFECTED PEOPLE

In simulating (running) the model, characters separated from an equation by at least one space are completely ignored by DYNAMO. (That is why no spaces are allowed *within* any DYNAMO equation.) One could begin all such definitions at a certain column, providing a model listing that is quite easy to follow.

Identifying individual simulation runs is made possible by the *RUN* statement. A descriptive heading such as BASE SIMULATION could be placed at the end of the model listing, after the SPEC, PRINT, and PLOT statements:

RUN   BASE SIMULATION

As with all DYNAMO statements, the word RUN identifying the form of this statement begins in the left-most space of the line. Following the word RUN can be any string of characters, including spaces, as long as it does not exceed 50. (In Mini-DYNAMO and DYNAMO F only the first 6 characters of the run label are printed.) The first simulation results obtained from a model containing the RUN statement given here would be labeled with the phrase BASE SIMULATION. The heading would automatically be dropped for subsequent runs of the model, or it can be changed to label each rerun appropriately (see Reruns, below).

Order of Statements

DYNAMO requires no particular order for the equations in a model. Arrange them in an order that is easy for a human reader to follow, and don't worry about the computer. (Section 4.5 contains some suggestions on ordering a model.)

Scaling Letters

If the coordinates on the y-axis scales are very large or very small, DYNAMO switches to an alphabetical abbreviation for scientific notation. The number 8.0A might appear on a scale, for example, to represent $8.0*10^{-3}$, or 0.008. Thus

Obtaining Output

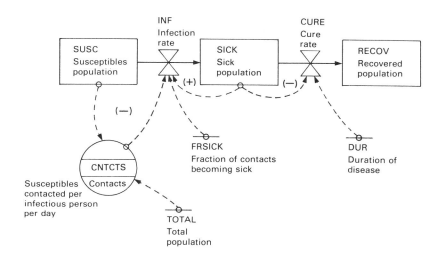

Figure 3.6: Flow diagram of epidemic model

the scaling letter A represents multiplication by 10-3. Other commonly encountered scaling letters are T (thousands) and M (millions). See the appendix for a complete list of scaling letters.

An Example

The model shown in figures 3.6 and 3.7 uses correctly all the aspects of DYNAMO discussed up to this point. The model captures in a very simplified form the dynamics of an epidemic such as influenza. Our concern here is for the form of DYNAMO statements, not the conceptualization supporting the design of the model. (The reader who wishes to understand the conceptual structure of the model would have to devote most of his attention to the formulation of the infection rate INF, but we will not do so here.) Note that the model contains three levels, two rates, and one auxiliary, given as a table function. The reader should check that each statement in the model fits the form required for it by DYNAMO. To insure understanding, one also ought to sketch from the table function data the graphical relationship specified in the auxiliary equation for CNTCTS (susceptibles contacted per infected person per day).

Figures 3.8 and 3.9 show the output resulting from a single simulation run of the model. This particular run was

```
*        SIMPLE EPIDEMIC MODEL
NOTE
L        SUSC.K=SUSC.J+DT*(-INF.JK)
N        SUSC=988
NOTE         SUSCEPTIBLE POPULATION  (PEOPLE)
R        INF.KL=SICK.K*CNTCTS.K*FRSICK
NOTE         INFECTION RATE  (PEOPLE PER DAY)
C        FRSICK=0.05
NOTE         FRACTION OF CONTACTS BECOMING SICK
NOTE         (DIMENSIONLESS)
L        SICK.K=SICK.J+DT*(INF.JK-CURE.JK)
N        SICK=2
NOTE         SICK POPULATION  (PEOPLE)
A        CNTCTS.K=TABLE(TABCON,SUSC.K/TOTAL,0,1,0.2)
NOTE         SUSCEPTIBLES CONTACTED PER INFECTED PERSON
NOTE         PER DAY  (PEOPLE PER PERSON PER DAY)
T        TABCON=0/2.8/5.5/8/9.5/10
NOTE         TABLE FOR CNTCTS
N        TOTAL=SUSC+SICK+RECOV
NOTE         TOTAL POPULATION (PEOPLE)
R        CURE.KL=SICK.K/DUR
NOTE         CURE RATE  (PEOPLE PER DAY)
C        DUR=10
NOTE         DURATION OF DISEASE  (DAYS)
L        RECOV.K=RECOV.J+DT*CURE.JK
N        RECOV=10
NOTE         RECOVERED POPULATION  (PEOPLE)
NOTE
SPEC     DT=0.25,LENGTH=50,PRTPER=5,PLTPER=1
PRINT    SUSC,SICK,RECOV,INF,CURE
PLOT     SUSC=W,SICK=S,RECOV=R/INF=I,CURE=C(0,200)
```

Figure 3.7: Listing of epidemic model

obtained using Mini-DYNAMO. The command to the computer was

MINDYN FLU

since the model was filed in the computer's memory under the name FLU. Your particular system command to activate DYNAMO may be different from "MINDYN _____"; consult the appropriate manuals or people. Figure 3.8 shows the output produced by including the PRINT statement and a nonzero PRTPER, while figure 3.9 shows the graphical output called for by the PLOT statement and the nonzero PLTPER specified in the model. Note that the graphical output is displayed sideways in order to show the TIME axis as the horizontal (x) axis and the different scales for the variables as the vertical (y) axis. DYNAMO plots variables across the page, with successive lines representing the increments in time specified

Obtaining Output   97

```
TIME      SUSC      SICK     RECOV     INF      CURE
E+00      E+00      E+00     E+00      E+00     E+00
 .0      988.00     2.00     10.00     1.00      .200
 5.      973.82    13.33     12.85     6.62     1.333
10.      881.54    86.83     31.63    42.13     8.683
15.      429.93   428.09    141.97   125.73    42.809
20.       62.18   531.31    406.50    23.13    53.131
25.       12.08   356.83    631.07     3.02    35.683
30.        4.30   220.86    774.82      .66    22.086
35.        2.30   134.63    863.05      .22    13.463
40.        1.57    81.69    916.71      .09     8.169
45.        1.25    49.48    949.24      .04     4.948
50.        1.09    29.95    968.93      .02     2.995
```

Figure 3.8: Printed output from the base run of the epidemic model.

by the plot period PLTPER. Note also that DYNAMO has selected two scales, as called for in the PLOT statement, showing each on the graphical output. One must read the appropriate scale for each variable. At TIME = 20 days, for example, the recovered population (RECOV, plotted R) is about 300 (people), while the cure rate (CURE, plotted C) plotted nearby is only about 55 (people per day). Finally, observe that DYNAMO shows the points where two or more characters are superimposed -- the letters along the side of the plot opposite the time scale indicate variables that would appear on the same point in the plot. The letters RI,SC opposite TIME = 2, for example, show that the graphs of RECOV (R) and INF (I) are superimposed at that time, and the graphs of of SUSC (S) and CURE (C) are also superimposed but at a different location in the plot.

Self-Test

No one is watching. Try answering the following questions by investigating the model listing in figure 3.7.

1. What kind of variable is SICK? Why?

2. Why does INF.JK appear in the level equation for SUSC with a negative sign?

3. Why doesn't DUR have a timescript?

4. What values of the variables SUSC, SICK, and RECOV are used to compute TOTAL?

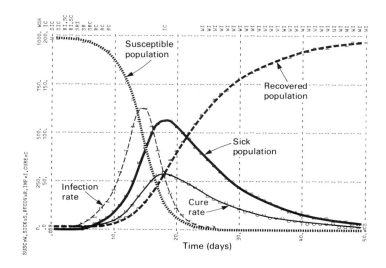

Figure 3.9: Plotted output from the base run of the epidemic model.

5. How many people had already had this disease and recovered at the time the simulation began?

6. What is the computation interval in this model?

7. Why are there only two scales for the five plotted variables?

8. Why does INF have the timescript KL in the rate equation in which it is computed, but has the timescript JK in the level equations where it is used?

9. Sketch the table function used in the auxiliary equation for CNTCTS.

10. Why are there 9 lines of printed output and 41 lines of plotted (graphical) output?

Suggested answers are contained in the appendix to this section.

Obtaining Output 99

Reruns

Once a DYNAMO model has been entered into a computer, using whatever editing and job control conventions are required, the model can be run. Printed or plotted output is produced, as in figures 3.8 and 3.9, and then DYNAMO shifts into rerun mode.

In response to DYNAMO's prompting phrase

TYPE RERUN

(some DYNAMO's say ENTER RERUN CHANGES), the user can change any constants or tables appearing in the model. If no reruns are desired, the user types QUIT to get out of DYNAMO and return to the computer's operating system.

To change in rerun mode the constant DUR (duration of disease) in the epidemic model to 5 days, for example, one types the following:

C   DUR=5

If that is the only change desired, the user then types the word RUN, followed, if desired, by a label identifying the run, just like a RUN statement placed at the end of the model listing. The statements

C   DUR=5
RUN DURATION=5

produce the following acknowledgment from DYNAMO,

        DUR
PRESENT 5.000
ORIGINAL 10.00

followed by the output shown in figures 3.10 and 3.11. DYNAMO then returns for another rerun of the model, printing TYPE RERUN. The constant DUR reverts back to its original value.

To change a constant for a number of successive reruns, one types CP instead of C. A constant changed by CP holds the newly assigned value permanently, until it is purposefully changed in a subsequent rerun.

Similar statements allow the alteration of the y-values in any table function in a model. Suppose one wished to change the table function for CNTCTS in the epidemic model and to

TIME	SUSC	SICK	RECOV	INF	CURE
E+00	E+00	E+00	E+00	E+00	E+00
.0	988.00	2.00	10.00	.997	.400
5.	977.27	8.42	14.31	4.185	1.684
10.	932.71	34.96	32.33	17.185	6.992
15.	757.53	137.18	105.28	62.978	27.437
20.	346.77	317.26	335.96	75.848	63.452
25.	118.13	245.00	636.86	20.259	49.000
30.	61.97	119.92	818.09	5.202	23.983
35.	46.16	52.21	901.61	1.687	10.442
40.	40.72	21.93	937.33	.625	4.385
45.	38.64	9.09	952.24	.246	1.818
50.	37.82	3.75	958.40	.099	.750

Figure 3.10: Epidemic model with DUR=5

suppress the printed numerical output. One could type the following in response to TYPE RERUN:

```
T     TABCON=0/4/7/8.8/9.5/1
C     PRTPER=0
RUN   NEW TABCON
```

The y-values in the new table for CNTCTS would be determined by the user; those given are merely examples. The zero value for PRTPER specifies that no tabular output should be produced. The results of such changes are shown in figure 3.12. These changes would revert to the original model values for the next rerun. Again, to change them permanently, one would type TP for the table and CP for the constant.

(The reader may wish to note in figure 3.12 how little the behavior of the model changed with the rather significant change in shape of the table. A useful exercise at this point would be to try to find a set of (reasonable) y-values for TABCON that do change the model's behavior. Can you, for example, prevent the epidemic? If you succeed in finding a table function that significantly alters the model's behavior, can you interpret what that function means in the real system?)

In addition to parameters occuring in C, T, and SPEC statements, the PRINT and PLOT statements may be changed in rerun mode simply by typing new PRINT or PLOT statements in the form previously described for them. (In Mini-DYNAMO and DYNAMO/F one can only PRINT or PLOT variables originally called for in PRINT or PLOT statements or reserved in computer memory for that purpose using the statement SAVE.)

Obtaining Output

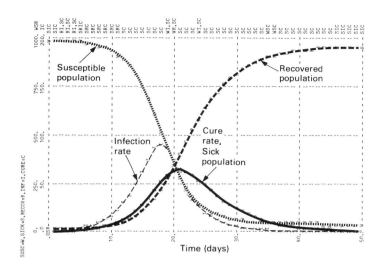

Figure 3.11: Epidemic model with DUR=5

Although N equations can not be changed in rerun mode, initial values of levels can be changed if they were written in the model with an N equation *and* a C equation. For example, the formulation for susceptibles in the epidemic model could be rewritten as

```
L    SUSC.K=SUSC.J+DT*(-INF.JK)
N    SUSC=SUSCN
C    SUSCN=988
```

Changing the constant SUSCN in rerun mode alters the initial value of the level SUSC.

Note that no other changes of model structure are permitted in the rerun mode. To alter actual equations, one must make the changes in the model outside of the rerun mode employing the editing capability used to enter the model in the first place. Clever uses of constants, however, can allow the modeler to test different equation formulations, but we reserve such subtleties for later.

Appendix - Answers to the Self-Test

1. SICK is a level variable, as indicated by its form and by the L in the left-most space on the line. SICK represents an accumulation of infected people.

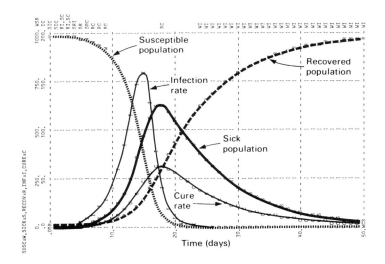

Figure 3.12: Output of the epidemic model in rerun mode, with PRTPER = 0 (no printed output) and new values for TABCON (the table for susceptibles contacted per infected person per day).

2. INF is the rate at which susceptible people are becoming infected. When infected, they move out of the susceptible population and into the sick population; hence INF is substracted in the equation for SUSC and added in the equation for SICK.

3. DUR is a constant, and constants do not have timescripts.

4. Their initial values are used: SUSC=988, SICK=2, and RECOV=10. Hence TOTAL equals 1000.

5. 10. The initial value of RECOV is 10.

6. DT = 0.5

7. The commas in the PLOT statement require DYNAMO to plot SUSC, SICK, and RECOV on one scale and INF and CURE on another. The slash in the PLOT statement allows DYNAMO to pick perhaps different scales for the two sets of variables.

Obtaining Output   103

8. Rates are computed in the present (time K) for the next time period (KL). Levels are computed in the present from their previous values (time J) and the values of the rates over the previous interval (JK). Hence rates are computed with subscripts KL and used with subscripts JK.

9. The x-axis ranges from 0 to 1 and represents values of the ratio SUSC.K/TOTAL. The graph contains the points (0,0), (.2,2.8), (.4,5.5), (.6,9.5), and (1,10). One would probably draw a smooth curve connecting these points, but DYNAMO actually computes along the straight-line segments joining them.

10. The print period PRTPER is set at 5, while the LENGTH of the simulation is 40. Hence there will be 40/5 or 8 intervals between the successive printed lines. With a first and last line, that means 9 lines of printed output. The same reasoning shows that the PLTPER value of 1 means 40 intervals in the plot and hence 41 lines.

3.5 DELAYS, SMOOTHS, AND AVERAGES

A model of a feedback system must trace the significant flows of material and information throughout the system. Sometimes delays exist. An order for widgets does not immediately result in a delivery of widgets, and they may not all arrive at the same time anyway. People contracting a disease may not immediately show its symptoms. Information about daily sales must accumulate before a picture of average sales per month can be obtained, and that accumulation takes time. Crops or trees planted cannot immediately be harvested. Delays are such an important part of the structure of feedback systems that DYNAMO has several delay functions preprogrammed for the convenience of the modeler.

Material Delays

Consider the simple epidemic model used as an illustration in section 3.4. Its basic rate/level structure is shown in figure 3.13. For certain policy experiments, it might more appropriately be seen as a four-level structure, the additional intervening level INC representing people in whom the disease is incubating but has not reached sufficient strength to cause symptoms to show. Figure 3.14 shows such a structure.

Figure 3.13: Epidemic model rate/level structure

To add such a level to the epidemic model listing in figure 3.7, we could include a level equation for the incubating population and a rate equation for its outflow rate, which we might call SYMP for the "rate of showing symptoms" or becoming symptomatic. The following are likely formulations:

```
L   INC.K=INC.J+DT*(INF.JK-SYMP.JK)
N   INC=TSS*INF
R   SYMP.KL=INC.K/TSS
```

where TSS stands for the time to show symptoms, which for influenza is about 3 days. The rate equation for SYMP is written just like the CURE rate. It means that 1/3 of the incubating population INC would become symptomatic every day, so that the average person would spend three days in the incubating population.

Precisely the same thing is accomplished by the following single DYNAMO equation:

```
R   SYMP.KL=DELAY1(INF.JK,TSS)
```

DELAY1

Described in all its glory, *DELAY1* is a *first-order exponential material delay.* A level, such as INC, is implicit in such an equation. The only thing lost is the accessibility of that level; one cannot print, plot or use the incubating population in this DELAY1 formulation because the level has been named internally by DYNAMO, not the modeler.

The rate/level diagram in figure 3.15 for the CURE rate and the SYMP rate help to show the structure DELAY1 always represents. Note the identical structure of the SYMP and CURE rates in the figure: they are both written as (level)/(time constant). For a first-order material delay the outflow rate is assumed to be the same sort of equation expressed as LEV.K/DEL, where DEL is the time of the delay and LEV is the internal level interposed by DYNAMO to generate the delay.

The level hidden away in a DELAY1 formulation must have an initial value, yet none is specified explicitly in the

Delays, Smooths, and Averages   105

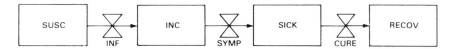

Figure 3.14: Epidemic model rate/level structure including a level for people in whom the disease is incubating

equation. DYNAMO cleverly initializes that internal level all by itself, with algebra that is equivalent to the N equation given above for INC in the epidemic model. Note that initially

```
SYMP    = INC/TSS
        = (INF*TSS)/TSS
        = INF.
```

Thus the INC level would be given an initial value that would make its inflow rate INF and its outflow rate SYMP momentarily *equal* at TIME=0. DELAY1 automatically initializes its internal level to put the inflow rate and its delayed version *in equilibrium*. No need for the modeler to worry about it (unless, of course, he wants the model to be out of equilibrium at TIME zero; for that nuance, see chapter 7).

DELAY3

There is no computational reason why the disaggregation of the infected population which resulted in a new level, the incubating population INC, needs to stop at just that one level. One could suggest that since the time to show symptoms TSS is three days, we could chop up the incubating population into those in their first day INC1, those in their second day INC2, and those in their third day INC3, as in figure 3.16.
The resulting delay between becoming infected (the INF rate) and showing symptoms (the SYMP rate) is known officially as a *third-order exponential material delay* and is represented in DYNAMO by the function DELAY3. One writes

```
R    SYMP.KL=DELAY3(INF.JK,TSS)
```

and DYNAMO automatically generates three internal levels between the INF rate and the SYMP rate, expressing their outflow rates as

```
LEVEL/(TSS/3)
```

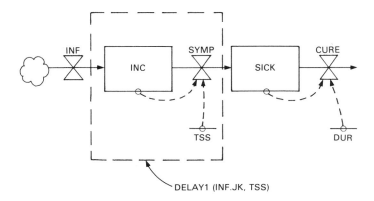

Figure 3.15: Rate/level diagram for DELAY1 in the epidemic model.

Thus the average time it takes an individual to pass through this third-order delay is still TSS (time to show symptoms). Note that the single DELAY3 equation is equivalent to three level equations, three N equations, and three rate equations--a considerable savings in typing and conceptual effort.

DELAYP

As with DELAY1, expressing the SYMP rate using DELAY3 does lose access to the incubating population, which in this case would be INC1 + INC2 + INC3. If one desires to express the SYMP rate as a third-order delay and wants to be able to plot or compute with the total incubating population INC, one merely needs to invoke another DYNAMO delay function denoted DELAYP, for delay with "pipeline" term.[7] One writes

    R   SYMP.KL=DELAYP(INF.JK,TSS,INC.K)

The inclusion of the third argument INC.K in the expression tells DYNAMO to give that name to the sum of the three internal levels in the delay structure, allowing the modeler

---

[7] DELAYP is not available in Mini-DYNAMO. The following, however, is equivalent and still saves a lot of typing:
    R   SYMP.KL=DELAY3(INF.JK,TSS)
    L   INC.K=INC.J+DT*(INF.JK-SYMP.JK)
    N   INC=INF*TSS

Delays, Smooths, and Averages 107

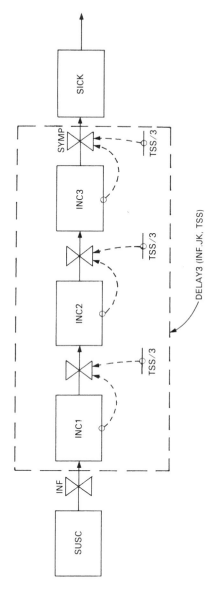

Figure 3.16: Further disaggregation of the incubating population INC

access to that value at all times. One could then plot INC or use it to compute other variables. Note that this one equation is now taking the place of ten: three levels, three initial values, three rates, and one auxiliary equation for the sum. DELAYP is a third-order delay; there is no corresponding pipeline formulation for a first-order delay. (If you wish to know the level in a first-order delay, write out the level and rate equations rather than use the DELAY1 function.)

Order of Material Delays

The meaning of the word "order" when applied to delays is probably clear from the preceding examples. It represents the number of levels contained in the delay. It is natural to ask what is gained by using a third-order delay instead of a first-order delay, and to note with some consternation that there doesn't seem to be any reason not to suggest fifth-order, thirteenth-order, or seventy-seventh-order delays. Some of the reasons for disaggregating and the reasons not to disaggregate too much are observable in figure 3.17.

Figure 3.17 shows how the outflow rate from a delay behaves when the inflow rate suddenly changes. The different curves show the response of first-, second-, third-, sixth-, and twelfeth-order delays. The behavior of the first-order delay is considerably different from the third-order delay: the latter is S-shaped, while the first-order delay shows simple, exponential goal-seeking behavior just like the coffee cooling model hand-simulated in section 3.2. The second-order delay shows the beginning of S-shaped behavior. The sixth- and twelfeth-order delays simply show more pronounced S-shaped curves.

As the order of the delay increases, the response to the step increase tends to look like a mimcing step, displaced in time by the specified delay time. Thus a substantial change in shape of response occurs from first to third order, but higher orders merely accentuate the S- shaped nature of the third-order response. For this reason DYNAMO includes DELAY1 and DELAY3 functions, but no single functions for higher orders. (If the modeler really feels the need for a higher order delay, he can string two third-order delays together to make a reasonably quick sixth-order delay, but such attention to order beyond three is seldom justifiable.)

To decide what order delay to use, the modeler must think about how the real system might behave in response to sudden changes. A delay between placing a bunch of orders and receiving the goods is probably best modeled as a third-order

Delays, Smooths, and Averages 109

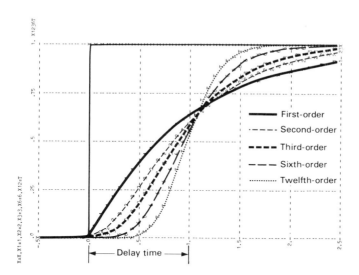

Figure 3.17: Behavior of first-, second-, third-, sixth-, and twelfth-order delays in response to a step increase

delay, because a sudden increase in orders placed would probably not immediately result in a growing receiving rate. The S-shaped response is justifiable because some of the orders would probably begin arriving before the delay time had elapsed, others would arrive late, while the greatest rate of arrivals (the steepest part of the S) would occur around the specified delay time. On the other hand, the incubating population in the simple epidemic model listed previously might just as well be captured in a first-order structure, since it is conceivable that some infected people might begin showing the symptoms quite soon after infection. (The reader may wish run the epidemic model with the two different structures for INF to verify that using a third-order delay structure results in slightly sharper, higher peaks in the graphs of INF, INC, and SICK, but the differences are almost too small to notice.)

"Smoothing" Information: the SMOOTH Function

No corporate officer would interpret a day's jump in sales as a permanent trend on which to base decisions about inventories, production, or employment. She would want to "smooth" out randomness from sales data sufficiently to detect real trends. Sales would be averaged over some period of time. Because averaging and smoothing processes abound in dynamic systems, DYNAMO provides a function for the purpose and calls it SMOOTH.

Smoothing or averaging information requires that it be accumulated. An average is, after all, a sum, an accumulation. Therefore, one expects that one or more levels ought to be involved. The DYNAMO formulation for a first-order smooth or information average is equivalent to the structure shown in figure 3.18. The equations intended by the flow diagram in that figure are the following:

```
L    SVAR.K=SVAR.J+DT*SRATE.JK
N    SVAR=VAR
R    SRATE.KL=(VAR.K-SVAR.K)/STIME
```

or, more simply,

```
L    SVAR.K=SVAR.J+(DT/STIME)(VAR.J-SVAR.J)
N    SVAR=VAR
```

One accomplishes precisely the same thing with the single auxiliary equation:

```
A    SVAR.K=SMOOTH(VAR.K,STIME)
```

DYNAMO takes care of the initialization, setting VAR and its smoothed value equal (in equilibrium) at TIME zero. The variable being smoothed (VAR) may be a level, rate, or auxliary. The quantity representing the smoothing time, STIME, is often a constant but may be a variable. Intuitively, the smoothing time represents the interval of time during which the values of the variable VAR are accumulated to produce an exponentially weighted moving average over time.

Why is such a formulation taken as a smoothing or an averaging of information? The simplest answer is that it is easy to formulate and it behaves properly. The equations are identical in structure to the coffee-cooling system hand-simulated in section 3.2 (compare figure 3.18 with figure 3.2). From that section we know that such a system shows exponentially goal-seeking behavior over time. Thus the smoothed variable SVAR will seek over time to equal VAR, in the same way as COFFEE approached the ROOM temperature. If VAR suddenly jumps and then holds steady, its smoothed value SVAR will head toward it, just as COFFEE temperature headed toward ROOM temperature, much as a moving average would approach VAR. If VAR were to jump back down before SVAR caught up to it, SVAR would not rise

Delays, Smooths, and Averages   111

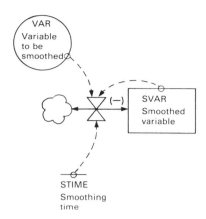

Figure 3.18: A first-order SMOOTH of a variable

as high as VAR did and would start back down in another exponentially goal- seeking pattern (see figure 3.19). That implies that, if VAR oscillates, its smoothed value SVAR will be delayed somewhat in time and will oscillate less dramatically (lower peaks, higher valleys). Thus the SMOOTH function will act to smooth out ups and downs in the variable being smoothed, just as an average would and just as we desired.

Readers familiar with smoothing formulations may want to observe that the level equation for a SMOOTHed variable can be written in the familiar weighted-average form for exponential smoothing. Suppose, for example, that ASR represents an average shipment rate computed in a first- order exponential smoothing formulation:

    A   ASR.K=SMOOTH(SR.JK,TASR)
where
    SR   = Shipment Rate (thingumabobs/week)
    TASR   = Time to Average SR (weeks).

Then from the definition of the SMOOTH function we know that

    ASR.K = ASR.J + (DT/TASR)(SR.JK-ASR.J)

which can be rearranged as follows:

ASR.K = ASR.J + (DT/TASR)(SR.JK) - (DT/TASR)(ASR.J)
      = (DT/TASR)(SR.JK) + (1-DT/TASR)(ASR.J)

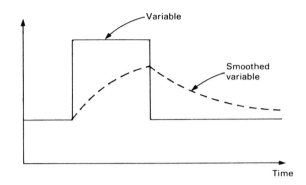

Figure 3.19: Behavior of the SMOOTH function in response to a step up and down

= ALPHA*SR.JK + (1-ALPHA)*ASR.J

where ALPHA is DT/TASR, a constant if the smoothing time TASR is constant. If ALPHA is between zero and one, such an expression is a weighted average of SR and ASR. The weighting depends upon the smoothing time constant TASR: a long averaging time places less weight on the most recent shipment rate and places more weight on the past average, causing ASR to respond more slowly to changes in SR, as desired. A short averaging time would make ALPHA large and place more weight on the most recent value of the shipment rate.

Technical Note on the Selection of DT

The weights in the last equation also depend upon DT, DYNAMO's computation interval. To keep ALPHA between zero and one, we must be sure to take DT less than TASR. If DT were set equal to TASR, then in one computation interval the value of ASR would jump to equal SR. The behavior would not be a sensible approximation to the smooth, exponentially goal-seeking pattern the formulation is trying to produce. Consequently, one must always take DT in a model to be smaller than such time constants. The observation holds for other time constants as well; recall the computational inaccuracies illustrated for the coffee- cooling loop at the end of section 3.2.

Consequently, as a simple rule of thumb, *take DT between one-half and one-tenth of the smallest time constant in the model.* Less than one-half will prevent extreme computational inaccuracies, while less than one-tenth is seldom necessary and requires a lot of computer time.

Information Delays

Just as material takes time to flow from one point in a system to another, so does information. The president of a company does not know today's sales rate for one of its branches; an economist does not know the gross national production rate for today; a New Yorker's perception of the job market in Juneau, Alaska, is bound to lag behind the true picture. Such delays are frequently an unavoidable part of system structure, which must be captured in equations.

The comments about the SMOOTH function reveal that averaging or smoothing information introduces delay. If a variable is rising, for example, its smoothed value is also rising but lags behind, as shown in figure 3.20. Consequently, the SMOOTH function is also frequently used to represent an information delay. When it is, it does not represent an intentional (policy) smoothing of information by actors in the system but rather a first-order exponential information delay.

DLINF3

One could string a series of first-order SMOOTHs together to create higher-order information delays, analogous to higher order material delays. The structure represented in figure 3.20 is a third-order information delay. Each successive level variable tries to track or "seek" the previous one, just as the COFFEE temperature sought after the ROOM temperature in the coffee-cooling loop. One could write out three level equations for this system to produce SV3, the third-order delayed version of VAR, but it is unnecessary to do so. The entire structure is captured by DYNAMO with the third-order information delay function DLINF3, and is written in a single equation:

A   SV3.K=DLINF3(VAR.K,STIME)

DYNAMO initializes the levels implicit in DLINF3 so that they all equal the initial value of the variable being delayed. The behavior of DLINF3 is identical to the behavior of the material delay DELAY3 except when the delay time is variable.[8] Nonetheless, the modeler should distinguish carefully between material and information delays and use the appropriate function. Note that a flow of material from a level reduces the level's size (other things being equal), while passing around information about a level in no way diminishes

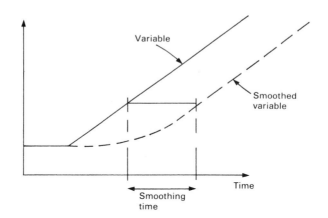

Figure 3.20: Behavior of the SMOOTH function in response to a steady increase, illustrating the delay introduced.

it. Only rates flow into material delays, while any kind of information--rate, level, or auxiliary--can be smoothed.

Diagraming Conventions for the Smooth and Delay Functions

Figure 3.22 shows the commonly accepted flow diagram symbols for delays and the SMOOTH function. SMOOTH is represented by a simple rectangle; the information inputs to the rectangle are understood to be the quantity being smoothed and the smoothing time. While the partitioned rectangle shown for the other delays is universally used, there is less agreement about what pieces of information ought to appear in each box. At a minimum, indicate the name of the delay function and the name of the output variable in the locations shown in the figure. If the accumulation in the delay pipeline is used in the model, use one of the boxes for its DYNAMO name and definition.

---

[8] The difference is suggested by the different initial value equations. In DELAY3 the equilibrium value of the internal levels is proportional to the delay time (see the N equation for INC in the epidemic model example). In DLINF3 each internal level in equilibrium has the value VAR. Hence, if the delay time changes, the equilibrium value of the levels in the pipeline of DELAY3 changes proportionally, but no change occurs in DLINF3.

Delays, Smooths, and Averages 115

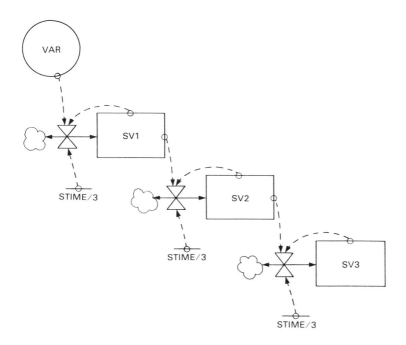

Figure 3.21: Third-order information delay structure

3.6 FUNCTIONS

DYNAMO provides a wide range of functions to aid the formulation and testing of dynamic models. We have met already the TABLE functions which perform linear interpolation and the DELAY and SMOOTH functions which do what their names imply. The remaining functions in DYNAMO may best be classified as mathematical, logic, and test functions, although the distinctions among them are not as tidy as these categories suggest.

Mathematical Functions

Five common mathematical functions are available in DYNAMO. They are listed here, together with their standard mathematical notations:

SQRT(X) = $\sqrt{x}$,   the non-negative square root of x,
SIN(X)   = sin x,   the sine of x,

116   Introduction to DYNAMO

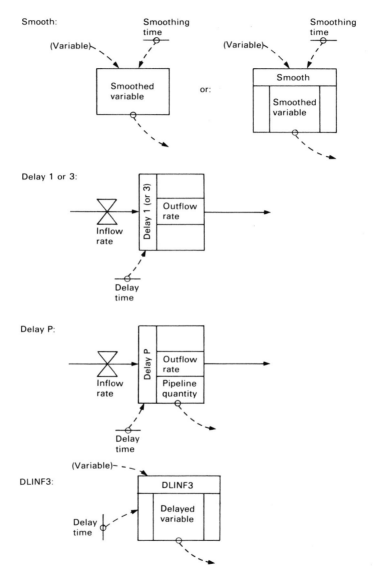

Figure 3.22: Flow diagram symbols for SMOOTH, DELAY1, DELAY3, DELAYP, and DLINF3.

Functions   117

COS(X)   = cos x,   the cosine of x,
EXP(X)   = e^x,   the exponential function, $(2.718..)^x$
LOGN(X) = $\log_e x$,   the natural logarithm of x

The uses of these functions in models of feedback systems are as varied as their applications anywhere. We leave it to readers to generate their own examples.

In some versions of DYNAMO (notably Mini-DYNAMO and DYNAMO II/370), to raise a number to a power one must use the exponential function EXP and its inverse, the natural logarithm function LOGN. The expression

EXP(A*LOGN(B))

is BA, B raised to the A power. To raise a number X to small integer powers such as 2 or 3, one saves some space and effort by writing repeated multiplication, X*X or X*X*X, but, to raise to higher powers or noninteger exponents, one must use the expression involving EXP and LOGN. (Other versions of DYNAMO recognize ** as the symbol for raising to a power, writing BA as B**A.)

One of the most important uses of the trigonometric functions SIN and COS -- representing oscillating influences outside the boundary of the model -- is discussed below under *Test Functions*.

Logic Functions

The functions in DYNAMO that perform logical operations like "if..., then..." are the MAX, MIN, CLIP, and SWITCH functions.

The expression MAX(A,B) takes the value of A or B, which ever is greater. Thus,

$$MAX(A,B) = \begin{cases} A, & \text{if } A \geq B \\ B, & \text{if } A < B \end{cases}$$

The MIN function is a companion to the MAX function, doing just the opposite:

$$MIN(A,B) = \begin{cases} B, & \text{if } A \geq B \\ A, & \text{if } A < B \end{cases}$$

The MAX function can be used to produce the *absolute value* of a number since

MAX(A,-A)

takes on the nonnegative value of A whether A itself is negative or not.

The MAX function is also occasionally used to prevent division by zero. The expression

A/MAX(B,0.01)

divides A by 0.01 if B happens to be smaller than that, so it effectively prevents division by zero (or negative) values of B. Such an expression is sometimes useful in a model as long as the quotient is meaningful for all positive values of B and the modeler does not wish to stop the simulation for occasional moments when B reaches zero.

It is tempting to miuse the MAX and MIN functions in the formulation of a model. The modeler needs to be especially careful in using these functions because they imply abrupt change in systems that are supposedly changing continuously through time. Always question whether the abrupt change a MAX or MIN would produce in model behavior adequately captures the behavior of the system. A table function smoothly changing the value of a variable might more accurately suit the structure and dynamics of the real system.

A common example occurs in controlling the outflow of a level such as an inventory. Suppose shipments SHIP deplete inventory INV by shipping at the rate of demand DMND, unless, of course, there isn't enough inventory left to satisfy demand. One might be prompted to write

R   SHIP.KL=MIN(DMND.K,INV.K)

making the shipment rate SHIP equal to demand DMND or to all of the remaining inventory, INV, whichever is smaller. When inventory is low, some demand goes unsatisfied. Thus the MIN function effectively prevents the inventory from going negative.

However, there are problems with this formulation. Most obvious, there is a problem with units in the equation. SHIP would be measured in something like widgets/month, while INV would be measured in widgets. But even if INV were divided by a factor of time, such a formulation in most applications would not capture the actual dynamics of the system. When inventory gets dangerously low, pressures usually develop in the system to constrain shipments before only one or two units of inventory are left. We ship to preferred customers only, or we try to smooth out shipments to keep the production department from having to lay off workers, and so on. The gradually increasing constraint on SHIP from low inventories is more pronounced, the more product lines INV is intended to represent in the aggregate. Shipments would be increasingly curtailed as the firm runs out of more and more product lines. In most cases there fore a more reasonable formulation would look like the following:

R   SHIP.KL=DMND.K*EIS.K

where EIS represents an effect of inventory on shipments and is expressed in a table function whose input (X) value is, perhaps, the ratio of inventory to desired inventory. Figure 3.23 shows some possible table functions for EIS.

The CLIP and SWITCH functions enable the modeler to change the values of quantities during a simulation. The form of the CLIP function is

CLIP(A,B,X,Y)

where A, B, X, and Y can be any constants or variables in the model. CLIP(A,B,X,Y) takes the value A as long as X is greater than or equal to Y; otherwise it takes the value B. Thus

$$CLIP(A,B,X,Y) = \begin{cases} A, & \text{if } X \geq Y \\ B, & \text{if } X < Y. \end{cases}$$

In most versions of DYNAMO this function has two names. The second name, FIFGE, is intended to describe what the function does: it takes the value of the First quantity IF the third is Greater than or Equal to the fourth. Thus CLIP(A,B,X,Y) and FIFGE(A,B,X,Y) mean the same thing. FIFGE might be preferable because its name is a mnemonic.

The form of the SWITCH function is similar but simply compares the third quantity to zero, not to a fourth quantity:

$$SWITCH(A,B,X) = \begin{cases} A, & \text{if } X = 0 \\ B, & \text{if } X \neq 0. \end{cases}$$

SWITCH also has a name intended to be a mnemonic: FIFZE, First *IF* third equal to *ZE*ro. Note SWITCH-alias-FIFZE does not separate values of X that are greater than zero from those less than zero: SWITCH(A,B,X) equals B for positive *and* negative values of X, taking the value A only when X = 0.

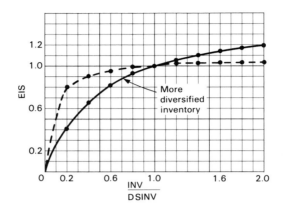

Figure 3.23: Table for EIS, effect of inventory on shipments

CLIP and SWITCH functions are usually used to test policy changes. They are generally not meaningful components of feedback structure. An example of a meaningful use of the CLIP function is the following, in which it is used to explore the effect of a change in the time required to bring an Inventory up to its desired level. To change Inventory Adjustment Time IAT from 4 to 6 weeks beginning at the 20th week of a simulation, one could write

A   IAT.K=CLIP(6,4,TIME.K,20)

Thus

$$IAT = \begin{cases} 6, & \text{if } TIME.K \geq 20 \\ 4, & \text{if } TIME.K < 20. \end{cases}$$

IAT will equal 4 until TIME reaches 20, when it will switch to 6.[9]

The SWITCH function can be cleverly used in the rerun mode to test alternative model equations. Recall that in the

---

[9] Computer round-off error can produce the situation in which TIME never exactly equals 20. In such a case this CLIP expression would make the switch from 4 to 6 when TIME equals (20+DT). If that bothers you, write the expression for IAT as CLIP(6,4,TIME.K-DT/2,20), or use the STEP function discussed under Test Functions. STEP has such a correction built in.

Functions

rerun mode parameters can be changed, but equations can not. Suppose a modeler wanted to test two proposed production policies, say,

```
A    A.K=(AVSALE.K)*(EFINV.K)*(EFBKLG.K) and
A    B.K=AVSALE.K+(DSINV.K-INV.K)/IAT+
X    (BKLG.K-DSBKLG.K)/BAT
```

(These DYNAMO variable names are intended to relate meaningfully to sales, inventories, and order backlogs, but for our purposes here the reader can treat them as strings of meaningless symbols.) Both variations could be written into the model, followed by

```
R    PROD.KL=SWITCH(A.K,B.K,SWB)
C    SWB=0
```

Until the constant SWB (switch B) is changed to a nonzero number in rerun mode, the production rate will be given by the A policy. Changing SWB to 1 in a rerun would alter the structure of the model, setting production according to the B policy.

A switch between two formulations can also be performed as follows:

```
R    PROD.KL=(1-SWB)*(A.K)+(SWB)*(B.K)
C    SWB=0
```

Notice that as long as SWB = 0, production PROD will be given by the variable A. If SWB is changed to one in a rerun, PROD will switch to the formulation given by the variable B. The complete expression here for PROD is more general than that given by the use of the SWITCH function, for it takes the form of a weighted average of A and B. One could, conceivably, set the value of SWA in this expression to values other than zero or one, setting PROD equal to some weighted combination of A and B. No such weighted average is possible from the SWITCH formulation.

Test Functions

A remarkable amount of information about a model and the feedback system it represents can be obtained by subjecting the model to several kinds of disturbances. A sudden increase in a variable, a steady decline, oscillations, and random disturbances all help to expose the relationships in the model between feedback structure and dynamic behavior. The purpose of such tests is understanding--both the computer model and ultimately the system it represents. DYNAMO provides several test functions useful for generating exogenous influences and disturbances, including STEP, RAMP, PULSE, the sine function SIN, and NOISE.

To illustrate what these test functions mean, how to use them, and what behavior one might expect, we shall apply them to a simplified inventory control model. The model assumes there is a delay between placing orders and receiving goods into inventory. The rate/level diagram for the model is shown in figure 3.24, and the listing of equations is shown in figure 3.25.

Without going into detail about the actual equations, note that the order rate ORDRS is equal to the average shipment rate AVSHIP plus an inventory adjustment term INVADJ. If INV is below desired inventory DI, INVADJ is positive; the model places more orders than the recent average shipment rate in order to bring inventory up to its desired level. Conversely, if INV is above DI, the inventory adjustment term INVADJ will be negative, reducing the orders placed below the recent average shipment rate, thereby trying to bring INV down. The tests we will perform with this model will reveal some omissions in such an inventory control scheme.

Note the form of the TEST equation:

```
A    TEST.K=TEST1*STEP(   )+
X    TEST2*RAMP(   )+
X    TEST3*PULSE(   )+
X    TEST4*AMP*SIN(   )+
X    TEST5*RANGE*NOISE(   )
```

Recall that the X in the left-most space in a line identifies the statement as a *continuation* of the previous line. The TEST equation is actually a long string of functions added together. The constants TEST1, TEST2, ..., TEST5 are all initially zero. To invoke a function, we will change in rerun mode one of the TEST's from zero to 1.

The STEP Function

The STEP function, as its name implies, is used to change a quantity abruptly at some point in time. The form of the STEP function is

   STEP(A,B)
where
   A represents the step height, and
   B represents the value of TIME at which the value of

Functions

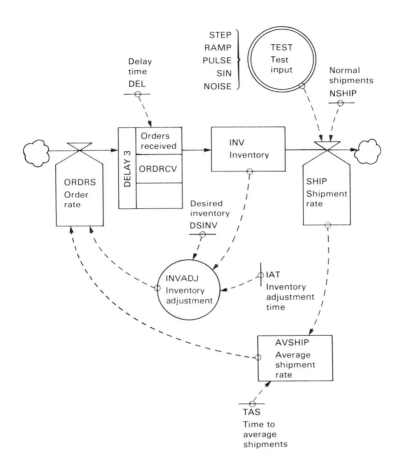

Figure 3.24: Flow diagram for a simple inventory control model

STEP changes.

Before time B, STEP equals zero; when TIME is greater than or equal to B, STEP takes the value specified by the quantity A. A and B can be represented by any quantity names one might use in a DYNAMO model. Their positions in the STEP function determine their meaning: the first argument is always the step height, and the second argument is

```
*   SIMPLE INVENTORY MODEL
NOTE
NOTE        SHIPMENTS
NOTE
R   SHIP.KL=NSHIP+TEST.K            SHIPMENT RATE (UNITS/WK)
C       NSHIP=100
A   TEST.K=TEST1*STEP(STH,STRT)+
X       TEST2*RAMP(SLP,STRT)+
X       TEST3*PULSE(HGTH,STRT,INTVL)+
X       TEST4*AMP*SIN(6.283*TIME.K/PER)+
X       TEST5*RANGE*NOISE()
C       TEST1=0/TEST2=0/TEST3=0/TEST4=0/TEST5=0
C       STH=10,STRT=2,SLP=20,HGTH=10,INTVL=200,AMP=10,
X       PER=5,RANGE=20
L   INV.K=INV.J+DT*(ORDRCV.JK-SHIP.JK)
N       INV=DSINV                   INVENTORY (UNITS)
R   ORDRCV.KL=DELAY3(ORDRS.JK,DEL)  ORDERS RECEIVED (UNITS/WK)
C       DEL=3 (WKS)                 DELAY IN RECEIVING ORDERS
NOTE
NOTE        ORDERS
NOTE
R   ORDRS.KL=AVSHIP.K+INVADJ.K      ORDERS PLACED (UNITS/WK)
A   AVSHIP.K=SMOOTH(SHIP.JK,TAS)    AVG SHIPMENT RATE (UNITS/WK)
C       TAS=2 (WKS)                 TIME TO AVERAGE SHIPMENTS
A   INVADJ.K=(DSINV-INV.K)/IAT      INVENTORY ADJUSTMENT (UNITS/WK
C       IAT=2 (WKS)                 INVENTORY ADJUSTMENT TIME
N   DSINV=DIC*NSHIP                 DESIRED INVENTORY (UNITS)
C       DIC=3 (WKS)                 DESIRED INVENTORY COVERAGE
NOTE
NOTE        CONTROL STATEMENTS
NOTE
SPEC    DT=.25/LENGTH=0/PRTPER=0/PLTPER=.5
PRINT   SHIP,TEST,INV,ORDRCV,ORDRS,AVSHIP,INVADJ
PLOT    SHIP=S/INV=I
```

Figure 3.25: Equations for the simple inventory control model

always the step time. The step height can be negative if one so desires, indicating a step down.

The behavior of the STEP function is shown in figure 3.26, along with the resulting behavior of the simple inventory model. In this run of the model,

```
A   TEST.K=STEP(HGHT,STRT)
C   HGHT=10
C   STRT=2
```

so the step height is 10, and the step occurs at TIME = 2 (weeks, in this model). As the normal value of shipments SHIP and ORDRS is 100 units per week in this model, this STEP change amounts to a sudden, permanent ten percent

Functions

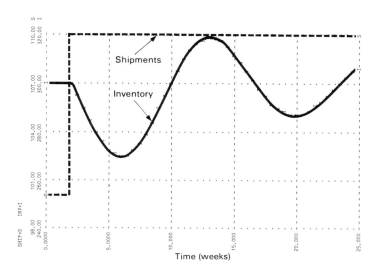

Figure 3.26: The STEP function and the resulting behavior of the inventory model

increase in shipments. In its effort to bring inventory back to its desired level (300 units), the model produces significant oscillations which appear to be damping out (deviating less and less over time from what will be the final equilibrium values of the variables). One might hope to design a more stable inventory control policy (and one can).

The RAMP Function

The RAMP function is a continuously growing or declining linear function of TIME. Its form in DYNAMO is

RAMP(A,B)
where
    A represents the slope of the linear function, and
    B represents the starting time for the ramp.

Before TIME equals B, the value of RAMP is zero. When TIME is greater than or equal to B, the value of RAMP is given by the linear function

    A * (TIME.K - B).[10]

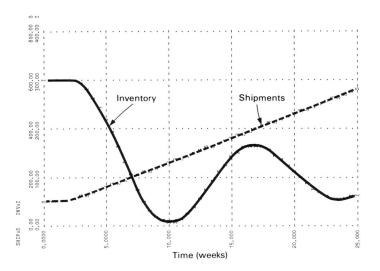

Figure 3.27: The RAMP function and the resulting behavior of the inventory model

The behavior of the RAMP function is shown in figure 3.27, as it influences the simple inventory model. In this run of the model

```
A    TEST.K=RAMP(SLP,STRT)
C    SLP=20
C    STRT=2
```

so the slope of the ramp is 20 (units per week), and the ramp begins at TIME.K = 2. The RAMP here represents a gradual linear increase in demand for shipments. Note that inventory drops almost to zero, with no effect at all on the shipment rate; a steeper ramp could presumably bring INV in this model to negative values. Clearly, some necessary negative feedback from inventory to the shipment rate has been omitted in this shipment rate formulation.

---

[10] This is true provided A and B are constant. RAMP is actually computed by DYNAMO as
```
L       RAMP.K=RAMP.J+DT*STEP(A.J,B.J)
N       RAMP=0
```

Functions

The PULSE Function

The PULSE function provides momentary jolts to a DYNAMO model. It is used to represent isolated changes in a variable, returning it immediately to its former value after each change. The form of the PULSE function in DYNAMO is

PULSE(A,B,C)
where
A represents the *height* of the pulse,[11]
B represents the *time of the first pulse*, and
C represents the *time interval between successive pulses*.

To PULSE a system just once, one merely makes the value of the third argument in the PULSE function very large, greater than any value of TIME encountered in the model run.

Figures 3.28 and 3.29 show the behavior of the PULSE function in the simple inventory control model. The TEST equation is

A   TEST.K=PULSE(HGHT,STRT,INTVL)
C   HGHT=10
C   STRT=2
C   INTVL= $\begin{cases} 200 & \text{(in figure 3.28)} \\ 5 & \text{(in figure 3.29)} \end{cases}$

Like the STEP, a single PULSE is sufficient to send this inventory control system into oscillations. Repeated pulses add little to our understanding of the behavior of the system; they serve merely to shift the curves to different positions on the oscillatory pattern shown in figure 3.28.

A PULSE occurs over the time interval of one DT, the simulation's closest approximation to instantaneously. In our inventory example the pulse raised the shipment rate SHIP 10 units per week for one DT, one 0.25 week period. Thus the total change in inventory caused by that pulse was (10 units per week)*(0.25 weeks) or 2.5 units. Note in figure 3.28 that INV dropped only 2.5 units, from 300 to 297.5. In general the total change produced in a level by a PULSE in one of its rates is the *area of the rectangle having base DT and height equal to the height of the PULSE*. In our example, that

---

[11] For those familiar with the mathematical impulse function, note that PULSE is different in that the *height* of the pulse is specified, not the area.

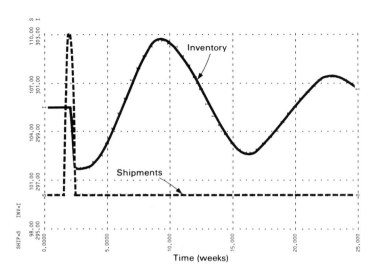

Figure 3.28: The PULSE function (interval = 200) and the resulting behavior of the inventory model

change is DT*HGHT. To add or subtract a given quantity Q to or from a level using PULSE, Q will have to equal DT*HGHT, so HGHT in the PULSE function must be Q/DT. Hence one must use PULSE(Q/DT,STRT,INTVL). A common use of such a formulation is to reduce some accumulation to zero regularly, as, for example, yearly production or monthly income. The following level equation shows a PULSE function being used to reduce the level LEV to zero every 52 weeks:

L   LEV.K=LEV.J+DT*(INRATE.JK-PULSE(LEV.J/DT,0,52))

The SIN Function

To test the response of a model to an oscillating (sinusoidal) variable, DYNAMO provides the sine of an angle in radians. In the expression

A*SIN(6.283*TIME.K/B)
where
A represents the amplitude of the oscillation,
B represents the period of oscillation, and
6.283 is an approximation to 2*pi.

Functions 129

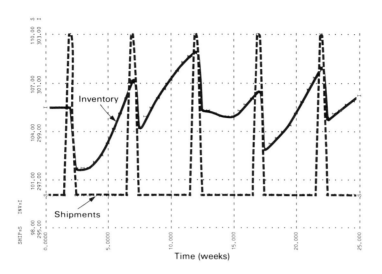

Figure 3.29: The PULSE function (interval = 5) and the resulting behavior of the inventory model

The amplitude A is the maximum deviation of this function from zero. The period B is the length of time between successive peaks or valleys. The function moves smoothly from zero to a maximum of A, back to zero, down to a minimum of -A, and then back to zero again every B units of time.

Figures 3.30 and 3.31 show the response of our simple inventory model to oscillations in the shipment rate, so that

A TEST.K=AMP*SIN(6.283*TIME.K/PER)
C AMP=10
C PER = $\begin{cases} 5 \text{ (in figure 3.30)} \\ 10 \text{ (in figure 3.31)} \end{cases}$

It is hardly surprising in figures 3.30 and 3.31 that in response to oscillating shipments the level of inventory oscillates. In figure 3.31, however, the amplitude of the oscillations in INV grows dramatically. A longer run would show that the oscillations are bounded. Nonetheless, a company with this inventory policy must hope that demand for shipments does not oscillate with a period near 10 weeks, or it has a big problem.

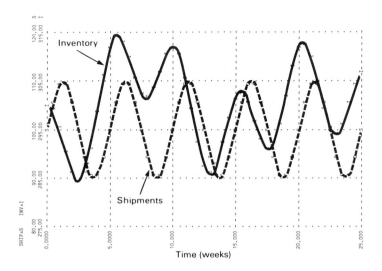

Figure 3.30: The Sine function SIN with period = 5, and the resulting behavior of the inventory model

The NOISE Function

DYNAMO provides a random number generator, a function called NOISE, whose values vary in a completely patternless, random way. The form of the NOISE function is, simply,

NOISE()

Note that nothing, not even a space, goes in the parentheses here. The parentheses help to identify the expression as a function producing many different values, rather than a constant, but otherwise they serve no purpose for the user. The values of the NOISE function range from -0.5 to +0.5 with a uniform or rectangular distribution. That is to say, any five-digit decimal between -0.5 and 0.5 is as likely to appear as any other one. To generate random numbers which vary over an interval of length A centered around the number B, one uses a DYNAMO expression in the form

A*NOISE()+B

The mean of such numbers would be B, and they would range from B - A/2 to B + A/2.

Functions 131

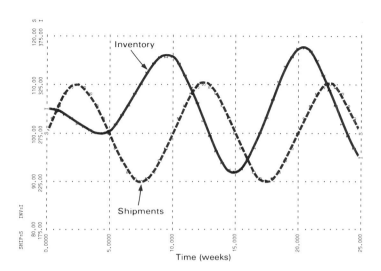

Figure 3.31: The Sine Function SIN with period = 10, and the resulting behavior of the inventory model

Figure 3.32 shows the effect of disturbing our simple inventory model with random jumps in shipments, as provided by the following test function:

```
A    TEST.K=RANGE*NOISE()
C    RANGE=20
```

Remarkably, the behavior of the model is far from random: it exhibits the same sort of oscillating behavior -- with nearly the *same period* -- as we observed when the model was STEPped and PULSEd. Feedback systems that have a natural period of oscillation tend to display it even when *randomly* disturbed. Since demand for real shipments contains some randomness, we are led to conclude that the inventory of any company with the ordering policy contained in our model would show oscillations, *for no other reason than the policy itself generates them.* The variability in demand is not the problem, the internal ordering policy is. It is for such revelations that one studies feedback systems.

There are significant subtleties in the use of NOISE in computer simulations. Note that a random number is selected every DT for each statement in a model containing the NOISE function. Thus, if one makes DT smaller, one puts more randomness into the system because more random numbers are

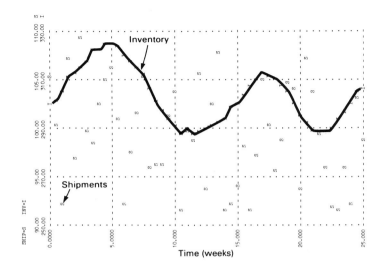

Figure 3.32: The NOISE function, and the resulting behavior of the inventory model.

being generated. Second, the number sequence generated by NOISE is only pseudo-random, in that the same patternless sequence is generated each time. Thus, if one makes no changes, two successive runs or reruns of a model containing NOISE will be exactly the same. Hardly what one would call truly random (but useful if one wishes to compare runs). One can change the random number sequence by specifying a different noise seed that starts the random sequence; see the reference manual for your particular version of DYNAMO. Finally, one may wish to simulate the random process known as "pink" noise, which has particular characteristics of some theoretical importance. See chapter 7 for details.

4 MODEL FORMULATION

Formulation is the process of translating model structure into equations. It is the transformation from an informal, conceptual view to a formal, quantitative representation. The purpose of formulation is to enable the model to be simulated (or solved analytically, if it were possible) to determine the dynamic behavior implied by the model's assumptions.

The stage of formulation is more than just a technical interlude between the more interesting stages of conceptualization and analysis of results. Formulated in equations, an informal causal-loop model sheds whatever ambiguity it had. Though no more accurate, in the sense of being a truer representation of reality, the formal model is much more precise. It has to be, for a computer to step it through time without getting confused. The necessary precision (unambiguity) of the formulation stage forces a clarity of thinking that improves the modeler's understanding of system structure.

The first section of this chapter discusses formulations in DYNAMO for common rate equations. The presentation is not intended to be a cookbook for rate equations but may help to save the reader from reinventing the wheel for certain common structures. Section 4.2 discusses auxiliary equations and describes the art of developing a table function. Levels are treated in section 4.3: though the equations for levels are trivial to master, deciding whether a concept should be represented by a level equation is sometimes subtle. Armed with the understandings of the first three sections, we return to the project model in section 4.4, formulating in DYNAMO equations the structure conceptualized in section 2.4. Sections 4.5, 4.6, and 4.7 use the model developed in 4.4 to discuss model documentation (including the use of DYNAMO's powerful documentor program), the thinking process involved in selecting parameters and initial values, and finding and correcting errors in model equations. The chapter ends in section 4.8 with a look back, summarizing a number of principles which help to guide the process of model formulation.

4.1 FORMULATING RATE EQUATIONS

Rate equations in a system dynamics model translate planning and system pressures into actions altering the state of the system. The entire process of model conceptualization and formulation is aimed toward a proper specification of rate equations.

This section discusses, in turn, rate equations which take the following forms:

CONST*LEVEL.K
LEVEL.K/LIFE
(GOAL.K-LEVEL.K)/ADJTM
AUX.K*LEVEL.K and LEVEL.K/AUX.K
NORM.K + EFFECT.K
NORM.K*EFFECT.K

Each pattern is illustrated in several different contexts, together with rationales for using such a formulation. At the completion of this section, the reader should be aware of some of the considerations involved in formulating rate equations and have a number of examples on which to base future formulations.

No list of patterns can exhaust the possible or necessary variations in rate equations, nor would such a list be advisable. In formulating a model, there can be no substitute for a fresh look at the way information combines to produce the change expressed in a given rate equation.

CONST*LEVEL.K

Figures 4.1 and 4.2 show two likely but simplified situations in which a rate equation takes the form

    R   RATE.KL=CONST*LEVEL.K

where CONST represents some system constant.

The structure in figure 4.1 is a classically simplified positive feedback loop which generates exponential growth over time. The interest payment rate, IPR, is formulated as a constant fractional annual interest rate, FAIR, times the bank balance, BAL:

Formulating Rate Equations

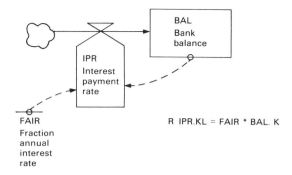

Figure 4.1: Interest payment rate formulated as a constant times the level of money in a bank balance.

    R   IPR.KL=FAIR*BAL.K
where
    IPR   = interest payment rate (dollars/year)
    FAIR = fractional annual interest rate
            (dollars/dollar/year, or fraction/year)
    BAL   = bank balance (dollars).

Note that the units in the equation correspond to the real world process the equation represents, and that they compute correctly:

    IPR = (fraction/year) * (dollars) = dollars/year.

The bank balance, increased by this compound interest, grows without bound, provided no one withdraws money from the account. As the balance grows, each successive computation of an interest payment is larger than the last, causing the growth pattern to curve upward, becoming increasingly steep as time goes on. One might encounter similar formulations in simple models involving growth, such as population (human or otherwise), the market for a company's sales, urban growth, and the like -- wherever the assumption of a constant fractional growth rate could be justified, given the purposes and scope of the model. (Note that common parlance calls the constant CONST the "fractional growth rate," (the interest rate, for example) causing potential confusion with our use of the word "rate.")

Figure 4.2 shows the production rate PR of a company formulated as the product of the workforce WF and a constant productivity PROD:

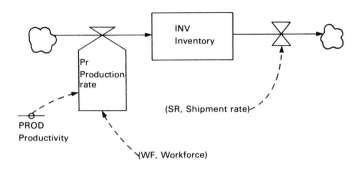

R PR.KL = PROD * WF.K

Figure 4.2: Production rate formulated as the product of the workforce and a constant productivity

```
R    PR.KL=PROD*WF.K
where
   PR    = Production (widgets/month)
   PROD  = Productivity (widgets/person/month)
   WF    = Workforce (people).
```

Again, note that the units agree:

$$PR = (\text{widgets/person/month}) * (\text{people})$$
$$= \frac{\text{widgets}}{\text{person*month}} * \text{people}$$
$$= \text{widgets/month}.$$

The assumption behind such a formulation is that the average productivity of workers in the firm is not changing over the time horizon of the problem addressed by the model: no overtime or undertime, no changes in the average workweek, and no effects of managerial pressure, or the like, on productivity.

While there are no doubt situations in which such formulations are appropriate, perhaps their most important functions here are to provide some simple beginnings that we will complicate later (for good reasons, of course) and to illustrate the use of dimensions as a check on equation formulation. The technique of dimensional analysis involves writing the dimensions of all the variables in an equation and simplifying the right-hand side algebraically, as if the words danced to the same tunes numbers do. In the production

Formulating Rate Equations 137

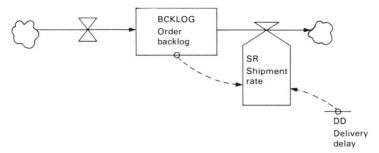

R SR.KL = BCKLOG.K/DD

Figure 4.3: Shipment rate from an order backlog expressed as LEVEL.K/LIFE.

example the ratio people/person is considered to be one, so people and person cancel, and the product PROD*WF.K is shown to have the units widgets/month, the proper units for the production rate PR.[1] Dimensional analysis cannot prove an equation is correct, but it neatly exposes the conceptual errors that produce inconsistencies in dimensions.

To produce a rate equation like either of these two examples, the modeler thinks of the process being modeled and asks if increases in the level result in proportional increases in its rate. If the process were to be represented by multiplication, what would the simplest possible equation be? The pattern of RATE.KL = CONST*LEVEL.K can later be made more flexible, if need be, by adding feedback to transform the constant CONST into a variable dependent on other quantities in the system.

LEVEL.K/LIFE

Many a rate equation depicting the outflow from a level is best thought of (initially at least) in the form

R   RATE.KL=LEVEL.K/LIFE

Figure 4.3 shows an example of a shipment rate depleting a backlog of orders for some product:

---

[1] Using the standard convention, we assume throughout that numbers or dimensions in the pattern A/B/C should be read (A/B)/C. Thus A/B/C = (A/B)/C = (A/B)*(1/C) = A/(B*C).

138   Model Formulation

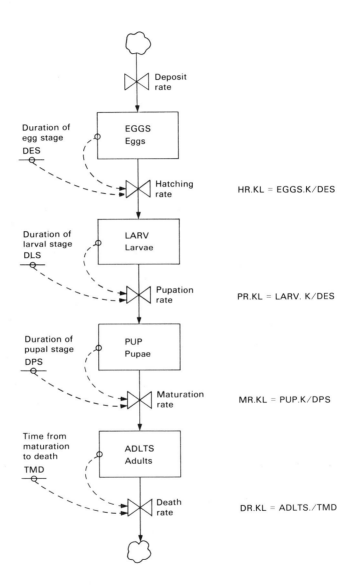

Figure 4.4: Stages of insect development, illustrating an aging chain with outflow rates in the form LEVEL.K/LIFE.

Formulating Rate Equations

```
   R    SR.KL=BCKLOG.K/DD
where
   SR       = shipment rate (widgets/week)
   BCKLOG   = order backlog (widgets)
   DD       = delivery delay (weeks).
```

The delivery delay is the *average time* an order remains in the order backlog before it is filled and shipped out. As in other outflow rate equations in this form, the denominator, DD, can be thought of as the average lifetime or the average dwell time of an item in the level. Note that the units agree: BCKLOG/DD is in widgets/week, as desired.

Figure 4.4 shows a cascade of four levels and their intervening rates. The figure shows an aging chain, here applied to an insect population but potentially useful in modeling different age groupings in any population, the aging of buildings in a city, stages of products in a manufacturing process, and so on. Each outflow is written in the form LEVEL.K/LIFE. In each case the denominator can be interpreted as the average time an item remains in the level before it flows on. (In this example, death rates from all but the last level have been omitted for simplicity.) (See section 5.3 for a discussion of when to disaggregate into such an aging chain and when not to.)

Expressing an outflow as the level divided by the average dwell time in the level has a number of important advantages. First, the constant in the denominator has a real-world meaning. It has meaningful units (time), not units contrived to enable the equation to pass the rigors of dimensional analysis. Second, the outflow rate can never deplete the level so far that the level becomes negative. The feedback from the level to its outflow rate guarantees that as the level heads toward zero so does the outflow rate. In contrast, if the shipment rate in figure 4.3 is the rate that depletes the inventory in Figure 4.2, there would be no feedback from inventory to its outflow rate, SR, so inventory could conceivably go negative if the order backlog got big enough. We will cure the problem for inventory later, but for the moment we merely note that, whenever we can formulate as we did for backlog, expressing the outflow from a level as the level divided by the average dwell time, we obviate the problem of levels going negative.

Certainly not all outflow rates should be formulated as LEVEL.K/LIFE, but it is fair to say that the pattern always ought to be considered as a promising candidate.[2]

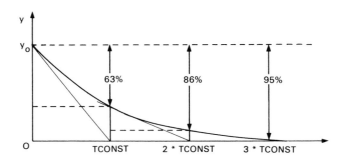

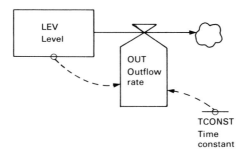

Figure 4.5: Behavior of a simple exponential decay structure, showing the role of the time constant

Technical Note

Figure 4.5 shows the dynamic behavior of a level influenced only by a single outflow rate expressed as

R   OUT.KL=LEV.K/TCONST

---

[2] Clearly, LEVEL/LIFE and CONST*LEVEL are equivalent if CONST = 1/LIFE. If one is appropriate, the other would compute just as well. Which form to use in any given situation is usually dictated by which feels the most natural. In which form is the constant most recognizable, or the units most meaningful?

Formulating Rate Equations 141

where
OUT = outflow rate (units/time unit),
LEV = level (units),
TCONST = time constant (time units).

The curve shows the same exponential decay pattern hand-computed for the coffee-cooling loop in section 3.2. The height of the curve as a function of time can be shown to be

$$y = y_o e^{-t/TCONST}.$$

From that equation one can compute that the curve drops 63 percent ($= 1-e^{-1}$) of the way toward zero during a time interval equal to TCONST. For a time interval twice that long, the curve drops 86 percent ($= 1-e^{-2}$) of the way from its initial value to zero.[3] In three time constants (3*TCONST) the curve drops 95 percent of the way from its initial value to zero.

The curve also has the tangency property illustrated in Figure 4.5: a tangent to the curve intersects the time axis after an interval of one time constant TCONST. This geometric property has a quantitative interpretation: if the level were depleted at a *constant* rate equal to its initial depletion rate, the level would be zero after one time constant TCONST.

Finally, there is a fixed relationship between TCONST and the time it takes for LEV to drop to *half* its original value. Setting $y = 0.5*y_o$ and solving

$$0.5*y_o = y_o e^{-t/TCONST}$$

for time t, the half-life of LEV can be shown to be $(\log_e 2)*TCONST$, or approximately 0.693*TCONST.

---

[3] The exponential function shown has the property that $y(t + TCONST) = e^{-1}*y(t)$, for any time t. Thus for *each* increment of time equal to TCONST, y(t) decreases by 63 percent ($= 1-e^{-1}$). Consequently, the 86 percent ($= 1-e^{-2}$) drop over two TCONST's can also be computed as an additional drop of 63 percent in the value of y from t=TCONST to t=2*TCONST: 0.63 + 0.63*0.37 = 0.86.

Actually, the half-life, or halving time, is the same at any point (t_o, y_o) on the exponential curve (a consequence of the fact that y_o cancels in solving for the half-life). Whenever an interval of time equal to 70 percent of TCONST elapses, the level drops about 50 percent.

In sum, there are four interpretations of the time constant TCONST. First, it represents the average lifetime or average dwell time of an item in the the level. Second, an elapsed time TCONST drops the level to 63 percent of its original value, from which one may conclude that after three time constants have elapsed the level is within 5 percent of zero. Third, the time constant represents the length of time it would take to exhaust the level if the initial outflow rate remained constant. Fourth, 70 percent of the time constant is approximately the half-life of the level. (The second, third, and fourth interpretations assume there is no inflow to the level.)

Similar comments apply to the simple exponential growth structure given by the pattern RATE.KL=CONST*LEVEL.K with CONST > 0 (exemplified in figure 4.1). An equation for the level in such a loop as a function of time can be written

$$y = y_o e^{CONST*t}.$$

The doubling time for the level, found by solving

$$2y_o = y_o e^{CONST*t}$$

for t, turns out to be

$$(\log_e 2)/CONST$$

or about 70 percent of (1/CONST). In the bank balance example, the balance doubles in about 0.7/FAIR, where FAIR is the fractional annual interest rate. (A compound annual interest rate of 7 percent doubles the account in about 10 years. On the other side of the coin: an inflation rate of 7 percent halves the value of our money in about 10 years.) If we call (1/CONST) the time constant for such formulations, we have exactly the same computation for doubling and halving times for exponential growth and decay: about 70 percent of the time constant involved.

These technical observations should help to characterize a time constant such as delivery delay, average product lifetime, or the reciprocal of the fractional annual interest

Formulating Rate Equations   143

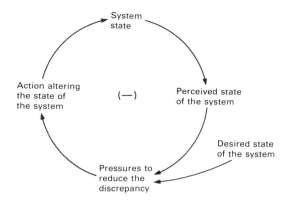

Figure 4.6: The generic negative feedback loop

rate. Furthermore, they can prove very useful in estimating time constants from real data: one can sometimes infer, for example, the half-life or doubling time of a variable from numerical data or a graph, and compute the corresponding time constant by dividing by 0.7. The data can also reveal whether it is reasonable to approximate the real structure by assuming a constant halving or doubling time (hence a constant time constant TCONST) or whether the structure must be complicated by treating the time constant as a variable.

(GOAL.K - LEVEL.K)/ADJTM

Perhaps the most classic negative feedback loop in human systems is the general structure striving to bring the actual state of a system closer to some desired or goal state. The loop is shown in general terms in figure 4.6. A rate equation in the form

    R   RATE.KL=(GOAL.K-LEVEL.K)/ADJTM

is a simple representation of such a negative loop structure. The constant labeled ADJTM stands for an adjustment time, a period over which the rate tries to close the gap between the level and its goal. The formulation represents a *net* rate, an aggregation of infows and outflows. It can be negative or zero as well as positive.

Figure 4.7 shows this pattern in a net hire/fire rate for a workforce. The equations for this piece of system structure are

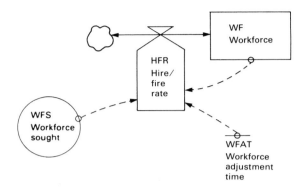

R HFR.KL = (WFS K-WF. K)/WFAT

Figure 4.7: A net hire/fire rate expressed in the pattern (GOAL.K-LEVEL.K)/ADJTM

```
L    WF.K=WF.J+DT*HFR.JK
R    HFR.KL=(WFS.K-WF.K)/WFAT
```
where
- WF = workforce (people)
- HFR = hire/fire rate (people/month)
- WFS = workforce sought (people)
- WFAT = workforce adjustment time (months).

The workforce sought is a quantity computed from other parts of the system, perhaps depending upon the firm's desired output or how much ahead or behind schedule it is. No matter how workforce sought is computed, the hire/fire rate attempts to adjust the actual workforce to equal the workforce sought. The adjustment takes some time, represented by the time constant WFAT in the denominator. Note that the units of the hire/fire rate check out to be people/month.

The hire/fire rate illustrated here is a net rate, capable of adding or subtracting people in the workforce, depending on how WF compares with WFS. If the workforce sought WFS is greater than the actual workforce WF, then (WFS.K-WF.K)/WFAT is positive, so people are hired. The workforce would rise. If WFS stayed constant for a while, WF would continue to rise until it equaled WFS, at which time the hire/fire rate would be zero, causing no further change in WF. Similarly, if WFS were less than WF, the hire/fire rate would be negative, lowering WF by subtraction until it

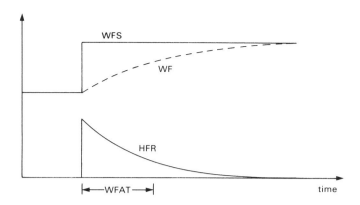

Figure 4.8: Behavior of the structure shown in figure 4.7 with a step in the workforce sought

reaches WFS. The system is clearly goalseeking; the workforce sought is the goal toward which WF is being adjusted.

The structure here is virtually identical to the coffee-cooling loop hand-simulated in section 3.2. It has very strong family ties to the outflow rate formulation LEVEL.K/LIFE discussed earlier: subtracting the rate LEVEL.K/LIFE can be thought of as adding the net rate (0 - LEVEL.K)/LIFE. Thus, the pattern LEVEL.K/LIFE is merely a special case of the net rate pattern (GOAL.K-LEVEL.K)/ADJTM. The goals implicit in the outflow rates in figures 4.3 and 4.4 are simply zero. The main difference between the outflow rate pattern shown in those figures and the net rate pattern illustrated in figure 4.7 is that the net rate can add or subtract to the level, while the outflow rates can only subtract.

As a consequence of the observations in the previous paragraph, we can infer, without any extra work, something of the behavior of structures involving the RATE.KL=(GOAL.K - LEVEL.K)/ADJTM pattern. Suppose the system shown in figure 4.7 had been in equilibrium for some time, WF = WFS, and the net HFR = 0. Then suppose there were a sudden, permanent increase in WFS, as shown in figure 4.8. The graph of WF would rise in an exponential, goal seeking pattern -- precisely the same shape as the graph in figure 4.5, merely flipped upside-down. The workforce adjustment time WFAT is the time constant of the level WF, analogous to LIFE and TCONST mentioned earlier. Thus we may conclude that the workforce rises halfway toward its new indicated value WFS in a time interval of about 0.7*WFAT. After an elapsed time of 3*WFAT, WF has closed 95 percent of the initial gap.

A sudden permanent drop in WFS would produce similar behavior, but in the negative direction.

Goal-seeking, negative feedback loop structures abound in real systems. Consequently, the simple representation of such structures by a net rate equation of the form RATE.KL = (GOAL.K-LEVEL.K)/ADJTM is potentially one of the most common general patterns for rate equations. Note, for example, that the rate of change assumed in the SMOOTH function appears in this pattern. In justifying the use of such a rate formulation, there are two considerations: structure and behavior. Does the negative loop, goal-seeking structure suit the situation, and would the actual behavior of the bit of reality being modeled look enough like figure 4.8 to warrant using a rate equation of the form (GOAL.K-LEVEL.K)/ADJTM?

AUX.K*LEVEL.K and LEVEL.K/AUX.K

Rate equations in the form

R   RATE.KL=AUX.K*LEVEL.K and
R   RATE.KL=LEVEL.K/AUX.K

are extensions or complications of the first two rate equation patterns discussed in this section, namely, CONST*LEVEL.K and LEVEL.K/LIFE. The constants in these latter expressions are replaced by variables. The thinking involved in formulating such equations is the same, whether constants or variables are used.

In the production example shown in figure 4.2, we could have written

R   PR.KL=PROD.K*WF.K
where
    PR   = production rate (widgets/month),
    PROD = productivity (widgets/person/month),
    WF   = workforce (people).

The only change is in viewing productivity PROD as a variable, as if undertime/overtime, changes in the workweek or changes in worker effectiveness are involved. The modeler adds a "dot K" (.K) to PROD.

The rate at which environmental pollutants are absorbed or neutralized could be represented (in the aggregate) by the pattern LEVEL.K/LIFE.K:

R   POLA.KL=POL.K/POLAT.K

Formulating Rate Equations 147

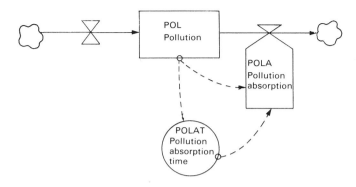

R POLA.KL = POL.K/POLAT.K

Figure 4.9: Pollution absorption formulated as POLAR.KL = POL.K/POLAT.K

where
  POLA = pollution absorption (pollution units/year),
  POL   = pollution (pollution units),
  POLAT = pollution absorption time (years).

The pollution absorption time POLAT represents the time it takes for an average pollutant to be broken down and absorbed by natural processes. It would presumably cover everything from tissue paper (six weeks?) to DDT (half-life in soil of about 10 years). In *World Dynamics,* Forrester chose to represent POLAT as a variable, rather than a constant, because

> "...as the amount of pollution increases, the pollution-absorption time is assumed to increase. This represents the poisoning and destroying of the pollution-cleanup mechanisms. Small amounts of pollution are dissipated quickly. But large amounts can have a cumulative effect by interfering with the natural processes of dissipation."[4]

Thus POLAT was formulated as a function of the pollution level POL, as shown in the flow diagram in figure 4.9. Deciding on the form of that functional relationship (not shown) requires some effort and expertise, but in the rate equation

---
[4] Forrester (1973), p. 58.

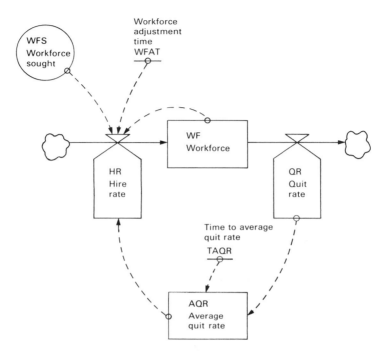

Figure 4.10: HR.KL = AQR.K + (WFS.K-WF.K)/WFAT

for POLA the only evidence of the change is the addition of ".K".

Thus nothing new has to be learned to formulate rate equations in these two patterns involving auxiliaries instead of constants, except how to formulate the equation for the auxiliary involved.

NORM.K + EFFECT.K

The previous patterns for rate equations can be considered the building blocks of most rate formulations. It is perhaps easiest to formulate a complicated rate as some normal or reference rate modified by one or more effects that add or multiply to adjust the reference rate to suit conditions in a model run. In this subsection we consider rate equations in which some normal rate is adjusted by the *addition* of one or more effects:

Formulating Rate Equations

    R   RATE.KL=NORM.K+EFFECT.K

The final subsection in 4.1 considers similar rate equations modified by multiplication.

Figure 4.10 shows a situation where a workforce is being increased by hires and decreased by quits. Management (in the model) has some idea of how large a workforce is indicated by conditions in the firm and tries to adjust its hiring to keep the workforce at or near the level indicated while at the same time replacing quits. The following hire rate does the job:

    R   HR.KL=AQR.K+(WFS.K-WF.K)/WFAT
where
    HR   = hire rate (people/month)
    AQR   = average quit rate (people/month)
    WFS   = workforce sought (people)
    WF   = workforce (people)
    WFAT   = workforce adjustment time (months).

(The formulation assumes there are no restrictions on the ability of the firm to hire as many people as it wants.)

In this formulation, the normal (reference) hiring rate is the average quit rate: management (in the model) normally hires to replace quits and maintain the size of the workforce. The second term in the equation represents the effect of changes in the desired size of the workforce.

All the parts of this hiring equation should look familiar even though the combination is new. The average quit rate could be a SMOOTH of the quit rate. Thus, if we were to represent it explicitly, even the rate of change of AQR in the SMOOTH function would be recognizable, since it would be in the pattern (GOAL.K-LEVEL.K)/ADJTM. The adjustment term trying to bring WF to equal WFS is in the same pattern. The complete rate equation states that the hiring rate is based upon the quit rate and is adjusted up or down depending on whether the workforce is at the level needed in the firm.

Perhaps a more readable version of the same equation would involve using an auxiliary for the adjustment term. Letting WFC represent the workforce correction, we would have

    R   HR.KL=AQR.K+WFC.K
where
    A   WFC.K=(WFS.K-WF.K)/WFAT
and
    A   AQR.K=SMOOTH(QR.JK,TAQR)

The result is a rate equation neatly in the form NORM.K + EFFECT.K.

Figure 4.11 shows a second example on this theme, but one with two adjustment terms. The context is managing an inventory of some sort. The firm takes some time to fill orders from inventory, so it has a backlog of orders. Inventory and the order backlog exist to smooth out variations in the stream of orders, so the firm can maintain production at a relatively constant pace. The firm tries to produce at a rate that will match the basic average shipment rate, raising or lowering production to adjust inventory and backlog to their desired levels. The result is a rate equation in the pattern NORM.K + EFECT1.K + EFECT2.K:

    R   PR.KL=ASR.K+IC.K+BC.K

where
    PR   = production rate (cars/month),
    ASR   = average shipment rate (cars/month),
    IC   = inventory correction (cars/month),
    BC   = backlog correction (cars/month).

The auxilary equations for IC and BC could be

    A   IC.K=(DI.K-I.K)/IAT and
    A   BC.K=(B.K-DB.K)/BAT

where
    DI   = desired inventory (cars)
    I   = inventory (cars)
    IAT   = inventory adjustment time (months)
    B   = order backlog (cars)
    DB   = desired order backlog (cars)
    BAT   = backlog adjustment time (months).

Note that, when there is too much inventory, production will be pulled below the average shipment rate by a negative value of IC. When the order backlog is too big, however, the backlog correction BC is positive, pushing up the production rate in an effort to fill more orders.

Evident in these examples is the requirement that the units of the terms in an additive equation must be the same. In the formulation for PR, for example, all three of the terms ASR, IC, and BC are measure in cars/month. One can not add apples and oranges in a system dynamics model, unless both are dimensioned as fruit.

(In some circumstances this formulation for a production rate could be seriously flawed: if I were big enough and/or B small enough, PR would become negative! It is more reasonable to view ASR + IC + BC as the desired production rate

Formulating Rate Equations

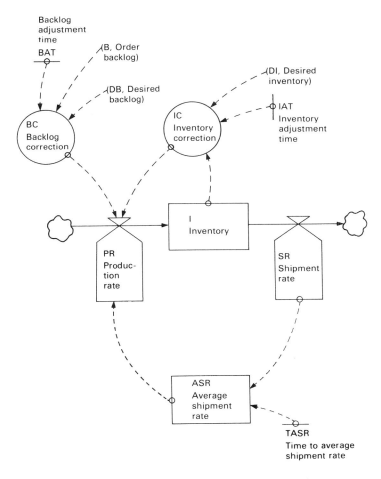

R PR.KL = ASR.K + IC.K + BC.K

Figure 4.11: A production rate formulated as a normal rate plus corrections for inventory and the backlog of orders

DPR, and to use DPR to determine the actual production obtainable from the firm's production capacity. Let us not worry about such complications at this point.)

It should be reasonably evident from these examples that one develops a rate equation in the form RATE.KL = NORM.K + EFFECT.K + ... from simple, well-understood pieces. We must emphasize that although the pattern here is helpful, it must not be applied in a cavalier fashion. Whether it fits a pattern or not, any rate equation has to represent the modeler's (and the client's) best understanding of the way change occurs at some particular point in the system--no more and no less.

NORM.K*EFFECT.K

The final pattern for rate equations that we will discuss is one of the most commonly used patterns for rate equations in system dynamics modeling. Like the previous pattern, NORM.K*EFFECT.K enables the building of complex rate formulations, incorporating, for example, factors outside the local subsystem in which NORM is computed. Here the rate is thought of as some normal rate of flow multiplied by one or more factors, raising or lowering the rate from its normal value. A multiplier that has no effect would have the value 1, an effect that raises the rate 30 percent above its normal value would have the value 1.3, a decrease of 15 percent is produced by a multiplier of 0.85, and so on. The tendency for people to see the world in terms of percentage changes makes rate equations in the form RATE.K = NORM.K*EFFECT.K relatively easy to conceptualize and frequently appropriate for modeling understandings of real processes.

For examples of an inflow rate and an outflow rate in this form, consider a changing urban population. Migration to and from the city is influenced among other factors by the job market in the city: the hope of employment draws people to the city, and persistently high unemployment can drive them away to greener pastures. Housing, cultural attractions, and other urban amenities also have their effects, but for the moment we consider only jobs.

Figure 4.12 shows a flow diagram of a city population in which migration is influenced by the job ratio--the number of jobs (filled and unfilled) in the city divided by the number of people in the potential laborforce (employed and unemployed). The immigration and emigration rates are formulated in the pattern NORM.K*EFFECT.K:

Formulating Rate Equations

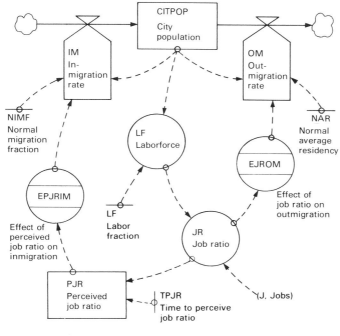

R IM.KL = (NIMF * CITPOP.K) * EPJRIM.K

R OM.KL = (CITPOP * K/MAR) * EJROM.K

Figure 4.12: Migration in a city, with effects of the job market captured in multipliers.

R   IM.KL=(NIMF*CITPOP.K)*EPJRIM.K
R   EM.KL=(CITPOP.K/NAR)*EJREM.K

where
- IM = immigration (people/year),
- NIMF = normal immigration fraction (fraction/year),
- CITPOP = city population (people),
- EPJRIM = effect of perceived job ratio on immigration (dimensionless),[5]

and
- EM = emigration (people/year),
- NAR = normal average residency (years),
- EJREM = effect of job ratio on emigration (dimensionless).

---

[5] Note that, in contrast to additive effects, multipliers are dimensionless.

each of the normal rates in this example are formulated using constants, in the patterns described at the beginning of this section. In the piece of model structure shown in figure 4.12, a job ratio is computed that directly influences emigration through the variable called EJREM. Immigration is assumed to be affected by a perceived job ratio, representing the delayed perceptions outsiders would have about the city before they moved to it.

The important thing to focus on in these formulations is the use of the multipliers EPJRIM and EJREM to change the normal migration rates as employment conditions change. One would expect that as the job ratio grows above 1, meaning there were more available jobs in the city than workers, there would be a tendency for immigration to increase (after a time) and emigration to decrease (all other effects being equal). Conversely, if the job ratio were to fall significantly below 1, one would expect a lower rate of immigration and a higher rate of emigration. The details of formulating variables like EPJRIM and EJREM will be discussed in section 4.2.

There is really no limit to the number of effects one could include in a rate equation using the pattern NORM.K*EFFCT1.K*EFFCT2.K*... Birth and death rates in Forrester's *World Dynamics* model, for example, contained effects of material standard of living, food, pollution, and crowding:

R   DR.KL=(DRN*P.K)*DRMM.K*DRFM.K*DRPM.K*DRCM.K
where
    DR   = death rate (people/year)
    DRN   = death rate normal (fraction/year)
    P   = population (people)
    DRMM   = death rate from material multiplier
              (dimensionless)
    DRFM   = death rate from food multiplier
              (dimensionless)
    DRPM   = death rate from pollution multiplier
              (dimensionless)
    DRCM   = death rate from crowding multiplier
              (dimensionless)

The reference death rate was set for global conditions in 1970, with DRN equal to 0.028 for an assumed normal (1970) global average lifespan of 1/0.028 or about 37 years. Thus each of these multipliers had to equal 1 for 1970 conditions of global average material standard of living, food per capita, pollution, and crowding. Considerable thought, of course, must go into the formulation of such multipliers, as we shall see in section 4.2.

Formulating Rate Equations   155

A Warning about Net Rates and Multiple Multipliers

When possible, it is nice to simplify the structure of a model by aggregating related inflows and outflows from a level and formulating the change in a net rate (see the hiring formulation discussed under the heading GOAL.K-LEVEL.K)/ADJTM, for example). However, multipling a net rate formulation by some effect in the pattern NORM.K*EFFECT.K can dangerous. EFFECT might have a value greater than one, as if it were trying to raise the level by increasing its net rate NORM, but if NORM happens at that time to be negative, the result will be the opposite.

This situation is likely to arise when more than one multiplier is involved. In the *World Dynamics* model, for example, Forrester might have tried to aggregate births and deaths into a single net birth rate equation:

```
R    NBR.KL=(NBRN*P.K)*NBRMM.K*NBRFM.K*NBRPM.K*
X    NBRCM.K
```

Yet in this formulation, a disastrous turn of world events that makes all the multipliers negative, indicating population should drop precipitously, would actually result in a *positive* net birth rate. The disaggregation into separate birth and death rate equations is forced by the multiple effects the model tries to capture.

More Examples

Any of the previously encountered rate equation patterns could be modified by multiplying by other factors. Recall the formulation for a shipment rate in figure 4.3:

```
R    SR.KL=BCKLOG.K/DD
```
where
```
    SR       = shipment rate (widgets/week)
    BCKLOG   = order backlog (widgets)
    DD       = delivery delay (weeks).
```

The formulation has the flaw that widgets would be shipped even if there were no widgets in inventory to ship! An improvement would modify the rate by an effect of inventory on shipments EIS:

```
R    SR.KL=(BCKLOG.K/NDD)*EIS.K
```

The most crucial property of EIS would be that it would have to be zero when inventory is zero, thus effectively shutting off shipments and increasing the actual delivery delay far above the normal delivery delay NDD. The resulting rate equation has the pattern NORM.K*EFFECT.K.

In a more complex example, a firm trying to expand production capacity (machines, buildings, trucks) will find itself under some conditions constrained in its ability to grow. It might *like* to place orders for capacity to replace discarded capacity and to adjust for changes in total desired changes in capacity, but it may find that banks do not like its current debt/equity situation. Hence its capacity ordering rate might be formulated:

    R   COR.KL=(ACDR.K+CC.K)*EDERCO.K

where
- COR = capacity order rate (capacity units/year)
- ACDR = average capacity discard rate (capacity units/year)
- CC = correction for capacity (capacity units/year)
- EDERCO = effect of debt/equity ratio on capacity orders (dimensionless).

The effect of debt/equity ratio on capacity orders could reduce the capacity order rate below that indicated by the amount needed for replacements and growth. It could even reduce orders to zero if banks simply refused all funding. Note that the pattern of this rather complex rate equation is (NORM.K + EFFCT1.K)*EFFCT2.K, a combination of the final two patterns discussed in this section.

Although the focus in this last example is on the use of the multiplier EDERCO, the reader might wish to complete the structure by suggesting simple formulations for ACDR and CC. Make use of previously encountered patterns for rate equations. Sketch the flow diagram, and write complete DYNAMO equations.

To Add or to Multiply?

And does it make a difference? Yes. Multiplication has the potential to shut off any rate, setting it to zero. Addition can not, except in the unlikely event that a negative term in the equation exactly counteracts a positive term with the same absolute value.

One might profitably think of addition as cooperating or contributing to the determination of a rate, while multiplication is potentially dominating. To decide which to use, then, one first tries to think out whether the various influences that combine to form a given rate cooperate or whether one could dominate the others. If one effect can dominate,

Formulating Rate Equations

then one must choose multiplication for that influence; if no effect dominates, then it is still an open question whether to add or multiply.

A glance back at some of the examples in this section serves to illustrate the point. Consider the hiring rate (figure 4.10) formulated as the sum of the average quit rate and a workforce correction term. The formulation assumes the personnel office (or whoever) simply adds the requests for new hires, replacements, and reductions in force to determine the positions to fill. No one of these considerations is assumed to have the potential to dominate the others, and together their net effect is the number of people personnel hires each month.

On the other hand, in the placing of orders for plant capacity, the banks had the last word. Internally, the company could decide to replace discarded capacity (ACDR) and adjust capacity up or down by adding a correction for capacity (CC), but its debt/equity ratio could be so big that no bank would loan them enough money to purchase a paper clip. Thus it was appropriate to model the influence of the debt/equity ratio as a multiplier, capable of sending to zero (dominating) the capacity ordering decision.

In the city migration example, it is reasonable to suggest that the job market could potentially reduce immigration to zero and speed emigration dramatically. Therefore, multiplying the effects of the perceived job ratios seems appropriate. Here, however, the decision is not as clear-cut as in the previous example. In the history of any city it might be hard to find a time when no new people were moving in, or in happier times, when none were moving out. Consequently, the modeler might be able to capture the dynamics of changing migration patterns using addition of effects--from jobs, housing, cultural amenities, and the like--rather than multiplication.

In such a case, a second consideration can be the deciding factor: in which form is it more natural to describe the various effects--as fractional or percentage changes, or as the addition or subtraction of actual quantities? Does management think of increasing productivity 10 percent or increasing output by 1000 widgets/month? Pick the rate formulation accordingly. Fractional changes suggest multiplication of effects, absolute changes suggest addition.

A final example, referring back to two production examples, can tie these perspectives together and lead us gracefully into section 4.2. In figure 4.2, a production rate was formulated as a product of the workforce WF and productivity PROD:

R   PR.KL=PROD.K*WF.K

(Here we have written PROD as a variable rather than a constant to allow for potential variations in productivity.) In figure 4.11, another production rate was written as a sum:

R   PR.KL=ASR.K+IC.K+BC.K

Is one of these right and the other wrong?

The formulation as a sum has the drawback that for sufficiently negative values of the inventory and backlog corrections IC and BC the production rate PR could become negative. Negative production is meaningless. Such an equation also claims that production will go on even if there are no machines and no people to work them. The one advantage the additive formulation has is that it states exactly what the managers in the firm have concluded (cooperatively) they want the firm to produce.

The product formulation does not suffer from the drawbacks of the additive formulation: if the workforce is zero, so is PR; and, if machines are needed and there aren't any, then presumably PROD would be zero, so again the formulation would behave correctly. The product formulation describes changes in the rate of output in fractional (percentage) terms, a bit less obvious, perhaps, than an absolute number of widgets.

The tidiest way of thinking of these two expressions is to view the multiplicative version as a sound equation for the actual production rate, while holding on to the additive version as the *desired* production rate. *If it could,* the firm would produce an amount sufficient to match the recent average shipment rate, adjusted up or down to put inventory and the order backlog in good shape. Thus DPR.K = ASR.K + IC.K + BC.K would be used to determine the size of the workforce WF the firm wants, and how productive it wants its current workers to be (overtime/undertime and the like).

In a sense, neither the additive nor the multiplicative formulation is really inappropriate as a production rate; one is more properly seen as desired rather than actual production. Their respective formulations suit the units in which changes in those variables are likely to be phrased: a new target for the *number* of widgets produced per month indicating that production should rise or fall by some *percentage.* In section 4.2 we shall show how desired and actual conditions such as DPR and PR can be linked.

Formulating Rate Equations 159

4.2 AUXILIARY EQUATIONS

A variable modeled as an auxiliary in a DYNAMO model represents *information* in the system. Since all information eventually traces back to the system levels, it is conceivable that one could write a model without using auxiliary equations, formulating all rate equations solely in terms of levels and constants. However, the resulting model listing would probably be nearly unreadable. Worse, some of the interesting information in the model would be inaccessable because DYNAMO can not print or plot quantities that have not been given variable names. Consequently, one captures some pieces of system information in auxiliary variables, which, as their name suggests, aid the formulation of rate equations.

In section 4.1 we repeatedly saw the need for auxiliaries: workforce sought WFS.K, pollution absorption time POLAT.K, inventory correction IC.K, productivity PROD.K, and multipliers such as EJREM.K, the effect of the job ratio on emigration. The diversity in just this small list gives some idea of the range of concepts auxiliaries might represent. Workforce sought, WFS, has units of people, making it perhaps sound like a level variable. The inventory correction IC has units of widgets/month, sounding potentially like a rate of flow. POLAT represents a length of time, productivity PROD might have units of widgets per person per week, and EJREM was conceived as a dimensionless multiplier. The range of meanings of auxiliaries is enormous, and clearly they can not be characterized by the types of dimensions involved.

Although an occasional auxiliary variable (such as workforce sought) may sound like it represents an accumulation in a system, auxiliaries have more in common with rate equations than with levels. Like rates and unlike levels, auxiliary equations represent algebraic computations,[6] not accumulations, and they have no standard form. An auxiliary variable can be computed from any combination of constants, levels, rates, or other auxiliaries--whatever the desired functional relationship dictates. The only limitations are con-

---

[6] An equation setting a variable equal to a first- or third-order information delay (the SMOOTH and DLINF3 functions) is also listed in a DYNAMO model as an auxiliary equation, even though these functions involve integration (levels). The reason is that DYNAMO creates a dummy variable in the subroutine for the function, computes with it, and then in a final *auxiliary* equation sets the model variable equal to the dummy.

ceptual--what is a meaningful algebraic representation of the way a quantity in the real system is determined?

For our purposes here we shall divide the discussion of auxiliaries into two parts, the first dealing with explicit algebraic computations and the second covering table functions. The division is more pedagogical than theoretical, for a table function is really nothing more than a simplified way of representing a potentially complicated algebraic function for which the modeler does not care to write a formula. There is an art to developing a table function, however, that warrants special attention.

A System Concept as an Algebraic Expression

The obvious task for the modeler is to translate a system conceptualization into quantitative terms. Deciding on level variables and the rates that affect them is the first step. The concepts that remain must be captured as auxiliaries, written as functions of the various pieces of information that converge (in a causal-loop diagram, for example) to form the concept.

The first key to developing an auxiliary equation is to be sure one is thinking of a concept as a quantity. *Identify the units of the concept.* Know what it means to say it rises or falls. If there is any doubt abouts its units, try getting some guidance by looking ahead to see where in the model the auxiliary will be used. Then *identify the units of the variables that converge to determine the concept.* The units of variables are one of the strongest indicators of possible functional forms for an auxiliary equation (or a rate equation, for that matter). Note that the units may have to be invented to meet the requirements of the model--it is possible that no one has ever measured the quantity before or even identified a scale by which it might be measured. A person's level of anxiety, management pressures, or the quality of a product in a model of corporate behavior are potentially very recognizable concepts. It is clear what it means for any of them to rise or fall. The modeler may wish to represent such a concept explicitly. To do so requires the invention of units and a measurement scale and consistent treatment throughout the model.

A second observation, although obvious, deserves to be stated: there is a distinction between a quantity and its effect(s) in the system. Formulate the variable first, and then focus on how to quantify its effects.

For examples, we will consider some of the concepts and variables scattered throughout the causal-loop diagrams in chapters 1 and 2. The formulation of the project model con-

Auxiliary Equations   161

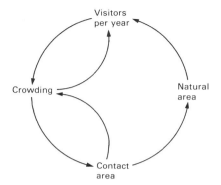

Figure 4.13: Crowding, from figure 1.5, "A public recreational area viewed as a feedback system"

ceptualized in section 2.4 will provide a more cohesive set of examples in section 4.4.

Consider the concept of crowding contained in the causal-loop diagram in figure 1.5 (repeated in figure 4.13). As conceived in the diagram, crowding is a function of the number of visitors per year and the contact area (the portion of the recreational area on which people walk or leave evidence of their presence). Crowding then influences visitors and the contact area. Distinguishing the concept from its effects, we note it is natural to think of crowding as measured by the visitor density, a ratio of people to area:

    A   VDENS.K=VPY.K/CA.K
where
    VDENS   = visitor density,
    VPY   = visitors per year (people/year),
    CA   = contact area (acres).

According to the computation, the units of VDENS would have to be (people/year)/acre. Although most of us would probably first think of crowding as measured by people/acre, the time units make sense because crowding is a concept involving both space and time--1000 visitors in one acre in a year is potentially a lot less crowded than 1000 visitors in one acre in 10 minutes. Note that the algebra of dimensions allows us to equate

    (people/year)/acre = people/(acre*year)
            = (people/acre)/year,

with the latter variation being perhaps the most intuitive expression for a measure of crowding.

The computation of natural area in figure 4.13 has a different character. It is apparently a function of only one variable, the contact area. The fact that both contact area and natural area have the same units suggests an additive (or subtractive) formulation, and one certainly exists in the system:

(contact area) + (natural area) = (total area).

So an auxiliary equation suggests itself for the natural area in this public recreational system:

    A   NA.K=AREA-CA.K
where
    NA   = natural area (acres),
    AREA = total area (acres),
    CA   = contact area (acres).

Thus, NA is actually a function of two quantities, one of which was omitted from the causal-loop diagram either for simplicity, because it was too obvious to bother to show, or because it was forgotten. Given the fixed bounds on such an area, it is likely that AREA would be a constant, so the modeler could focus the most attention on fomulating the contact area, CA. (For reasons discussed in section 4.3, the contact area would then be modeled as a level. The perceptive reader will see that there is a certain amount of arbitrariness here, in that one could have conceptualized the system in such a way that NA is the level and CA is computed from it in an auxiliary.)

One should note from these two examples that if an auxiliary is a function of one or more quantities and all their units agree, then an additive (or subtractive) formulation is likely. If the units of an auxiliary are different from the variables that determine it, then multiplication (or division) must be involved somehow. The observation is trivial, but it exposes the thinking that must take place to formulate auxiliaries.

Consider the salesman loop in figure 1.7, reproduced as figure 4.14. The loop summarizes the process by which salesmen generate orders, which produce revenue and help to determine the budget, and eventually indicate the number of salesmen the budget can support. If indicated salemen is greater than salesmen, then the loop implies that salesmen would be hired and the variables in the loop would grow. If the number of indicated salesmen is less than the current

Auxiliary Equations 163

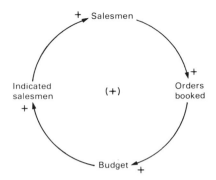

Figure 4.14: The salesmen loop (from figure 1.7)

number of salesmen, then some of the salesforce would be let go, and the variables in the loop would decline.

How should orders booked be formulated? Its units should either be widgets or widgets per month (for example), depending upon whether one interprets the variable as representing the accumulated backlog of orders or the flow of orders into the backlog. The intent of the causal-loop diagram is that orders booked should rise as the number of salesmen rises (ceteris paribus), and that the budget should rise as a result of an increase in orders booked. Because a budget is a "per time" concept (a monthly or an annual budget), we may conclude that orders booked was intended to represent the flow of orders, not the stock.

Since the units of salesmen (people) and orders booked (say, widgets) are different, multiplication or division must be involved, so one is prompted to write

    A   OB.K=S.K*OBPS

or, alternatively,

    A   OB.K=S.K*OBPS.K,
where
    OB   = orders booked (widgets/month),
    S   = salesmen (people),
    OBPS = orders booked per salesman
            (widgets/month/person).

OBPS could be a constant or another variable (depending on considerations not shown here), but whichever is chosen, the multiplicative form for OB is clearly indicated by the units of the quantities involved.

For practice, the reader might try formulating auxiliary equations in a similar fashion for the budget and the number of indicated salesmen in figure 4.14. Assume for simplicity that orders are shipped and paid for as soon as they are booked. The enthusiastic reader may want to complete the loop by formulating a net hire/fire rate equation (implicit in the causal-loop diagram), making use of the understandings in section 4.1.[7]

Examples of a number of other auxiliary equations are contained in section 3.3, together with enough information about the meanings of variables to allow the reader to verify the formulations. Turning now to table functions, we postpone further discussion of explicit algebraic auxiliaries until the formulation of the project model in section 4.4.

Formulating Table Functions

A table function is a simple computational technique for specifying a relationship graphically in a model. The DYNAMO expressions for table functions were presented in section 3.3. Here we investigate the thinking required to formulate conceptually meaningful table functions.

Consider the portion of a simplified urban model shown in causal-loop and flow diagrams in figure 4.15. The structure depicts the construction and demolition of business structures BS in a city of fixed land area, showing an important negative feedback loop constraining construction as the city land fills with buildings, roads, parks, parking lots, and the like. The reader should be agile enough at this point to write the level equation for BS, the rate equations for BC and BD, and the auxiliary equation for LFO, which are implied by the flow diagram.

---

[7] There will be two more quantities like OBPS in the loop, invoked to formulate the auxiliaries: net revenue per sale and salesman salary. Show that if the product of two of them divided by the third is greater than 1, the salesforce grows, while if it is less than 1, the salesforce declines. That conclusion is equivalent to saying that the salesmen level in the positive feedback loop grows only if the *gain* of the loop is greater than 1. For a more complete discussion of these ideas, see Forrester (1968a).

Auxiliary Equations

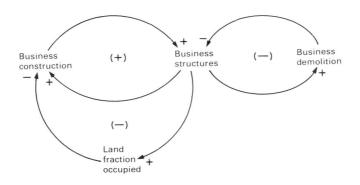

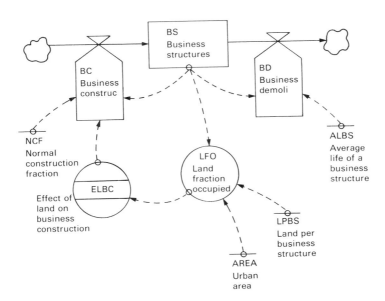

Figure 4.15: A portion of a simplified urban model containing an effect of land on business construction

Our focus is on the formulation of the business construction rate BC. Making use of the pattern of NORM.K*EFFECT.K, we can write the rate as follows:

R   BC.KL=NCF*BS.K*ELBC.K
where
    BC   = business construction (structures/year),
    NCF   = normal construction fraction (fraction/year),
    BS   = business structures (structures),
    ELBC = effect of land on business construction (dimensionless).

This equation states that business structures are constructed at a certain normal fraction per year of existing structures, unless the land available for construction becomes scarce. Although the term "normal" is common in such formulations, it is more appropriate to view NCF as a reference value, not a normative one. The flow diagram shows the effect of land on business construction as an auxiliary variable, a function of the land fraction occupied LFO. Our task is to formulate a table function for ELBC.

Formulating a Table Function for ELBC

LFO ranges from 0 (the city area before people arrive) to 1 (a completely built-up city with no further room for construction). The most crucial attribute of ELBC is that it must be zero when LFO reaches one. When there is no more available land, construction must cease until land becomes available through demolition of existing structures. Therefore we know one point on the table for ELBC, the point (1,0).

Because ELBC multiplies in the equation for BC, we are interested also in the value(s) of LFO for which ELBC is 1. Those points represent the no effect points, or the normal points, the values of LFO at which business construction proceeds at its reference rate, determined by the value of NCF. Thus, there is a connection between NCF, LFO, and ELBC: *the value(s) of LFO for which ELBC = 1 implicitly define(s) the normal or reference condition of the city, in which the fractional construction rate is NCF.*

We shall *define* the reference conditions to be when 60 percent of the city land is built-up, when land fraction occupied LFO is 0.6. From city records we then should be able to find the actual fractional construction rate at that point in the city's history and set NCF equal to it. Let us assume we found NCF to be 0.7 (fraction/year) when LFO was 0.6. (We might have found the number of business structures doubling about every ten years when the land was about 60 percent built up.) We now know another point on the table, the point (0.6,1), a so-called normal, or reference, point.

There are two other major considerations: the *slope* of the table and its *shape*. The effect of an increase in land fraction occupied should eventually be to decrease business construction: as land becomes scarcer, the more desirable parcels are taken up, zoning is introduced, building permits become required and with time become ever harder to obtain, land becomes outrageously expensive, and so on. Therefore, the causal link from LFO to business construction is negative. Consequently, the slope of the table for ELBC should be negative. For the shape of the table, consider the curves sketched in Figure 4.16, which meet all the requirements for ELBC that we have proposed thus far.

In alternative 1 of figure 4.16, ELBC is 1 in the region where LFO is less than 0.6. The implication is that throughout the early history of the city growth proceeded at a constant fractional rate (all other potential effects on construction not changing). As LFO increases beyond 0.6, ELBC becomes increasingly steep as LFO approaches 1. Because of the curvature, everywhere on this curve a ten-percent increase in LFO brings ever greater decreases in the value of ELBC as LFO increases. The curvature thus implies that every increase in LFO brings ever more severe constraints on business construction until eventually all construction is shut off when LFO equals 1.

Alternative 2 begins above 1 for low values of LFO and declines throughout the entire range of LFO. The table implies that the fractional growth rate of the city was above NCF until the reference point, where LFO = 0.6. The S-shape of alternative number 2 has an interesting interpretation. Unlike alternative 1, alternative 2 becomes steepest near the middle of the range of values for LFO, when LFO is about 0.7. Therefore a ten-percent change in LFO brings the greatest percentage drop in business construction in the range where LFO is near 0.7. In alternative 1, the greatest percentage drop occurred at the end of the interval. Thus, while alternative 1 implies that constraints on construction pile up more and more rapidly as LFO approaches 1, alternative 2 says the most rapid burgeoning of constraints occurs around the time when about 70 percent of the land is built up. Both tables choke off construction when LFO reaches 1, but alternative 2 throttles down sooner.

The modeler (and the client) would select which of these scenarios best suits their understandings of the effect of land on business construction and write the DYNAMO statements

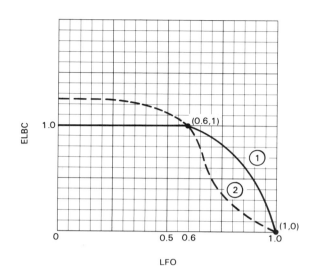

Figure 4.16: Alternative table functions for ELBC, the effect of land on business construction

for ELBC accordingly. If we decided upon alternative 2, we would write something like the following:

```
A    ELBC.K=TABLE(TELBC,LFO.K,0,1,0.1)
T    TELBC=1.3/1.28/1.25/1.22/1.18/1.1/1/.7/.3/.1/0
```
where
    ELBC   = effect of land on business construction (dimensionless),
    TELBC = table for ELBC.

(See section 3.3 if this notation is unfamiliar.) The x-axis of the table was chopped up here in increments of 0.1 in order to obtain a relatively smooth broken line approximation to the curve we drew. A larger x-increment--fewer subintervals and y-values in the table--could be used. It would mean sharper corners between the line segments along which DYNAMO interpolates, but they would probably not be noticeable in the behavior of the model. The y-values listed in TELBC are most easily read off a sketch of the table on graph paper, but one should remember that exact values are meaningful for only two of the points, (0.6,1) and (1,0).

A rather different shape for ELBC is shown in figure 4.17. The table first rises and then falls. It has a positive

Auxiliary Equations 169

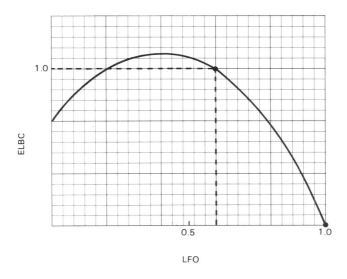

Figure 4.17: A table function for the effect of land on business construction which expresses both positive and negative effects.

slope up to LFO = 0.4 and a negative slope thereafter. Such a table is trying to capture two different effects--a positive link and a negative link--in one formulation. In its positive slope range, the table implies that, as the city becomes more built up, its annual fractional growth rate *increases.* The assumption is that in the early history of the city, when little land is built upon, business structures make it easier to build more business structures: a brick yard facilitates the construction of brick buildings; paved roads allow materials to be transported more easily and quickly; water, gas, and electric facilities attract new construction; and so on. The negative region of the table expresses the same constraint on construction from available land discussed above.

A table which tries to capture effects with opposite polarities is not particularly bad practice, but one must be very clear about the distinct natures of the two effects. For example, in the positive slope range, the curve for ELBC uses land fraction occupied as a proxy for infrastructures that facilitate early growth in the city. The diminishing availability of land is not really speeding growth; it is what is on the land in those early years that accelerates growth. In an urban model that does not separate out kinds of business structures (such as roads vs. buildings, construction companies vs. opera companies), such a surrogate formulation may be necessary because the real determinates of accelerating growth are below the level of aggregation in the model.

Formulating an Undertime/Overtime Effect

Consider the problem of incorporating undertime and overtime in a production sector. Assume that within the model there is a well-defined notion of desired output DOUT, intended to match sales and, if necessary, bring inventory and the order backlog to their desired conditions. The following production rate has been proposed:

    R   PR.KL=NPROD*WF.K*EUTOT.K
where
    PR   = production rate (thingumabobs/week),
    NPROD   = normal productivity
                (thingumabobs/person/week),
    WF   = workforce (people),
    EUTOT   = effect of undertime/overtime (dimensionless).

The rate equation is in the form of (normal rate)*effect. The task is to formulate a table for EUTOT as a function of desired output DOUT.

The task is greatly simplified if we "normalize" or reference the input to the table, that is, if we formulate EUTOT as a function of the *ratio* of DOUT to some normal production rate NPR, defined to be the output rate if there were no undertime or overtime. Referencing the input to the table function changes the scale of the x-axis of the table so that it ranges around 1, where 1 represents normal desired output conditions. Hence, we propose to formulate

    A   EUTOT.K = f(DOUT.K/NPR.K)
where
    EUTOT   = effect of undertime/overtime,
    f( )   represents a table function,
    DOUT   = desired output (thingumabobs/week),
    NPR   = normal production rate (thingumabobs/week).

NPR would be, by definition,

    A   NPR.K=NPROD*WF.K

the normal production rate that would pertain if no undertime or overtime were in effect.

Auxiliary Equations

Proceeding as before for ELBC, we consider the slope of the table, its shape, and some specific points. Because the input to the table was normalized, the no effect point for the multiplier is (1,1). The point (1,1) corresponds to the situation in which DOUT/NPR = 1 and EUTOT = 1, so the production rate PR = (NPROD*WF)*EUTOT = (NPR)(1) = DOUT, just as desired in normal conditions. The slope of the table is also rather obvious: desired output should have a positive effect on production, so the slope of the table for EUTOT should be positive.

Figure 4.18 shows the beginnings of a table for EUTOT. The (1,1) point is drawn in, and the y = x line is sketched as a *reference line*. The reference line has the property that at every point on it EUTOT = DOUT/NPR. That is, the undertime/overtime effect represented by the y = x line is just what is needed to set the production rate equal to desired output, as the following algebra verifies:

PR = (NPROD*WF)*EUTOT = (NPR)*(DOUT/NPR) = DOUT.

Why not draw the table function for EUTOT right along that reference line?

Such a table function assumes there is no limit to the amount that production can be raised or lowered by overtime or undertime. But limits certainly exist. Few workers would be willing to put in more than, say, a 55-hour workweek; the union might have a contract that limits overtime to no more than 50 hours per week and requires that all workers be paid for at least, say, 35 hours a week; the machines or whatever that workers use in production might require a certain amount of downtime for maintainence and repairs. These limitations force the table for EUTOT to depart from the reference line and to *saturate* for both high and low values of DOUT/NPR, that is, to become horizontal. Values of DOUT/NPR above a certain number will result in no further increases in overtime; values of DOUT/NPR below some other number will produce no further decreases in production from undertime.

One can estimate the maximum and minimum (saturation) values of EUTOT from workweek limitations. If the maximum average workweek conceivable in the firm is, say, 50 hours per week and the normal is 40, a 50-hour average workweek represents a fractional change in production of 50/40 or 1.25. Hence, the maximum achievable value of EUTOT would be 1.25, and would represent a 25 percent increase in productivity resulting from overtime. If the minimum workweek allowed by union restrictions or whatever is 35 hours, then the minimum value of EUTOT would be 35/40, or 0.875.

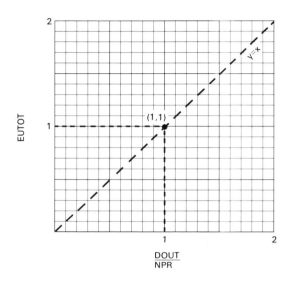

Figure 4.18: The normal point (1,1) and a reference line for EUTOT.

Consequently, for values of DOUT/NPR between 0.875 and 1.25, EUTOT can follow rather closely to the reference (y = x) line, but outside that range, EUTOT must be horizontal. Figure 4.19 shows such a table function.

The table function shown in figure 4.19 curves smoothly away from the reference line as it nears the extreme possible values. A smooth curve without sharp corners represents the likely response of an *aggregate* system. Not all workers will go on overtime and undertime at the same time or to the same extent. As a few start on overtime, the aggregate curve will pull away from the ideal path lying along the reference line. As the numbers of workers on overtime and the average amount each puts in approach their maximums, the table function approaches the saturation limit.

The DYNAMO statements for EUTOT shown in figure 4.19 could be the following:

```
A    EUTOT.K=TABHL(TEUTOT,DOUT.K/NPR.K,.7,1.4,.1)
T    TEUTOT=.875/.89/.92/1/1.1/1.18/1.22/1.25
```

The use of the high-low table function TABHL notifies DYNAMO that the modeler does not wish to be told when DOUT/NPR falls outside the range of the x-values specified.

Auxiliary Equations 173

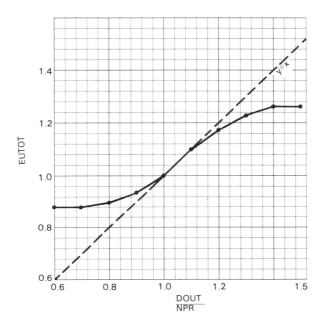

Figure 4.19: A table function for EUTOT showing the saturation effects of maximum allowable amounts of undertime and overtime.

The function TABLE produces a message each time the x-value first falls outside the range of the table and when it later reenters the range. (See section 3.3.) The high-low table is useful only with a table function that saturates, and even then the use of TABLE is advised until the modeler is certain the table range is properly specified.

Guidelines for Formulating Table Functions

Embedded in the analyses of the effect of land on business construction and the effect of undertime/overtime are a number of important principles for formulating table functions in system dynamics models:

1. There are four potential considerations in formulating a table function: slope, shape, one or more specific points, and one or more reference lines.

2. Set the slope of the table to match the polarity of the effect it represents: a negative link is represented by a table function with negative slope, and a positive link by a positive slope.

3. Select the shape of the curve, paying attention to its slope and curvature at both extremes and in the middle of the table. The flattening out of a table function corresponds to a weakening or saturating effect. Steepening a table represents the strengthening of an effect.

4. If possible, normalize or reference the table by formulating it as a function of the *ratio* of the input to its reference value. The reference point is then clearly indicated as the one with x-coordinate equal to 1. (Occasionally, referencing the input to a table by *subtracting* the reference value is more appropriate; in such a case the reference point has x-coordinate zero.)

5. Determine the locations (coordinates) of as many of the following points as possible:

    - where the y-value of the table is 0,
    - where the y-value of the table is 1,
    - where the x-value of the table is 0,
    - where the x-value of the table is 1,
    - the extreme x- and y-values shown in the table,
    - any others.

6. For a table representing a multiplier, the point(s) for which the table (y) value is 1 are the no-effect, or reference, points. They *define* normal, or reference, conditions, including perhaps the values of constants in the formulation. For a table function that adds to a formulation, the no-effect points are those for which the table value is 0.

7. Reference lines, such as $y = x$, $y = y_o$, or $y = x_o$, can be useful to represent desired or ideal conditions, saturation limits, or other referents for a table.

Summary Comments on Auxiliary Equations

Quantifying pieces of information around feedback loops in a model requires a clear conception of the way those pieces are determined in the system being modeled. The arrows that converge in a causal-loop diagram give some idea of an appropriate computation for a system concept as a model variable, but often the modeler must make a deliberate shift from thinking conceptually to thinking algebraically. The most powerful guidance in formulating auxiliaries comes from the *units* of the quantities involved. Dimensional analysis--the manipulation of units as if they behaved like algebraic quantities--is a critically necessary skill.

Formulating and interpreting table functions is something of an art. To some, it may look too much like an art to have a justifiable role in serious quantitative modeling. Yet, analyzing the possibilities for the slope and shape of such a function, sketching some reference lines, and locating two or three points leave remarkably little leeway for the remainder of the function. Moreover, table functions, or the algebraic functions they represent, are extremely useful for formulating the nonlinear pressures and effects that abound in societal problems. Often a table function must be invoked to represent a link believed to exist in the real system but not discernable in real system data.

Because of the flexibility of table functions, there are potential pitfalls in their use. As we have seen, there is considerable subtlety to the inferences one can make about the nature of an effect from such things as the curvature of the table function used to represent it. The ease with which a table can be changed in DYNAMO's rerun mode can lead to experiments that violate some of those subtle assumptions. If the experimenter is not careful he can easily change a table in such a way that it no longer passes through a certain reference point or along a particular reference line or no longer saturates at a reasonable value.[8]

A second temptation that stems from the simplicity of tables is to use a table function to quantify an idea when further thought would have produced a more explicit formulation. Consider, for example, the following two formulations for a desired production rate:

A   DPR.K=ASR.K*EIDP.K and

---
[8] A number of critical analyses of system dynamics models have fallen prey to some of the pitfalls of table function formulation. See, for example, Garn and Wilson, 1972, p. 155.

A   DPR.K=ASR.K+(DI.K-INV.K)/IAT
where
   ASR   = average shipment rate (brooms/month),
   EIDP = effect of inventory on desired production,
   DI   = desired inventory (brooms),
   INV   = inventory (brooms),
   IAT   = inventory adjustment time (months).

EIDP would be a table function, probably formulated in terms of INV.K/DI.K, trying to represent pressure to raise or lower production relative to its basic rate, the average shipment rate. The correction term in the second equation expresses a more explicit decision rule (not necessarily a better one but more explicit). The speed of response to conditions in which INV is not equal to DI, for example, is only implicitly represented in the first equation, in the steepness of the table. In the second equation, it is exposed in the constant IAT. In general, explicit formulations are preferable. They are easier for the modeler to control and can be compared more reliably to the reality they try to represent. The tendency to overuse table functions is most damaging when it results in a formulation that ignores accumulations or delays in the system.

4.3 WHAT IS A LEVEL?

The concept of a level was introduced in chapter 2. In chapter 3 the form of a DYNAMO level equation was developed. From these discussions we know that a level variable is an accumulation over time, a storage device for material, energy, or information. Rates of flow increase or decrease the level accumulated, and the DYNAMO equation for any level has a consequently predictable form. All this is clear, because the notion of accumulation is rather simple. What is perhaps not clear at this point is that while all levels are accumulations, not all accumulations should be levels.

The purpose of this section is to help the reader decide when a system concept should be modeled with a level equation and when it would be more appropriate to model the same concept as a constant or as an auxiliary. Is price, for example, best viewed as a level or an auxiliary, or perhaps even a constant? The answer is, it depends, and this section attempts to show on what.

The Snapshot Test

A simple mental experiment can help to identify the potential levels in a system. Imagine stopping time in the system,

What is a Level? 177

freezing all flows instantaneously, as if one took an all encompassing photograph of the system capturing intangible and invisible characteristics as well as physical processes. The potential level variables are those that still exist and have meaning in the snapshot. One would still be able to measure the extent of accumulations even if time were stopped, while flows would be stilled, perhaps visible in the photograph but not measurable.

Thus potential levels in a system include such obvious accumulations as people, inventory, production capacity, serum cholesterol, and bank balances. Included as well, however, are such things as cultural traditions, average sales rates, habits, and perceptions, for these too would not disappear if time were stopped. In the snap shot test, the current flow of sales would be stopped, but the average level of sales over the past year would continue to exist and be knowable. The fact that the units of average sales (for example, widgets per month) make the variable sound like a rate does not alter the fact that averaging involves an accumulation over time, implying that average sales is more like a level than a rate. *Units are no help in selecting level variables.*

Yet not all quantities that continue to exist in a snapshot test ought to be modeled as levels. An obvious set of exceptions are constants. One might have an essentially constant workforce in a production setting, for example, and model the workforce by simply assigning it a value in a DYNAMO C equation. A snapshot would show the workforce, and it might certainly be thought of as net accumulation of prior hires and fires, yet it would not be modeled as a level.

Now one might argue that the constant workforce counterexample is spurious because the accumulation involved took place outside the time frame of the model. Yes, and that is the beginning of an important insight about system levels. There is a conceptual connection between the time horizon of the model and the time over which kinds of variables change. Some constants, for example, might be thought of as accumulations that change only very slowly over the time frame of the problem. By analogy, one might suggest that an accumulation that changes very rapidly -- over a time interval much shorter than the time frame of the model -- could be properly modeled as an auxiliary. Before exploring the suggestion, we shall develop a deeper view of the different ways levels and auxiliaries contribute to the dynamics of a system.

Levels Transform Behavior

A level variable in a causal-chain has the capability of changing the pattern of behavior of coming into it. Turn back to figures 4.7 and 4.8 for a simple illustration of the fact. In those figures an abrupt step increase in the size of the workforce sought WFS produced a smoothly rising pattern in the workforce WF as new people were hired according to the net hire/fire rate equation

HFR.KL=(WFS.K-WF.K)/WFAT

The behavior of WFS was transformed by the accumulation process of the level equation for WF, changing an abrupt step increase of the auxiliary into the smooth, exponential goal-seeking behavior of the level.

Examples of the capability of a level to transform the shape of an input abound. Suppose the inflow to a level is constant over time, its graph over time a horizontal straight line. In the absence of any outflows, the level will show a *rising* straight-line pattern, whose slope equals the constant inflow rate. Or if the inflow shows a rising straight-line pattern, the level would show a graph over time that curves upward in a parabolic arc. An oscillating pattern in the inflow would result in a similar oscillating pattern in the level, but it would be shifted significantly along in time. (The reader may recall, or wish to show graphically, that the shift is always to the right a length of time equal to one-fourth of the period of the oscillations.)

The algebraic computations in auxiliaries (and rates) behave quite differently. In contrast to the behavior of a level, if the inputs to an auxiliary equation are constant, then so is the value of the auxiliary computed from them. If an input oscillates, then so will the auxiliary, but there will be no shift along in time of the sort that a level equation would produce. There are no delays: a change in a quantity in an auxiliary equation is immediately transmitted algebraically to the auxiliary variable itself. A step rise, for example, in a quantity in an auxiliary equation produces an immediate step change in the auxiliary --no transformation of the pattern over time except perhaps to stretch it, shrink it (changes of scale), or flip it upside down (multiplying by a negative).[9]

---

[9] The auxiliary might not change at all, if, for example, the changing input is in a saturation (horizontal) region of a table function.

What is a Level?

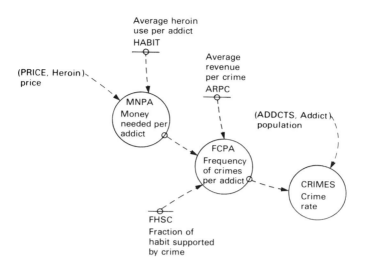

Figure 4.20: A chain of auxiliary variables from a model dealing with the problem of heroin-related crime

Figure 4.20 shows an instructive example of a string of auxiliaries from a model dealing with heroin-related crime (see also figure 2.7). The equations represented by figure 4.20 are the following:

```
A    MNPA.K=PRICE.K*HABIT
A    FCPA.K=(MNPA.K*FHSC)/ARPC
A    CRIMES.K=FCPA.K*ADDCTS.K
```
where
- MNPA = money needed per addict (dollars/addict/week),
- PRICE = heroin price (dollars/bag),
- HABIT = average heroin use per addict (bags/addict/week),
- FCPA = frequency of crimes per addict (crimes/addict/week),
- FHSC = fraction of habit supported by crime (dimensionless),
- ARPC = average revenue per crime (dollars/crime),
- CRIMES = heroin-related crime rate (crimes/week),
- ADDCTS = addict population (addicts).

This string of auxiliaries provides nice practice in dimensional analysis, but our purpose is to note that the behavior over time of these variables is very tightly linked. In spite of the apparent length of the causal chain here, a change in PRICE over time, for example, is immediately passed along to produce a corresponding change in CRIMES having the same shape over time. The coupling is in the nature of the algebra, for one could substitute backward algebraically and write a single equation for CRIMES:

A   CRIMES.K=PRICE.K*(HABIT*FHSC/ARPC)*ADDCTS.K

Clearly, a change in PRICE (or in any of these other quantities) in this model brings an immediate, corresponding change in CRIMES.

Thus, if the modeler believed that heroin price and heroin-related crime had a bit of slack between them in the real system, with changes in one not immediately or exactly duplicated in the other, the modeler would have to interpose a level variable somewhere in the information path from PRICE to CRIMES. One would have to think about where in this causal-chain a delay, in the form of some accumulation, intervenes. (Perhaps the most likely spot would be in the frequency of crimes per addict: it might take a bit of time to alter the pattern of average heroin-related criminal behavior.)

A general lesson for model formulation appears from this discussion: if variables along a causal-chain are very tightly linked in their behavior over time, they may be modeled as auxiliaries. If one can vary in a different pattern from another in the chain, then the presence of a level variable is clearly indicated. The level would serve to decouple the ends of the causal-chain of auxiliaries, allowing them to behave more independently over time--more like a fluid coupling than a gear chain. In a sense, the decoupling property of a level variable is simply a restatement of the transforming property, but it deserves a separate emphasis.

A Level as a Decoupler of Rates

If one were ask a business manager what the purpose of an inventory is, the person would likely say that it allows a firm to respond rapidly to fill an order without abruptly changing production. Instead of tying production directly to orders (the way it's done with ice cream cones and blast furnaces), an inventory is interposed between production and orders so that one can be relatively smooth and predictable even if the other bumps around. An inventory thus decouples production and shipment rates, meaning it allows them to vary over time somewhat independently.

What is a Level?

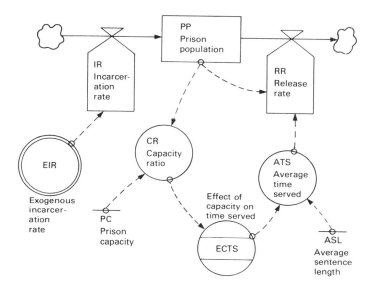

Figure 4.21: A simple prison population model illustrating the decoupling of rates by a level.

Any level variable with inflows and outflows has this decoupling capability. There are times when the capability is weakened, as when an inventory gets very low, but in general the capacity to decouple rates is one of the defining characteristics of level variables. It is inherent in the accumulating nature of the computation in a level equation. By rising or falling when necessary, the level variable absorbs any inequalities among its inflow and outflow rates.

A very instructive example, with some potential policy implications, is provided by the accumulation of prisoners in a prison, as diagrammed in figure 4.21. The prison population is increased by incarcerations and decreased as prisoners complete their sentences or are released on parole. The flow diagram assumes that the incarceration rate is determined exogenously, outside the boundary of this little system, and that the release rate is formulated simply (in a pattern suggested in section 4.1) as the prison population divided by the average time served. Average time served is expressed as an auxiliary variable, the product of some constant average sentence length and an effect of capacity on time served, ECTS. The structure assumes that as the prison fills up to capacity and beyond, prisoners would be paroled sooner to make way for new inmates, so ECTS makes the average time served a smaller and smaller fraction of average sentence length as the capacity ratio approaches and then exceeds 1.

For practice, the reader may wish to write equations for this simple model and formulate an appropriate table function for ECTS. While we shall give those equations shortly, our focus at the moment is on a thought experiment: How would such a system behave in response to a rise in the incarceration rate, caused, for example, by an increase in police action, more rapid handling of court cases, or elimination of plea bargaining?

There are two cases to consider, and the distinctions between them are the point of this discussion. Suppose first that there is considerable excess prison capacity, with the capacity ratio perhaps as low as 50 percent. A sudden rise in incarcerations would cause the level of the prison population to begin to rise, reaching as high as, say, 75 percent before leveling off. The release rate would also begin to rise as the prison population rises. It would, in fact, show the same pattern of behavior over time as the prison population curve because as long as the capacity ratio stays low the release rate RR is essentially just proportional to the prison population:

R   RR.KL=PP.K/ATS.K
where
   RR   = release rate (people/year),
   PP   = prison population (people),
   ATS = average time served (years).

In this case ATS is essentially equal to the constant average sentence length ASL when there are no shortening effects due to strained prison capacity.

These patterns are illustrated in the graph in figure 4.22.

Consider now the case in which the prison is significantly overcrowded. Capacity is exceeded by perhaps as much as 40 percent, meaning the capacity ratio CR is 1.4. The behavior of the system to a step increase in incarcerations is now considerably different. In an overcrowded system, there is little ability of prisons to absorb an inequality between the incarceration and release rates. With no place to put additional prisoners, the higher inflow rate produces very little increase in the prison population but instead immediately raises the outflow rate. The lack of excess prison capacity effectively *couples* the admittance of new prisoners to the release of a nearly equal number of old prisoners. The level of the prison population no longer decouples its inflow and

What is a Level?

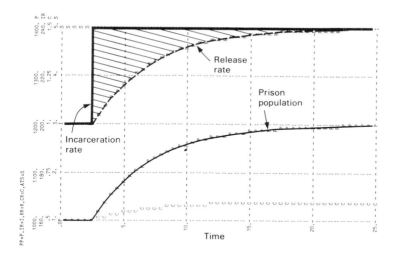

Figure 4.22: The response of a prison system to a sudden, permanent rise in incarcerations, assuming the system is operating well below capacity

outflow rates. The graph in figure 4.23 shows such behavior.[10]

The point of this prison example is to emphasize the decoupling characteristic of levels. Yet it also has potential policy implications for managing a criminal justice system: some policies intended to improve the system can be completely counteracted by a lack of excess prison capacity. Increasing the number of judges to reduce court backlogs, using television testimony to speed the hearing of cases, increasing police forces and enforcement activity, and plea bargaining

---

[10] The amount of the rise in the prison population in both figures 4.22 and 4.23 is equal to the the *area* between the graphs of the incarceration rate and the release rate. (That conclusion is really a version of the fundamental theorem of calculus, but one can see intuitively that it is reasonable by noting that the dimensions of a square unit of area between the rate curves are people/year (vertical) and years (horizontal), yielding area units of people.) The closer coupling of the rates in the second figure results in a much smaller area between the rate curves and hence a much smaller rise in the prison population.

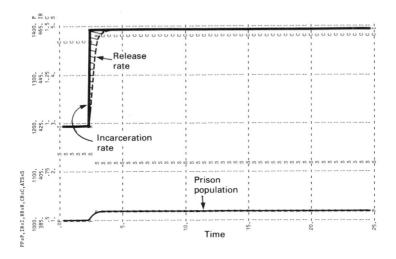

Figure 4.23: The response of a prison system to a sudden permanent rise in incarcerations, assuming the system is operating at 140 percent of capacity

are policies that attempt to speed the flow of criminals into prison. In situations where prison capacity is hopelessly overloaded, such policies can only produce a corresponding increase in the release of prisoners to make room for the larger inflow. The average sentence length must lower, and whatever reasons society has for placing people in prisons would be subverted.

Modeling Note

For those readers interested in the DYNAMO model that produced figures 4.22 and 4.23, a listing of it follows with brief explanatory notes:

```
*       SIMPLE PRISON POPULATION MODEL
L       PP.K=PP.J+DT*(IR.JK-RR.JK)
N       PP=IPP
C       IPP=1000
NOTE        PRISON POPULATION  (PEOPLE)
R       IR.KL=NIR+STEP(SH,ST)
NOTE        INCARCERATION RATE  (PEOPLE/YEAR)
N       NIR=PP/ATS
NOTE        NORMAL INCARCERATION RATE  (PEOPLE/YEAR)
C       SH=40
NOTE        STEP HEIGHT FOR IR
C       ST=2.5
NOTE        STEP TIME FOR IR
```

What is a Level?

```
R     RR.KL=PP.K/ATS.K
NOTE        RELEASE RATE  (PEOPLE/YEAR)
A     ATS.K=ASL*ECTS.K
NOTE        AVERAGE TIME SERVED  (YEARS)
C     ASL=5
NOTE        AVERAGE SENTENCE LENGTH  (YEARS)
A     ECTS.K=TABLE(TECTS,CR.K,0,2,.2)
NOTE        EFFECT OF CAPACITY ON TIME SERVED  (DIMENSIONLESS)
T     TECTS=1/1/1/1/1/.94/.7/.5/.3/.2/.15
NOTE        TABLE FOR EFFECT OF CAPACITY ON TIME SERVED
A     CR.K=PP.K/PC
NOTE        CAPACITY RATIO  (DIMENSIONLESS)
C     PC=2000
NOTE        PRISON CAPACITY  (PEOPLE)
NOTE
NOTE        CONTROL STATEMENTS
NOTE
SPEC  DT=.5/LENGTH=25/PLTPER=.5
PLOT  PP=P(1000,1400)/IR=I,RR=R(160,240)/CR=C(.5,1.5)/ATS=S(1,5)
RUN   FIG 4.22
C     PC=700
PLOT  PP=P(1000,1400)/IR=I,RR=R(385,465)/CR=C(.5,1.5)/ATS=S(1,5)
RUN   FIG 4.23
```

A Level as Potential for Disequilibrium

The previous paragraphs note that it is characteristic of level variables to have the potential to be fluid couplers between causal chains and to decouple inflow and outflow rates. As a consequence of these characteristics, adding a level to a model adds an additional possibility for *disequilibrium behavior.* Equilibrium in a dynamic system is the situation in which all the inflows exactly balance their corresponding outflows. Since a level can decouple its rates of flow, adding a level to a model creates an additional opportunity for an inequality between inflows and outflows to exist.

Consider a model with just one level. When it reaches equilibrium, its inflows and outflows balance, and the level ceases to change. Since there are no other levels in the system, every auxiliary and every rate in the system trace their values back to this one level. If the level does not change, then none of the auxiliaries in the model change, and consequently the inflow and outflow rates do not change. Hence all the quantities in the model remain constant from the moment the inflows and outflows to the one level become equal.

If a second level is added to the model, then, to be in equilibrium, it requires two sets of inflow and outflow rates to be simultaneously in balance. One level could be in a state of equilibrium, with its inflows and outflows equal, but the other level in the system could be out of equilibrium. The disequilibrium forces from the second level could feed back to pull the first level out of equilibrium. The result might be oscillations, as, for example, in a simple ecosystem in which predator and prey populations move out of phase with each other.

One can correctly conclude from these observations that a model containing just one level variable can not oscillate, while a model containing two (or more) levels can. The reason is that the addition of the second level adds the possibility that one part of the system can be in disequilibrium when another part is in equilibrium. Moreover, a nonsensical system with no levels could show no dynamics at all, for, once all the parameters in the model were set and auxiliaries were computed, nothing could change. Levels add potential for disequilibrium behavior and, in a sense, are the source of all dynamic behavior in system dynamics models.

Very Fast and Very Slow Accumulations

In discussing the snapshot test, we suggested that there is a conceptual connection between the time horizon of a model and the time over which the different kinds of variables change. Obviously, constants do not change at all over a simulation. They represent quantities that remain essentially fixed over the time frame of the model. Yet the quantity a constant represents may well change over a longer time frame. A model addressing oscillations in a production- distribution chain over a four-year time horizon might well assume the production plant is constant over that time. Yet, if the model were addressing dynamics over a twenty-year period, the plant capacity would almost certainly have to be modeled as a level variable. Thus, a quantity is modeled as a constant not because it never changes but because *in the time frame of the model* it essentially does not change. Some constants are simply accumulations that are changing too slowly to be of any dynamic significance.

---

[11] The following discussion illustrates a formulation technique that is occasionally applicable, but the intent of the discussion is more to reinforce the conceptual differences between auxiliaries and levels.

What is a Level?

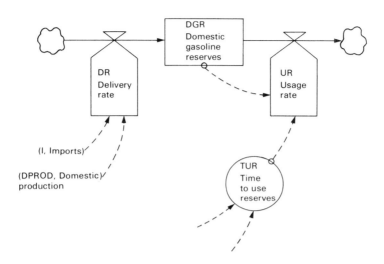

Figure 4.24: Formulating domestic gasoline reserves as a level variable

Accumulations that change too rapidly to be modeled easily or usefully as levels also crop up.[11] Consider the aggregate inventory of gasoline residing in gas stations, pipelines, and land storage facilities in the United States. Suppose one were interested in policy questions associated with the life cycle of United States petroleum usage (see figure 2.3, for example). The time horizon of the model might be on the order of 200 years. Should the U. S. inventory of gasoline be represented in the model as a level variable?

If it were, the aggregate inventory structure could look something like that shown in figure 4.24, where the usage rate is intended to be written as:

R   UR.KL=DGR.K/TUR.K
where
    UR   = usage rate (gallons/year),
    DGR = domestic gasoline reserves (gallons),
    TUR = time to use reserves (years).

The form of this rate equation is LEVEL.K/LIFE.K: the time to use reserves is the variable "time constant" for the level of reserves. (Recall section 4.1.) It indicates how long the reserves would hold up if used at a constant rate. In 1979 the United states held in its tanks and pipelines a level of reserves sufficient to cover about 30 days of gasoline use. Thus TUR would be in the range of 1/12 of a year.

A time constant of 1/12 year is very short compared to a time frame of 200 years. Any dynamic patterns generated by the gasoline inventory would be of such short duration that they would largely be swamped by the patterns of exploration, discovery, and depletion dominating the life cycle of petroleum in the 200-year span. Furthermore, if one graphed the inflow and outflow rates to the reserves level over such a long time frame, the two curves would probably lie essentially on top of one another. The small inequalities absorbed by changes in the reserve inventory would not be visible on graphs covering 200 years of model behavior.

In fact, over such a long time frame, it is more appropriate to model the quantity of domestic gasoline reserves as an auxiliary rather than a level (assuming one wants to see such a variable plotted out by the model). One technique for doing so makes use of the fact that the inflows and outflows to a level with such a relatively short time constant can be viewed as essentially equal, treats the rates and the level as auxiliaries, and solves for the value of the inventory DGR in terms of the inflow rate:

DR.KL ≃ UR.KL = DGR.K/TUR.K

so

A   DGR.K=DR.K*TUR.K
where
   DGR = domestic gasoline reserves (gallons),
   DR   = delivery rate (gallons/year),
   TUR = time to use reserves (years).

Thus the auxiliary variable for gasoline reserves would be computed as a product. Given the approximate value of TUR, DGR would be in the neighborhood of 1/12 of the inflow to domestic reserves. Such a computation would be diagramed as in figure 4.25.

Thus an accumulation which is very "fast" relative to the time horizon of the model could be modeled as an auxiliary. One advantage of doing so is avoiding massive amounts of computation resulting from a tiny solution interval in the model. Recall the rule of thumb for DT that suggests a value between one-half and one-tenth of the smallest time constant in the model. In the petroleum model with domestic gasoline reserves as a level, DT would have had to be as small as 1/25 year, resulting in 200/(1/25) or 5,000 iterations through the computations in the model for every run. The miniscule

What is a Level?

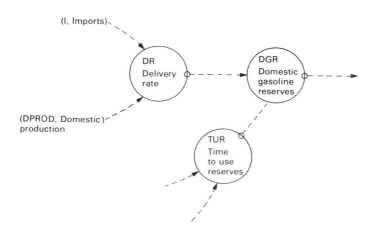

Figure 4.25: Domestic Gasoline Reserves as an auxiliary variable

dynamics gained by representing reserves as a level would not be worth the compuational cost or even the time to wait for the output.

Summary

This section has presented a number of ways of viewing how a level acts in a system dynamics model. All the views are related. In fact, one may justly feel that each is a restatement of some single underlying characteristic of levels that stems from the process of accumulation all level variables somehow reflect. We have seen that levels change the shape over time of an input, they decouple inflow and outflows allowing them more or less independent movement, and (consequently) they bring to a model potential for disequilibrium behavior.

Whether to represent a concept as a level or not depends first upon whether or not the concept can be viewed as some sort of stock or accumulation over time. The snapshot test is sometimes difficult to apply, however, because it must cover such accumulation processes as perceiving, averaging, and forming traditions, as well as the more mundane stocks such as inventories and people. Then the experienced modeler also questions whether a rough estimate of the time constant associated with an accumulation suggests that the accumulation be modeled as a level, as an auxiliary if it is very fast, or as a constant if it is very slow. The question of whether an accumulation could be modeled as an auxiliary is rather advanced, however, and should not be allowed to get in the way of one's early modeling efforts. At first, the guiding principle in model formulation should be conceptual simplicity. As one fine old scholar-teacher used to say, before you split hairs, you must grow some.

4.4 FORMULATING A PROJECT MODEL

The understandings of rate, auxiliary, and level equations discussed in the preceding three sections only become owned when they are applied. This section attempts to provide the experience of completely formulating a simple model, the project model conceptualized in section 2.4. The model formulated here will provide a focus for the discussions in chapters 5 and 6.

The model developed in this section is a quantitative representation of the feedback structure diagrammed in figure 2.28 (repeated as figure 4.26, showing levels and rates explicitly). We will first locate quantities that will be modeled as levels. Then the rates that affect them will be identified. The model will then be developed around these levels and rates by considering in turn each of the following small sectors:

- real progress

- undiscovered rework

- perceived progress

- effort perceived remaining

- hiring

- scheduling

About 23 equations from now, we will have a model that is capable of reproducing the reference modes for the problem of overruns described and defined in section 2.4.

Reading about equations is no substitute for trying to to write them. We strongly urge the reader, before reading what follows, to try formulating equations for the quantities and concepts contained in the hybrid causal-loop diagram in

Formulating a Project Model

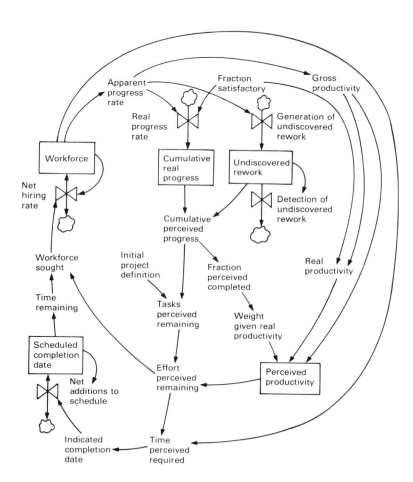

Figure 4.26: Structure of the project model, showing levels and rates.

figure 4.26. Where our equations and the reader's differ, there will be opportunities for some new understandings: the reader can analyze the differences to decide whether one is an improvement over the other for some identifiable reason(s), or whether both formulations are in fact defensible. Where our equations and the reader's match, not much new will be learned, but there will be the internal reward of knowing that at least we all agree.

As we proceed, there will be a number of points at which a range of alternative formulations for a given structure are conceivable. To prevent the discussion from foundering in a sea of "could be's," we will ignore alternatives on the first pass. At the end of this section, there will be some discussion of how one might justifiably have done things differently.

Levels and Rates in the Project Model

The most obvious accumulations in the structure diagrammed in figure 2.27 appear to be the workforce, cumulative real progress, and undiscovered rework. Each of these passes the snapshot test for levels. Moreover the flows into and out of these accumulations are rather clear. The workforce is influenced by the hiring rate and the rates at which people quit or are transferred. The flow affecting cumulative real progress is the current rate at which real progress is made. The inflow to the level of undiscovered rework corresponds to real progress and is the rate at which faulty or unsatisfactory "progress" is being made. Finally, unlike cumulative real progress, the level of undiscovered rework has an outflow rate, the rate at which such work is detected. Figure 4.26 shows these rates and levels superimposed on the causal-loop structure developed in chapter 2.

As shown in figure 4.26, there are two additional concepts in the model which will be modeled here as level variables: the scheduled completion date and perceived productivity. While not as obvious an accumulation as, say, wine in a wine glass, the scheduled completion date does pass the snapshot test. If time were stopped in the system, outside observers would still be able to see the date at which the project was currently being planned to finish. Furthermore, being a policy variable, the scheduled completion date requires some planning and adjustment time. It would not respond immediately to changes in the completion date currently indicated by the size and productivity of the workforce and the number of man-months of effort perceived remaining.

Similar comments hold for perceived productivity, the perception of people within the system about how many tasks per

Formulating a Project Model 193

month the average worker in the project is able to complete satisfactorily. The perception of productivity tends to persist through time and to change neither exceedingly rapidly nor exceedingly slowly. Thus it satisfies a number of the characteristics we have attributed to levels. Figure 4.26 omits an explicit rate of change for the perceived productivity level, not because it does not have one but because the intent is to model perceived productivity as a SMOOTH. The rate equation for a smoothing or exponential averaging formulation is frequently omitted from diagrams for visual simplicity.

To put some order into the equation writing that follows, the discussion will be divided into the six small functional areas listed at the beginning of this section. We begin in the upper left of figure 4.26 with equations dealing with real and apparent progress and plan to move roughly clockwise down through the center of the figure, and end by coming back up the left side.

Real Progress

We shall assume that the project to be completed consists of a number of tasks. The task unit will become defined when we assign numbers to such variables as the initial project definition (a number of tasks) and normal worker productivity (tasks completed per month).

Progress is made as workers on the project perform tasks, some fraction of which are completed satisfactorily and the rest of which require rework. Thus it seems reasonable to formulate the real progress rate as the product of some apparent progress rate and a fraction satisfactory. The remaining fraction of tasks accomplished per month would be (1 - fraction satisfactory), and those would be tasks eventually requiring rework.

The workforce (people) determines the apparent progress rate APPRG (tasks/month), so we shall formulate APPRG as the product of the workforce and a total or gross productivity factor (tasks/person/month), representing both satisfactory and unsatisfactory work. The units show such a product formulation makes sense. Hence, we have the following equations:

```
A     APPRG.K=WF.K*GPROD
NOTE     APPARENT PROGRESS RATE  (TASKS/MONTH)
C     GPROD= ?
NOTE     GROSS PRODUCTIVITY
NOTE     (TASKS/PERSON/MONTH)
R     RPRG.KL=APPRG.K*FSAT
```

```
NOTE    REAL PROGRESS RATE  (TASKS/MONTH)
C   FSAT= ?
NOTE    FRACTION SATISFACTORY  (DIMENSIONLESS)
```

WF denotes the workforce, which will be computed in a level equation.

At some point we will have to give values to the constants GPROD and FSAT, but for the moment we will leave them unspecified in order to concentrate on the formulation of auxiliary, rate, and level equations. Note that while we have formulated GPROD and FSAT as constants here, nothing prevents us from returning to these concepts later and reformulating them as variables (presumably auxiliaries). Whether to do so or not will depend upon whether we eventually come to believe that writing them as constants ignores some significant or interesting feedback effects in the system. For example, overtime and undertime effects would push GPROD up and down, and FSAT might be affected by how close the project is to completion. For now, however, for simplicity they are constants.

Real progress accumulates in the level called cumulative real progress:

```
L    CRPRG.K=CRPRG.J+DT*RPRG.JK
N    CRPRG=0
NOTE    CUMULATIVE REAL PROGRESS  (TASKS)
```

The level has only an inflow, because real progress, once made, is assumed not to decay. The N equation for CRPRG gives the initial value of cumulative real progress, which ought to be zero at the beginning of a project.

Before continuing on to formulate equations relating to undiscovered rework, the reader should note the forms of the rate, auxiliary, and level equations thus far. The rate equation for RPRG is in the form AUX.K*CONST, not a pattern mentioned in section 4.1. However, substituting the quantities from which APPRG is computed shows RPRG to be computed as WF.K*GPROD*FSAT, in the common rate equation form LEVEL.K*CONST. For the various auxiliary equations, check that the units agree. Finally, one might note that level equations are no exception to the rule that dimensions must agree: DT*RPRG.JK has units of (months)*(tasks/month), or tasks, which are the units of CRPRG.

Undiscovered Rework

The fraction of the apparent progress rate that is unsatisfactory is termed the generation of undiscovered rework. Its

Formulating a Project Model 195

equation has the same form as that of the real progress rate RPRG:

```
R    GURW.KL=APPRG.K*(1-FSAT)
NOTE    GENERATION OF UNDISCOVERED REWORK
NOTE    (TASKS/MONTH)
```

RPRG computes the fraction of APPRG that is satisfactory, and GURW simply computes the remainder.

The generation of undiscovered rework accumulates in the level of undiscovered rework, an essential part of our dynamic hypothesis of why there may be overruns in projects:

```
L    URW.K=URW.J+DT*(GURW.JK-DURW.JK)
N    URW=0
NOTE    UNDISCOVERED REWORK  (TASKS)
```
where
    DURW = detection of undiscovered rework (tasks/month).

Again, at the beginning of the project there should be no undiscovered rework yet, so the N equation sets URW initially to zero.

The realization that undiscovered rework has an outflow, detection of undiscovered rework, brings us to our first real departure from the causal-loop structure conceptualized in section 2.4. GURW was not identified there. We have to invent a formulation for it from scratch, considering what it should depend on and then what the form of the equation should be. Following the advice of section 4.1, we consider a rate equation for the outflow in the form LEVEL.K/LIFE.K:

```
R    DURW.KL=URW.K/TDRW.K
NOTE    DETECTION OF UNDISCOVERED REWORK
NOTE    (TASKS/MONTH)
```

The time to detect rework TDRW can be interpreted as the average time between the apparent completion of a task and when it is discovered that it must be redone. This time constant TDRW could be formulated as a constant or as an auxiliary if there were reason to believe that over the course of a 40-month project the time would vary.

It seems reasonable to assume that TDRW becomes shorter as the project appears to near completion. As final reports are being written and mock-ups of the product are put in final form, previous errors should become evident more quickly than when the project first began. In addition, the kinds of work done on a project toward the end (some as simple as typing) involve errors that are inherently quicker to perceive, and these are aggregated with much more significant unsatsifactory work in the level of undiscovered rework. Hence, we are led to suggest that TDRW ought to formulated as a function of the fraction of the project perceived to be completed:[12]

    A   TDRW.K = f(FPCOMP.K),
where
    TDRW   = time to detect rework (months),
    f( )   = a table function,
    FPCOMP = fraction perceived completed (dimensionless).

FPCOMP ought to be simply the ratio of the number of tasks believed to be completed satisfactorily to the number of tasks initially believed required. It ought to vary from zero to 1, and as it nears 1, we wish to assume TDRW decreases. The assumptions we wish to incorporate in the function f suggest formulating TDRW as a table function:

    A   TDRW.K=TABLE(TTDRW,FPCOMP.K,0,1,.2)
    NOTE   TIME TO DETECT REWORK (MONTHS)
where
    TTDRW   = table for time to detect rework,
    FPCOMP = fraction perceived completed (dimensionless).

Note that the table goes from FPCOMP = 0 to FPCOMP = 1. Increments of 0.2 are shown in the table function expression, but one could choose a finer partition if a smoother curve appears necessary.

To formulate the table function for TDRW one must consider slope, shape, reference points, and any reference lines that might be known. The slope matches the polarity of the influence, so it is negative: as the fraction perceived completed rises toward 1, the time to detect rework should decline. The shape should be roughly horizontal near FPCOMP = 0, reflecting the assumption that increases in the fraction perceived completed bring little change in TDRW near the beginning of the project. The curve should steepen near FPCOMP = 1 to show the influence strengthening, reducing TDRW more and more as FPCOMP approaches 1. Thus, the

---

[12] It is a thorny question whether to use the actual fraction completed instead of the fraction perceived completed. Here we are swayed by the thought that the kinds of activities undertaken toward the perceived completion of the project will tend to uncover areas needing rework.

Formulating a Project Model 197

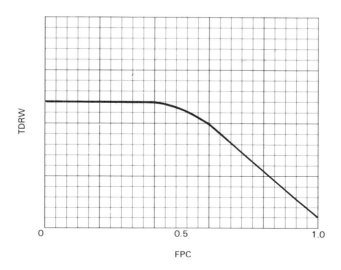

Figure 4.27: General shape of the table function for time to detect rework as a function of fraction perceived completed

table should look like one of the alternatives in figure 4.27. The table might begin dropping down a bit sooner or a bit later, and it could conceivably show a saturating effect as it approaches some limiting minimal value as FPCOMP nears 1.

Now what might the values of TDRW be along such a curve? Our best approach would be to ask our client for estimates. We would be interested mainly in bounds for TDRW, estimates of an upper limit and a lower limit, for we have just thought out how the variable ought to behave in between. Lacking any i[formation about TDRW from our client or from a literature search (as might happen), we would have to guestimate. In a forty-month, nearly four-year project, TDRW must be less than forty months throughout, and it is hardly conceivable that it could drop below, say, two weeks even at the end. It might be reasonable to give it a value around one year for perhaps half the project, then dropping with increasing steepness to, say, half a month at the end. Such as set of values are shown in the following T equation:

T   TTDRW=12/12/12/10/5/.5
NOTE   TABLE FOR TIME TO DETECT REWORK

Simulations of the model will show just how sensitive the behavior of the model is to different formulations of this table. We can experiment with formulating TDRW to be a constant, by setting TTDRW equal to a string of 12's or 3's (or whatever) in DYNAMO's rerun mode. Alternatively, we can explore the effects of assuming higher or lower upper bounds or a saturating shape for the table near FPCOMP = 1.

Perceived Progress

Cumulative perceived progress was conceived to be the sum of cumulative real progress and undiscovered rework:

```
A    CPPRG.K=CRPRG.K+URW.K
NOTE     CUMULATIVE PERCEIVED PROGRESS (TASKS)
```
where
    CRPRG = cumulative real progress (tasks),
    URW   = undiscovered rework (tasks).

Knowing the number of tasks perceived to be completed and the number of tasks initially thought to make up the project, one can easily compute the fraction perceived completed as a ratio:

```
A    FPCOMP.K=CPPRG.K/IPD
NOTE     FRACTION PERCEIVED COMPLETED
NOTE     (DIMENSIONLESS)
C    IPD= ?
NOTE     INITIAL PROJECT DEFINTION (TASKS)
```

We shall assume that the number of tasks in the initial project definition does not change throughout the project, and give IPD a value later. As noted earlier, one might alternatively assume that one of the causes of overruns in a project is underestimating the size of the job at the beginning. Such a hypothesis could be built into our model (and will be in chapter 5) by formulating in terms of a current project definition which grows as the project proceeds.

Figure 4.28 shows a flow diagram of the structure of the model up to this point. Not yet formulated are the areas of the model dealing with effort perceived remaining, hiring, and scheduling.

Effort Perceived Remaining

We assume that policy planning about projects like ours is phrased in terms of man-months of effort expended and

Formulating a Project Model   199

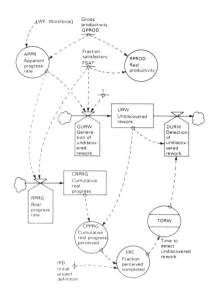

Figure 4.28: Flow diagram showing the formulations for real progress and undiscovered rework in the project model

remaining. Planning for adjustments in the workforce (people) or the schedule (months) requires some idea of the man-months remaining on the project. The model must translate from progress perceived in task units to man-month units for effort perceived remaining. The units of productivity, tasks/person/month, are precisely the appropriate translation factor, as the following bit of dimensional analysis shows:

(tasks)/(tasks/person/month) = (tasks)/[tasks/(man*month)]
                             = (tasks/tasks)/(1/(man*month))
                             = man*months.

Therefore, to formulate effort perceived remaining, we first compute the number of tasks perceived remaining and then divide by perceived productivity:

```
    A     EPREM.K=(IPD-CPPRG.K)/PPROD.K
    NOTE      EFFORT PERCEIVED REMAINING
    NOTE      (MAN*MONTHS)
where
    IPD    = initial project definition (tasks),
    CPPRG  = cumulative perceived progress (tasks),
    PPROD  = perceived productivity (tasks/person/month).
```

This formulation illustrates rather vividly how the technical skill of dimensional analysis can contribute to conceptual clarity.

In section 2.4 we envisioned perceived productivity as a combination of gross productivity and real productivity, with more weight being given to real productivity as the project appeared to near completion. At the beginning of the project, estimates of productivity would have to be based on previous project experience, if any. As the project develops and the need for rework begins to appear, people in the project will be increasingly able to perceive how productive the workforce has actually been. A simple formulation that can have these characteristics is a *weighted average:*

```
    WTRP.K*RPROD.K + (1-WTRP.K)*GPROD
where
    WTRP   = weight given real productivity,
    RPROD  = real productivity (tasks/person/month),
    GPROD  = gross productivity (tasks/person/month).
```

The weighting factor WTRP would have to be formulated so that it moves from essentially 0 at the beginning of the project to 1 as the project appears to near completion. Formulating WTRP as a table function of the fraction perceived completed FPCOMP would make it easy to give values to such a weighting factor.

Under the assumption that the adjustment of perceptions of productivity takes some time, we shall write the equation for perceived productivity as a SMOOTH of the weighted average described above. To simplify the equations, let us compute the weighted average in a separate equation, calling it "indicated" productivity:

```
    A      PPROD.K=SMOOTH(IPROD.K,TPPROD)
    NOTE     PERCEIVED PRODUCTIVITY
    NOTE     (TASKS/PERSON/MONTH)
    C      TPPROD= ?
    NOTE     TIME TO PERCEIVE PRODUCTIVITY (MONTHS)
    A      IPROD.K=WTRP.K*RPROD.K+(1-WTRP.K)*GPROD
    NOTE     INDICATED PRODUCTIVITY
    NOTE     (TASKS/PERSON/MONTH)
```

The SMOOTH function is described in detail in section 3.5. Suffice to say here, it is exactly equivalent to writing PPROD as a level equation in which the rate of change is expressed as (GOAL.K-LEVEL.K)/ADJTM, for example,

```
    L      PPROD.K=PPROD.J+DT*(IPROD.J-PPROD.J)/TPPROD
```

Formulating a Project Model

N   PPROD=IPROD

Whichever way it is written, PPROD will adjust over time toward the weighted average of RPROD and GPROD. The rapidity of the adjustment will depend upon the time constant TPPROD, to which we will assign a value later.

Real productivity, measured in tasks per person per month, is simply the fraction of gross productivity that is satisfactory:

A   RPROD.K=GPROD*FSAT
NOTE   REAL PRODUCTIVITY
NOTE   (TASKS/PERSON/MONTH)

To anticipate later model refinements, we are computing RPROD in an auxiliary equation as if it were a variable, even though it is given as a product of constants and could therefore be computed in an N equation.

It remains to construct the table function for the weighting factor WTRP. We know it should be a function of the fraction perceived completed which is near zero when FPCOMP = 0 and is probably one when FPCOMP = 1. The slope must be positive, indicating that as FPCOMP increases more weight is given real productivity. It would seem most reasonable to assume an S-shape for the function, because such a shape would produce saturating effects at the beginning and the end of the project. As the project appears to be 90 percent completed, for example, the perception of productivity would probably be dominated almost completely by real productivity, with little room for further increases in WTRP as FPCOMP approaches 100 percent. Figure 4.29 shows a likely table function for the weighting factor WTRP.

The DYNAMO equations for the table in Figure 4.29 are:

A   WTRP.K=TABLE(TWTRP,FPCOMP.K,0,1,.2)
NOTE   WEIGHT GIVEN REAL PRODUCTIVITY
T   TWTRP=0/.1/.25/.5/.9/1
NOTE   TABLE FOR WTRP
where
    FPCOMP = Fraction Perceived Completed (dimensionless).

The particular values assumed here imply the most rapid change in the perception of productivity (the steepest part of the table) occurs when the project appears to be between 40 percent and 70 percent completed. It will be interesting to explore how the model behaves when different assumptions about the weighting factor WTRP are simulated. A string of one's for TWTRP, for example, assumes perfect perception of

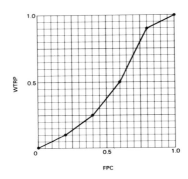

Figure 4.29: Suggested table function for the weight given real productivity, WTRP

productivity throughout the project. While not testing a really tenable assumption, such a simulation will add to our understanding of WTRP and the dynamics of perceived productivity in the model.

As figure 4.26 suggests, the remaining two sectors, hiring and scheduling, have great similarities. Consequently, we are very close to being done with the formulation of the entire project model. Have courage.

Hiring

The workforce has been conceived as a level variable influenced by a net hiring rate:

```
L    WF.K=WF.J+DT*HR.JK
NOTE      WORKFORCE (PEOPLE)
N    WF=WFN
C    WFN= ?
NOTE      INITIAL VALUE FOR WORKFORCE
```

We are making no distinction here between people newly hired and people fully experienced in the project. Because of differences in productivity, the disaggregation of the workforce into a workforce in training and an experienced workforce will add some vitally important dynamics, which we will explore later in chapter 5. The real question for us here is how to formulate the hiring rate HR. Following a suggestion of section 4.1, we consider a rate equation in the pattern (GOAL.K-LEVEL.K)/ADJTM:

```
R    HR.KL=(WFS.K-WF.K)/WFAT
NOTE      NET HIRING RATE (PEOPLE/MONTH)
```

Formulating a Project Model 203

where
- HR = net hiring rate (people/month),
- WFS = workforce sought (people),
- WF = workforce (people),
- WFAT = workforce adjustment time (months).

The workforce sought would be the number of people required to finish the project in the scheduled time remaining. Presumably, WFS would be easy to calculate, knowing the man-months of effort perceived remaining and the months of time remaining.

Such a hiring rate formulation assumes that some time elapses between the determination of how many people are needed on the project and the acquisition (or laying off) of a requisite number of people. If WFS is above WF, people will be hired or transferred into the project; if WFS is less than WF, people will be reassigned to other work or let go; and if WF is equal to WFS, no changes in the workforce will take place because the it contains precisely the right number of people to complete the project on schedule. Assuming that it takes about the same amount of time to bring people into the project as it does to transfer them out of it, the formulation appears to capture neatly the dynamics required.

Throughout most of the project the workforce sought, WFS, ought to be precisely the number of people believed to be necessary and sufficient to complete the project on time. Having a perception of the number of man-months of effort remaining, and knowing how many months remain before the scheduled completion date, project management ought to be able to compute the desired workforce by dividing

$$EPREM.K/TREM.K$$

where
- EPREM = effort perceived remaining (man*months),
- TREM = time remaining (months).

Note that the units of these terms help to show that division is the correct arithmetic: (man*months)/months = people.

Workforce sought, however, should not be set directly equal to the quotient EPREM.K/TREM.K. Toward the end of the project, for example, there would be considerable reluctance to bring on new people, even though the time and effort perceived remaining imply more people are needed. It would take too much time to acquaint new people with the mechanics of the project, integrate them into the project team, and train them in the necessary technical areas. We shall assume that our project managers reach a point, say within six months of the perceived end of the project, at which they become unwilling to change the size of the workforce, no matter how many people EPREM.K/TREM.K indicates are required to finish on time.

Therefore we are led to our second departure from the basic structure outlined in chapter 2 and diagramed in figure 4.26. We shall compute the quantity EPREM.K/TREM.K as an indicated workforce, and formulate the workforce sought as a weighted average of IWF and the current workforce WF:[13]

```
A    WFS.K=WCWF.K*IWF.K+(1-WCWF.K)*WF.K
NOTE    WORKFORCE SOUGHT (PEOPLE)
A    IWF.K=EPREM.K/TREM.K
NOTE    INDICATED WORKFORCE (PEOPLE)
```

The willingess to change workforce, WCWF, must be a number that moves from 1 in the early stages of the project to 0 near the end. When WCWF equals 1, WFS = IWF = EPREM/TREM, and the project will try to adjust its workforce to the number perceived required. When WCWF eventually becomes 0, WFS will equal WF, and so the hiring rate will fall to zero, as desired.

Figure 4.30 shows a potential policy for the willingness to change workforce. The table assumes that, when the time remaining TREM is perceived to be at least 21 months, there is total willingness to adjust the size of the workforce to suit the needs of the project. As the number of months perceived remaining drops from 21 to 6, the table shows increasing reluctance to change the size of the workforce. Within 6 months of the end of the project, the table assumes the project will not alter its workforce. If the project is behind schedule with less than 6 months to go, project managers will cope by pushing back the scheduled completion date. (Note that in this simple view of project management, variations in the intensity of effort, such as overtime, are being ignored. We will add such complexities later.) The DYNAMO equations for the table shown in figure 4.30 are:

```
A    WCWF.K=TABHL(TWCWF,TREM.K,0,21,3)
NOTE    WILLINGNESS TO CHANGE WORKFORCE
NOTE    (DIMENSIONLESS)
T    TWCWF=0/0/0/.1/.3/.7/.9/1
NOTE    TABLE FOR WCWF
```

---

[13] TREM approaches zero as the project nears completion, so we might wish to formulate IWF as EPREM.K/MAX(TREM.K,.001) to prevent division by zero.

Formulating a Project Model

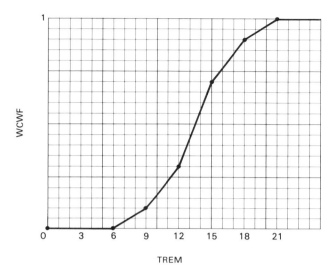

Figure 4.30: Table function for the willingness to change workforce, WCWF, showing a likely project management policy.

It is important to realize that the variable WCWF is an expression of a *policy* for managing projects. A rather wide range of functions are possible here, capturing a number of different strategies for how to balance workforce and schedule adjustments throughout the project to minimize overruns and costs.

Scheduling

It makes sense to formulate the scheduled completion date SCD, not as an actual date, but as a number of months from the beginning of the project. Then simply by subtracting the current value of TIME (the number of months elapsed in the simulation), we can determine the scheduled time remaining TREM, used above to indicate the size of the workforce desired.

```
A     TREM.K=SCD.K-TIME.K
NOTE     TIME REMAINING (MONTHS)
where
   SCD     = scheduled completion date (months),
   TIME    = DYNAMO's variable for elapsed simulation time
```

(months).

We need not include an equation for TIME (in fact we'd get an error message if we did) because DYNAMO automatically computes TIME for every model in what amounts to a level equation, TIME.K = TIME.J + DT. TIME in this model is in months because we have selected DT to be in months.

The level equation for scheduled completion date accumulates the changes in the schedule:

```
L    SCD.K=SCD.J+DT*NAS.JK
NOTE    SCHEDULED COMPLETION DATE (MONTH)
N    SCD=SCDN
C    SCDN= ?
NOTE    INITIAL VALUE OF SCD
```

NAS, net additions to schedule, is the rate of change of SCD, representing decisions by project management to change the scheduled completion date. We shall assume that project management accurately computes the completion date indicated by current perceived conditions in the project, taking account of the size of the workforce being sought, and gradually adjusts the schedule accordingly. NAS is easily formulated in the pattern

(GOAL.K - LEVEL.K)/ADJTM,

where the goal is an indicated completion date:

```
R    NAS.KL=(ICD.K-SCD.K)/SAT
NOTE    NET ADDITIONS TO SCHEDULE
NOTE      (MONTHS/MONTH)
C    SAT= ?
NOTE    SCHEDULE ADJUSTMENT TIME (MONTHS)
```

The adjustment will not happen instantaneously but will take place over a time interval SAT, yet to be determined. Taken together, the rate equations for hiring and for net additions to schedule guarantee that management will adjust either the workforce, the scheduled completion date, or both to keep the project under control and on some targeted path toward completion.

At any point in the project the indicated completion date is the number of months from the present it would take to complete the project. Thus,

```
A    ICD.K=TIME.K+TPREQ.K
NOTE    INDICATED COMPLETION DATE (MONTH)
```

Formulating a Project Model 207

where
- TIME = DYNAMO's variable for elapsed simulation time,
- TPREQ = time perceived required (months).

The time perceived required TPREQ represents the remaining time required to complete the project, given its current condition. In a computation analogous to the equation for the workforce sought, the time perceived required can be determined from the size of the workforce sought and the man-months of effort perceived remaining:

```
A     TPREQ.K=EPREM.K/WFS.K
NOTE       TIME PERCEIVED REQUIRED  (MONTHS)
```
where
- EPREM = effort perceived remaining (man*months),
- WFS = workforce sought (people).

Computing in terms of WFS rather than WF means that we are assuming that schedule adjustments are made with full awareness of the hiring decisions being made. Note that with the exception of a translation from potentially sexist to neutral terminology the units check as desired.

The model is complete! Figure 4.31 shows a flow diagram for the sectors of the model dealing with effort perceived remaining, hiring, and scheduling. Together with figure 4.28, it gives a complete picture of the project model.

Getting the Model to Run

Table 4.1: Initial values of levels

WFN	2	people
CRPRG	0	tasks
URW	0	tasks
SCDN	40	tasks

Table 4.2: Values of constants

GPROD	1	task/person/month
FSAT	0.7	(dimensionless)
IPD	1200	tasks
TPPROD	6	months
WFAT	3	months
SAT	6	months

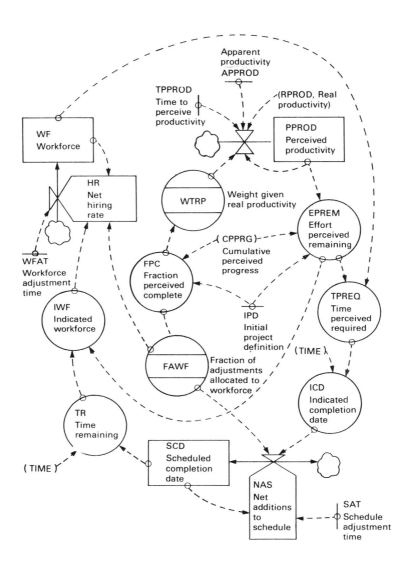

Figure 4.31: Flow diagram showing the formulations in the model for effort perceived remaining, hiring, and scheduling

Formulating a Project Model

Before simulating the model, we would need to assign initial values to the level variables and pick values for the constants. Since our focus here is on the formulation of equations, we will postpone a detailed discussion of parameter values in the model until section 4.6. But it would be a shame not to be able to get some output here after all this work, so without justification we will tentatively adopt the values shown in tables 4.1 and 4.2. Note that GPROD and IPD *define* the meaning of the task unit: whatever tasks are, one of them is normally completed by a person every month and 1200 of them can make up a 40-month project.

Next, a number of parameters controlling details of the simulation need to be specified, including DT, LENGTH, PLTPER, and PRTPER.

The smallest time constant in table 4.2 is three months, so by our rule of thumb (see section 3.5) DT should be less than one-half of three and will likely not need to be taken less than one-tenth of three. Setting DT equal to 0.5 months seems quite reasonable. However, it is important to note that some time constants in the model are hidden in table functions and other variables affecting the system levels in their rate equations. The time to detect rework TDRW, for example, can go as low as 0.5 months at the very end of the simulation. Perhaps it would be better to take DT equal to 0.25.

The values of PLTPER and PRTPER can be selected more to suit the whim of the modeler. Fifty lines of output fit nicely on a computer page, and the project is supposed to take about 40 months (see SCD above) so taking the plot period PLTPER equal to 1 month would probably give a nice plot. For the time being, the PRTPER will be set equal to zero to suppress printed, tabular output.

The determination of LENGTH in this model is unusually difficult. Usually, one sets the length of a simulation to some constant value. Here, however, we wish the model to run until the project is completed, and since we hope that our model shows overruns in the project we don't know how much over 40 months to set the LENGTH parameter. A way out of the problem is provided by DYNAMO's logic function, the CLIP. We will use the following equation for LENGTH:

```
A    LENGTH.K=CLIP(TIME.K,MAXLEN,FCOMP.K,1)
C    MAXLEN=100
```

FCOMP denotes the fraction of the project actually completed at any given point in time. The CLIP function sets LENGTH equal to MAXLEN as long as FCOMP is less than 1. When FCOMP equals or exceeds 1, LENGTH drops to equal the cur-

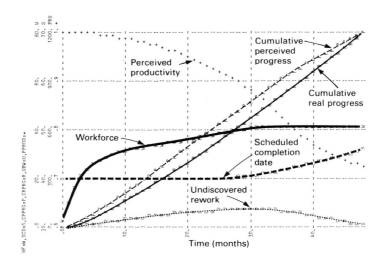

Figure 4.32: Base run of the project model.

rent value of TIME, which is the cue for DYNAMO to halt the simulation. In this fashion the simulation will always run until the project is actually completed, unless, of course, it takes longer than 100 months. The following auxiliary computes FCOMP, the fraction completed:

```
A    FCOMP.K=CRPRG.K/IPD
```
where
  CRPRG = cumulative real progress (tasks),
  IPD   = initial project definition (tasks).

The model is now ready to run. A complete listing of equations, with explanatory NOTEs, a PLOT statement, and equations for DT, LENGTH, PLTPER, and PRTPER, is given at the end of this section.

Figure 4.32 shows the base run of our model. It does indeed show overruns in its schedule and, as a bit of analysis will show, in the size of the workforce one would have expected to employ to complete a project of the size assumed.

Looking Back

The focus of this section has been on formulating equations for a simple but meaningful model addressing a set

Formulating a Project Model

of real problems. We have seen the selection of level variables in the model and the formulation of five rate equations and thirteen auxiliaries, including three table functions. Many of the patterns in these equations should have been familiar, given the discussions in the first three sections of this chapter. Others, such as the weighted averages in PPROD and WCWF, are new and deserve special attention because of their potential for representing a wide range of concepts.

The model developed thus far is one of many possible models of project dynamics. It is oversimplified in places, because its primary purpose is the illustration of model formulation, not really the investigation of project dynamics. The reader may have wished to see other equations for certain variables; certainly, there were other ways of formulating parts of the model, even given the rather precise structure diagramed in chapter 2. The current model has been an introduction and will be a point of departure for us later. After using the model as a medium for discussing model documentation (section 4.5), parameter selection (section 4.6), and debugging (section 4.7), we shall devote some attention in chapter 5 to alternative formulations and additional structure.

Listing of the Project Model

```
*       SIMPLE PROJECT MODEL.
NOTE
NOTE        REAL PROGRESS
NOTE
A       APPRG.K=WF.K*GPROD
NOTE        APPARENT PROGRESS RATE  (TASKS/MONTH)
C       GPROD=1
NOTE        GROSS PRODUCTIVITY (TASKS/PERSON/MONTH)
R       RPRG.KL=APPRG.K*FSAT
NOTE        REAL PROGRESS RATE  (TASKS/MONTH)
C       FSAT=0.7
NOTE        FRACTION SATISFACTORY  (DIMENSIONLESS)
L       CRPRG.K=CRPRG.J+DT*RPRG.JK
N       CRPRG=0
NOTE        CUMULATIVE REAL PROGRESS (TASKS)
NOTE
NOTE        UNDISCOVERED REWORK
NOTE
R       GURW.KL=APPRG.K*(1-FSAT)
NOTE        GENERATION OF UNDISCOVERED REWORK  (TASKS/MONTH)
L       URW.K=URW.J+DT*(GURW.JK-DURW.JK)
N       URW=0
NOTE        UNDISCOVERED REWORK  (TASKS)
R       DURW.KL=URW.K/TDRW.K
NOTE        DETECTION OF UNDISCOVERED REWORK  (TASKS/MONTH)
A       TDRW.K=TABLE(TTDRW,FPCOMP.K,0,1,0.2)
NOTE        TIME TO DETECT REWORK (MONTHS)
T       TTDRW=12/12/12/10/5/.5
NOTE        TABLE FOR TIME TO DETECT REWORK
NOTE
```

Model Formulation

```
NOTE        PERCEIVED PROGRESS
NOTE
A     CPPRG.K=CRPRG.K+URW.K
NOTE        CUMULATIVE PERCEIVED PROGRESS (TASKS)
A     FPCOMP.K=CPPRG.K/IPD
NOTE        FRACTION PERCEIVED COMPLETED (DIMENSIONLESS)
C     IPD=1200
NOTE        INITIAL PROJECT DEFINITION (TASKS)
NOTE
NOTE        EFFORT PERCEIVED REMAINING
NOTE
A     EPREM.K=(IPD-CPPRG.K)/PPROD.K
NOTE        EFFORT PERCEIVED REMAINING (MAN*MONTHS)
A     PPROD.K=SMOOTH(IPROD.K,TPPROD)
NOTE        PERCEIVED PRODUCTIVITY (TASKS/PERSON/MONTH)
C     TPPROD=6
NOTE        TIME TO PERCEIVE PRODUCTIVITY (MONTHS)
A     IPROD.K=WTRP.K*RPROD.K+(1-WTRP.K)*GPROD
NOTE        INDICATED PRODUCTIVITY (TASKS/PERSON/MONTH)
A     RPROD.K=GPROD*FSAT
NOTE        REAL PRODUCTIVITY (TASKS/PERSON/MONTH)
A     WTRP.K=TABLE(TWTRP,FPCOMP.K,0,1,0.2)
NOTE        WEIGHT GIVEN REAL PRODUCTIVITY
T     TWTRP=0/.1/.25/.5/.9/1
NOTE        TABLE FOR WTRP
NOTE
NOTE        HIRING
NOTE
L     WF.K=WF.J+DT*HR.JK
NOTE        WORKFORCE (PEOPLE)
N     WF=WFN
C     WFN=2
NOTE        INITIAL VALUE FOR WORKFORCE (PEOPLE)
R     HR.KL=(WFS.K-WF.K)/WFAT
NOTE        NET HIRING RATE (PEOPLE/MONTH)
C     WFAT=3
NOTE        WORKFORCE ADJUSTMENT TIME (MONTHS)
A     WFS.K=WCWF.K*IWF.K+(1-WCWF.K)*WF.K
NOTE        WORKFORCE SOUGHT (PEOPLE)
A     IWF.K=EPREM.K/TREM.K
NOTE        INDICATED WORKFORCE (PEOPLE)
A     WCWF.K=TABHL(TWCWF,TREM.K,0,21,3)
NOTE        WILLINGNESS TO CHANGE WORKFORCE (DIMENSIONLESS)
T     TWCWF=0/0/0/.1/.3/.7/.9/1
NOTE        TABLE FOR WCWF
NOTE
NOTE        SCHEDULING
NOTE
A     TREM.K=SCD.K-TIME.K
NOTE        TIME REMAINING (MONTHS)
L     SCD.K=SCD.J+DT*NAS.JK
NOTE        SCHEDULED COMPLETION DATE (MONTH)
N     SCD=SCDN
C     SCDN=40
```

Formulating a Project Model

```
NOTE       INITIAL VALUE OF SCD
R      NAS.KL=(ICD.K-SCD.K)/SAT
NOTE       NET ADDITIONS TO SCHEDULE  (MONTHS/MONTH)
C      SAT=6
NOTE       SCHEDULE ADJUSTMENT TIME  (MONTHS)
A      ICD.K=TIME.K+TPREQ.K
NOTE       INDICATED COMPLETION DATE  (MONTH)
A      TPREQ.K=EPRFM.K/WFS.K
NOTE       TIME PERCEIVED REQUIRED  (MONTHS)
NOTE
NOTE       CONTROL STATEMENTS
NOTE
SPEC   DT=.25,MAXLEN=100,PLTPER=1,PRTPER=0
A      LENGTH.K=CLIP(MAXLEN,0,1,FCOMP.K)
A      FCOMP.K=CRPRG.K/IPD
PLOT   WF=W(0,80)/SCD=S(30,70)/CPPRG=P,CRPRG=R,URW=U(0,1200)/PPROD=+(.6,1)
```

4.5 DOCUMENTATION

One of the most difficult tasks facing a person building a quantitative model is the communication of model structure and implications. A first step toward clear communication is thorough documentation, indicating what model variables mean, where they appear in the model, and how they interrelate. The purpose of documentation is to make the model as accessable as possible to readers who might wish to understand it, verify findings, use it for further investigations, or modify it for their own uses.

In section 3.4 three kinds of statements in DYNAMO were introduced that facilitate the documentation of a model and its simulation runs: the * statement, the NOTE statement, and the RUN statement. The models we have listed have used these appropriately as labeling devices. We used the * statement as the first line in a listing to identify the model and label output.[14] NOTE statements were used to separate the model into meaningful sectors, define variable names, and add whatever comments might be helpful in the equation listing. We followed some RUN statements with text, such as BASE SIMULATION, to identify the computer output for those runs.

---

[14] Major sectors of larger models can be identified with additional * statements, with NOTE statements used to divide major sectors into subsectors. Multiple * statements have a special significance to the DOCUMENTOR: see Dividing into Sectors.

This section illustrates the capabilities of the DYNAMO DOCUMENTOR, a computer program that dramatically simplifies the task of communicating model structure. We shall use it to produce a complete documentation of the project model, including a fully annotated equation listing, a list showing where each variable in the model is used, a dictionary of variable names in the model, and a reformatted version of the model.

Using the DOCUMENTOR

The DOCUMENTOR is computer program separate from the DYNAMO compiler. The specific commands that call up the DOCUMENTOR and apply it to a DYNAMO model vary depending upon the computer system on which the software is installed. See the appropriate references for your system. By way of example, we note that the runs of the DOCUMENTOR that follow in this section were produced by the simple system command

DOC PM2

where DOC stands for DOCUMENT and PM2 is the name we had given to the file containing the project model shown in section 4.4. There are a number of ways of preparing the model for documentation and using the DOCUMENTOR: a separate file of definitions of model variables can be used, a DOPT statement may be included to specify particular options, and so on. However, for the moment we postpone the details of model preparation in order to focus on the results.

Description of Output

In response to the simple computer cue representing the command "document this particular DYNAMO model," the DOCUMENTOR produces the first three listings shown in the appendix to this section.

The first listing is a fully documented, numbered equation listing, containing definitions of all the variables as they were given in the original NOTE statements in the model and showing at the right where each variable is used in the model. The equations are numbered according to type: the active equations (L, R, and A equations) are numbered with integers in the order they appear in the original listing, and the inactive equations (N, C, and T equations) are numbered with decimals (such as 2.1) to identify the active equations after which they are located (and to which, one hopes, they pertain).

This fully documented, numbered equation listing provides a complete statement of the model, as readable as the original NOTE statements in the model allow. It is particularly useful cut up and inserted in an equation-by- equation description of a model, in much the same fashion as the presentation in section 4.4.

The second listing is a dictionary of the names of all the variables and functions appearing in the model. In the column headed WHR-CMP (meaning where computed) the alphabetical list shows the number of the equation in which the quantity is assigned its values. At the right in this dictionary are the definitions that appeared in appropriately placed NOTE statements in the original model listing (or in a separate definition file). Much like any dictionary, this alphabetical list provides the user quick access to the meanings of variables and functions appearing in the model. At the bottom of the list the DOCUMENTOR notes the names of terms appearing in the model that were not defined.

The third listing from this run of the DOCUMENTOR is another "Where Used" list, arranged alphabetically and showing all the variables and functions appearing in the model. Beside each term is a list of all the places in the model where the term appears. Each equation in which a term is used is identified by the name of the variable computed, the type of equation, and its number in the model listing. This separate Where Used list repeats the information contained on the right side of the documented equation listing in a form which fits on standard 8 1/2 by 11 pages.

The reader may wish to refer back to one or more of these listings in following the discussions in chapters 5 and 6.

Preparing a Model for the DOCUMENTOR

Although the DOCUMENTOR is a remarkably sophisticated program, it has no intelligence: it cannot discern meanings, so it can pair up variables and definitions only by placement. As a result, a model and its definitions of terms must be written in particular forms before the DOCUMENTOR can produce the various listings shown above.

Placement of Definitions

To pair definitions with variable names there are three options:

1. Each variable can be defined *on the same line in which it is computed*. Simply leave one or more spaces after the equation, and type the definition. DYNAMO uses the spaces as a cue that the equation is finished and a definition follows. For example,

    A   APPRG.K=WF.K*GPROD APPARENT PROGRESS

2. Each variable can be defined *in a NOTE statement in the line immediately following the equation for the variable*. This is the format we used for the project model in section 4.4, which produced the DOCUMENTOR output shown above.

3. A *separate definition file* can be created that simply lists each term in the model followed on the same line by one or more spaces and the definition of the term. Such a file for the project model would include, for example,

    APPRG APPARENT PROGRESS (TASKS/MONTH)
    GPROD GROSS PRODUCTIVITY (TASKS/MAN/MONTH)

    The first option has the advantage of creating a relatively compact equation and definition listing, but long definitions sometimes spill over onto the next line, requiring the use of a NOTE card to continue the definition.[15] For example, one might have to write

R   GURW.KL=APPRG.K*(1-FSAT) GENERATION OF
NOTE   UNDISCOVERED REWORK (TASKS/MONTH)

    The second option has a very uniform appearance but creates a rather long model listing. The third option is useful for big models and models existing in a number of versions for which one list of common definitions suffices. The definitions in all of these variations are vastly improved by the inclusion of the units of each variable in parentheses.

Dividing into Sectors and Subsectors

    The DOCUMENTOR recognizes a blank NOTE card as a separation between groups of equations. A sequence such as

---

[15] In DYNAMO II and III the word NOTE can be replaced by a space in column 1, producing a less egregious continuation.

```
NOTE
NOTE    PERCEIVED PROGRESS
NOTE
```

serves to separate the next group of equations from what preceeded them and to head them with the title PERCEIVED PROGRESS. We used such headings in the project model in section 4.4, and they appeared as sector headings in the first listing produced above by the DOCUMENTOR.

Major sectors in a large model may be headed by * statements. A new * statement signals the DOCUMENTOR to begin a new sector on a new page and labels every subsequent page with the new heading the * statement contains. NOTE statements can then be used as above to divide major sectors into subsectors.

Ordering Equations

The DOCUMENTOR numbers and presents equations in the order of the original model listing. Take care to list equations in a natural order for a reader. Tastes vary--a natural order for one may seem a hodgepodge to another--but a few general guidelines appear to be prerequisites to an orderly listing:

- Place important variables in conspicuous places, and order the listing so that less important variables point toward them.

- Complete small groups of equations before moving on. For example, place constant, table, and initial value equations immediately after the active equations to which they pertain.

- In general, list equations in a way that tries to minimize the need for a reader to look ahead for the meaning or computation of a variable.

DOCUMENTOR Options and the DOPT Statement

To instruct the DOCUMENTOR to print out particular listings and formats, the user may place one or more DOPT

---

[16] Consult your computer facility, or write Pugh-Roberts Associates, Inc., 5 Lee St., Cambridge, Mass. 02139.

statements at the top of the equation listing. For a complete discussion of all the options available, consult the *Documentor User Manual*.[16] Here we merely note that options exist for the user to

- control the basis for grouping equations,
- set the basis for numbering equations,
- specify the source of definitions,
- change the length, width, or margin of the listings,
- specify the device on which the output is to appear,
- suppress unwanted listings,
- request a reformatted model.

Reformatting

The Reformat option RFRMT can be used to rewrite a model listing in a variety of ways. To illustrate the possibilities, we note that running the DOCUMENTOR with

DOPT RFRMT

at the top of the equation listing produces the model listing shown at the end of the apendix to this section. Definitions no longer follow their respective equations but appear instead at the right, each beginning neatly in column 35.

The RFRMT option produces output in this general form -- equations on the left, definitions on the right. It is frequently used for cleaning up the listings of models already in this general format, starting equations and definitions in standard columns, renumbering equations, and numbering the pages of the listing.

Summary of Documentation

The purpose of documentation is accessability. Computer models by their very nature are accessable (readable) only to a small minority of people. An undocumented model reduces that small group almost to zero. Exceptionally clear documentation has the chance of actually increasing the number of people who can understand the assumptions of a model to the extent that they can work with it or contribute to the discussion of the problems it addresses.

Documentation

Whether one uses the DYNAMO DOCUMENTOR or not, one must accompany any model with documentation that supplies the sort of information that we have presented here for the project model. An undocumented model is sure to wind up on some intellectual scrap heap.

DOCUMENTOR output for the simple project model

PAGE 1   SIMPLE PROJECT MODEL.   6/17/80   6/23/81

    REAL PROGRESS

APPRG.K=WF.K*GPROD   A,1   >RPRG,R,2/GURW,R,4
GPROD=1   C,1.1   >APPRG,A,1/IPROD,A,12/RPROD,A,13
    APPRG   - APPARENT PROGRESS RATE (TASKS/MONTH) <1>
    WF   - WORKFORCE (PEOPLE) <15>
    GPROD   - GROSS PRODUCTIVITY (TASKS/PERSON/MONTH) <1>

RPRG.KL=APPRG.K*FSAT   R,2   >CRPRG,L,3
FSAT=0.7   C,2.1   >RPRG,R,2/GURW,R,4/RPROD,A,13
    RPRG   - REAL PROGRESS RATE (TASKS/MONTH) <2>
    APPRG   - APPARENT PROGRESS RATE (TASKS/MONTH) <1>
    FSAT   - FRACTION SATISFACTORY (DIMENSIONLESS) <2>

CRPRG.K=CRPRG.J+DT*RPRG.JK   L,3   >CPPRG,A,8/FCOMP,A,27/PLOT,28
CRPRG=0   N,3.1
    CRPRG   - CUMULATIVE REAL PROGRESS (TASKS) <3>
    RPRG   - REAL PROGRESS RATE (TASKS/MONTH) <2>

    UNDISCOVERED REWORK

GURW.KL=APPRG.K*(1-FSAT)   R,4   >URW,L,5
    GURW   - GENERATION OF UNDISCOVERED REWORK (TASKS/MONTH)
              <4>
    APPRG   - APPARENT PROGRESS RATE (TASKS/MONTH) <1>
    FSAT   - FRACTION SATISFACTORY (DIMENSIONLESS) <2>

URW.K=URW.J+DT*(GURW.JK-DURW.JK)   L,5   >DURW,R,6/CPPRG,A,8/PLOT,28
URW=0   N,5.1
    URW   - UNDISCOVERED REWORK (TASKS) <5>
    GURW   - GENERATION OF UNDISCOVERED REWORK (TASKS/MONTH)
              <4>
    DURW   - DETECTION OF UNDISCOVERED REWORK (TASKS/MONTH)
              <6>

PAGE 2   SIMPLE PROJECT MODEL.   6/17/80   6/23/81

DURW.KL=URW.K/TDRW.K   R,6   >URW,L,5
    DURW   - DETECTION OF UNDISCOVERED REWORK (TASKS/MONTH)
             <6>
    URW   - UNDISCOVERED REWORK (TASKS) <5>
    TDRW   - TIME TO DETECT REWORK (MONTHS) <7>

TDRW.K=TABLE(TTDRW,FPCOMP.K,0,1,0.2)   A,7   >DURW,R,6
TTDRW=12/12/10/5/.5   T,7.1   >TDRW,A,7
    TDRW   - TIME TO DETECT REWORK (MONTHS) <7>
    TTDRW - TABLE FOR TIME TO DETECT REWORK <7>
    FPCOMP - FRACTION PERCEIVED COMPLETED (DIMENSIONLESS) <9>

          PERCEIVED PROGRESS

CPPRG.K=CPPRG.K+URW.K   A,8   >FPCOMP,A,9/EPREM,A,10/PLOT,28
    CPPRG - CUMULATIVE PERCEIVED PROGRESS (TASKS) <8>
    CRPRG - CUMULATIVE REAL PROGRESS (TASKS) <3>
    URW   - UNDISCOVERED REWORK (TASKS) <5>

FPCOMP.K=CPPRG.K/IPD   A,9   >TDRW,A,7/WTRP,A,14
IPD=1200   C,9.1   >FPCOMP,A,9/EPREM,A,10/FCOMP,A,27
    FPCOMP - FRACTION PERCEIVED COMPLETED (DIMENSIONLESS) <9>
    CPPRG - CUMULATIVE PERCEIVED PROGRESS (TASKS) <8>
    IPD   - INITIAL PROJECT DEFINITION (TASKS) <9>

          EFFORT PERCEIVED REMAINING

EPREM.K=(IPD-CPPRG.K)/PPROD.K   A,10   >IWF,A,18/TPREQ,A,24
    EPREM - EFFORT PERCEIVED REMAINING (MAN*MONTHS) <10>
    IPD   - INITIAL PROJECT DEFINITION (TASKS) <9>
    CPPRG - CUMULATIVE PERCEIVED PROGRESS (TASKS) <8>
    PPROD - PERCEIVED PRODUCTIVITY (TASKS/PERSON/MONTH) <11>

Documentation   221

222 Model Formulation

```
PAGE 3      SIMPLE PROJECT MODEL.     5/17/80                      6/23/81

PPROD.K=SMOOTH(IPROD.K,TPPROD)                                     A,11          >EPREM,A,10/PLOT,28
TPPROD=6                                                           C,11.1        >PPROD,A,11
    PPROD   - PERCEIVED PRODUCTIVITY (TASKS/PERSON/MONTH)   <11>
    IPROD   - INDICATED PRODUCTIVITY (TASKS/PERSON/MONTH)   <12>
    TPPROD  - TIME TO PERCEIVE PRODUCTIVITY (MONTHS)        <11>

IPROD.K=WTRP.K*RPROD.K+(1-WTRP.K)*GPROD                            A,12          >PPROD,A,11
    IPROD   - INDICATED PRODUCTIVITY (TASKS/PERSON/MONTH)   <12>
    WTRP    - WEIGHT GIVEN REAL PRODUCTIVITY                <14>
    RPROD   - REAL PRODUCTIVITY (TASKS/PERSON/MONTH)        <13>
    GPROD   - GROSS PRODUCTIVITY (TASKS/PERSON/MONTH)       <1>

RPROD.K=GPROD*FSAT                                                 A,13          >IPROD,A,12
    RPROD   - REAL PRODUCTIVITY (TASKS/PERSON/MONTH)        <13>
    GPROD   - GROSS PRODUCTIVITY (TASKS/PERSON/MONTH)       <1>
    FSAT    - FRACTION SATISFACTORY (DIMENSIONLESS)         <2>

WTRP.K=TABLE(TWTRP,FPCOMP.K,0,1,0.2)                               A,14          >IPROD,A,12
TWTRP=0/.1/.25/.5/.9/1                                             T,14.1        >WTRP,A,14
    WTRP    - WEIGHT GIVEN REAL PRODUCTIVITY                <14>
    TWTRP   - TABLE FOR WTRP                                <14>
    FPCOMP  - FRACTION PERCEIVED COMPLETED (DIMENSIONLESS)  <9>

            HIRING

WF.K=WF.J+DT*HR.JK                                                 L,15          >APPRG,A,1/HR,R,16/WFS,A,17/PLOT,28
WF=WFN                                                             N,15.1
WFN=2                                                              C,15.2        >WF,N,15.1
    WF      - WORKFORCE (PEOPLE)                            <15>
    HR      - NET HIRING RATE (PEOPLE/MONTH)                <16>
    WFN     - INITIAL VALUE FOR WORKFORCE (PEOPLE)          <15>
```

```
PAGE 4      SIMPLE PROJECT MODEL.    6/17/80                          6/23/81

HR.KL=(WFS.K-WF.K)/WFAT                                               R,16       >WF,L,15
WFAT=3                                                                C,16.1     >HR,R,16
    HR    - NET HIRING RATE (PEOPLE/MONTH) <16>
    WFS   - WORKFORCE SOUGHT (PEOPLE) <17>
    WF    - WORKFORCE (PEOPLE) <15>
    WFAT  - WORKFORCE ADJUSTMENT TIME (MONTHS) <16>

WFS.K=WCWF.K*IWF.K+(1-WCWF.K)*WF.K                                    A,17       >HR,R,16/TPREQ,A,24
    WFS   - WORKFORCE SOUGHT (PEOPLE) <17>
    WCWF  - WILLINGNESS TO CHANGE WORKFORCE (DIMENSIONLESS)
            <19>
    IWF   - INDICATED WORKFORCE (PEOPLE) <18>
    WF    - WORKFORCE (PEOPLE) <15>

IWF.K=EPREM.K/TREM.K                                                  A,18       >WFS,A,17
    IWF   - INDICATED WORKFORCE (PEOPLE) <18>
    EPREM - EFFORT PERCEIVED REMAINING (MAN*MONTHS) <10>
    TREM  - TIME REMAINING (MONTHS) <20>

WCWF.K=TABHL(TWCWF,TREM.K,0,21,3)                                     A,19       >WFS,A,17
TWCWF=0/0/0/.1/.3/.7/.9/1                                             T,19.1     >WCWF,A,19
    WCWF  - WILLINGNESS TO CHANGE WORKFORCE (DIMENSIONLESS)
            <19>
    TWCWF - TABLE FOR WCWF <19>
    TREM  - TIME REMAINING (MONTHS) <20>

            SCHEDULING

TREM.K=SCD.K-TIME.K                                                   A,20       >IWF,A,18/WCWF,A,19
    TREM  - TIME REMAINING (MONTHS) <20>
    SCD   - SCHEDULED COMPLETION DATE (MONTH) <21>
```

PAGE 5   SIMPLE PROJECT MODEL.   6/17/80   6/23/81

SCD.K=SCD.J+DT*NAS.JK   L,21
SCD=SCDN   N,21.1
SCDN=40   C,21.2   >TREM,A,20/NAS,R,22/PLOT,28
   SCD   - SCHEDULED COMPLETION DATE (MONTH) <21>   >SCD,N,21.1
   NAS   - NET ADDITIONS TO SCHEDULE (MONTHS/MONTH) <22>
   SCDN   - INITIAL VALUE OF SCD <21>

NAS.KL=(ICD.K-SCD.K)/SAT   R,22   >SCD,L,21
SAT=6   C,22.1   >NAS,R,22
   NAS   - NET ADDITIONS TO SCHEDULE (MONTHS/MONTH) <22>
   ICD   - INDICATED COMPLETION DATE (MONTH) <23>
   SCD   - SCHEDULED COMPLETION DATE (MONTH) <21>
   SAT   - SCHEDULE ADJUSTMENT TIME (MONTHS) <22>

ICD.K=TIME.K+TPREQ.K   A,23   >NAS,R,22
   ICD   - INDICATED COMPLETION DATE (MONTH) <23>
   TPREQ - TIME PERCEIVED REQUIRED (MONTHS) <24>

TPREQ.K=EPREM.K/WFS.K   A,24   >ICD,A,23
   TPREQ - TIME PERCEIVED REQUIRED (MONTHS) <24>
   EPREM - EFFORT PERCEIVED REMAINING (MAN*MONTHS) <10>
   WFS   - WORKFORCE SOUGHT (PEOPLE) <17>

       CONTROL STATEMENTS

SPEC DT=.25,MAXLEN=100,PLTPER=1,PRTPER=0   25

LENGTH.K=CLIP(MAXLEN,0,1,FCOMP.K)   A,26
   FCOMP - FRACTION ACTUALLY COMPLETED (DIMENSIONLESS) <27>

FCOMP.K=CRPRG.K/IPD   A,27   >LENGTH,A,26
   FCOMP - FRACTION ACTUALLY COMPLETED (DIMENSIONLESS) <27>
   CRPRG - CUMULATIVE REAL PROGRESS (TASKS) <3>
   IPD   - INITIAL PROJECT DEFINITION (TASKS) <9>

PAGE 6   SIMPLE PROJECT MODEL.   6/17/80

PLOT WF=W(0,80)/SCD=S(30,70)/CPPRG=P,CRPRG=R,URW=U(0,1200)/
PPROD=+(.6,1)
    WF   - WORKFORCE (PEOPLE) <15>
    SCD   - SCHEDULED COMPLETION DATE (MONTH) <21>
    CPPRG - CUMULATIVE PERCEIVED PROGRESS (TASKS) <8>
    CRPRG - CUMULATIVE REAL PROGRESS (TASKS) <3>
    URW   - UNDISCOVERED REWORK (TASKS) <5>
    PPROD - PERCEIVED PRODUCTIVITY (TASKS/PERSON/MONTH) <11>

RUN BASE RUN

```
PAGE 1         SIMPLE PROJECT MODEL.  6/17/80                              6/23/81

*     SIMPLE PROJECT MODEL.  6/17/80                                       00000000
NOTE                                                                       00000001
NOTE         REAL PROGRESS                                                 00000002
NOTE                                                                       00000003
A     APPRG.K=WF.K*GPROD           APPARENT PROGRESS RATE (TASKS/MONTH)    00000010
C     GPROD=1                      GROSS PRODUCTIVITY (TASKS/PERSON/       00000011
                                     MONTH)                                00000012
R     RPRG.KL=APPRG.K*FSAT         REAL PROGRESS RATE (TASKS/MONTH)        00000020
C     FSAT=0.7                     FRACTION SATISFACTORY                   00000021
                                     (DIMENSIONLESS)                       00000022
L     CRPRG.K=CRPRG.J+DT*RPRG.JK                                           00000030
N     CRPRG=0                      CUMULATIVE REAL PROGRESS (TASKS)        00000031
NOTE                                                                       00000032
NOTE         UNDISCOVERED REWORK                                           00000033
NOTE                                                                       00000034
R     GURW.KL=APPRG.K*(1-FSAT)     GENERATION OF UNDISCOVERED REWORK       00000040
                                     (TASKS/MONTH)                         00000041
L     URW.K=URW.J+DT*(GURW.JK-DURW.JK)                                     00000050
N     URW=0                        UNDISCOVERED REWORK (TASKS)             00000051
R     DURW.KL=URW.K/TDRW.K         DETECTION OF UNDISCOVERED REWORK        00000060
                                     (TASKS/MONTH)                         00000061
A     TDRW.K=TABLE(TTDRW,FPCOMP.K,0,1,0.2)                                 00000070
                                   TIME TO DETECT REWORK (MONTHS)          00000071
T     TTDRW=12/12/12/10/5/.5       TABLE FOR TIME TO DETECT REWORK         00000072
NOTE                                                                       00000073
NOTE         PERCEIVED PROGRESS                                            00000074
NOTE                                                                       00000075
A     CPPRG.K=CRPRG.K+URW.K        CUMULATIVE PERCEIVED PROGRESS           00000080
                                     (TASKS)                               00000081
A     FPCOMP.K=CPPRG.K/IPD         FRACTION PERCEIVED COMPLETED            00000090
                                     (DIMENSIONLESS)                       00000091
C     IPD=1200                     INITIAL PROJECT DEFINITION (TASKS)      00000092
NOTE                                                                       00000093
NOTE         EFFORT PERCEIVED REMAINING                                    00000094
NOTE                                                                       00000095
A     EPREM.K=(IPD-CPPRG.K)/PPROD.K EFFORT PERCEIVED REMAINING (MAN*       00000100
                                     MONTHS)                               00000101
A     PPROD.K=SMOOTH(IPROD.K,TPPROD) PERCEIVED PRODUCTIVITY (TASKS/PERSON/ 00000110
                                     MONTH)                                00000111
C     TPPROD=6                     TIME TO PERCEIVE PRODUCTIVITY           00000112
                                     (MONTHS)                              00000113
A     IPROD.K=WTRP.K*RPROD.K+(1-WTRP.K)*GPROD                              00000120
                                   INDICATED PRODUCTIVITY (TASKS/PERSON/   00000121
                                     MONTH)                                00000122
A     RPROD.K=GPROD*FSAT           REAL PRODUCTIVITY (TASKS/PERSON/        00000130
                                     MONTH)                                00000131
A     WTRP.K=TABLE(TWTRP,FPCOMP.K,0,1,0.2)                                 00000140
                                   WEIGHT GIVEN REAL PRODUCTIVITY          00000141
T     TWTRP=0/.1/.25/.5/.9/1       TABLE FOR WTRP                          00000142
NOTE                                                                       00000143
NOTE         HIRING                                                        00000144
NOTE                                                                       00000145
L     WF.K=WF.J+DT*HR.JK           WORKFORCE (PEOPLE)                      00000150
N     WF=WFN                                                               00000151
C     WFN=2                        INITIAL VALUE FOR WORKFORCE (PEOPLE)    00000152
R     HR.KL=(WFS.K-WF.K)/WFAT      NET HIRING RATE (PEOPLE/MONTH)          00000160
C     WFAT=3                       WORKFORCE ADJUSTMENT TIME (MONTHS)      00000161
A     WFS.K=WCWF.K*IWF.K+(1-WCWF.K)*WF.K                                   00000170
```

```
PAGE 2        SIMPLE PROJECT MODEL.  6/17/80                        6/23/81

                                    WORKFORCE SOUGHT (PEOPLE)       00000171
A   IWF.K=EPREM.K/TREM.K            INDICATED WORKFORCE (PEOPLE)    00000180
A   WCWF.K=TABHL(TWCWF,TREM.K,0,21,3)                               00000190
                                    WILLINGNESS TO CHANGE WORKFORCE 00000191
                                    (DIMENSIONLESS)                 00000192
T   TWCWF=0/0/0/.1/.3/.7/.9/1       TABLE FOR WCWF                  00000193
NOTE                                                                00000194
NOTE        SCHEDULING                                              00000195
NOTE                                                                00000196
A   TREM.K=SCD.K-TIME.K             TIME REMAINING (MONTHS)         00000200
L   SCD.K=SCD.J+DT*NAS.JK           SCHEDULED COMPLETION DATE (MONTH) 00000210
N   SCD=SCDN                                                        00000211
C   SCDN=40                         INITIAL VALUE OF SCD            00000212
R   NAS.KL=(ICD.K-SCD.K)/SAT        NET ADDITIONS TO SCHEDULE (MONTHS/ 00000220
                                    MONTH)                          00000221
C   SAT=6                           SCHEDULE ADJUSTMENT TIME (MONTHS) 00000222
A   ICD.K=TIME.K+TPREQ.K            INDICATED COMPLETION DATE (MONTH) 00000230
A   TPREQ.K=EPREM.K/WFS.K           TIME PERCEIVED REQUIRED (MONTHS) 00000240
NOTE                                                                00000241
NOTE        CONTROL STATEMENTS                                      00000242
NOTE                                                                00000243
SPEC DT=.25,MAXLEN=100,PLTPER=1,PRTPER=0                            00000250
A   LENGTH.K=CLIP(MAXLEN,0,1,FCOMP.K)                               00000260
A   FCOMP.K=CRPRG.K/IPD             FRACTION ACTUALLY COMPLETED     00000270
                                    (DIMENSIONLESS)                 00000271
PLOT WF=W(0,80)/SCD=S(30,70)/CPPRG=P,CRPRG=R,URW=U(0,1200)/PPROD=+(.6, 00000280
X      1)                                                           00000281
RUN BASE                                                            00000290


PAGE 3        SIMPLE PROJECT MODEL.  6/17/80                        6/23/81

            LIST OF SYMBOLS WITH NO DEFINITIONS

CLIP        DT        LENGTH    MAXLEN    PLTPER    PRTPER    SMOOTH    TABHL
TABLE       TIME
```

PAGE 1   SIMPLE PROJECT MODEL. 6/17/80   6/23/81

LIST OF VARIABLES

SYMBOL	T	WHR-CMP	DEFINITION
APPRG	A	1	APPARENT PROGRESS RATE (TASKS/MONTH) <1>
CPPRG	A	8	CUMULATIVE PERCEIVED PROGRESS (TASKS) <8>
CRPRG	L	3	CUMULATIVE REAL PROGRESS (TASKS) <3>
	N	3.1	
DT	S	25	
DURW	R	6	DETECTION OF UNDISCOVERED REWORK (TASKS/MONTH) <6>
EPREM	A	10	EFFORT PERCEIVED REMAINING (MAN*MONTHS) <10>
FCOMP	A	27	FRACTION ACTUALLY COMPLETED (DIMENSIONLESS) <27>
FPCOMP	A	9	FRACTION PERCEIVED COMPLETED (DIMENSIONLESS) <9>
FSAT	C	2.1	FRACTION SATISFACTORY (DIMENSIONLESS) <2>
GPROD	C	1.1	GROSS PRODUCTIVITY (TASKS/PERSON/MONTH) <1>
GURW	R	4	GENERATION OF UNDISCOVERED REWORK (TASKS/MONTH) <4>
HR	R	16	NET HIRING RATE (PEOPLE/MONTH) <16>
ICD	A	23	INDICATED COMPLETION DATE (MONTH) <23>
IPD	C	9.1	INITIAL PROJECT DEFINITION (TASKS) <9>
IPROD	A	12	INDICATED PRODUCTIVITY (TASKS/PERSON/MONTH) <12>
IWF	A	18	INDICATED WORKFORCE (PEOPLE) <18>
LENGTH	A	26	
MAXLEN	S	25	
NAS	R	22	NET ADDITIONS TO SCHEDULE (MONTHS/MONTH) <22>
PLTPER	S	25	
PPROD	A	11	PERCEIVED PRODUCTIVITY (TASKS/PERSON/MONTH) <11>
PRTPER	S	25	
RPRG	R	2	REAL PROGRESS RATE (TASKS/MONTH) <2>
RPROD	A	13	REAL PRODUCTIVITY (TASKS/PERSON/MONTH) <13>
SAT	C	22.1	SCHEDULE ADJUSTMENT TIME (MONTHS) <22>
SCD	L	21	SCHEDULED COMPLETION DATE (MONTH) <21>
	N	21.1	
SCDN	C	21.2	INITIAL VALUE OF SCD <21>
TDRW	A	7	TIME TO DETECT REWORK (MONTHS) <7>
TPPROD	C	11.1	TIME TO PERCEIVE PRODUCTIVITY (MONTHS) <11>
TPREQ	A	24	TIME PERCEIVED REQUIRED (MONTHS) <24>
TREM	A	20	TIME REMAINING (MONTHS) <20>
TTDRW	T	7.1	TABLE FOR TIME TO DETECT REWORK <7>
TWCWF	T	19.1	TABLE FOR WCWF <19>
TWTRP	T	14.1	TABLE FOR WTRP <14>
URW	L	5	UNDISCOVERED REWORK (TASKS) <5>
	N	5.1	
WCWF	A	19	WILLINGNESS TO CHANGE WORKFORCE (DIMENSIONLESS) <19>
WF	L	15	WORKFORCE (PEOPLE) <15>
	N	15.1	
WFAT	C	16.1	WORKFORCE ADJUSTMENT TIME (MONTHS) <16>
WFN	C	15.2	INITIAL VALUE FOR WORKFORCE (PEOPLE) <15>
WFS	A	17	WORKFORCE SOUGHT (PEOPLE) <17>
WTRP	A	14	WEIGHT GIVEN REAL PRODUCTIVITY <14>

Documentation

PAGE 2   SIMPLE PROJECT MODEL.   6/17/80   6/23/81

              WHERE-USED LIST

```
SYMBOL   WHERE-USED

APPRG    GURW,R,4/RPRG,R,2
CLIP     LENGTH,A,26
CPPRG    PLOT,28/EPREM,A,10/FPCOMP,A,9
CRPRG    PLOT,28/FCOMP,A,27/CPPRG,A,8
DURW     URW,L,5
EPREM    TPREQ,A,24/IWF,A,18
FCOMP    LENGTH,A,26
FPCOMP   WTRP,A,14/TDRW,A,7
FSAT     RPROD,A,13/GURW,R,4/RPRG,R,2
GPROD    RPROD,A,13/IPROD,A,12/APPRG,A,1
GURW     URW,L,5
HR       WF,L,15
ICD      NAS,R,22
IPD      FCOMP,A,27/EPREM,A,10/FPCOMP,A,9
IPROD    PPROD,A,11
IWF      WFS,A,17
MAXLEN   LENGTH,A,26
NAS      SCD,L,21
PPROD    PLOT,28/EPREM,A,10
RPRG     CRPRG,L,3
RPROD    IPROD,A,12
SAT      NAS,R,22
SCD      PLOT,28/NAS,R,22/TREM,A,20
SCDN     SCD,N,21.1
SMOOTH   PPROD,A,11
TABHL    WCWF,A,19
TABLE    WTRP,A,14/TDRW,A,7
TDRW     DURW,R,6
TIME     ICD,A,23/TREM,A,20
TPPROD   PPROD,A,11
TPREQ    ICD,A,23
TREM     WCWF,A,19/IWF,A,18
TTDRW    TDRW,A,7
TWCWF    WCWF,A,19
TWTRP    WTRP,A,14
URW      PLOT,28/CPPRG,A,8/DURW,R,6
WCWF     WFS,A,17
WF       PLOT,28/WFS,A,17/HR,R,16/APPRG,A,1
WFAT     HR,R,16
WFN      WF,N,15.1
WFS      TPREQ,A,24/HR,R,16
WTRP     IPROD,A,12
```

4.6 PARAMETERS AND INITIAL VALUES

To simulate a system dynamics model, one must first assign values to all the constants, table functions, and level variables appearing in the equation listing. Auxiliaries and rates are computed from levels and constants, so they do not usually require separate initial value computations. This section addresses the question, Where do the numbers come from? Our discussion here will be divided into two parts, the first focusing on parameter selection and the second on the process of determining the initial values of system levels.[17]

The question of parameter selection is perhaps the most often asked and the most often misunderstood inquiry about the system dynamics approach. A number of claims about parameter values in system dynamics models are intially rather disturbing. It has been asserted that feedback models are relatively insensitive to parameter changes, that parameters need not be estimated with statistical confidence intervals in order to have confidence in the policy implications of a model, and--the key claim on which the others rest--that the behavior of a system dynamics model is much more a consequence of its structure than its parameter values. To people used to the numerical sensitivity of statistical models, who may have devoted considerable efforts to statistical estimation and validation, such claims are disturbing.

Necessary Accuracy

A modeler should estimate parameter values only to the degree of accuracy required. Such a statement is hardly controversial: most would agree it would be rather silly to spend time and effort (and money) obtaining some value accurate to within plus or minus 1 percent if plus or minus 50 percent would suffice. The question is, Accuracy required for what purpose?

There is a range of possible purposes and therefore a range of accuracies required. One might be striving for a model that tracked certain historical time paths to within some specified tolerance. Or one might be seeking to understand or improve upon some general pattern of behavior, such as inventory oscillations. In the latter example one might well care more for an approximate fit of periodicities and amplitudes than historically correct values of inventory at particular points in time. For a client to believe in the formu-

---

[17] The discussion of parameter selection draws on material contained in Graham (1980) and Graham (1976), pp. 248-252.

Parameters and Initial Values 231

lation of a model, still other levels of parameter accuracy might be required.

But in the last analysis, the purpose of most system dynamics studies is policy analysis. *If the policy implications of a model do not change when its parameters are varied plus or minus some percent, then from the modeler's point of view the parameters do not need to be estimated any more accurately than that.*

Thus to decide how much effort to put into estimating a given parameter value, one ought to know how sensitive the behavior of the model is to the value of that parameter. Yet to know that, one must run the model, and that requires parameter values. To resolve the dilemma, the modeler picks some values rather quickly and simulates the model. Initial estimates are made carefully, to be sure, with as much concern for accuracy as can be easily mustered, but keeping in mind that one can always go back and estimate more carefully if it makes a difference. In setting parameter values it pays to remember that, if it doesn't make any difference, then *it really doesn't make any difference.*

Kinds of Parameters

A glance over the various equations littered throughout the book shows a number of different types of parameters. There are constant *measures* such as AREA in an urban model, *conversion factors* such as a constant for worker productivity PROD, *reference parameters* such as a normal delivery delay, *growth* or *aging factors* such as a fractional annual interest rate FAIR or an average lifespan, and *adjustment times* such as the constant for the time to perceive productivity TPPROD in the project model. Different types of parameters may call for different estimation strategies.

Within any of these types a given parameter may be a *definitional* parameter--one whose assigned value contributes to the definition of a concept or measurement unit in the model. In the project model formulated in section 4.4, gross productivity GPROD was such a definitional parameter. Setting GPROD equal to 1 task/person/month implicitly contributes to the defintion of the task unit: a task is large enough to require 1 man-month of effort. Setting GPROD equal to 20 or 21 tasks/person/month would imply that a task was something that a person could complete in a single work day (there are about 21 working days in a month). Giving values to definitional parameters is made simple by the realization that, as long as consistency throughout the model is maintained, values of definitional parameters are, within limits, essentially arbitrary.

Whatever the parameter type, the system dynamics approach insists that any parameter (and any variable) in a system dynamics model should have a clear correspondence to a real quantity or concept. In some sense, the quantity should be observable, although there are situations in which the term intuitable is more appropriate. Specifying the units of a parameter helps to guarantee such observability, by flagging parameters with awkward units contrived more to suit some mathematical formulation than the structure of the real system. The insistence on realism and observability helps to ease the task of parameter selection.

Kinds of Parameter Estimates

In general parameters can be estimated three ways: from firsthand knowledge of a process, from data on individual relationships in a model, and from data on overall system behavior. Parameters that represent *measurements* tend to be simple to estimate because they are easy to know firsthand. A city has an area, for example, that can be found in an atlas or estimated from a map; a prison has a given projected capacity, which a warden is likely to know, or which can be estimated by counting cells or bunks; average sentence length in a criminal justice system could be computed from court data; the maximum production capacity in a firm is undoubtedly known by people responsible for planning production; and so on. To obtain a value for such a parameter, one simply consults the appropriate source: observation of the real system itself, data, literature (scholarly or otherwise), or people. Don't forget people.

Conversion factors such as land per housing unit, productivity, or revenue per sale translate from one kind of unit to another, for example, from buildings to acres, people to widgets per month, or thingumabobs to dollars. Again, the use of realistic parameters and knowledge of the real system should make estimating such parameters reasonably straightforward.

Care must be taken, however, to be sure that the data used matches the meanings of the quantities in the model. In a population model, for example, does the birth rate factor represent births per year per person, per adult, per female, or per woman of child-bearing age? The meaning in the model will make a difference in the data sought and the way it is used in estimating the parameter. Land per housing unit in an urban model provides a slightly subtler example. Unless the model explicitly represents roads, parks, and the like, they must be considered to be aggregated with the buildings of the city. Hence, land per housing unit is more than just

Parameters and Initial Values 233

the land on which the physical building sits. It must include some portion of the land taken up by the infrastructure required to support a housing unit.

The lesson here is that aggregation in a model affects the meaning of a parameter, and it is the meaning *in the model* that determines how data should be used for assigning parameter values.

Normal or *reference parameters* often appear in rate equations in the pattern NORM.K*EFFECT.K and require knowledge or data on individual relationships in a model. Consider, for example, the following formulation for the shipment rate SR from an inventory:

R   SR.KL=(BCKLOG.K/NDD)*EIS.K

where
- SR = shipment rate (baubles/week),
- BCKLOG = order backlog (baubles),
- NDD = normal delivery delay (weeks),
- EIS = effect of inventory on shipments (dimensionless).

EIS (discussed in more detail in section 4.1) has the properties that it takes the value 1 when inventory equals some desired level of inventory and approaches 0 as inventory approaches 0. To assign a value to NDD the equation can be rewritten (ignoring timescripts):

NDD = (BCKLOG*EIS)/SR,

showing that NDD is the value of BCKLOG/SR when EIS equals 1, that is, when inventory is at its desired level.

Hence, to estimate NDD for a model of a given firm, one obtains data on the order backlog and the average shipment rate during periods of inventory adequacy, and then simply computes ratios. Note that, since NDD is formulated as a constant, one hopes that all such computations produce essentially the same result. If considerable variability appears, the formulation of the original rate equation is suspect. Perhaps another effect on SR is present and was ignored.

The strategy described here for NDD is required to estimate any parameters appearing in rate equations of the form

   rate = (normal fraction) * level * multipliers

or

   rate = (level / normal life) * multipliers.

The value of such a normal parameter is the appropriate ratio of the level and rate when the multipliers are all 1.

On the far end of the scale of difficulty are parameters that represent *adjustment times,* such as perception delays. Estimating such parameters requires thinking about the time required to form habits, change perceptions, and adjust traditions to new circumstances. Unfortunately, nature is not likely to perform a nice step increase experiment that would allow us to read off a time constant from an exponentially goal-seeking response of something like perceived productivity. When data is lacking, we have to resort to some mental experiments.

Bounds

In the absence of data one can frequently estimate reasonable upper and lower bounds for the value of a parameter and select a trial value somewhere in between. An example occurred in the development of the table function for the time to detect rework, TDRW, in the formulation of the project model (see section 4.4). Hunting for the length of time a mistake or a misdirection in the project could remain until it became obvious, we observed that in a 40- month project the value could not be as large as 40 months (without dire consequences). It also seemed unlikely that it could be as small as a single month. So we sought an initial value for TDRW between 1 and 40, finally agreeing on 12 months as a reasonable guess. The resulting number is a shot in the twilight: it is an educated guess that could be improved still further if one can find arguments that help to reduce the interval of reasonable values.

Applying such a strategy to estimate the time to perceive productivity in the project model, one would try to imagine how long it would take for an abrupt, permanent change in productivity to become perceived as a new norm. One first must know who is supposed to be doing the perceiving, workers responsible for the productivity or project managers considerably more distant from it. In our model the notion of perceived productivity is used to determine hiring and scheduling, so it is most appropriately interpreted as a management perception. It seems unlikely that in a project covering nearly four years project management would completely change their perception of worker productivity in, say, less than one month. Nor would they take as long as 40 months to become convinced a new standard of productivity had emerged. It might take as much as 18 months for a complete change of perception, and the half-life of the perception could easily be 3 to 6 months. Because of the properties of

Parameters and Initial Values

the exponential averaging formulation inherent in the SMOOTH, PPROD would show 50 percent of a permanent step increase in productivity in 0.7*TPPROD months. (See sections 3.5 and 4.1.) If we guessed a half-life of PPROD of 4 months, we would be led to suggest a value for TPPROD around 6 months. The exponential averaging in PPROD also guarantees that after three time constants, or 18 months, have elapsed, PPROD will be within 5 percent of the new value of productivity assumed in this mental step increase experiment.

Estimating from Process and from Behavior

Detailed knowledge of a process represented in the aggregate in a model can be used to estimate a parameter value. Alternatively, the overall behavior of the model or of a piece of it can be used. The meaning of these two approaches and their relative merits are best presented in an example. Consider the graph in figure 4.33 representing the irruption of a deer population that occurred when all their natural predators were removed by bounty hunters.

Suppose that the graph in figure 4.33 is a reference mode for a modeling study in which the equations relating to the deer population growth are the following:

```
L    DP.K=DP.J+DT*(DNGR.JK-DPR.JK)
NOTE    DEER POPULATION
R    DNGR.KL=DGRF.K*DP.K
NOTE    DEER NET GROWTH RATE (DEER/YEAR)
```
where
    DPR   = deer predation rate,
    DGRF   = deer growth rate factor.

We are interested in determining the maximum value of DGRF. The maximum value would be the growth rate fraction observable in the case of a deer population not constrained by food or predation.

One way to proceed would be to quantify the process by which a deer population grows, thinking of the quantities involved in births and deaths and then aggregating to come up with a maximum value of DGRF. Suppose we think in terms of a deer birth rate factor DBRF and a deer death rate factor DDRF. Then

DNGR = DBRF*DP - DDRF*DP = (DBRF - DDRF)*DP, so the deer growth rate factor is (DBRF - DDRF). To compute the maximum value of DGRF we want to find the maximum value of DBRF and the minimum value of DDRF and subtract.

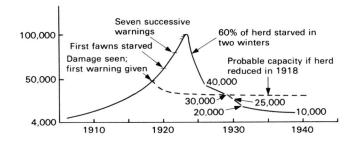

Figure 4.33: The effect of removal of natural predators on the deer population on the Kaibab Plateau on the north rim of the Grand Canyon, Arizona. (Source: Kormondy (1969), p. 96. See also Goodman (1974), p. 377-388.)

The deer birth rate factor could be computed as the product of a number of factors: the number of fawns per litter, the fraction that survive, the number of litters per adult female per year, the fraction of adult deer that are female, and the fraction of deer that are adult. One grand multiplication of these times the deer population DP would result in units of deer per year, as desired. One could go to a zoological reference, but let us do the best we can with what we have in our heads.

Like horses and cows, deer probably rarely have more than one fawn at a time, so the number of deer per litter ought to be 1. Something less than 100 percent survive to adulthood; let us say 90 percent in good times. Each adult female seems to give birth each spring, so the number of litters per adult female per year is likely to be 1. One can safely assume that about half the adult deer population is female, so we acquire a factor of 0.5. To figure the fraction of deer that are adult, we could assume that the deer population might be uniformly distributed from 1 year of age up through some average lifespan, say, 6 or 7 years so that the fraction adult would be about (7-1)/7 or 6/7. The result of these estimates is a proposed value for DBRF of 1*.9*1*.5*(6/7), or about 0.39.

DDRF must be 1/7 to be consistent with our estimate of an average deer lifespan of 7 years in healthy conditions. Hence we conclude that the maximum deer net growth factor DGRF is about 0.39 minus 0.14, or about 0.25. Its units are deer/deer/year, as the reader might wish to check. The "rule of 70" for exponentially growing quantities (see the Technical Note in section 4.1) suggests a maximum deer population doubling time of about 0.7/0.25 or 2.8 years.

Parameters and Initial Values

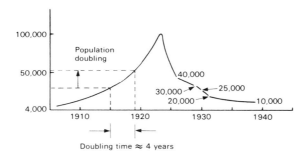

Figure 4.34: Deer population doubling time

The estimation procedure used here made use of data or estimates for a host of parameters aggregated within the single model parameter DGRF. Such procedures have been termed estimation "below the level of aggregation in the model." They focus on detailed knowledge or estimates about the processes the aggregate parameter represents.

Alternatively, one might use aggregate data to estimate DGRF. The reference to doubling times suggests how. Selecting a period in the graph in figure 4.33 in which neither predators nor food appear to inhibit growth, one can pick a population value, double it, and read off the corresponding doubling time from the horizontal axis. The lines drawn on the graph in figure 4.34 show one such experiment, from which one can estimate a doubling time of about 4 years. Again using the rule of 70, a doubling time of four years implies DGRF equals 0.7/4 or about 0.18.

This second approach involves estimating a parameter from knowledge of behavior rather than from knowledge of underlying processes. One approach produced an estimate of the maximum value of DGRF of 0.25, while the other yielded an estimate of 0.18. Given the rather dramatic differences in the two approaches (and the armchair estimates involved in one of them), it is remarkable that they are as close as they are.

What might account for differences between the estimates? Faulty guesstimates in the first procedure are a likely source. Lower values of the survival fraction or the average lifespan are well within likely ranges. Just changing the estimate of the average lifespan from 7 to 6 years drops the estimate of DGRF to less than 0.21. (The average lifespan is the most sensitive number in the computation because it influences both birth and death rate factors.) And perhaps deer do not bear fawns until their third season? A second potential source of inaccuracies are omissions of factors in the process computation. One might easily forget, for example, that that some fraction of young ones do not make it through the winter, or that using a single level for the deer population aggregates fawns and adults, so that the fraction adult must be included in the birth rate computation. A third potential source of differences is the possibility that growth rate evident in the graph in figure 4.34 is less than the maximum possible growth rate, because of some factors not taken into con sideration. It is, after all, aggregate data from a multi-loop system, and we are using it to try to estimate a single number.

Parameters and Validity

Where possible it is preferable to estimate parameters below the level of aggregation in the model, as we did in the first example in this section, using knowledge of the process a parameter represents in the aggregate. Such parameter estimates tend to be relatively quick, they make use of the maximum information available about the system, and each estimate stands alone for scrutiny.

Estimating parameters from behavior, on the other hand, has two drawbacks. First, it requires making assumptions about equations in the model. In the deer population example, the second estimation procedure had to assume an equation for deer net growth rate of the form DNGR = DGRF*DP. More assumptions mean more potential sources for error. Better to limit assumptions as much as possible by estimating parameters without making use of equations in the model. Second, estimating from behavior interferes with an important test of model validity. The modeler and the client are fundamentally interested in whether the model generates the reference behavior modes of the study. If the parameters are estimated to fit model behavior to the reference modes, then no one gains any additional confidence in the model by noting that the fit exists. In contrast, if parameters are estimated one at a time from knowledge of underlying processes, and *then* the resulting model as a whole generates the reference behavior, everyone's confidence in the model justifiably increases. Further discussion of these issues is contained in section 5.4.

Estimating Parameters Statistically

Statisticians emphasize repeatedly that correlation coefficients do not imply causal connections. It is well known that jogging and healthy heart-lung systems are positively correlated, but it is also well known that the correlation does not imply that either one causes the other. But, even if a causal connection is assumed for other reasons, a correlation coefficient can give the wrong impression about the polarity of the connection.

The connection between bank loan interest rates and demand for loans provides an example. Since the interest rate amounts to the cost of borrowing, as the interest rate rises, there should be an inhibiting effect on the demand for loans, a negative link in a system dynamics model. Yet real data show a *positive* correlation between interest rates and the demand for loans. The reason for the apparent contradiction is that there is another causal-link between loan demand and interest rates: as the demand for funds rises, banks raise interest rates to maintain a balance between demand and the supply of loanable funds. Thus the correlation coefficient is positive because of this positive link in a negative feedback loop connecting interest rates and loan demand.

Similar confusions of polarity could occur with correlations computed in the context of any negative feedback loop. The problem is that correlations made from total system behavior do not measure ceteris paribus causal relationships. Nature tends not to hold everything else constant while we collect data for a correlation. Yet a single link assumed in a system dynamics model is a ceteris paribus assumption. The lesson is clear: be wary of correlational data in setting parameters.[18]

Multiple regression procedures to estimate a parameter can suffer from similar difficulties.[19] Procedures with the greatest promise for statistical estimation of parameters in system dynamics models appear to be those involving full-information maximum likelihood via optimal filtering.[20] DYNASTAT, a package of computer routines designed for DYNAMO models

---

[18] There are some large potential pitfalls involved in using real world, multi-loop data to determine parameter values involved in a single causal link. See, for example, Nordhaus (1973), pp. 1160-1163, and the response in Forrester, Low, and Mass (1974), pp. 171-177.

[19] For discussions of some of the problems of ordinary and generalized least-squares estimation for feedback models, see Senge (1980), Mass and Senge (1978), and Richardson (1981).

[20] See Peterson (1980).

[21] Pugh-Roberts Associates, Inc., 5 Lee St., Cambridge, Mass. 02139.

and capable of this sort of estimation, has recently been developed.[21]

Suffice to say, experience with feedback models will convince the reader that model behavior really *is* more a consequence of structure than parameter values. One should therefore be more concerned with developing the arts of conceptualization and formulation than finding ultimate parameter selection methods. Our advice for beginners would be to estimate parameters with good Statistics (data) but not statistics (mathematical methods). In the system dynamics context the latter are a collection of power tools that just might cut off your intuition.[22]

Initial Values of Levels

In selecting the initial values of the levels in a model, there are three general situations. The modeler must decide whether to match some historical situation, or to intialize the model in equilibrium or for some particular path of growth (or decline). Each of these presents its own particular set of problems.

The project model falls into the first category. It requires initial values for four explicit levels: the workforce, cumulative real progress, undiscovered rework, and the scheduled completion date. The fifth level in the model, implicit in the SMOOTHed equation for perceived progress, is automatically assigned its initial value by DYNAMO. The initial values ought to match the situation at the beginning of the work on a project. Therefore cumulative real progress and undiscovered rework should have initial values of zero, since no work would have been accomplished yet. The initial value chosen for the workforce depends upon whether we wish the model to pass through a start-up period in which the basic project work force is hired or whether we wish to assume the model runs begin with a full staff in place. For the base run shown at the end of section 4.4, an initial value of 2 was used, and the run began with the rapid hiring of a basic project workforce. The initial value of the scheduled completion date must reflect how long the project is initially assumed to take. Given the problem definition, a value of 40 months is appropriate. The initial value chosen for SCD should affect only

---

[22] For excellent examples of the use of data for a small but significant system dynamics model, see Meadows (1970). Data is assembled to select parameter values to fit a general commodity model to the four-year pork production cycle and the sixteen-year beef cycle.

Parameters and Initial Values 241

the length of the run, not the basic behavior of the model, since it represents a definition of the expected duration of the project.

The ease with which the initial values for the project model can be selected is deceptive. Frequently, initializing a model is annoyingly difficult and technically taxing. The first set of technical skills to acquire are those that enable a model to be started and run in equilibrium.

Initializing for Equilibrium

A large number of models, dealing with such things as inventory dynamics, business cycles, and predator-prey interactions, can be classified as equilibrium models, because their dynamics involve deviations from equilibrium conditions. One acquires the greatest understanding of such models by formulating them in equilibrium and then disturbing them to generate their particular dynamic patterns. Such analyses reveal the essential relationships between system structure and behavior. Once understood, those relationships can shed light on system behavior in the context of general growth or decline.

A model is in equilibrium when the inflows and outflows to each level balance. Each system level is unchanging, and as a consequence all model variables are constant over time. Information and material still flow throughout the system, but the result of the flows is no net change in any level, so all variables remain constant.

The key to initializing a model in equilibrium is the balancing of inflow and outflow rates for each level in the model in turn. Consider first the simple case of one level influenced by a single exogenous inflow rate and an outflow rate in the form LEVEL.K/LIFE:

```
L    LEV.K=LEV.J+DT*(IN.JK-OUT.JK)
N    LEV= ?
R    IN.KL=EXOG.K
R    OUT.KL=LEV.K/TCONST
```

where EXOG represents a constant or a variable determined as a function of time and TCONST is the time constant associated with the level LEV. To find the value of LEV that will put the model in equilibrium, set the outflow rate equal to the inflow rate and solve for the required value of LEV as follows:

```
OUT            = IN
LEV/TCONST     = EXOG
```

LEV   = EXOG*TCONST

Then, to set this simple structure in equilibrium initially, write the N equation for LEV:

N   LEV=EXOG*TCONST

In the N equation no timescripts are used, even if some of the quantities are variables, because there are no dynamics involved in the computation, just initial values. For a computation like this one, the modeler could do the arithmetic and type in a particular value in place of EXOG*TCONST, or he could type the line as shown to require DYNAMO to do the computation. The advantage of the latter is that the model would always begin in equilibrium for any values of EXOG and TCONST.

An extended example of this sort of equilibrium computation arises in a structure illustrated in section 4.1. Figure 4.4 showed an aging chain for an insect population in which the hatching, pupation, maturation, and death rates (HR, PR, MR, DR) were all expressed in the form LEVEL.K/LIFE. To set such an aging chain in equilibrium initially, we require

HR = PR = MR = DR,

that is,

EGGS/DES = LARV/DLS = PUP/DPS = ADLTS/TMD,

where the denominators represent the durations of the stages and the time from maturity to death. Picking values for the time constants and one of the level variables determines all the other initial values in the chain. If, for example,

N   ADLTS=100000,

then

N   PUP=DPS*ADLTS/TMD
N   LARV=DLS*ADLTS/TMD
N   EGGS=DES*ADLTS/TMD

put the model into initial equilibrium, provided the rate of depositing eggs equals ADLTS/TMD as well. The equations verify the common sense intuition that the stages that last longer have larger equilibrium populations.

A varient of the simple inventory model listed in section 3.6 provides an example of initialization in a slightly more

Parameters and Initial Values   243

complicated setting. In that model, goods are shipped out of
inventory, orders are placed to replace shipments and keep
inventory at its desired level, and those orders are received
into inventory after a production delay. The equations from
that model which are relevant to the discussion here are:

    R   SHIP.KL=NSHIP+TEST.K
    A   AVSHIP.K=SMOOTH(SHIP.JK,TAS)
    A   DSINV.K=AVSHIP.K*NIC
    A   INVADJ.K=(DSINV.K-INV.K)/IAT
    R   ORDRS.KL=AVSHIP.K+INVADJ.K
    R   ORDRCV.KL=DELAY3(ORDRS.JK,DEL)
    L   INV.K=INV.J+DT*(ORDRCV.JK-SHIP.JK)
    N   INV= ?

(Consult section 3.6 if you wish to see definitions of model
variables and the complete set of equations.) Here we have
modified the model slightly by formulating desired inventory
DSINV as a variable, a function of the average shipment rate
AVSHIP and a number of weeks of normal inventory coverage
NIC.
   To set the model in initial equilibrium, the inflows and outflows to the system levels must balance. In this case, that
requirement is equivalent to setting the three rates of flow
equal:

    ORDRS = SHIP = ORDRCV.

Since DELAY3 automatically sets ORDRCV = ORDRS at TIME =
0, we only need to consider ORDRS and SHIP. Their
equations imply

    AVSHIP + INVADJ = NSHIP + TEST.

Now consider what each of these ought to be in the initial
equilibrium situation. NSHIP is the normal shipment rate,
and the variable TEST is intended to be initially zero, until
any one of several exogenous disturbances are introduced.
Next, it is only reasonable to assume that the inventory
adjustment term INVADJ ought to be zero in equilibrium,
indicating that inventory is at its desired level.[22] Finally,

---

[22] If ORDRS were based only on INVADJ, then INVADJ could
not be zero in equilibrium. Thus inventory would not be
equal to desired inventory in equilibrium, an unhappy state
of affairs. When formulating rate equations, keep in mind the
implications for steady state conditions.

the SMOOTH function sets AVSHIP equal to SHIP initially. These relationships imply that the following hold in initial equilibrium:

```
SHIP    = NSHIP
ORDRS   = AVSHIP = SHIP
INV     = DSINV
```

Therefore the initial value for Inventory INV satisfies:

INV = DSINV = AVSHIP*NIC = SHIP*NIC = NSHIP*NIC.

We could set INV to any one of these. For simplicity, we shall use

    N   INV=DSINV

Because DYNAMO is very thorough and knows the initial values of *all* of its computations, one could actually use any one of the four initial value equations implied by the previous string of equalities. DYNAMO knows, for example, that if DSINV appears in an N equation, the value called for is the initial value of DSINV, and from other computations it knows that that value is NSHIP*NIC. A very forgiving language.

    As in the first example, the modeler could either type the N equation as a computation as shown above, or compute by hand and write

    N   INV=300

(or whatever INV figures out to be, given selected values of NSHIP and NIC). Again, the advantage of the formula is that it places the model in equilibrium for any values NSHIP and NIC, in case they might be changed in a rerun.

    The simple prison population model used as an illustration in section 4.3 shows an initialization in which the level variable is picked in advance and the parameter determining a *rate* is computed in an N equation to start the model in equilibrium. The scheme eliminates the need to know what value a particular table function will produce initially. The relevant equations are

```
L    PP.K=PP.J+DT*(IR.JK-RR.JK)
N    PP=1000
R    IR.KL=NIR+STEP(SH,ST)
N    NIR= ?
R    RR.KL=PP.K/ATS.K
```

Parameters and Initial Values

ATS, average time served, was computed as the product of a constant average sentence length and a variable effect of capacity on time served, expressed as a table function of the ratio of the prison population to prison capacity. (Consult section 4.3 for details, if necessary.) The model will be in equilibrium if

   IR = RR.

Substituting,

   NIR + STEP(SH,ST) = PP/ATS or
   NIR = PP/ATS,

since the initial value of the STEP will be zero.

Therefore the model will initially be in equilibrium if the constant normal incarceration rate NIR is computed in an N equation as the initial value of the prison population PP divided by the initial value of the average time served ATS. Note that since ATS is a rather complicated variable, it might be difficult to find its equilibrium value by hand. It would require computing the initial value of the capacity ratio and looking up the corresponding value of ECTS in the table for the effect of capacity on time served, perhaps having to perform linear interpolation in the process. Fortunately, DYNAMO simply traces through the model, computes the initial value of ATS, and uses its value to compute the initial value of NIR as PP/ATS. The model is automatically started off in equilibrium.

For more complex models the process of initializing for equilibrium is the same as in these examples, repeated over and over, tracing values around the model. With many level variables, however, or with more than one inflow and outflow rate for some of the levels, the process can be quite difficult. When all else fails, one can run the model until equilibrium results, printing out the values of the levels to determine those needed for equilibrium. Such an approximation is not as desirable as an analytic procedure, partly because a change in any parameter requires reinitializing the model, partly because one learns more about the model by analyzing the equilibrium condition, and partly because it's not very esthetically pleasing.

Initializing for Growth

Some problems do not lend themselves to simple steady state initialization. The dynamics of personnel management and capacity acquisition for a growth company would undoubtedly require a model initialized in a growth mode. Perhaps the initial conditions should be formulated to start the model off in steady linear or exponential growth. Such conditions are difficult to formulate. The equations for hiring and orders for production capacity, for example, must contain explicit growth terms (or alternatively, the levels for personnel and capacity must differ from their desired values). Understanding the process of initializing for growth will yield some insight into the subtleties involved.

Suppose the reference mode for a study shows population initially growing at 2.3 percent per year, and you wish to initialize the model to match that as closely as possible. A simple identity is the required key:

(Inflows - Outflows)/Level = Fractional change

Consider a simple formulation in which the birth rate is a birth rate factor BRF.K times population POP and the death rate is POP divided by average lifespan:

```
L   POP.K=POPL.J+DT*(BRTH.JK-DTHS.JK)
R   BRTH.KL=BRF*POP.K
R   DTHS.KL=POP.K/ALIFE.K
```

The identity becomes

(BRF*POP - POP/ALIFE)/POP = FC

which simplifies to BRF - 1/ALIFE = FC. FC represents the fractional change in the level of population. Thus if ALIFE is known, we could write

```
N   BRF=FC+(1/ALIFE)
C   FC=.023
```

to set population growing initially at 2.3 percent per year.

In the population example the level POP dropped out of the computations. Frequently, it does not. Consider the problem of initializing a workforce to grow initially at, say, 30 percent per month, and suppose the rate of change of the workforce level is written as a single net hiring rate HR just as in the project model. The fractional change identity becomes

HR/WF = FC.

Substituting,

Parameters and Initial Values   247

[(WFS - WF)/WFAT]/WF = FC.

Solving for WF produces the initial value of the workforce desired:

WF = WFS/(FC + 1/WFAT).

Writing this expression in an N equation for WF will initialize the workforce so that it begins every model run growing at the fractional rate determined by FC. For a 30 percent per month growth rate, set FC = 0.3. For a 10 percent per month decline, set FC = -0.1.

The fractional change identity also holds if FC = 0. The condition in that case guarantees equilibrium. One can actually initialize a model using one or more fractional growth constants FC1, FC2, ..., and then be able to start a model either in equilibrium (FC1 = FC2 = ... = 0) or with various levels growing at any desired rates, simply by changing the values of the FC's in reruns.

The population example at the beginning of this section has some interesting implications when FC = 0. In order for the population system to be in equilibrium, FC = 0, and the equilibrium condition must be

BFR = 1/ALIFE;

the birth rate factor must be the reciprocal of the average lifespan. Therefore, when the population of an area, a country, or the globe comes into equilibrium, as it eventually must (if only momentarily before a decline), this condition must hold. Current global birth rates of 4 percent to 5 percent annually imply an equilibrium average lifespan of 1/0.04 or 1/0.05--a mere 25 to 20 years. One might rather hope for an equilibrium in which the fractional birth rate is between 1 percent and 2 percent, but it is not clear that hope is enough.

4.7 DEBUGGING

Few models run perfectly the first time they are simulated. No one really expects them to. Models are developed in pieces and in an iterative process that seeks to improve model formulation on a number of successive passes. But even the pieces or the stages seldom run perfectly the first time because of coding errors--bugs--in the DYNAMO equations in the model. Finding bugs in DYNAMO models--a process known to computer buffs as debugging--is the subject of this section. After a brief description of DYNAMO's error- checking routines, the section focuses on an important class of model flaws known as simultaneous equations, and ends with a discussion of finding errors that DYNAMO can not detect.

The processes of finding and correcting errors is facilitated by the extensive error-checking capabilities of DYNAMO. The language can flag more kinds of errors than most of us can make, and it can tell us about them in annoyingly pointed English. In most error messages DYNAMO prints out the offending line and marks with a "V" the probable location of the bug. (Truly creative errors can confuse it, however, and can occasionally produce misplaced accusations or messages that don't accurately describe the real mistake.) Some errors, such as simultaneous equations, are so well handled that in complex models the user is better off leaving the checking and identification of such bugs entirely to DYNAMO.

Three Classes of Errors

DYNAMO divides the errors it finds in a model into three classifications: warnings, serious errors, and fatal errors. Warnings identify errors that DYNAMO can recognize but are not likely to affect the simulation results. A ".J" was written as a ".K," for example. DYNAMO attempts to think for itself about an error rating a warning, substitutes what it thinks the user really intended, and runs the model. Usually, that works out very well. Serious errors, on the other hand, will probably affect the results of the simulation. In time-sharing systems DYNAMO halts the simulation if a serious error turns up. In batch systems DYNAMO will run the simulation anyway, hoping that the user can get something out of the run in return for all the time spent setting it up. Fatal errors are fatal to the run, usually not the user. DYNAMO is unable to recover from fatal errors. The simulation is aborted before the model run begins.

To find errors in a model DYNAMO searches through the list of equations three times. On the first pass, it hunts for errors in the form of equations. Timescripts are one target. One might type a ".K" in the wrong place by accident or in an attempt to force DYNAMO to compute strangely. Such a transgression rates a warning. Algebraic misformulations are also noted, such as parentheses that do not pair up properly or ambiguous expressions such as A/B+C. For the latter DYNAMO warns that it will compute the expression as if it were written (A/B)+C and proceeds.

On the second pass, DYNAMO uncovers such things as quantities used in the model but not defined, quantities defined but not used, and level equations without initial values. The first and third of these are serious errors, while the second rates only a warning informing the user of what might be an oversight.

The third pass locates simultaneous auxiliary equations and simultaneous initial value (N) equations. Both of these kinds of errors are fatal, but simultaneous auxiliaries are more significant to the modeler because they indicate a conceptual error.

If a model makes it through the three passes without serious or fatal flaws, it is simulated, but it is not yet free from scrutiny. DYNAMO flags a variety of run-time errors, such as the wrong number of entries in a table, division by zero, floating point overflows (numbers bigger than the largest expressible in the computer's arithmetic), impossible computations such as square roots or logs of negative numbers, and so on. The error messages for such flaws and the necessary cures are reasonably clear.

The final chapters of the *DYNAMO User's Manual* and the *Mini-DYNAMO User's Guide* discuss every DYNAMO error message and the unusual error conditions that might produce them.[23] Rather than amble through a general treatment of errors, we prefer to focus here on one important class of errors--simultaneous equations.

Simultaneous Equations

We have observed that DYNAMO computes initial values of all variables whose initial values are not specified, including initial values of the levels implicit in smoothing and delay formulations. Computing the initial values of the variables in a feedback loop requires starting with a known value and computing successive values around the loop. If it happens that no variable in the loop has an assigned or computable initial value, then none of the initial values in the loop can be asigned. Such a situation commonly occurs in a loop containing nothing but auxiliaries and one or more SMOOTHs or DELAYs to be initialized by DYNAMO. DYNAMO reports "simultaneous N equations" and lists the variables involved.

A related but more serious error generates the message "simultaneous active equations." If a chain of auxiliary equations forms a feedback loop without a level variable interposed somewhere, the auxiliaries in the loop are termed simultaneous. Each is supposed to be computed in the present (.K) from predecessors in the loop. Yet computing in such a fashion around the loop would eventually require com-

---

[23] See Pugh (1976) and Shaffer (1979).

puting an auxiliary as a function of itself, in the present. Thus simultaneous auxiliaries create an impossible computation task for DYNAMO (and for us if we were to do the work by hand). Their existence in a model suggests any one of several conceptual and technical mistakes that require serious attention.

In conceptualizing and formulating a model, the modeler should always be on the look out for closed loops of active equations that do not contain at least one level or SMOOTH. Besides the computational problem they create, they raise the conceptual (and philosophical) questions of whether loops in the real system behave like sets of simultaneous equations. But in the case of N equations it may be difficult to anticipate simultaneities. Some initial values are embedded in the DELAY and SMOOTH functions and others are created by DYNAMO from auxiliary and rate equations. It is vastly easier, and not irresponsible, to worry about the problem of simultaneous initial values only after DYNAMO has explicitly identified it.

An Example of Simultaneous Initial Value (N) Equations

Consider the simple inventory model discussed in sections 3.6 and 4.6. As some of the tests in section 3.6 revealed, it is missing a feedback link that would restrain shipments as inventory heads toward zero. The inventory level in that model can become negative. Suppose that a modeler, ignoring our cautions on using the MIN function in section 3.5, tries to cure the problem by replacing

    A   SHIP.K=NSHIP+TEST.K

with

    A   SHIP.K=MIN(NSHIP+TEST.K,INV.K/DT)

(See figure 4.35 for a complete listing of the model with this addition.)

We still do not advocate it, but this new expression will indeed prevent SHIP from exhausting the inventory level INV, and the modeler has even sidestepped an earlier criticism we made. By dividing by DT, he has ensured that the units of SHIP are always widgets/week rather than just widgets. But a problem still exists, and it is not really the fault of the MIN or the unusual use of DT. When the model is run, we receive an error message:

Debugging

```
*  REVISED INVENTORY MODEL WITH INITIALIZATION ERROR
NOTE
NOTE       SHIPMENTS
NOTE
R   SHIP.KL=MIN(NSHIP+TEST.K,INV.K/DT)   SHIPMENT RATE (UNITS/WK)
C       NSHIP=100
A   TEST.K=TEST1*STEP(STH,STRT)+
X       TFST2*RAMP(SLP,STRT)+
X       TEST3*PULSE(HGTH,STRT,INTVL)+
X       TEST4*AMP*SIN(6.283*TIME.K/PER)+
X       TEST5*RANGE*NOISE()
C       TEST1=0/TEST2=0/TEST3=0/TEST4=0/TEST5=0
C       STH=10,STRT=2,SLP=20,HGTH=10,INTVL=200,AMP=10,
X       PER=5,RANGE=20
L   INV.K=INV.J+DT*(ORDRCV.JK-SHIP.JK)
N      INV=DSINV                        INVENTORY (UNITS)
R   ORDRCV.KL=DELAY3(ORDRS.JK,DEL)   ORDERS RECEIVED (UNITS/WK)
C       DEL=3 (WKS)                   DELAY IN RECEIVING ORDERS
NOTE
NOTE       ORDERS
NOTE
R   ORDRS.KL=AVSHIP.K+INVADJ.K       ORDERS PLACED (UNITS/WK)
A   AVSHIP.K=SMOOTH(SHIP.JK,TAS)     AVG SHIPMENT RATE (UNITS/WK)
C       TAS=2 (WKS)                   TIME TO AVERAGE SHIPMENTS
A   INVADJ.K=(DSINV-INV.K)/IAT       INVENTORY ADJUSTMENT (UNITS/WK)
C       IAT=2 (WKS)                   INVENTORY ADJUSTMENT TIME
A   DSINV.K=DIC*AVSHIP.K             DESIRED INVENTORY (UNITS)
C       DIC=3 (WKS)                   DESIRED INVENTORY COVERAGE
NOTE
NOTE       CONTROL STATEMENTS
NOTE
SPEC    DT=.25/LENGTH=0/PRTPER=0/PLTPER=.5
PRINT SHIP,TEST,INV,ORDRCV,ORDRS,AVSHIP,INVADJ
PLOT  SHIP=S/INV=I
```

Figure 4.35: Equations for the simple inventory control model, with attempted reformulation of the shipment rate equation.

SIMULTANEOUS EQNS IN N EQNS
INVOLVED ARE: SHIP, AVSHIP#4, AVSHIP, DSINV, INV

Comparing the variables involved with the listing, we discover that only one has an explicit N equation:

N   INV=DSINV

So to compute the initial value of INV, DYNAMO moves to DSINV and interprets its auxiliary equation as an N equation:

N   DSINV=DIC*AVSHIP

Thus the initial value of DSINV depends on a constant, DIC, and AVSHIP, another auxiliary without an N equation. But AVSHIP is computed by the SMOOTH function, which includes an N equation setting the smoothed value equal to the variable being smoothed:

N   AVSHIP=SHIP

Again we find a variable without an N equation, the rate SHIP. As with auxiliaries, DYNAMO repeats the active equation as an N equation:

N   SHIP=MIN(NSHIP+TEST,INV/DT).

Thus, the initial value computations would form the string:

N   INV = DSINV = DIC*AVSHIP = DIC*SHIP
       = DIC*MIN(NSHIP+TEST,INV/DT).

That is,

N   INV = f(INV).

Since the initial value of INV depends on itself, we have a loop of simultaneous N equations.

(The list of variables in DYNAMO's error message includes a strange variable with a "#" as part of its spelling, AVSHIP#4. The variable was created by DYNAMO to assist in generating the SMOOTH function for AVSHIP. The message advertises the fact that that variable internal to the SMOOTH function is part of the loop of variables that cannot be assigned initial values without additional information. See section 7.2 for details about variables like AVSHIP#4.)

The simplest way to cure the problem of simultaneous N equations is to pick one of the variables in the loop and assign it an appropriate initial value in a separate N equation. For example, we could add

N   INV=1000.

Alternatively, we could initialize one of the variables in terms of other variables, for example,

N   SHIP=NSHIP+TEST,

under the assumption that NSHIP and TEST are assigned their initial values in ways that do not depend on quantities in our troublesome loop. Either of these would serve to start

Debugging

off the chain of initial value computations without confusion. The model would then be able to compute successive values of all the variables over time: INV.K would be computed from SHIP.JK, and SHIP.KL would then be computed from INV.K, and so on through time.

An Example of Simultaneous Active Equations

Curing simultaneous N equations requires little conceptual effort. However, when a simultaneity persists in the model even after initial conditions have been properly specified, the modeler has to do some thinking. To illustrate the notion of simultaneous active equations, and to show how they are likely to arise in ordinary modeling efforts, we shall invent a likely formulation to add to the project model. Although technically flawed, the structure we will add is intended to capture a very real component of project overruns.

One of the hypotheses about the source of overruns in large projects is that the people involved fail to realize the size of the task at the beginning, and they maintain a somewhat distorted view until near the end of the project. To add that hypothesis to our model requires rethinking the role of the quantity called the initial project definition IPD.

Let us introduce two new concepts: a final project definition FPD (a constant), and a variable current project definition CPD. Like IPD, each of these is phrased as a number of tasks to be completed in the project. The final project definition is not knowable to the actors in the system, but they approach a more and more accurate picture of it toward the end of the project. One way to formulate the current project definition is as a weighted average of the initial and the final project definitions:

A   CPD.K=WTFPD.K*FPD+(1-WTFPD.K)*IPD
NOTE   CURRENT PROJECT DEFINITION (TASKS)

As in the formulations for WTRP and WCWF, it makes sense to express WTFPD, the weight given the true project definition, as a function of the fraction perceived completed. At the beginning of the project, when FPCOMP is near zero, CPD would be dominated by the initial project definition (which would presumably be a good deal smaller than the final project definition). As project managers perceived that the project was nearing completion, they would have to continually revise their definition of the project as more unanticipated tasks came to light. Hence a likely formulation is the table function shown in figure 4.36, represented in DYNAMO by

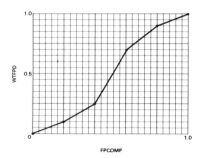

Figure 4.36: Table function for the weight given the final project definition, used in the formulation of the current project definition

```
A     WTFPD.K=TABLE(TWTFPD,FPCOMP.K,0,1,.2)
NOTE      WEIGHT GIVEN THE FINAL PROJECT
NOTE      DEFINITION
T     TWTFPD=0/.1/.25/.7/.9/1
NOTE      TABLE FOR WTFPD
```

Completing the changes necessary is a new equation for the fraction perceived completed, using the current project definition instead of the initial project definition used originally:

```
A     FPCOMP.K=CPPRG.K/CPD.K
NOTE      FRACTION PERCEIVED COMPLETED
```
where
   CPPRG = cumulative perceived progress (tasks).

Summarized in brief, the proposed changes are as follows:
Add:

```
A     CPD.K=WTFPD.K*FPD+(1-WTFPD.K)*IPD
A     WTFPD.K=TABLE(TWTFPD,FPCOMP.K,0,1,.2)
T     TWTFPD=0/.1/.25/.7/.9/1
C     FPD=1200
```

Change:

	From	To
A	FPCOMP.K=CPPRG.K/IPD	FPCOMP.K=CPPRG.K/CPD.K
A	FCOMP.K=CRPRG.K/IPD	FCOMP.K=CRPRG.K/FPD
C	IPD=1200	IPD=800
A	EPREM=(IPD-CPPRG.K)/ PPROD.K	EPREM.K=(CPD.K-CPPRG.K)/ PPROD.K

The reader should try to find the flaw(s) in these new formulations before reading on.

Running the project model with these changes prompts DYNAMO to respond[26]

```
SIMULTANEOUS EQNS IN A EQNS
    INVOLVED ARE WTFPD#2,WTFPD,CPD,FPCOMP
SIMULTANEOUS EQNS IN N EQNS
    INVOLVED ARE WTFPD#2,WTFPD,CPD,FPCOMP
```

Studying the variables involved in the simultaneous equations, we see the reason for the error message. To compute the fraction perceived complete, FPCOMP, one needs to know CPD. But to compute CPD, WTFPD is necessary, and that depends upon FPCOMP. The model thus contains a closed loop of variables without any intervening level.

Three cures are possible. They apply in different situations and, unlike simultaneous N equations, they require a reconceptualization of the faulty loop.

First, a level can be incorporated in the loop of simultaneous auxiliaries. Was some accumulation or smoothing process in the loop overlooked? In our case that is a possibility, since it could easily be argued that the current project definition passes the snapshot test for levels, tending to persist through time and adjust moderately slowly. CPD might be formulated as a SMOOTH of the weighted sum of FPD and IPD.

Second, the simultaneity might be solved algebraically to create several auxiliaries without feedback. Is the feedback process the loop tries to capture really a part of the system? If so, does it react too quickly to justify the inclusion of a level in the loop? If the answer to both questions is yes, one might try replacing the loop with a string of equivalent auxiliaries. In our case the table function makes this nearly impossible. However, if we felt it was justified, we might approximate the weighting factor WTFPD by a straight line function (WTFPD = FPCOMP) from 0 to 1 to produce a simple formula for CPD, namely, CPD = FPD*FPCOMP + (1-FPCOMP)*IPD. The simple formula would allow us to substitute for CPD in the auxiliary equation for FPCOMP

---

[26] Because of limitations imposed by small computers, Mini-DYNAMO reports not just the variables in the simultaneous loop but also all those variables connected to them without an intervening level. To find the loop, the user must refer to a flow diagram of the variables listed.

FPCOMP = CPPRG/[FPD*FPCOMP + (1-FPCOMP)*IPD]

and solve for FPCOMP to produce a non-linear auxiliary equation for FPCOMP in terms of CPPRG, IPD, and FPD. The result would have the essential characteristics the feedback loop of auxiliaries was trying to capture. (The goal of the algebraic process just given is the expression of FPCOMP as a function of the variable CPPRG, cumulative perceived progress. Realizing that in advance, one could instead sketch FPCOMP as a table function of CPPRG which makes FPCOMP grow from 0 to 1 as CPPRG increases from 0 to the value of FPD, the final project definition. A table function for FPCOMP provides greater latitude for the functional specification.)

The third cure for simultaneous equations is to reconsider the variables involved in the formulation. Can some equation in our loop be reformulated in terms of variables that do not depend on it? Substituting FCOMP, fraction complete, for FPCOMP in the equation for WTFPD is one candidate. The original formulation states that the estimate of project size depends on the *perceived* fraction complete. The assumption was that project personnnel cannot believe that they are entering the final phases of a project without realizing the true magnitude of the job. The alternative formulation with *true* fraction complete states that progress toward the real goal dictates one's ability to see that goal. It assumes that project personnel could believe they are done when, in fact, they are not. In the early phases of a project the first formulation may be more defensible; in latter (apparent) phases the second may make more sense. Since neither argument is overwhelming, we might chose the reformulation because it eliminated a nasty simultaneity problem.

In deciding among these three alternatives, focus more on conceptualization than technique. Solve the problem of simultaneous active equations by adding levels or manipulating algebra only if a serious reconsideration of the real system suggests that is the appropriate structure.

Finding Bugs DYNAMO Can't

Suppose the first simulation of the project model had pro-

---

[28] To save space, the run shown in figure 4.37 has been shortened from what it might have been by setting the maximum length, MAXLEN, equal to 40. With MAXLEN equal to its original value of 100 months, the simulation shows exactly the same patterns merely drawn out to greater extremes.

Debugging

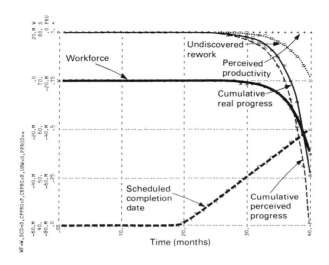

Figure 4.37: Simulation of the project model containing an undetected error

duced output like that shown in figure 4.37.[28] The graphs have nothing to do with the reference modes for the study, and there are negative ranges in some of the scales. The workforce, for example, apparently becomes negative somewhere in the simulation. Note also that some of the scale numbers are enormous: the scaling letters T and M represent thousands and millions in DYNAMO--hardly appropriate ranges for the fraction perceived completed, the workforce, and cumulative progress. Clearly, there must be errors, yet the only error messages from DYNAMO are three notifications that table ranges have been exceeded:

```
BELOW RANGE OF TABLE
    AT TIME  .75000   IN EQUATION FOR  WTRP
BELOW RANGE OF TABLE
    AT TIME  .75000   IN EQUATION FOR  TDRW
```

Perhaps the most useful strategy in such a situation is to shorten LENGTH and *print* values of the variables *every DT*, in order to see in slowest possible detail what the model is doing. Figure 4.38 shows the output of the model when the following instructions were typed in rerun mode:

```
C MAXLEN=3
```

TIME	WF	HR	SCD	CPPRG	CRPRG	URW	FPCOMP
E+00	E+00	E+00	E+00	E+00	E+00	E+00	E-03
.0	2.000	-9.333	40.000	.000	.000	.000	.000
.25	-.333	-10.170	40.000	.500	.350	.150	.417
.5	-2.876	-11.082	40.000	.414	.292	.122	.345
.75	-5.646	-12.076	40.000	-.308	-.212	-.096	-.257
1.	-8.665	-13.160	40.000	-1.717	-1.200	-.518	-1.431
1.25	-11.955	-14.341	40.000	-3.873	-2.716	-1.157	-3.227
1.5	-15.540	-15.629	40.000	-6.838	-4.808	-2.029	-5.698
1.75	-19.448	-17.033	40.000	-10.680	-7.528	-3.153	-8.900
2.	-23.706	-18.564	40.000	-15.477	-10.931	-4.546	-12.897
2.25	-28.347	-20.233	40.000	-21.308	-15.080	-6.229	-17.757
2.5	-33.405	-22.053	40.000	-28.265	-20.040	-8.225	-23.554
2.75	-38.918	-24.037	40.000	-36.445	-25.886	-10.559	-30.371
3.	-44.928	-26.201	40.000	-45.955	-32.697	-13.258	-38.296

Figure 4.38: Printing every DT to locate bugs in the project model.

```
C   PRTPER=.25
C   PLTPER=0
PRINT WF,HR,SCD,CPPRG,CRPRG,URW,FPCOMP
```

A first glance at the printed output reveals that the hiring rate HR is negative from the very beginning of the run! People are being let go, even though there are only two people in the workforce to begin with and there is a lot of work for them to do in the next 40 months. In the next DT, the workforce has been pulled negative, and the hiring rate becomes even more negative. The next variables to go distinctly awry are cumulative perceived progress CPPRG, cumulative real progress CRPRG, undiscovered rework URW, and fraction perceived completed FPCOMP. All rise a bit at first and then drop below zero after three computation intervals. Recalling that apparent progress in the model is a product involving the workforce, we can reason that progress accumulates positive increments as long as the workforce is positive and accumulates negatives when WF becomes negative. The bug (if there is only one) definitely appears to be in the hiring and workforce area of the model.

So the modeler would focus on the hiring sector of the model, which in this flawed version of the project model contains the following equations:

```
NOTE        HIRING
NOTE
L    WF.K=WF.J+DT*HR.JK
N    WF=WFN
C    WFN=2
NOTE        WORKFORCE (PEOPLE)
```

Debugging

```
R    HR.KL=WCWF.K*(WF.K-WFS.K)/WFAT
NOTE     HIRING RATE (PEOPLE/MONTH)
C    WFAT=3
NOTE     WORKFORCE ADJUSTMENT TIME
A    WFS.K=WCWF.K*IWF.K+(1-WCWF.K)*WF.K
NOTE     WORKFORCE SOUGHT (PEOPLE)
A    WCWF.K=TABLE(TWCWF,FPCOMP.K,0,1,0.1)
NOTE     WILLINGNESS TO CHANGE WORKFORCE
T    TWCWF=0/0/0/.1/.3/.7/.9/1
NOTE     TABLE FOR WCWF
A    IWF.K=EPREM.K/TREM.K
NOTE     INDICATED WORKFORCE (PEOPLE)
```

where EPREM represents the man-months of effort perceived remaining and TREM stands for time remaining. So where might the error(s) be?

The negative hiring rate at TIME = 0 suggests something is wrong in the HR equation. Staring at the hiring rate equation long enough should reveal that the factor involving WF and WFS was intended to be in the form

(GOAL.K-LEVEL.K)/ADJTM,

but was erroneously written as (LEVEL.K-GOAL.K)/ADJTM. The reversal of terms reverses the sign of the hiring rate, so negatives result. The sign change changes the polarity of the hiring loop from negative to positive: a gap between and WFS generates negative hiring (transfers?) which feeds back to increase the gap. A serious bug has been found. In this particular case it is the only bug in the model; when the equation is rewritten correctly, the model is bug-free and produces the output shown at the end of section 4.4. In buggier situations, one might have to correct the error and repeat the process to find still more errors.

The example illustrates the potential power of printing model variables every DT to understand model quirks. Occasionally, model misbehavior might not occur near the beginning of the run, and printing every DT out as far as necessary would consume huge amounts of time and paper. In such a case, write the print period PRTPER as a *variable*, which is zero for a while and then becomes equal to DT during the time interval of interest. The following auxiliary equation does the trick:

```
A    PRTPER.K=STEP(DT,PSTART)
```

where PSTART is the print starting time. Assign a value to PSTART in a C statement, and set LENGTH to halt the simulation after an appropriate number of lines have been printed out.

Common Bugs

   A number of typing errors can cause bugs in a model that are hard to find. Beware of the following:

1. Blanks within column 1: Some DYNAMOs allow the user to type a blank in the first column in place of the word NOTE, to signal that what follows in the line is to be ignored in the computations of a simulation. An accidental blank in column 1 can therefore wipe an equation out of the model.

2. Oh's (O) instead of zero's: Some terminals and teletypes make very little distinction between the letter O and the digit 0.

3. Blanks in a line: Everything following a space in a DYNAMO line is ignored in the computations, assuming whatever follows the space is a comment. A space in the middle of an equation chops off everything following it.

4. Line too long: Everything beyond column 72 is ignored, but some computer systems will not report that a line is too long. Use continuation statements (X) for lengthy equations.

One other general class of error deserves mention: taking too large a value for DT. A value of DT less than one-half the smallest time constant in a model usually guarantees that DT will not influence model behavior. Sometimes the time constant associated with a level is not obvious, however, and the modeler may occasionally take too large a value for DT. Symptoms of a DT too large include unexpected oscillations, oscillations increasing in amplitude, and jerky behavior overtime. The wise modeler checks the value by trying a simulation with a value one-half or one-quarter as big, to see if changes in model behavior result. If they do, then a smaller DT is called for.

Summary of Debugging Techniques

   DYNAMO's error messages and the technique of printing every DT are the most reliable debugging tools available. But there is a third tool that should not be ignored in the debugging process: documentation. Be sure to annotate and save every model run and every model change made in the debugging stage, at least until the model behaves properly. Bugs have a way of breeding more bugs before they are finally eliminated. The wise exterminator keeps a record of the nests encountered.

This section has focused mainly on finding technical flaws in a model. The more subtle arts involved in exploring a model for conceptual flaws will be discussed in chapter 5.

4.8 PRINCIPLES OF MODEL FORMULATION

This chapter has focused on the formulation of equations for simulating a dynamic feedback system over time. The discussion has been largely technical, under the assumption that prior to the formulation stage a problem had been well-defined and an appropriate structure conceptualized. Embedded within the examples and discussions throughout this chapter are a number of biases about what constitutes good system dynamics modeling practice. As a summary for the chapter, this section lists those biases as a set of principles for system dynamics modeling.[29]

The principles that follow are intended as quick guidelines for system dynamics model formulation. At one level they may be regarded as a set of technical do's and don'ts, rigid proscriptions for acceptable models. A more enlightened view will see each as a modeling aid. Some flag potential technical mistakes. Others crystalize philosophical assumptions. Some can, in fact, be violated (we will try to note the likely ones). But as with most violations of good practice, the violator must be aware of the principle, aware of the affront, and able to justify it in that particular modeling situation. It takes a good acrobat to be a clown.

The section is organized as a list of principles, each followed by one or more paragraphs of brief explanatory comments and references to earlier sections. The principles should look familiar, even obvious, or should trigger a feeling of the sudden recognition of an understanding that has remained unconscious and implicit to this point. Those principles that seem at all surprising deserve most of the reader's attention.

---

[29] The original presentation of many of the prinicples in this section is in Forrester (1968b).

Principles of Model Structure

Conserved Flows and Information Links

- All level variables represent accumulations. They can be changed only by moving their contents between levels, sources or sinks.

System dynamics models distinguish carefully between physical flows and information links. Physical quantities must be conserved: widgets can not be shipped out, for example, without being produced or without depleting a stock of widgets. A conserved flow from A to B must necessarily decrease A and increase B. In contrast,

- Information is not a conserved flow. Information from a single source can be transmitted to other variables in the system without diminishing the source.

The necessary distinction between conserved flows and information flows gives rise to the solid and dotted lines in flow diagrams. Rate and auxiliary variables, representing information in the system, are not conserved through time as level variables are.

A "conserved subsystem" comprises the entire flow of *one kind* of material, the rates that control the flow, and all the levels in which it accumulates. The units in every level in a conserved subsystem are the same. (Figure 2.11 shows an example of a large conserved subsystem; figure 4.10 shows a highly aggregated one.)

- In any conserved flow subsystem, the rate and level variables must alternate.

This principle is implicitly reflected in aging chains such as figure 4.4. It is related to the principle that

- Levels are changed only by rates.

No level directly affects the value of another level. A level is computed from its past value and the rates of flow affecting it over the intervening time interval. The present value of a

---

[30] Occasionally, for convenience, a level equation is written in terms of other levels, but such a formulation always implicitly embodies a rate equation that can be written separately.

Principles of Model Formulation

level can be computed without the present or previous values of any other level variables.[30]

- Only information links can connect between conserved subsystems.

No physical quantity can flow between different conserved subsystems: a workforce can control (determine) the flow of products into an inventory, but people do not flow into (become) widgets in inventory. Thus it is not material flows but rather information about levels in one conserved subsystem that can control rates rates of flow in other subsystems.

- Rates depend, in principle, only on levels and constants.

The only inputs to rates are information links, and all information can be traced ultimately back to the system levels and parameters. Auxiliary variables organize pieces of information for output or for convenience in writing rate equations; in principle a model could be written without them (but hardly anyone should ever want to prove it in practice). This modeling principle is a consequence of the premise that

- Most rates are not instantaneously knowable by most actors in the system. There can be, in principle, no rate-to-rate connections in a model.

To be used as information affecting other subsystems, a rate of flow should be averaged over a period of time. No rate can, in principle, control another rate without an intervening level. Examples reflecting this principle appeared in figures 4.11 and 4.12.

Two exceptions to the principle of no rate-to-rate connections should be noted. In bookkeeping situations a physical rate frequently determines directly a rate of change of some paper quantity. The rates of change affecting a book value of inventory, for example, must depend directly on the rates of flow increasing and decreasing actual inventory. Or in a model of an economy, a hiring rate could simultaneously increase the level of employed people while decreasing the level of job vacancies. The second exception is more pragmatically based: in practice a rate may be expressed directly in terms of another rate when the time constant of the intervening levels are very small relative to the other time constants in the system. (It is because of the no rate-to-rate principle that DYNAMO will respond with an error message to a rate equation in the form

R   RATE2.KL=RATE1.KL.

DYNAMO will accept

R   RATE2.KL=RATE1.JK,

setting RATE2 equal to the prevous value of RATE1 but the equation should not be deliberately used to introduce a delay of length DT. The solution interval DT is a simulation parameter having no real system meaning whatsoever.) Necessary and Sufficient Structure

- Every feedback loop in a model must contain at least one level.

Without a level, rate-to-rate connections or simultaneous auxiliaries result. Simultaneities create philosophical problems of causal sequence as well as impossible computation tasks.

- Level variables completely describe the system condition.

The values of auxiliaries and rates can be traced back to be expressed in terms of the system levels. Once the values of constants and the initial values of level variables are assigned, all other variables can be computed and the system simulated over time.

Principles of Model Building

Units

- In every model equation the units of measure must be consistent. Terms added together or subtracted must have identical units.

Dimensional analysis cannot prove an equation is correct, but it can certainly prove some equations to be incorrect. (See sections 4.1, 4.2, and 4.4 for examples.)

- Levels and rates cannot be distinguished by their units of measure.

The fact that a variable has units such as widgets per month does not guarantee that the variable should be formulated in a rate equation. The variable might be the recent average shipment rate, involving an accumulation (summing) process that suggests the variable is appropriately modeled as a

Principles of Model Formulation

level. In the project model formulated in section 4.4, there are levels with units as diverse as people (the workforce), tasks per person per month (perceived productivity), and even time itself--months (the scheduled completion date).

- Within any subsystem of conserved flows, all levels have the same units of measure and all rates are measured in those same units divided by time.

Parameters

- Like every variable in a model, every parameter should have a meaningful interpretation or counterpart in the real system.

Parameters should have recognizable meanings to people closest to the real system. Never invoke meaningless proportionality coefficients and conversion factors solely to adjust dimensions.

Decisions

- All rate equations should make sense even under extreme or unlikely conditions.

Rates represent decisions for action and change. They must be robust, capable of responding reasonably in the face of strange, unanticipated values of the inputs to the decision. Implausible model behavior can never be justified as a consequence of the model operating outside its normal operating region.

- Distinguish between desired conditions and actual conditions.

This principle helps to ensure the observance of the previous one. Inventory might be so high, for example, that managers might desire negative production to reduce inventory, but the actual condition could obviously go no lower than zero.

- Distinguish between actual conditions and perceived conditions.

If there is reason to believe that actual and perceived conditions may be different, decisions for action can only be based on perceptions, not the real conditions themselves. Rate equations representing decisions must take account of whatever perception delays and bias exist in the real system.

Hiring and scheduling decisions in the project model, for example, were based upon effort *perceived* remaining and *perceived* productivity, both of which are significantly different from the corresponding true quantities in the model. More generally,

- Decisions can be based only on available information.

Rate equations representing decisions for action in the real system cannot be influenced in the model by information not accessible to the decision-makers. Usually, only a small fraction of the information in the system is available at a point of decision making, and it may be biased, distorted or delayed.

5 MODEL TESTING AND FURTHER DEVELOPMENT

The system dynamics modeling process is iterative, passing through a number of sequences of conceptualization, formulation, testing (simulating), reconceptualization, and refinement. The iterative approach to the formulation of a final model is usually forced by the complexity of the problem being addressed. But, even when it is possible to build a model all at once, it is advisable to assemble it in pieces and test the behavior of the growing structure in stages. An iterative approach to model formulation generates insights into the relationships between system structure and behavior, and those insights are the heart of the reasons for building a model.

This chapter gives the flavor of the cyclic, iterative nature of the system dynamics approach by returning to the project model a number of times. Section 5.1 illustrates how insights about structure and behavior can be drawn from a model. The important question of model sensitivity and robustness is addressed in section 5.2. The project model is used to illustrate how sensitivity testing can be performed and which of the different kinds of sensitivities one needs to pay attention to. Section 5.3 uses the project model to discuss refining and reformulating a model and makes some suggestions for knowing when to stop. Section 5.4 is an introduction to the thorny area of model validity.

By the end of this chapter, the reader will have been exposed to a number of technical skills and philosophical issues in the area of modeling that might best be termed model progress, and we will have a more refined project model with which to discuss policy analysis in chapter 6.

5.1 UNDERSTANDING MODEL BEHAVIOR

By exploring the behavior generated by individual feedback loops and by various combinations of loops in the model, the modeler learns about structure and behavior. One need not guess at the cause of, say, oscillations in a particular variable. Simulation experiments isolating and combining the effects of suspected factors can precisely pinpoint the structure responsible.

The notion of causation in feedback models requires comment. At the level of individual causal links in a model, we are comfortable with assuming "as this rises, that rises (or falls) by so much," and we assume relationships that we express in equations and table functions. But at the level of the feedback loop, causality is muddled. Does population cause births, or do births cause population? Neither alone is an adequate view. Each variable is a cause-effect of the other. The feedback view antiquates the notion of a simple, linear, left-right causality. Chickens and eggs are not a causal dilemma if one focuses on what they cause together, namely, exponential growth in the barnyard.

So, in hunting for the causes of model behavior, we seek feedback structures, not isolated variables. While a single factor can change the strength of a feedback loop and affect its dominance in the rest of the model, it is more useful to see the loop, not the factor, as the causal agent in the system. To analyze the relationships between structure and behavior, this section considers changes in parameters and table functions which *deactivate* feedback loops. The role of any feedback loop in a model can be inferred from the behavior of the model with and without the loop.

Changing Structure with Extreme Parameter Values

Recall in the project model the parameter FSAT, the fraction of work completed satisfactorily. In the original model run, FSAT was set equal to 0.7, as a reference against which other values and model behavior can be compared. Any values of FSAT greater than 0 and less than 1 result in the generation of a mix of real progress and undiscovered work. But when FSAT equals 1, its maximum possible value, the structure of the model changes. All of the feedback structure on the right side of figure 4.26 is eliminated, leaving a system with essentially only three levels (workforce, cumulative real progress, and the scheduled completion date) and the simple structure first conceptualized in figure 2.25.

The structural change is evident in the equations in the model involving FSAT:

Understanding Model Behavior

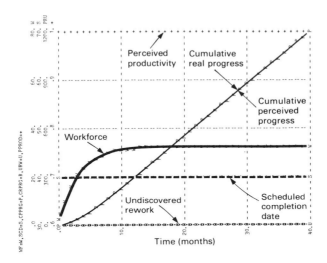

Figure 5.1: Simulation of the basic project model with fraction satisfactory FSAT equal to 1

```
R    RPRG.KL=APPRG.K*FSAT
NOTE      REAL PROGRESS (TASKS/MONTH)
A    RPROD.K=GPROD*FSAT
NOTE      REAL PRODUCTIVITY
NOTE      (TASKS/PERSON/MONTH)
R    GURW.KL=APPRG.K*(1-FSAT)
NOTE      GENERATION OF UNDISCOVERED REWORK
NOTE      (TASKS/MONTH)
```

(where APPRG = apparent progress and GPROD = gross productivity). When FSAT equals 1 in these equations, all of apparent progress APPRG is real. No undiscovered rework is generated, and GURW = 0. In addition, the equation for indicated productivity shows that when FSAT equals 1, IPROD is a constant, not a variable:

```
A    IPROD.K=WTRP.K*RPROD.K+(1-WTRP.K)*GPROD
         = WTRP.K*GPROD*FSAT + (1-WTRP.K)*GPROD
         = WTRP.K*GPROD + (1-WTRP.K)*GPROD
         = GPROD
```

Since perceived productivity is a SMOOTH of IPROD, we see that, when FSAT equals 1, PPROD = GPROD throughout the run.

Changing the parameter FSAT to 1 thus wipes out a number of feedback links and loops and transforms a number of variables in the model into constants. The resulting model behavior is shown in figure 5.1. It simulates a flawless project with a transient start-up period during which the necessary workforce is hired.

The experiment performed in figure 5.1 illustrates the way deactivating pieces of model structure can locate the structure essential for certain behavior modes. In this case our dynamic hypothesis for schedule and manpower overruns centered on FSAT and the accumulation of undiscovered rework, so we are not surprised by the results of the experiment. However, we should be reinforced that our dynamic hypothesis really does have the power to cause the problematic behavior. In a larger, more complex model, uncovering the structures responsible for different behavior modes is more difficult, but in principle the method of deactivating different loops is still the key.

A very general technique for deactivating loops applies to net rate equations of the form (GOAL.K-LEVEL.K)/ADJTM, which appear in formulations involving exponential averages, smooths, perception delays, or adjustments of various kinds. Making the time constant ADJTM in such formulations huge sets the rate essentially to zero. The level the rate affects is transformed into a constant, never changing from its initial value throughout the run. As a result, whatever feedback paths the level is in are interrupted and feedback loops are broken.

In the project model there are three potential examples, involving the workforce and schedule adjustment times WFAT and SAT and the time to perceive productivity TPPROD. Figure 5.2 shows a simulation in which TPPROD was set equal to one billion months (1E9 in DYNAMO notation). PPROD shows up as a constant, as expected, equal to its initial value, gross productivity GPROD.

This experiment thus wipes out all the feedback loops that pass through PPROD. In the experiment setting FSAT equal to 1, those loops were also deactivated, but in addition the entire undiscovered rework structure was removed. Here, the dynamic hypothesis involving undiscovered rework is still active in the model, and the behavior is very similar to the base run.

By comparing runs in which overlapping parts of a model are neutralized, the modeler can locate with certainty the structures essential for various behavior modes. Here we can observe, for example, that holding PPROD constant at its initial value (and therefore overestimating productivity) results in a slightly *smaller* workforce and an extention of the project

Understanding Model Behavior   271

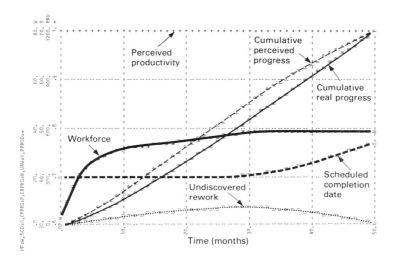

Figure 5.2: Simulation of the simple project model with perceived productivity fixed as a constant by setting TPPROD equal to one billion months (1E9).

duration from 48 to 50 months. Thus continually misperceiving productivity can increase the schedule overrun by a small amount, but the dominant influence on the pattern of overruns appears to be undiscovered rework.

Replacing Table Functions with Constants

Changing a variable to a constant in order to deactivate feedback loops is a simple task for variables expressed as table functions. In the project model the time to detect rework TDRW was expressed as a table function of the fraction perceived completed:

```
A   TDRW.K=TABLE(TTDRW,FPCOMP.K,0,1,.2)
T   TTDRW=12/12/12/10/5/.5
```

The time to detect rework can be changed to a constant of, say, 6 months by replacing TTDRW in a rerun with a string of 6's:

```
T   TTDRW=6/6/6/6/6/6
```

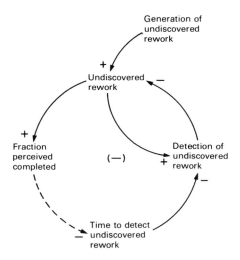

Figure 5.3: The feedback loop in the project model deactivated by making the time to detect rework TDRW a constant

The change eliminates a negative feedback loop shown in figure 5.3 by removing the dotted causal link.

The results of this change in TDRW are shown in the simulation in figure 5.4. The run is no different from the base run (figure 4.32) until near the end. In the test run the level of undiscovered rework continues to rise and remain high throughout the run. Workers on the project perceive the project is completed about three months before it actually is. It is conceivable that work on the project would cease even though the final product still contains a large component of unsatisfactory work.

The use of such a change in model structure to generate insights is more dramatically illustrated by a simulation in which willingness to change workforce, WCWF, is set equal to a constant, thereby deactivating the feedback loops that pass through WCWF. In the real system the change represents a shift in the way project managers make decisions adjusting the workforce and the schedule. Setting WCWF equal to 0.5 constantly throughout the run means that any adjustments required to put the project on target for completion will be handled by changes in both the workforce and the scheduled completion date. The reader should try to sketch the behavior of variables in the model that would be produced by this change.

Understanding Model Behavior

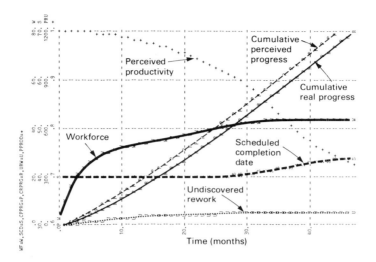

Figure 5.4: Simulation of the project model with the time to detect rework set equal to a constant (TTDRW=6/6/6/6/6/6)

Figure 5.5 shows the behavior of the model when the table values for WCWF are changed as follows:

Original:
 TWCWF=0/0/0/.1/.3/.7/.9/1
Change:
 TWCWF=.5/.5/.5/.5/.5/.5/.5/.5

For this run the plot period PLTPER was changed to 1.5 months to reduce the output to a managable size, for, as the reader should note, this run shows a project that takes 73 months!

The reason for this extreme schedule overrun can be traced to the start-up period. Starting with only two people initially, management (in the model) sees that the initial effort perceived remaining is too great to be completed by the tiny initial workforce in the initial time scheduled, so it has to adjust. With WCWF equal to 0.5, management adjusts both SCD and WF with equal fervor, with the unhappy result that the scheduled completion date rises in the start-up period to 60 months while only about 20 people are hired to do the work. The product

 (20 people)*(60 months)*(1 task/person/month)

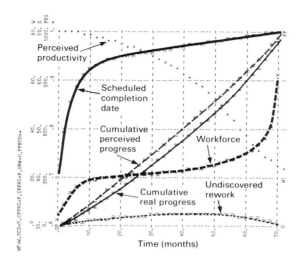

Figure 5.5: Simulation of the project model with the willingness to change workforce WCWF held constant at 0.5 (TWCWF=.5/.5/.5/.5/.5/.5/.5/.5)

does equal 1,200 tasks, the initial project definition, but few would manage a project in such a fashion. We conclude that it is essential in this model to require WCWF to be 1 throughout the start-up period of the project, so that the model will place the project on target by hiring the necessary number of people instead of pushing off the completion date.

The behavior of the run at the end of the project in figure 5.5 exposes another essential feature of our model's structure. Perceiving they are behind, and willing to hire workers right up to the very end, project managers in this run triple the workforce in the last quarter of the project. Few projects would be managed this way, and even if one were, the dramatic rise in the workforce would not immediately lead to increased productivity. The new people would have to be brought up to date on the work accomplished so far. We conclude that it is essential in the model to bring the willingness to change workforce, WCWF, to 0 sometime before the end of the project. Furthermore we are led to suggest disaggregating the workforce into new and experienced workers, so that the differences in their productivities can be captured in the model.

Understanding Model Behavior

Using Test Functions

In section 3.6 a small, mildly faulty inventory control model was used to illustrate the meaning of a variety of test functions in DYNAMO, including STEP, RAMP, PULSE, SIN (the trigonometric function, sine), and NOISE. Such functions are particularly useful in understanding the behavior of pieces or sectors of a model, and sometimes they are appropriate whole-model tests. Usually, these functions are used to disturb a model from equilibrium. The consequences of such ideal disturbances applied to a sector of a model shed light on the behavior of that sector when coupled with others.

The main conceptual and technical tasks for the modeler are the building of a submodel (a sector) from the larger model under construction, the choice of variables in the submodel to disturb with test functions, and the choice of tests. The test inputs should be placed in equations representing inputs to the sector from other sectors in the larger model, or from outside the system boundary. Rate equations, or auxiliaries that feed directly into rates, are the common choices for test inputs: business cycle oscillations in orders placed, random variation in the rate of infection in an epidemic, a step decrease in the productivity of workers suddenly slowing the production rate, and so on.

For examples of the use of test functions and the insights about structure and behavior one may derive from them, the reader should refer back to section 3.6.

Learning from Simulation Experiments

Simulations that test pieces of model structure are powerful tools for generating insight and understanding, but there is a danger in them. They are seductively easy to perform. See the words TYPE RERUN, enter a few lines and say "RUN," and more model output flows dutifully from the machine. But insights do not come from random experiments. Typing, as the computer buffs like to say, is no substitute for thinking, and the warning applies especially to reruns.

To learn the most from simulations that explore the implications of model structure, plan a sequence of simulations that focus on a particular element of structure. Analyze in advance the model behavior expected for each experiment. Write down your hypotheses, and compare them to observed model behavior. If unexpected behavior appears, analyze why. Think out key experiments with the model that can test competing hypotheses. Keep a notebook in which every detail of model structure, analysis, conjecture, and observation is recorded. Don't rely on memory to store model-based understandings. In the end, the simulation model is a laboratory tool. As in all science, a key requirement of lab experiments is *reproducibility*.

It is particularly edifying to reason through the behavior of the model in each simulation, explaining each rise, dip, turning point and phase relationship as much *in terms of the real system* as the model allows. The following causal explanation/description of the base run of the project model is an example, written for the run shown in figure 4.32. It is based upon understandings of model structure, model behavior, and a number of test simulations:

> At the beginning, the project is scheduled to take 40 months. Only two people are working on it initially, presumably project directors. Based on past experience, which may apply to this project only approximately, project managers assume that worker productivity will be about one task per person per month, and they estimate the number of tasks in the project to be 1,200. These numbers indicate a workforce of 30 is required, so in the first eight to ten months (the start-up period) workers are hired rapidly. The workforce rises a bit above 30 by ten months because a small amount of time was lost in the start-up period which must be made up by hiring a few more workers.
>
> Between 10 and 20 months, the second quarter of the project, the project is perceived to be almost on schedule. However, people on the project are beginning to perceive that productivity is not as high as it was initially thought to be: PPROD begins to drop in this quarter. As a result the project is perceived to be behind schedule. The hiring rate, which dropped steadily throughout the start-up period, begins to rise slightly, and more workers are added to the project workforce.
>
> In the third quarter, months 20 to 30, perceived productivity is dropping steadily as project personnel take note of the number of man-hours expended and the number of tasks perceived completed and see that it is no longer reasonable to assume a productivity of 1 task per person per month. Based upon current estimates of actual productivity and the effort remaining, additional people are hired and the decision is made to let the scheduled completion date slip.
>
> By month 30, supposedly three-quarters through, the project is perceived to be less than 65 percent completed. Everyone knows the project is well behind schedule; even the officially scheduled completion date has been pushed back from 40 to 41 months. The reasons for

being behind schedule are becoming clearer, as the need comes to light for redoing work thought to be completed satisfactorily. Undiscovered rework peaks about month 30. As the project heads toward completion, it becomes easier to spot aspects of the project that require rework. People are diverted from new tasks to redoing old ones, so perceived progress slows down.

After month 30, management refuses to hire more people, assuming that the project is so close to being done that it is preferable to push back the scheduled completion date if necessary to complete the project. By month 40 productivity is perceived to be close to what it actually was throughout the project, since it is easy to look back over the 40-month experience and get a reasonably accurate view. Rework is being discovered and attended to more and more rapidly, with the result that cumulative perceived progress slows its rise and cumulative real progress speeds up toward the end.

The project ends about eight months after its initial scheduled completion date, with a workforce of about 41 people, 11 more than would have been initially projected.

Such a story, phrased in real terms and enlightened by understandings of model structure and behavior, helps to bridge the gap between model and reality. It helps to ensure that causal structure in the model matches real structure. Where it exposes questions or alternative explanations, it helps to identify other simulations necessary for clear understandings. It may even expose faulty structure. Our story above, for example, describes the dynamics of perceived productivity in a way that suggests it should be computed from cumulative perceived progress and the number of man-months of cumulative effort expended on the project. Perhaps the weighted average formulation for PPROD can be improved upon.

Most important, repeatedly explaining model behavior in real terms approaches the fundamental purpose of system dynamics modeling--insight and understanding about a real problem. Understanding model structure and behavior is only a means to an end, the understanding of real system structure and behavior. The ultimate goal is to increase the likelihood that beneficial policies for the real system can be found and implemented.

5.2 SENSITIVITY

To place faith in model-based analyses and policy recommendations, the modeler has to know the degree to which those analyses might change as reasonable alternative assumptions are built into the model. How sensitive is the model to changes in parameter values and apparently minor variations in the formulation of equations? If reasonable changes in the model force changes in model-based conclusions about the real system, then those conclusions are not warranted.

Three Kinds of Sensitivity

The discussion of model sensitivity can be clarified by the observation that there is a hierarchy of three different kinds: numerical, behavioral, and policy sensitivity. A model is numerically sensitive if a parameter or structural change results in changes in numerical values computed in the course of the simulation. All quantitive models exhibit numerical sensitivity.

Behavioral sensitivity is a concern for dynamic simulation models. It refers to the degree to which the behavior exhibited by the model changes when a parameter value is changed or an alternative formulation is used. Recall that we use the term "behavior" to stand for *patterns or shapes of graphs over time.* We shall soon see that the behavior of system dynamics models tends to be rather insensitive to parameter changes. The observation is important and controversial, so it will require a close look.

Policy sensitivity is potentially the most damaging. Indeed the question of whether model-based policy conclusions change with reasonable changes in the model is the main reason for considering model sensitivity. No parameter in a model will have a unique value that fits the real system; a range of meaningful values is usually possible. If a model's behavior is so sensitive to its parameters over their reasonable ranges that the model can not help to assess the merits of competing policies, then the model is useless as a policy tool.

If model-based policy conclusions are sensitive to a particular parameter in the model, the modeler must first try to determine if the sensitivity is an artifact of the model or an accurate reflection of the real world. If analysis suggests the sensitivity is an artifact of the formulation, then the model must be reformulated to remove the sensitivity. The structure of the real system must be reanalyzed and the model reconceptualized, probably in a more detailed, disaggregated form. If, however, the structure seems suitable and consistent with the real system, then the modeler must try to estimate the parameter as accurately as possible. Because the real system in this case may have the same sensitivity, it would be a good idea to seek other beneficial policies that do not depend so on the value of the parameter.

There is a happy side to policy sensitivity, namely, insensitivity. If policy analyses hold up as parameters are varied over selected ranges, then those parameter values need not be estimated with any greater accuracy than those ranges. If the purpose is policy analysis, then it makes no difference if a model is numerically sensitive (they all are), or even behaviorally sensitive (system dynamics models tend not to be), as long as the policy conclusions are robust in the face of parameter changes. Esthetic or theoretical considerations may press for insensitivity of all three kinds, but in the last (policy) analysis only policy insensitivity really counts.

Parameters and Structure

In the discussions that follow, the emphasis is on testing a model for parameter sensitivities, but there are situations in which sensitivity to alternative structural formulations should be tested as well. Usually, careful analysis of the real system suggests a well-defined feedback structure that in turn leads to a particular set of model equations. Sometimes, however, one or more alternative formulations suggest themselves, and modeler and client find it hard to prefer one over the others. Each must be tested in the model. If all the choices cause the model to behave in essentially the same ways in the same scenarios, then any of them will do; pick one, and ignore the rest. If the model behaves differently, then comparison with the behavior of the real system should determine which is the more suitable formulation.

An example of a structural sensitivity test in the project model was suggested in section 5.1 in the description of the behavior of perceived productivity. There we suggested that perhaps a more appropriate formulation would base perceived productivity on the same computation actors in the system would use. In the early stages of an unfamiliar project, productivity would be estimated from past experience, the constant GPROD in our model, as in our original weighted average formulation. As man-months are expended and cumulative progress is perceived, actors in the system would rely more and more on an estimate of productivity actually indicated by the amount of cumulative perceived progress and the number of man-months of cumulative effort. The following equations would be required:

```
A    IPROD.K=WTCP.K*(CPPRG.K/CUMEFF.K)+(1-WTCP.K)*
X    GPROD
NOTE      INDICATED PRODUCTIVITY
```

```
NOTE      (TASKS/PERSON/MONTH)
A    WTCP.K=TABLE(TWTCP,FPCOMP.K,0,1,.2)
          WEIGHT GIVEN COMPUTED PRODUCTIVITY
T    TWTCP=0/.1/.9/1/1/1
NOTE      TABLE FOR WTCP
L    CUMEFF.K=CUMEFF.J=DT*WF.J
N    CUMEFF=0.0001
NOTE      CUMULATIVE EFFORT (MAN*MONTHS)
```

CUMEFF is a new variable, but WTCP is simply a revision of the weight given real productivity WTRP used in the original formulation. The table function values are selected to ensure that indicated productivity is initially GPROD and is completely determined by the computed average productivity, CPPRG.K/CUMEFF.K, by the time 60 percent of the project is perceived completed. Other similar weightings are certainly possible. Note the tiny nonzero initial value given to CUMEFF to prevent division by zero at the start of the simulation.

Figure 5.6 shows a simulation of the project model with these changes. It requires a very close look to see the differences between this run and the original base run of the model shown in figure 4.32. The time paths of the workforce, the scheduled completion date, undiscovered

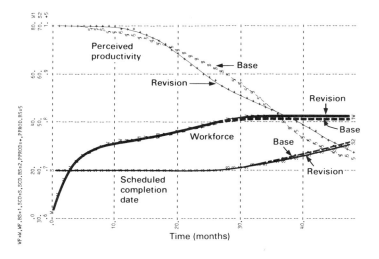

Figure 5.6: Simulation of the project model with a revised structure for perceived productivity PPROD

rework, and cumulative progress are virtually identical in both graphs. Subtle differences between the two figures can be seen in the graph of perceived productivity. In the test run perceived productivity eventually appears to lock into a linear path about halfway through the run. In the base run perceived productivity follows an S-shaped path. Both runs end with PPROD a little more than 0.7.

What accounts for the differences in the runs? The linear pattern in PPROD appearing halfway through the test run results from the fact that the two variables from which it is computed show basically linear patterns over time. The S-shaped pattern in PPROD in the base run is a simple consequence of the S-shaped table function for the weighting factor WTRP. Make WTRP linear and PPROD in the base run would be linear.

In spite of these tiny differences the two runs exhibit virtually identical overruns in workforce and schedule, and for the same reasons. Our observations tend to support the conclusion that the structural change in PPROD does not significantly change the behavior of the model. The behavior of the model is largely insensitive to this change. In the versions of the model which follow, we shall return to the original formulation.

It is tempting to assert that system dynamics models are insensitive to minor structural changes and deservedly quite sensitive to major ones, but the statement tends to be circular. We would likely define minor and major in terms of the extent of the effects of the changes on model behavior. Suffice to say, model structure is a powerful determinant of model behavior. Vary structure in any but the most trivial ways, and it is likely that behavior will change significiantly. Model structure is analogous to the assumptions in a logical argument: change the assumptions, and the conclusions almost always will have to change.

The most significant structural changes are those that alter the model boundary; such changes can have a large impact on model behavior. Care in setting the boundary at the start of a study is essential. A reconsideration of it near the end is advised in section 6.3.

Insensitivity to Parameter Changes

Parameter values do not have as powerful an effect on model behavior as the feedback, rate, and level structure of the model has. It frequently happens that a parameter change has virtually no effect on model behavior. Numerical values will change, to be sure, but the patterns over time generated by the model will usually remain fundamentally the

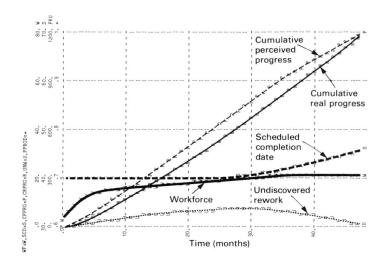

Figure 5.7: Simulation of the project model with the value of gross productivity GPROD raised from 1 to 2

same. We will note some examples from the project model and then try to reason out why feedback models exhibit this sort of robustness.

Figure 5.7 shows a run of the original project model in which the value of gross productivity GPROD is raised from 1 to 2 tasks per person per month--a change of 100 percent. Strangely enough, the project still takes about 48 months! The test run and the base run show the same patterns of behavior in all variables. There are important numerical differences: the workforce is about half as big everywhere in the test run as in the base run, while perceived productivity is twice as big in the test run. A bit of reflection shows these results are obvious ramifications of the decision structure of the model. Believing that workers are twice as productive, project managers in the model hire only half as many workers while still designing the project to take 40 months. The overruns result, just as in the base run, from the various effects of undiscovered rework.

Analogous results would be obtained if we halved GPROD, setting it equal to 0.5 tasks per person per month. We conclude that the model exhibits numerical sensitivity to changes in the value of GPROD, but the model's behavior is insensitive to them. (Dramatically different results would be obtained if we doubled the value of GPROD midway through a simulation. Such a run would not be a sensitivity test, however; it would represent a hypothetical scenario in which people suddenly became more productive, perhaps in response to a wildly successful incentives program.)

Consider a run of the project model in which the initial project definition, IPD, is raised from 1,200 tasks to 1,800 tasks, a 50 percent increase. Without bothering to show the computer output, we claim that the behavior of the model would be unchanged, just as in the previous simulation. The reasons the graphs over time would show exactly the same patterns as in the base run is again the decision structure postulated for project managers in the model. The project would still be scheduled for 40 months, so to complete more tasks in the same amount of time more workers would be hired. The size of the workforce would change, but the pattern of its graph over time would not. The project would still run on about 8 months too long because of undiscovered rework. Reducing IPD produces similar results with a reduced workforce. The behavior of the model is simply insensitive to changes in the size of the initial project definition.

These examples of parameter insensitivity may have seemed startling at first, although understandable in retrospect when analyzed in terms of the structure assumed in the model. Part of our initial surprise may be due to the belief that a smaller project or greater productivity really *should* shorten the work, even if the project is scheduled to take 40 months. One might well assume that as the workforce grows in size the number of man-hours spent each week in organization and communication tasks would grow, leaving less time for the actual, productive, project work. The model ignores, perhaps improperly, any effect of the size of the workforce on productivity. While these two simulations show parameter insensitivity, perhaps we should also take from them the hint that a feedback link has been left out.

To motivate one final observation about parameter sensitivity, consider a run of the project model in which the fraction satisfactory FSAT is changed from 0.7 to 0.5, roughly a 30 percent drop. As shown in figure 5.8, the project is completed in about 58 months, rather than the 48 months of the base run, more than double the schedule overrun and an increase in project duration of about 20 percent. Other than stretching the project out in time, the parameter change has no influence on the patterns generated over time by the model.

There is a significant difference between dropping FSAT to 0.5 and raising FSAT to 1 (compare figure 5.8 with 5.1). When FSAT was raised to 1, the behavior of the model did

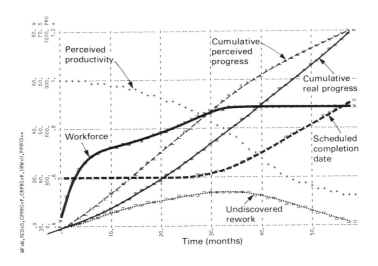

Figure 5.8: Simulation of the project model with the fraction of work satisfactory FSAT reduced from 0.7 to 0.5

change. Overruns were eliminated, for the obvious reason that no rework was required. We could not say the behavior of the model is insensitive to a change in FSAT to 1. There is a difference in sensitivity here because one of these changes actually changes the feedback, rate, and level structure of the model, while the other does not. Raising FSAT to 1 wipes out the level of undiscovered rework and the rates and feedback links associated with it. It is this structural change which produces the change in model behavior.

The lesson in these observations is that a discovery of a sensitive parameter may actually be a discovery of a parameter change that alters model structure. Know the parameter ranges that preserve model structure, and recognize when a parameter value falls outside its meaningful range.

Understanding Parameter Insensitivity

The preceding sensitivity tests provide evidence for the claim that system dynamics models tend to be relatively insensitive to parameter changes. In spite of such evidence the claim is controversial. Most of us are most familiar with situations in which numbers matter a lot. Some see the claim as a weak attempt to justify the down-playing of statistical procedures for estimating confidence intervals for parameters.

Sensitivity 285

Some, realizing that all quantitative models are numerically sensitive, view the stress on behavioral insensitivity as a sidestepping of a fundamental sensitivity issue.

Yet the evidence persists that patterns of behavior in system dynamics models tend not to be significantly changed by parameter changes. Furthermore the emphasis on behavior--patterns of graphs over time--is understandable if the problem being addressed is captured as a set of dynamic patterns. Project overruns, for example, are a mix of patterns over time which, in fact, remain essentially unchanged by most of the parameter changes explored in this section. To use the project model as a policy analysis tool, we really do not need to know with any greater accuracy the values of the parameters tested so far. The *amount* of an overrun and the relative costs associated with it will eventually be critical to our policy recommendations. The *pattern* of overruns is our current concern and it has remained remarkably persistent throughout these sensitivity tests.

There are two primary reasons why system dynamics models tend to be less sensitive to changes in parameters than changes in structure, and why they are less sensitive to numerical values in general than certain other types of quantitative models. The first of these concerns *dominant feedback loops:* in a feedback structure, a loop that is primarily responsible for model behavior over some time interval is known as a dominant loop. When a population is growing, for example, the positive loop involving births dominates the negative loops trying to halt growth. Changing a parameter in a feedback loop that is dominated by other loops in the system will have little effect on the behavior of the system.

Complicating the picture is the fact that loop dominance usually shifts among a number of loops in the course of a simulation. Changing a parameter may thus have an effect over part of the run but not in another part. Changing the workforce adjustment time WFAT in the project model, for example, may have some effect in the early part of a simulation but will have little or no effect toward the end when hiring is dominated by the feedback loop involving the willingness to change workforce.

The second deals with *compensating feedback loops:* a parameter change may weaken or strengthen a feedback loop, but the multi-loop nature of a system dynamics model naturally strengthens or weakens other loops to compensate. The result is often little or no overall change in model behavior. In the elementary example presented earlier, changing GPROD in the project model changed the strength of the production power in the model, but the hiring loop simply compensated, leaving the behavior (but not the numbers) unchanged.

Reflecting on the origins of the notion of feedback shows why feedback systems tend to compensate for parameter changes. Engineering feedback control systems were developed for precisely the purpose of creating systems that are insensitive to parameter changes. The purpose of the governor of a steam engine is to maintain engine speed even if the load changes or friction in the bearings increase. Societal systems have the same sort of feedback control structures, even though they are not engineered for commonly accepted goals.

Thus compensating feedback is a property of real systems, as well as system dynamics models, and is the reason real systems tend to be resistant to policies designed to improve behavior. Reality, one might say, is relatively insensitive to parameters. A good model capturing the feedback control mechanisms of a real system should exhibit the same characteristic.

Changing Initial Values

To complete these examples of parameter sensitivity tests, we would like to experiment with different initial values for the workforce and the scheduled completion date. Since DYNAMO is designed to allow C and T statements to be changed in rerun mode, without rewriting and recompiling the model, we have written the initial values as follows:

```
N    WF=WFN
C    WFN=2
NOTE     INITIAL VALUE OF WORKFORCE
N    SCD=SCDN
C    SCDN=40
NOTE     INTIAL VALUE OF SCHEDULED
NOTE      COMPLETION DATE
```

To change the initial value of the workforce level in a rerun, to, say, 30, simply write

```
C  WFN=30
RUN
```

when DYNAMO asks for rerun changes. Whenever an initial value of a level is set as a number rather than determined in a computation, it is advisable to use N and C statements in this fashion to allow flexibility in the choice of initial values.

Sensitivity

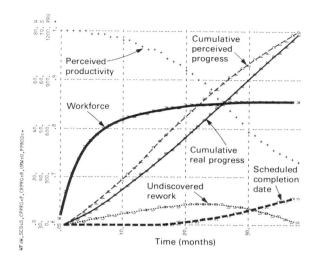

Figure 5.9: Simulation of the project model with the initial scheduled completion date lowered from 40 to 30 months (SCDN=30)

Figures 5.9, 5.10, and 5.11 show simulations of the project model varying the initial values of SCD and WF. Figure 5.9 shows that reducing the initial value of SCD to 30 shortens the run considerably, completing the project in about 38 months for an schedule overrun of, again, 8 months. More people are hired than in the base run to get the project done quicker, but the behavior of the model is, once again, unaffected by the parameter change. Figure 5.10 shows what happens when the workforce starts off at 30 people, the number expected to be required to complete 1,200 tasks in 40 months. The start-up period is eliminated in this run; presumably, it occurred before the start of the simulation. Otherwise, no significant differences.

Finally, figure 5.11 tries starting with a workforce of 100, many more than would be expected to be necessary. Predictably, the project management realizes 100 is too many and starts the project off by letting people go. The workforce declines initially, settling to about 28. Before it declines, however, the large initial workforce gets the project off to a headstart, so the project is undoubtedly perceived to be ahead of schedule. Soon, however, undiscovered rework takes its toll, and project management reverses itself, hiring or transferring back some of the people it had let go. Even-

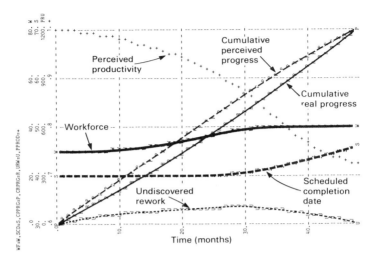

Figure 5.10: Simulation of the project model with the initial workforce set at 30 people instead of 2 people (WFN=30)

tually, in spite of the head start, the project still takes 47.5 months to complete, almost as much as the base run.

It is fair to conclude that the behavior of the model is somewhat sensitive to changes in the initial value of the workforce WF, but insensitive to changes in the initial value of the scheduled completion date SCD.

Testing Alternative Table Formulations

Table functions can be viewed as strings of parameters, or perhaps as approximations of polynomials whose coefficients would be parameters. Whatever their categorization, any table function, like a parameter, has a range of meaningful alternatives. Different choices would have the same general slope and shape and would pass through some of the same reference points but depart from the chosen table function in other, minor ways. Figure 3.23, for example, showed two alternatives for an effect of inventory on shipments, differing mainly in the rapidity with which shipments would be constrained as inventory gets small. Figures 4.16 and 4.17 showed a more extensive range of alternatives for an effect of land fraction occupied on the rate of business construction in an urban model. For the same reasons parameter sensitivity

Sensitivity

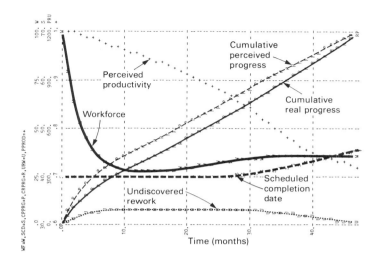

Figure 5.11: Simulation of the project model with the initial workforce raised to 100 people (WFN=100)

must be investigated, the effects of reasonable alternative table functions should be tested in the model.

The procedure is straightforward. Compare the behavior of the model in its base run with its behavior in a simulation with an alternative table. Test at first only the extremes of the likely alternatives. If no significant change in model behavior results, then one can conclude that the behavior of the model is relatively insensitive to changes in that table, within the range of reasonable alternatives. However, if model behavior changes in ways that seem important for the purposes of the study, then the table is a cause for concern and attention, leading perhaps to reformulation of the model.

For conceptual clarity it is important to distinguish between those table changes that are extreme enough to change model structure and those that test sensitivity without changing the feedback, rate, and level structure of the model. As we noted at the end of section 4.2, it is too easy to change fundamental assumptions in a model by changing a few numbers in a table function and then to conclude the model is sensitive to it.

Figure 5.12 shows two alternative table functions for the effect of land on business construction from an urban model, which differ only in that one goes to zero when land fraction occupied is 1 while the other stops at some positive number.

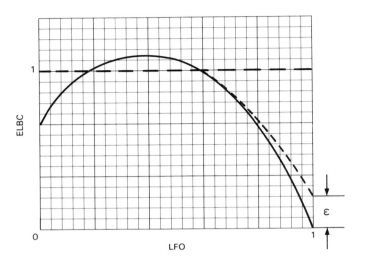

Figure 5.12: Alternative table functions for the effect of land on business construction ELBC from an urban model (see also figures 4.16 and 4.17.)

The tables look almost identical, yet express fundamentally different assumptions: one of them assumes construction can continue forever when the city land is completely built over--that cleared land is not required for construction--an untenable assumption. Predictably, an urban model simulated with these two tables would show dramatically different behavior patterns.[1] Yet a wide variety of tables for ELBC holding to the assumption that construction goes to zero when the land is filled up produce almost no change in the patterns of behavior in existing system dynamics models of urban land use.[2] The proper conclusion is that such models are very sensitive to the assumption of zero construction when land fraction occupied is 1 but essentially insensitive to variations in ELBC which do not violate that assumption.

For examples of a table sensitivity tests in the project model, consider the table for the time to detect rework TDRW. Figure 5.13 shows the original table and two likely

---

[1] For examples, unfortunately with an inappropriate interpretation, see Garn and Wilson (1972).
[2] See Alfeld and Graham (1976), pp. 41-45.

Sensitivity 291

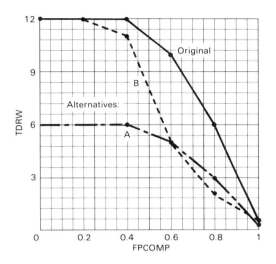

Figure 5.13: Alternative table functions for the time to detect rework TDRW in the project model

alternatives. Alternative A assumes the same shape but reduces the time initially required from 12 months to 6 months. The change essentially tests the sensitivity of the model to a change in scale in this table. Alternative B starts with TDRW equal to 12 months but drops sooner and tests the significance of the shape of the tail of the table.

Figures 5.14 and 5.15 show the behavior of the project model with alternatives A and B, respectively. The reader should draw his own conclusions about the sensitivity of the model to these changes.

The Significance of Model Sensitivity

If simulations uncover a parameter or a table function that has a significant effect on the behavior of the model, there are three plausible responses, each reasonable under different circumstances. First, the sensitivity may indicate that the parameter or table in question must be estimated with great care. If model-based conclusions hang on which of several narrow ranges of a parameter truly reflect the real system, then one must somehow estimate the parameter closely enough to be able to determine which range is appropriate.

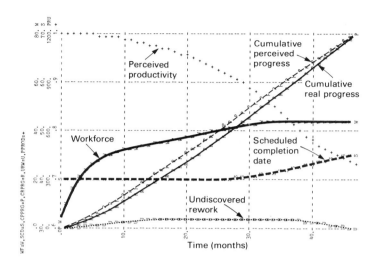

Figure 5.14: Simulation of the project model with the time to detect rework TDRW expressed as Alternative A in figure 5.13. (TTDRW=6/6/6/5/3/.25)

A second response is to reformulate the model to capture in more detail the structure of the real system which was aggregated in the parameter or table. The reformulation would likely involve changing a constant to a variable embedded in some new feedback structure, or replacing a table with a feedback structure involving one or more levels and rates. The effort is to remove sensitivities that are due solely to the artificial simplifications that make up the modeling process. Because of the properties of loop dominance and compensation, the incorporation of more detailed feedback structure appropriate to the real system should help to desensitize an inappropriately sensitive model.

A third and potentially important response is to interpret the sensitivity as an indicator of a point of leverage in the real system. When model structure is appropriate to the problem and parameters and tables most carefully done, a sensitivity in the model may merely reflect a sensitivity in the real system, a place where intervention would have a significant effect. Interpreting model sensitivities as policy indicators is one of the aspects of model-based policy analysis discussed in chapter 6.

Sensitivity 293

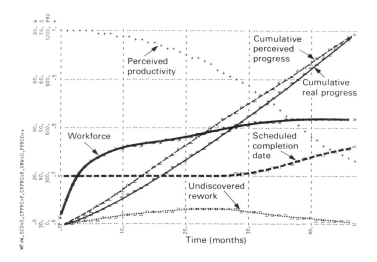

Figure 5.15: Simulation of the project model with the time to detect rework TDRW expressed as Alternative B in figure 5.13 (TTDRW=12/12/11/5/2/.5)

5.3 REFINEMENT AND REFORMULATION

The iterative model-building process that we have been urging involves a number of passes through the stages of conceptualization, formulation, simulation, and evaluation. At each successive pass, parts of the model are reformulated and refined, perhaps even deleted, and other structure is added. The purpose of the iterative process is eventually to produce a model highly consistent with the real system, well suited for its purposes, and well understood. This section discusses several considerations in model refinement and reformulation, including disaggregation, enriching the dynamic hypothesis, adding feedback structure, turning constants into variables, incorporating policy evaluation criteria, and most important, when to stop. The project model will provide examples: in one pass (usually many would be required) we will produce a refined and reformulated model that will be used as a basis for the discussions of model-based policy analysis in chapter 6.

Incorporating Criteria for Evaluating Policies

The model should contain within it the measures of policy effectiveness that would be used in the real system. Costs, profits, revenue growth, market share, and the like will influence policy choices in corporate settings, so they should be generated endogenously by a corporate model. Other models--macroeconomic, environmental, psychological, biological--yield analogous measures of policy effectiveness. Their inclusion in the model facilitates comparisons among different policy simulations and helps to sharpen the problem focus of the study.

In the project model the concerns are overruns in workforce and schedule. In addition project managers are no doubt concerned with costs. The inclusion of a variable for project cost in the model would enable us to chose cost-effective policies and also to observe some of the nasty tradeoffs involved in project management. Trying to hold costs down by skimping on the workforce hired will probably lead to large overruns in the schedule, while hiring a lot of people to be sure to finish on time will run up costs.

Consequently, we add a simple cost computation to the project model, first computing the number of man-months of cumulative effort and then translating that into a dollar figure by simply multiplying by a constant cost per man-month:

```
L     CUMEFF.K=CUMEFF.J+DT*WF.J
N     CUMEFF=0
NOTE     CUMULATIVE EFFORT (MAN-MONTHS)
A     COST.K=CUMEFF.K*CPMM
NOTE     PROJECT COST (DOLLARS)
C     CPMM=3000
NOTE     COST PER MAN-MONTH
NOTE     (DOLLARS/PERSON/MONTH)
```

DT is the time interval from J to K (0.25 months in our model), and WF.J is approximately the number of people working on the project from time J to time K. Hence the product DT*WF.J computes the additional number of man-months of effort expended on the project in the time interval from J to K. The level equation for CUMEFF accumulates those products and thus, by the end of the project, gives the total number of man-months of effort involved. We shall assume for simplicity that the cost per man-month is constant; in particular, even when we add an effect of schedule pressure on productivity we shall assume that people are not paid at a higher rate for overtime hours.

Refinement and Reformulation 295

With productivity at 1 task/person/month and a cost per man-month of $3,000, a 1,200-task project with no flaws would require 1,200 man-months and cost $3.6 million.

Disaggregating a Level

When should a level variable be reformulated as a sequence of two or more levels? Examples encountered in the simple epidemic model (sections 3.4 and 3.5), aging chains (as in figure 4.1-4), and the discussion of the order of material delays (section 3.5) show the disaggregation of one level into several connected by rates of flow. Since such a disaggregation is almost always possible technically, we need to ask when is such a refinement advisable?

There are two (and only two) considerations for reformulating a level as a sequence of two or more levels: policy analyses and model behavior. First, *is the disaggregation required in order for the model to be able to address particular policy issues?* Figure 2.12 showed the disaggregation of a heroin addict population, to include levels of addicts in methadone programs and addicts in community rehabilitation programs. The purpose of explicitly representing these subsets of the total addict population is to enable the model to address policy issues associated with methadone maintenance and nonmethadone rehabilitation programs. If those policy issues were judged not to be a part of the purpose of the model, those levels might have been omitted from the model and the populations they represented aggregated into a single level of, say, ex- addicts, or perhaps moved outside the system boundary and represented as a sink.

The second reason for disaggregating a level involves the dynamics of the system. *Does the disaggregation of a level into two or more levels have the potential to change significantly the behavior of the model?* If this appears likely, then the disaggregation should be tested, and the degree of change balanced against the increase in model complexity. The final arbiter should be model-based policy analyses. If the change in behavior has the potential to alter policy conclusions, then the disaggregation is essential.

Note that in the refinement-reformulation stage, as in the original formulation stages, the modeler should avoid the temptation to disaggregate a level simply to make the model look more like the real system. If model behavior and policy analyses are not affected, then the supposed refinement is just a decoration, and one that has the potential to distract the eye from what is important.[3]

Disaggregating the Project Workforce

As an example of disaggregating a level variable, consider replacing the workforce level in the project model with two levels, one representing workers newly brought into the project, and the other representing workers experienced on the project.

The change is suggested both by policy considerations and model behavior. Workers new to the project could be assumed to be less productive than workers experienced on the project. Thus a rise in workers new to the project would not fully increase the rate of progress until the new workers become assimilated into the experienced workforce. The dynamics caused by the delay in assimilating workers new to the project could conceivably have significant effects on model-based conclusions. The disaggregation is also suggested by the new policy levers it could add to the model. With a new workforce level and an experienced workforce level, one could test the effects of training programs to speed assimilation of new workers and increase their productivity.

Therefore, we shall add the structure shown in figure 5.16. The DYNAMO equations for the new levels and rates for the workforce are the following:

```
A     WF.K=EXPWF.K+NEWWF.K
NOTE     WORKFORCE (PEOPLE)
L     EXPWF.K=EXPWF.J+DT*WFAR.JK
NOTE     EXPERIENCED WORKFORCE (PEOPLE)
N     EXPWF=EXPWFN
C     EXPWFN=2
NOTE     INITIAL VALUE OF EXPWF
R     WFAR.KL=NEWWF.K/ASMT
NOTE     WORKFORCE ASSIMILATION RATE
NOTE       (PEOPLE/MONTH)
C     ASMT=6
NOTE     WORKFORCE ASSIMILATION TIME
NOTE       (MONTHS)
L     NEWWF.K=NEWWF.J+DT*(HR.JK-WFAR.JK)
NOTE     NEW WORKFORCE (PEOPLE)
N     NEWWF=NEWWFN
```

---

[3] A client may require such a refinement in order to believe in model-based conclusions, but such a consideration is more political than conceptual or technical, and we leave it for the modeler to decide his own course. Some of the problems associated with building confidence in a model are discussed in section 5.4.

Refinement and Reformulation

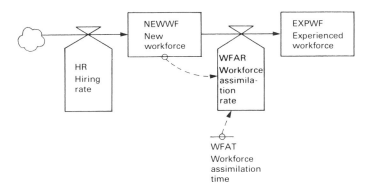

Figure 5.16: Flow diagram of the project workforce disaggregated into two levels, new workforce and experienced workforce.

```
C    NEWWFN=1
NOTE    INITIAL VALUE OF NEWWF
```

The hiring rate HR and the workforce sought WFS on which HR depends remain as they were in previous listings.

The equation for EXPWF contains no explicit outflow rate. The only conditions in which EXPWF can decline in this simple formulation is if the assimilation rate WFAR becomes negative, in other words, only if we have negative people in the NEWWF level! That weakness in the formulation will not trouble us, however, because the model never seems to get enough workers. Furthermore, if WFS ever does happen to drop below WF, making the hiring rate negative, the model will lower the workforce by removing people from the NEWWF level; just as in the real world, the newly hired will be the first ones transferred out if they are no longer needed. If concerns about a negative NEWWF level arise, however, the modeler could formulate a simple outflow rate in the form (fraction of EXPWF transferred per month)*(EXPWF.K). The inclusion of the outflow rate would add the possibility of representing feedback effects to and from quits and transfers.

Consequences of Disaggregating a Level

Separating a level into two levels often almost doubles the number of equations associated with the original level. In the project model, the explicit representation of workers new to the project suggests that worker productivity-- represented previously by the constant GPROD--should be reformulated. One technically simple approach is to provide two constants for worker productivity, one for NEWWF and the other for EXPWF, form products like the previous GPROD*WF.K, and sum the two terms. We shall illustrate a different approach, one that involves fewer equations but adds considerable subtlety to the productivity formulation.

We shall leave the formulation for apparent progress as it was, the product of the total workforce WF and gross productivity GPROD, but the latter will become a variable, GPROD.K, computed as the product of a constant normal gross productivity and an effect of experience on gross productivity. The latter effect, EEXPGP, will be expressed as a table function of EXPWF/WF, the fraction of the workforce that is experienced. The equations follow, and the table function for EEXPGP is shown in figure 5.17.

```
A       GPROD.K=NGPROD*EEXPGP.K
NOTE        GROSS PRODUCTIVITY
NOTE          (TASKS/PERSON/MONTH)
C       NGPROD=1
NOTE        NORMAL GROSS PRODUCTIVITY
NOTE          (TASKS/PERSON/MONTH)
A       EEXPGP.K=TABLE(TEEXPG,FEXP.K,0,1,.2)
NOTE          EFFECT OF EXPERIENCE ON GROSS
NOTE          PRODUCTIVITY
T       TEEXPG=.5/.55/.65/.75/.87/1
NOTE          TABLE FOR  EEXPGP
A       FEXP.K=EXPWF.K/NEWWF.K
NOTE          FRACTION OF WORKFORCE EXPERIENCED
```

The table function for EEXPGP is remarkably subtle in its implications. We will discuss them briefly as a kind of post-graduate course in the art of formulating table functions presented in section 4.2. First note that, if the average new worker is half as productive as the average experienced worker, and no effects other than numbers are considered, then the table function should be a straight line from (0,0.5) to (1,1) (shown in figure 5.17 as reference line A).[4] Our table incorporates two additional effects that oppose each other but in the net pull productivity below reference line A.

---

[4] Productivity would simply be the weighted average [(0.5)*NEWWF + (1)*EXPWF]/WF, or (0.5)(1-FEXP) + (1)(FEXP), which simplifies to the obviously linear function 0.5 + (0.5)(FEXP).

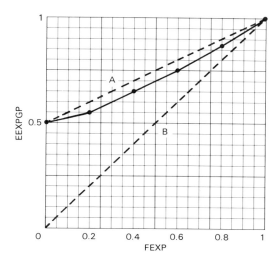

Figure 5.17: Table function for the effect of experience on gross productivity, showing reference lines on which the function is based

Figure 5.17 assumes that people new to the project are not only less productive than experienced people, but to some extent *inhibit* the effectiveness of experienced workers. New people need to be brought up to speed on the project. They must be assimilated into the project team. They have to learn their roles and whatever project specific skills are required, and experienced workers must take some of their project time to teach or supervise them. That time taken away from the project reduces the effectiveness of the experienced workforce, so the graph of EEXPGP drops below the weighted average formulation represented by reference line A.

On the other hand, supervision of people new to the project ought to *raise* their productivity above the value assumed for an unsupervised NEWWF (namely 0.5). Such a consideration argues for lifting EEXPGP *above* reference line A. We have two competing influences. The table function we have drawn in figure 5.17 assumes that the loss of productivity of experienced workers is greater than the corresponding gain in productivity of new workers. (A likely alternative would be to assume EEXPGP lies above the reference line when FEXP nears 1 and drops below the reference line when FEXP drops sufficiently below 1.)

An additional subtlety in the table for EEXPGP is reflected in reference line B (figure 5.17). This is the weighted average formulation for productivity under the assumption that workers new to the project have *zero* productivity (the explanation is analogous to the previous footnote). Thus, if EEXPGP drops below reference line B for any values of FEXP, the result is tantamount to assuming that those new workers on the project have, in the net aggregate average, *negative* productivity. Such values for EEXPGP may be justifiable for certain ranges of the ratio of new to experienced workers, if, for example, new workers did not contribute much at first and experienced workers had to spend almost all their time training them. But unless we wish to make the assumption of a net negative contribution of new workers for some values of FEXP, we have to take care that the table function for EEXPGP nowhere falls below reference line B. Table functions are indeed subtle.

Enriching the Dynamic Hypothesis

One of the strengths of a system dynamics model is its ability to incorporate simultaneously more than one hypothesis about the causes of a problem and also to compare their potentials for generating the problem behavior. For simplicity it is wise to take advantage of the iterative approach to model building: begin with one clear dynamic hypothesis, formulate the model and play with it, and then in later refinement-reformulation stages add other hypotheses that appear to have the potential to be responsible for some of the problem behavior.

In our first pass at formulating a project model, we chose to formulate our dynamic hypothesis solely in terms of the notion of undiscovered rework. We ignored a potentially competing or contributing hypothesis, namely, that complex projects by their nature involve more tasks than project managers can anticipate initially. So as an example of enriching the dynamic hypothesis we introduce a variable current project definition CPD, formulated as a function of FCOMP, the fraction of the project completed:

```
A     CPD.K=TABLE(TCPD,FCOMP.K,0,1,.2)
NOTE     CURRENT PROJECT DEFINITION  (TASKS)
T     TCPD=800/830/900/1000/1140/1200
NOTE     TABLE FOR  CPD
N     FPD=TABLE(TCPD,1,0,1,.2)
NOTE     FINAL PROJECT DEFINITION  (TASKS)
A     FCOMP.K=CRPRG.K/FPD
NOTE     FRACTION COMPLETED (DIMENSIONLESS)
```

Refinement and Reformulation 301

where CRPRG represents cumulative real progress.

The result of this change in the model is an *underestimation* of the size of the project in its early phases. The definition of the project starts at 800 tasks and then rises in a sigmoid pattern, eventually reaching the 1,200 tasks assumed in the simple project model. The table formulation allows the testing of a wide variety of assumptions about the early perceptions of the size of the project. For comparisons, however, the final project definition FPD should remain fixed at 1,200 tasks. Note the neat use of the DYNAMO table function expression for FPD--making the independent variable of the function 1 causes the expression to pick out the final y-value in the table as the initial value of FPD, just as desired. The expression is handier than assigning FPD the value 1,200 in a C equation, because it would adjust FPD automatically if the table for CPD is varied.

Adding Feedback Structure

Adding additional feedback effects between existing model variables is perhaps the most common kind of model reformulation. The conceptualization and formulation of new feedback links in an existing model follow the patterns involved in developing the original model. The most likely stumbling blocks are simultaneous equations (see section 4.7). When a new link is added, it is easy to close accidentally a loop that contains no level variable. Some experienced modelers try to anticipate simultaneities and prevent them before they show up in an error message, while others charge ahead and worry about simultaneities only when DYNAMO flags them. The choice is yours.

As with other complicating refinements, it is usually advisable to add additional feedback links sequentially rather than all at once. Understandings gained from seeing the incremental effects of successive refinements probably save the modeler time in the long run. Furthermore, adding new links one or a few at a time makes it easier to track down and correct any simultaneities that crop up.

For examples of refining a model by adding new feedback links, we will add effects of experience and effects of schedule pressure on gross productivity and the fraction of work satisfactory. EEXPGP was discussed in detail earlier; the others are contained in the revised equations that follow.

```
A     APPRG.K=WF.K*GPROD.K*ESPGP.K
NOTE      APPARENT PROGRESS (TASKS/MONTH)
A     ESPGP.K=TABLE(TESPGP,ICD.K/SCD.K,.9,1.2,.05)
NOTE      EFFECT OF SCHEDULE PRESSURE ON
```

```
NOTE      GROSS PRODUCTIVITY
T    TESPGP=.9/.92/1/1.1/1.18/1.23/1.25
NOTE      TABLE FOR ESPGP

A    FSAT.K=NFSAT*EEXPFS.K*ESPFS.K
NOTE      FRACTION SATISFACTORY
NOTE      (DIMENSIONLESS)
C    NFSAT=0.7
NOTE      NORMAL FRACTION SATISFACTORY
A    EEXPFS.K=TABLE(TEEXPF,FEXP.K,0,1,.2)
NOTE      EFFECT OF EXPERIENCE ON
NOTE      FRACTION SATISFACTORY
T    TEEXPF=.5/.6/.7/.8/.9/1
NOTE      TABLE FOR EESPFS
A    ESPFS.K=TABLE(TESPFS,ICD.K/SCD.K,.9,1.2,.05)
NOTE      EFFECT OF SCHEDULE PRESSURE ON FRACTION
NOTE      SATISFACTORY (DIMENSIONLESS)
T    TESPFS=1.1/1.06/1/.96/.9/.83/.75
NOTE      TABLE FOR ESPFS
```

Rather than discuss these in detail, we merely note their rationales and leave the details for the reader. EEXPFS, like EEXPGP, is a function of the fraction of the workforce experienced and represents the assumption that inexperienced workers produce more unsatisfactory work than experienced workers. The effects of schedule pressure are functions of the ratio ICD/SCD, the indicated completion date over the scheduled completion date. Ratios greater than one mean the project is behind schedule. ESPGP and ESPFS assume that the resulting schedule pressures *increase* gross productivity and *decrease* the fraction of work performed satisfactorily. ESPGP is really assumed to represent an overtime effect. (A more pessimistic assumption for ESPGP would show a peak and *decline* for high values ICD/SCD, indicating that lots of schedule pressure has a debilitating effect in spite of overtime hours. A more sophisticated formulation would add an explicit fatigue effect dependent upon an accumulation of ESPGP that could only dissipate over time. But our purposes are best served by simplicity.)

Interpreting the effect of schedule pressure on gross productivity ESPGP as an overtime effect requires including ESPGP in our previous computation for cumulative effort:

```
L    CUMEFF.K=CUMEFF.J+DT*(WF.J*ESPGP.J)
```

The idea here is that a fixed workforce working, say, 48-hour weeks rather than 40-hour weeks is putting in 48/40 or 120 percent of normal effort and would contribute to the measured cumulative effort just as much as a workforce 1.2 times as big. Including ESPGP in CUMEFF guarantees that the overtime increases the cost of the project (see the cost equation given at the beginning of this section).

Note that computing COST as CUMEFF times a constant cost per man-month means that people will be paid for their overtime, but at the same rate as their regular work. A formulation for time-and-a-half would require raising CPMM whenever ESPGP rises above 1. We won't bother, but the reader might want to see how such a pay scheme influences the simulations.

Turning Constants into Variables

Reformulating a constant as a variable is often merely a slightly different way of phrasing the addition of a feedback link to a model. A constant could be reformulated as a level variable or as an auxiliary, but in general this sort of reformulation presents no new difficulties. As an example, consider making the normal fraction satisfactory NFSAT a function of FCOMP, the fraction of the project complete:

```
A    NFSAT.K=TABLE(TNFSAT,FCOMP.K,0,1,.2)
NOTE    NORMAL FRACTION SATISFACTORY
NOTE    (DIMENSIONLESS)
T    TNFSAT=.5/.55/.63/.75/.9/1
NOTE    TABLE FOR NFSAT
```

(Recall that NFSAT is multiplied by EEXPFS and ESPFS to compute FSAT in the reformulated model.)

The rationale for making NFSAT a variable in this fashion is the realization that tasks near the beginning of a large project are much different from those near the end. Near the beginning the project is being defined, different approaches are being explored, ideas are on the drawing board. The work toward the middle of the project involves prototypes and mock-ups. Near the end the tasks look more like finishing touches, assembling documentation, typing reports, and so on. The likelihood of performing work that must be redone is much greater at the beginning of a project than at the end. (Note that in the formulation of TDRW, the time to detect rework, the model already assumes that tasks toward the end of the project are of a different character, with the need for rework easier to discern as the project is being brought to a close.)

Printing Summary Information

Some pieces of information about a simulation are more telling in numerical, rather than graphical, form. In the project model, it would be useful to print out the initial and final values of such variables as the size of the workforce, the scheduled completion date, the number of man-months of cumulative effort, the cost of the project, the fraction of the project perceived completed, and the fraction actually completed. Thus we include the following PRINT statement:

PRINT WF,SCD,CUMEFF,COST,FPCOMP,FCOMP

Usually, to print the initial and final values in a simulation, one would set the constant PRTPER equal to LENGTH in an N equation. In the project model, however, the length of the simulation varies:

A   LENGTH.K=CLIP(TIME.K,MAXLEN,FCOMP.K,1)

The run stops whenever FCOMP reaches 1, and that could happen almost anytime depending upon conditions in the particular scenario being simulated. Hence, we need to formulate PRTPER as a variable also, using an auxiliary equation:

A   PRTPER.K=LENGTH.K

What to Plot?

The general rule is plot a lot. Plot important feedback effects even if they are not observable in the real system, as well as key system variables. To keep the plots reasonably simple, call for several plots by including more than one plot statement, each with no more than, say, six variables. Once the likely ranges of plotted variables have been determined from a number of runs, fix the scales in the plots so that any run is easily comparable to any other.

For the reformulated project model we shall use the following PLOT statements:

```
PLOT    WF=W(0,60)/SCD=S(30,70)/CPPRG=P,CRPRG=R,
X       URW=U(0,1200)/PPROD=+(.6,1)
PLOT    ESPGP,EEXPGP,ESPFS,EEXPFS(.5,1.5)/FEXP=X(0,1)
```

Together with the summary printed information including the man-months of cumulative effort and the total cost of the project, these plots should give us a good understanding of each

Refinement and Reformulation

policy simulation and allow easy comparisons of the effectiveness of different policies.

When to Stop Refining and Reformulating?

There is no simple answer. If you believe that the model has reached a stage at which it is well suited for its purposes and highly consistent with reality, stop there. If you can perceive when the marginal benefits of further refinements begin to be outweighed by their marginal costs (effort, time, dollars), stop there. If the model in its present form helps people to generate what appear to be real insights about the problem addressed, by all means stop there. If the model has become so large and complex that you can no longer understand it, you have gone too far too fast; drop back.

The trouble with these suggestions is that all except the last require wisdom to apply. Know that the tendency is almost always to go too far. No model is the final, perfect model. The process is iterative and could conceivably go on until the modeler totters from old age. Keeping that firmly in mind may make it easier to say, "this far, but no further."

Beyond the pragmatic issues of limited time, energy, and computer accounts, the issue of when to stop is intimately tied to the question of model validity, the subject of the next section. Before turning to it, we present the base run of the reformulated project model, and a fully documented model listing.

Printed summary of initial and final values in the base run of the revised project model:

TIME	WF	SCD	CUMEFF	COST
E+00	E+00	E+00	E+00	E+03
0.00	3.0	40.0	0.0	0.0
61.50	38.0	59.0	2082.5	6247.4

Appendix: The Revised Project Model

```
*       REVISED PROJECT MODEL.
NOTE
NOTE        REAL PROGRESS
NOTE
L       CRPRG.K=CRPRG.J+DT*RPRG.JK
N       CRPRG=0
NOTE        CUMULATIVE REAL PROGRESS   (TASKS)
R       RPRG.KL=APPRG.K*FSAT.K
NOTE        REAL PROGRESS RATE   (TASKS/MONTH)
A       APPRG.K=WF.K*GPROD.K*ESPGP.K
NOTE        APPARENT PROGRESS RATE  (TASKS/MONTH)
A       GPROD.K=NGPROD*EEXPGP.K
NOTE        GROSS PRODUCTIVITY (TASKS/PERSON/MONTH)
```

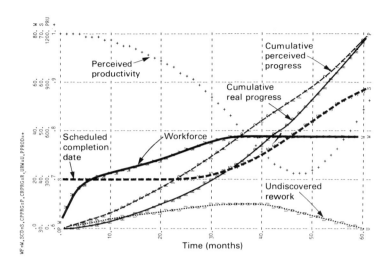

Figure 5.18: The behavior of key variables in the base run of the reformulated project model

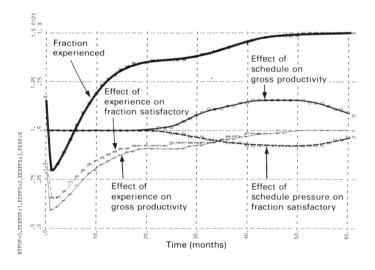

Figure 5.19: The behavior of new feedback effects in the base run of the reformulated project model

Refinement and Reformulation

```
C     NGPROD=1
NOTE        NORMAL GROSS PRODUCTIVITY  (TASKS/PERSON/MONTH)
A     FSAT.K=NFSAT.K*EEXPFS.K*ESPFS.K
NOTE        FRACTION SATISFACTORY
A     NFSAT.K=TABLE(TNFSAT,FCOMP.K,0,1,.2)
NOTE        NORMAL FRACTION SATISFACTORY
T     TNFSAT=.5/.55/.63/.75/.9/1
NOTE        TABLE FOR  NFSAT
NOTE
NOTE        EFFECTS OF EXPERIENCE AND SCHEDULE PRESSURE
NOTE
A     EEXPGP.K=TABLE(TEEXPG,FEXP.K,0,1,.2)
NOTE        EFFECT OF EXPERIENCE ON GROSS PRODUCTIVITY
NOTE        (DIMENSIONLESS)
T     TEEXPG=.5/.55/.65/.75/.87/1
NOTE        TABLE FOR  EEXPGP
A     FEXP.K=EXPWF.K/WF.K
NOTE        FRACTION OF WORKFORCE EXPERIENCED  (DIMENSIONLESS)
A     EEXPFS.K=TABLE(TEEXPF,FEXP.K,0,1,.2)
NOTE        EFFECT OF EXPERIENCE ON FRACTION SATISFACTORY
NOTE        (DIMENSIONLESS)
T     TEEXPF=.5/.6/.7/.8/.9/1
NOTE        TABLE FOR  EEXPFS
A     ESPFS.K=TABLE(TESPFS,ICD.K/SCD.K,.9,1.2,.05)
NOTE        EFFECT OF SCHEDULE PRESSURE ON FRACTION SATISFACTORY
NOTE        (DIMENSIONLESS)
T     TESPFS=1.1/1.06/1/.96/.9/.83/.75
NOTE        TABLE FOR  ESPFS
A     ESPGP.K=TABLE(TESPGP,ICD.K/SCD.K,.9,1.2,.05)
NOTE        EFFECT OF SCHEDULE PRESSURE ON GROSS PRODUCTIVITY
NOTE        (DIMENSIONLESS)
T     TESPGP=.9/.92/1/1.1/1.18/1.23/1.25
NOTE        TABLE FOR  ESPGP
NOTE
NOTE        UNDISCOVERED REWORK
NOTE
R     GURW.KL=APPRG.K*(1-FSAT.K)
NOTE        GENERATION OF UNDISCOVERED REWORK  (TASKS/MONTH)
L     URW.K=URW.J+DT*(GURW.JK-DURW.JK)
N     URW=0
NOTE        UNDISCOVERED REWORK  (TASKS)
R     DURW.KL=URW.K/TDRW.K
NOTE        DETECTION OF UNDISCOVERED REWORK  (TASKS/MONTH)
A     TDRW.K=TABLE(TTDRW,FPCOMP.K,0,1,0.2)
NOTE        TIME TO DETECT REWORK  (MONTHS)
T     TTDRW=12/12/12/10/5/.5
NOTE        TABLE FOR  TDRW
NOTE
NOTE        PERCEIVED PROGRESS
NOTE
A     CPPRG.K=CRPRG.K+URW.K
NOTE        CUMULATIVE PERCEIVED PROGRESS  (TASKS)
A     FPCOMP.K=CPPRG.K/CPD.K
NOTE        FRACTION PERCEIVED COMPLETED  (DIMENSIONLESS)
A     CPD.K=TABLE(TCPD,FCOMP.K,0,1,.2)
NOTE        CURRENT PROJECT DEFINITION  (TASKS)
T     TCPD=800/830/900/1000/1140/1200
NOTE        TABLE FOR  CPD
N     FPD=TABLE(TCPD,1,0,1,.2)
NOTE        FINAL PROJECT DEFINITION  (TASKS)
```

```
A      FCOMP.K=CRPRG.K/FPD
NOTE        FRACTION COMPLETED  (DIMENSIONLESS)
NOTE
NOTE        EFFORT PERCEIVED REMAINING
NOTE
A      EPREM.K=(CPD.K-CPPRG.K)/PPROD.K
NOTE        EFFORT PERCEIVED REMAINING  (MAN*MONTHS)
A      PPROD.K=SMOOTH(IPROD.K,TPPROD)
NOTE        PERCEIVED PRODUCTIVITY  (TASKS/MONTH)
A      IPROD.K=WTRP.K*RPROD.K+(1-WTRP.K)*NGPROD
N      IPROD=NGPROD
NOTE        INDICATED PRODUCTIVITY  (TASKS/PERSON/MONTH)
C      TPPROD=6
NOTE        TIME TO PERCEIVE PRODUCTIVITY
A      WTRP.K=TABLE(TWTRP,FPCOMP.K,0,1,0.2)
NOTE        WEIGHT GIVEN REAL PRODUCTIVITY
T      TWTRP=0/.1/.25/.5/.9/1
NOTE        TABLE FOR  WTRP
A      RPROD.K=NGPROD*FSAT.K
NOTE        REAL PRODUCTIVITY  (TASKS/PERSON/MONTH)
NOTE
NOTE        HIRING
NOTE
A      WF.K=EXPWF.K+NEWWF.K
NOTE        WORKFORCE  (PEOPLE)
L      EXPWF.K=EXPWF.J+DT*WFAR.JK
NOTE        EXPERIENCED WORKFORCE  (PEOPLE)
N      EXPWF=EXPWFN
C      EXPWFN=2
NOTE        INITIAL VALUE OF  EXPWF
R      WFAR.KL=NEWWF.K/ASMT
NOTE        WORKFORCE ASSIMILATION RATE  (PEOPLE/MONTH)
C      ASMT=6
NOTE        ASSIMILATION TIME  (MONTHS)
L      NEWWF.K=NEWWF.J+DT*(HR.JK-WFAR.JK)
NOTE        NEW WORKFORCE  (PEOPLE)
N      NEWWF=NEWWFN
C      NEWWFN=1
NOTE        INITIAL VALUE OF  NEWWF
R      HR.KL=(WFS.K-WF.K)/WFAT
NOTE        HIRING RATE  (PEOPLE/MONTH)
C      WFAT=3
NOTE        WORKFORCE ADJUSTMENT TIME  (MONTHS)
A      WFS.K=WCWF.K*IWF.K+(1-WCWF.K)*WF.K
NOTE        WORKFORCE SOUGHT  (PEOPLE)
A      WCWF.K=TABHL(TWCWF,TREM.K,0,21,3)
NOTE        WILLINGNESS TO CHANGE WORKFORCE
T      TWCWF=0/0/0/.1/.3/.7/.9/1
NOTE        TABLE FOR  WCWF
A      IWF.K=EPREM.K/TREM.K
NOTE        INDICATED WORKFORCE  (PEOPLE)
NOTE
NOTE        SCHEDULING
NOTE
A      TREM.K=SCD.K-TIME.K
NOTE        TIME REMAINING  (MONTHS)
L      SCD.K=SCD.J+DT*NAS.JK
N      SCD=SCDN
NOTE        SCHEDULED COMPLETION DATE  (MONTH)
```

Refinement and Reformulation

```
C     SCDN=40
NOTE        INITIAL VALUE OF  SCD
R     NAS.KL=(ICD.K-SCD.K)/SAT
NOTE        NET ADDITIONS TO SCHEDULE   (MONTHS/MONTH)
C     SAT=6
NOTE        SCHEDULE ADJUSTMENT TIME   (MONTHS)
A     ICD.K=TIME.K+TPREQ.K
NOTE        INDICATED COMPLETION DATE  (MONTH)
A     TPREQ.K=EPREM.K/WFS.K
NOTE        TIME PERCEIVED REQUIRED   (MONTHS)
NOTE
NOTE        INDICATORS
NOTE
L     CUMEFF.K=CUMEFF.J+DT*(WF.J*ESPGP.J)
N     CUMEFF=0
NOTE        CUMULATIVE EFFORT   (MAN*MONTHS)
A     COST.K=CPMM*CUMEFF.K
NOTE        PROJECT COST   (DOLLARS)
C     CPMM=3000
NOTE        COST PER MAN*MONTH   (DOLLARS)
NOTE
NOTE        CONTROL STATEMENTS
NOTE
SPEC  DT=.25,MAXLEN=100,PLTPER=1.25
A     LENGTH.K=CLIP(TIME.K,MAXLEN,FCOMP.K,1)
A     PRTPER.K=LENGTH.K
PLOT  WF=W(0,80)/SCD=S(30,70)/CPPRG=P,CRPRG=R,URW=U(0,1200)/
X     PPROD=+(.6,1)
PLOT  ESPGP,EEXPGP,ESPFS,EEXPFS(.5,1.5)/FEXP=X(0,1)
PRINT WF,SCD,CUMEFF,COST,FCOMP,FPCOMP
```

5.4 THE QUESTION OF VALIDITY

> No model has ever been or ever will be thoroughly validated. ... "Useful," "illuminating," or "inspiring confidence" are more apt descriptors applying to models than "valid."
>
> --Martin Greenberger et al. (1976)

The modeler, the clients, and any audience for a model-based study want to know how much trust to place in analyses based upon the model. The formal processes that lead people to place confidence in a model are frequently referred to as the validation of the model. Among the many different modeling methodologies, there is little agreement about what good validation is or ought to be.

Validity and Model Purpose

We begin with an important premise: it is meaningless to try to judge validity in the absence of a clear view of model purpose. The observation is frequently acknowledged and yet all-too-often forgotten in the heat of a discussion of model validity. We have stressed that a system dynamics model addresses a problem, not a system, and is designed to answer a reasonably well-defined set of questions. The confidence we place in the model to help us analyze that problem and answer those questions should not depend upon whether the model can address other problems or answer other questions. Statements of model purpose and the goals of a modeling effort thus serve two purposes: they focus a study initially and aid in judging the validity of the results.

The dangers of ignoring goals when assessing model validity are irreverently suggested by Lewis Carroll's observation that a clock that never runs at all is more accurate than a clock that loses a minute a day: the stopped clock is exactly right twice every day, while the merely slow one is right only about once every two years! Which is the more valid time piece? The answer must depend upon whether its purpose is to be exactly right or to help the owner get to appointments reasonably on time. The distinction may be expressed more generally as whether the purpose is point prediction or behavior prediction. The example takes on more significance when it is noted that it concerns judging the validity of a model of an oscillating system (true period = 12 hours) by considering whether it exhibits the *behavior* of the real system (prediction of behavior) or how well it predicts the true state of the system (point prediction).[5]

The Question of Validity

Critics of a modeling study must therefore keep in mind the purposes for which the model was constructed and confine their evaluations to those purposes. In like manner believers in the strength of a model must confine their applications of it to questions it is capable of addressing. Operators and observers alike can err by ignoring model purpose.

Validation as Innoculation

There is a tendency to think of validation as a process similar to warding off the measles: a model, susceptible to contagious criticism, gets validated and becomes immune to further attack. Often some measure of its immunity is sought--a goodness-of-fit test, a multiple-correlation coefficient (R^2)--much like a count of antibodies in a mathematical bloodstream. We tend toward such a view of validation because for some models it may be appropriate, because it is comforting, and because it relieves observers of the arduous task of repeating a validation process for themselves. Unfortunately, validation as innoculation is not really an appropriate way to think of the process of building confidence in a system dynamics model.

As it is usually phrased, the question is, Is the model valid?, or, Has the model been validated? The implication of these questions, and perhaps even our own characterization of the modeling process (see section 1.3), is that validation is a one-time process that takes place after a model is built and before it is used for policy analyses. In the system dynamics approach validation is an on-going mix of activities embedded throughout the iterative model-building process. No single test suffices to validate a system dynamics model or provide a measure of its degree of validity. An observer wishing to make a judgement about the validity of a system dynamics study must follow much the same path as the modeler. There is no royal (or Student's) road to validation in the system dynamics approach.

Validity as Suitability and Consistency

For our purposes the question of validity is more useful when translated into two questions:

---

[5] Forrester (1961), pp. 126-127, discusses an analogous example involving oscillating inventories and points out that the statistician's least-squares criterion would lead to the conclusion that the better fit (point prediction) is obtained from a model showing *constant* inventories, rather than a model showing oscillations with incorect period.

- Is the model *suitable* for its purposes and the problem it addresses?

- Is the model *consistent* with the slice of reality it tries to capture?

The word validity connotes an absolute, as if a valid model were true. Yet no science claims to be able to prove truth, merely to try to get better at disproving some falsehoods. Suitability and consistency of a model are the most we can hope for, and at best we can claim only to be able to recognize their absence. The questions are perhaps best phrased in the negative: In what ways is the model unsuitable for its purposes or inconsistent with observed reality?

Rephrasing the question of validity into these two questions helps to reinforce the importance of model purpose in validation. Without taking into account the goals of a modeling study and the purposes of a model, it is impossible to decide whether or not the model is suitable. Suitable for what?

The system dynamicist (and other modelers, we submit) must test assiduously for suitability and consistency, revising the model until both are satisfactory. Subjectivity is involved, of necessity. If a perfect match between model and reality can never be achieved, how much consistency is sufficient? How much inconsistency is acceptable? Even the objective statistician relies on a subjectively chosen standard, such as the 95% confidence interval or significance level.

The system dynamicist seeks a shared consensus about the suitability and consistency of a model. If the goal of a study is the implementation of improved policies, then modeler and clients must together create a policy consensus, an inherently subjective (and fragile) meeting of minds.

Model Utility and Effectiveness

The final judgement of a model ought to be its utility and effectiveness. (That is not to say it is, but that in our opinion it ought to be.) How effective is the model in achieving the purposes of the study? Can the model or its results be used? Utility and effectiveness are related to suitability and consistency, for a model unsuitable for its purposes can hardly be expected to be effective in achieving them, but the notion of effectiveness goes beyond suitability and consistency and the usual interpretations of validation. A model's utility and effectiveness depend in addition on the degree to which the model communicates, helps to generate insights,

The Question of Validity

enhances understanding, and in general reaches and influences its audience.

The ultimate test of a policy-oriented model would be whether policies implemented in the real system consistently produce the results predicted by the model. But implementation requires confidence, which must be built by other means, and after all, waiting to see how the real system responds to a policy defeats the purpose of modeling. The penultimate judgement, a combination of a host of evaluations ranging from the technical to the subjective, should most properly be the model's utility and effectiveness.

Tests for System Dynamics Models

Although subjective in the last analysis, the process of judging the suitability and consistency of a system dynamics model includes a number of reasonably objective tests which have been touched upon previously and which we now bring together.[6] They divide naturally into four groups, as illustrated at the top of table 5.1. Any one of these tests by itself is certainly inadequate as an indicator of model validity. Taken together, they are a formidable filter, capable of trapping and weeding out weaker models and allowing passage only to those most likely to reflect something close to truth.

Tests of the Suitability of Structure

- *Dimensional consistency:* Do the dimensions of the variables in every equation of the model agree with the computation?

A system dynamics model containing equations whose dimensions do not balance properly is unsuitable for any purposes, except perhaps to serve as a bad example. Dimensional consistency contributes to model validation provided it is coupled with the requirement that all parameters have a meaningful real-world interpretation and are not to be invoked merely to fix dimensional inconsistencies.

- *Extreme condition tests in equations:* Does every equation in the model make sense even if subjected to extreme but possible values of its variables?

---

[6] The discussion that follows is adapted from Forrester and Senge (1980). See also Forrester (1961), chapter 13.

Table 5.1: Summary table of tests for building confidence in system dynamics models.

	Focusing on STRUCTURE	Focusing on BEHAVIOR
Testing SUITABILITY for purposes (tests focusing inward on the model)	Dimensional consistency Extreme conditions in equations Boundary adequacy -important variables -policy levers	Parameter (in)sensitivity -behavior characteristics -policy conclusions Structural (in)sensitivity -behavior characteristics -policy conclusions
Testing CONSISTENCY with reality (tests comparing the model with information about the real system)	Face validity -rates and levels -information feedback -delays Parameter values -conceptual fit -numerical fit	Replication of reference modes (boundary adequacy for behavior) -problem behavior -past policies -anticipated behavior Surprise behavior Extreme condition simulations Statistical tests -time series analyses -correlation & regression
Contributing to the UTILITY & EFFECTIVENESS of a suitable, consistent model	Appropriateness of model characteristics for audience -size -simplicity/complexity -aggregation/detail	Counter-intuitive behavior -exhibited by model -made intuitive by model-based analyses Generation of insights

The Question of Validity

The modeler should mentally test each equation (particularly each rate equation) to be sure it remains meaningful under extreme conditions, even if unlikely. Knowing that each equation by itself is robust relieves an observer of any anxiety that the model may at some time be operating in regions where one or more equations behave unreasonably. The extreme condition test can expose the need for nonlinearities (such as table functions or saturation effects) and distinctions between desired and actual conditions, and other model shortcomings.

- *Boundary adequacy tests for structure:* Does the structure of the model contain the variables and feedback effects necessary to address the problem and suit the purposes of the study?

Too small a boundary means important variables have been omitted, some policy levers of interest may be lacking, and/or some active feedback links have been ignored. Confidence in the ability of such a model to address its purposes is not justified. *Care must be taken to overcome the tendency to see system behavior as a consequence of external or exogenous events.* The boundary must be drawn large enough to capture important influences endogenously, as parts of information feedback processes. On the other hand, too large a boundary means that the model can obscure the central relationships between structure and dynamics in much the same way that the complexity of the real system obscures them. Before declaring a model invalid because more should have been included with the model boundary, be sure that the proposed additional structure is necessary, given the purposes of the model.

Tests of the Suitability of Model Behavior

- *Parameter (in)sensitivity:* Is the behavior of the model sensitive to reasonable variations in parameter values? That is, do the modes (patterns) of behavior exhibited by the model change with minor parameter changes? More critically, do the policy conclusions stemming from model-based analyses change with reasonable variations in parameter values?

It is possible to build confidence in a model showing parameter sensitivities, by justifying parameter values beyond a shadow of a doubt, but it is a lot easier to build confidence if the model tends to be insensitive to reasonable changes in parameter values.

- *Structural (in)sensitivity:* Is the behavior of the model sensitive to reasonable alternative formulations? That is, do the modes of behavior exhibited by the model change when equally likely alternative formulations are tested in the model?

Again, the more critical question is whether model-based policy conclusions vary with minor changes in structure. A model that exhibits such policy sensitivity tends to be judged unsuitable for policy analysis.

Tests of the Consistency of Model Structure with the Real System

- *Face validity:* Does the model's structure look like the real system? Is the model a recognizable picture of the real system? Are those who know the system most closely convinced that a reasonable fit exists between the rate/level/feedback structure of the model and the essential characteristics of the real system?

If the structure does not appear to fit the real system, then even if the behavior of the model is judged appropriate it is hard to place high confidence in analyses based upon the model.

- *Parameter values:* First, are the parameters themselves recognizable in terms of the real system, or are some of them contrived to make the units in some equations balance? Second, are the values selected for the parameters consistent with the best information available about the real system?

If data exists, was it used in a sensible way to justify values? If data is not available, what is the quality of the information used to make estimates? Whatever way the parameter values were estimated, do they make sense to those most knowledgable about the system?

Tests of the Consistency of Model Behavior with the Real System

- *Replication of reference modes* (boundary adequacy tests focusing on behavior): Does the model endogenously reproduce the various reference behavior modes that initially defined the study, including the problematic behavior, any observed responses to past policies, and any

The Question of Validity 317

conceptually anticipated behavior arising from hypothetical situations?

A system dynamics model that cannot reproduce its reference behavior modes is invalid.

- *Surprise behavior:* Does the model under some test circumstances produce dramatically unexpected behavior, not observed in the real system?

There are two possible conclusions: the model is incorrect and must be revised, or the model has correctly identified a mode of behavior that may occur in the real system under the same circumstances. If careful analysis leads to the conclusion that the mechanisms producing the anomalous behavior are real and the behavior meaningful, then confidence in the model justifiably increases, even though the behavior has yet to be observed in the real system.[7]

- *Extreme condition simulations:* Does the model behave reasonably under extreme conditions or extreme policies, even ones that have never been observed in the real system?

What is reasonable in such tests is subjective, but the tests are nonetheless extremely powerful in building confidence in a system dynamics model. A model that does not behave reasonably under extreme conditions is suspect, because one may not be certain when aspects of extreme conditions may crop up in ordinary runs.

- *Statistical tests:* Does the model output behave statistically like data from the real system?

While we have not emphasized statistical procedures for estimating parameters, and we have claimed that a single statistical measure of the validity of system dynamics model is not available, there is, nonetheless, a role for formal statistical procedures in testing how consistent the behavior of the model is with the behavior of the real system. Any statistical

---

[7] The lift in confidence that comes from validating surprise behavior is greater than if the model had been constructed with that behavior as an additional reference mode. A theory that can predict new information is more inspiring than one that was constructed to be consistent with that information to begin with, even though both fit the information equally well.

tests that can be performed on data from the real system can be replicated using data obtained from a representative run of the model under the influence of some randomness (using the NOISE function appropriately). If the tests on noisy model data produce essentially the same statistical results as in the real system, then confidence in the consistency of the model with the real system justifiably increases. Note that we are emphasizing statistical tests on the output of the whole model. The judgement of whether the results are essentially the same statistically is subjective rather than statistical.

Additional Characteristics Contributing to Model Utility and Effectiveness

- *Appropriateness of structure for audience:* Is the size of the model, its simplicity or complexity, and its level of aggregation or richness of detail appropriate for the audience for the study?

To be most effective, a model may have to trade off some richness of detail to be tractable to its audience. Conversely, an audience most influenced by an apparently exhaustive model may lead the modeler, striving for an effective model, to the inclusion of more detail than is strictly necessary for the structure, behavior, and policies under study. Taking account of audience characteristics will not enhance model validity, but it may enhance the audience's perception of model validity. (Is that politics or just good communication?)

- *Counterintuitive behavior:* In response to some policies, does the model exhibit behavior that at first contradicts intuitions and later, with the aid of the model, is seen as a clear implication of the structure of the system?

It has been observed that complex feedback systems tend to exhibit counterintuitive behavior.[8] Long-term reponses are characteristically the opposite of short-term responses, for example, and one's intuitions are often based on one or the other perspective, seldom on both. There is nothing like a

---

[8] Forrester (1971). Reprinted in Meadows and Meadows (1973) and Forrester (1975). The latter reference also includes excerpts from the discussion that followed the initial presentation of the paper before the Subcommittee on Urban Growth of the Committee on Banking and Currency in the House of Representatives.

The Question of Validity

little counterintuitive behavior to enhance the effectiveness of a model. (Is that showmanship or just the nature of complex feedback systems?)

- *Generation of insights:* Is the model capable of generating new insights, or at least the feeling of new insights, about the nature of the problem addressed and the system within which it arises?

A truly effective model will be a fertile source of new ideas and understandings, and perhaps new questions that go beyond the model's capabilities. Insights come, if they come at all, as much from the the model-building process as from a finished model. If somewhere along the way in a modeling study the feeling grows that new insights have been obtained, the model can hardly be judged invalid.

The Validity of the Project Model

Our development of the project model has addressed at various times most of the questions raised in this section. The model's adequacy under some tests, such as dimensional consistency, can be asserted at this stage without additional investigation. Its sensitivity to some parameter changes and alternative formulations has been explored, but more might have to be done to establish real confidence in those areas. Certainly, its boundary is adequate to generate the reference modes of the study, but it is unlikely that the model includes all the dynamic hypotheses capable of generating schedule and cost overruns. It has face validity--variables and their relationships sound and look like a real project--although it would be hard to recognize what particular kind of project it most resembles.

Some of the tests raise questions about the model. It may fail some extreme condition tests at this point. For example, it the project were to start with far too large a workforce, the level called NEWWF can become negative (the reader should check to see how this could happen). Fortunately, for values of NEWWF that are not too negative, the impact on the behavior of the model is not at all strange or silly: the table function for the effect of the experience on gross productivity EEXPGP makes sense even if the fraction experienced, FEXP, takes on (meaningless) values above one or below zero. Progress always remains positive, and the model will rather quickly right itself. But for some model purposes, the hiring and assimilation rates should be reformulated (and transfer out of the project added) to prevent a negative new workforce from ever arising.

Finally, some of the tests suggested above have not been applied to the project model. At this point, we leave them for the consideration of the reader.

6 POLICY ANALYSIS

Model-based policy analyses involve the use of the model to help investigate why particular policies have the effects they do and to identify policies that can be implemented to improve the problematic behavior of the real system. The goal is an understanding of what policies work and why. To draw understanding from model-based analyses, the modeler needs all the intuitions and skills discussed up to this point in the book. The only new technical skills required are the ability to translate real-world policy alternatives into changes in the model, and, on the flip side, the ability to interpret model changes meaningfully as policy options in the real system.

Policy alternatives in the real system correspond to one or a mixture of two kinds of model manipulations: parameter changes (including minor variations in table functions) and structural changes (changes in the form or number of equations). Both involve changing how decisions are made. Sensitive policy parameters in a model suggest leverage points in the real system -- places where a change in existing influences in the system would improve matters. Model changes involving new feedback structure suggest new ways of manipulating information in the real system to improve behavior. This chapter considers each type in turn. The project model will provide us with illustrations, but we will stop far short of a full investigation of the policy implications of the model. The final section of the chapter returns to the issue of validity, focusing on the question of the reliability of model-based policy recommendations.

6.1 PARAMETER CHANGES AS POLICY ALTERNATIVES

Changing the value of a parameter in rerun mode, running the model, and comparing the resulting model behavior with the base run can be interpreted as a test of the model's sensitivity to that parameter (as discussed in section 5.2). Sometimes, however, it can also be interpreted as a test of a policy change in the real system. Some parameters in the model can be classified as *policy parameters*--numbers whose values are to some extent within the control of actors in the real system. Testing the model's sensitivity to the value of a policy parameter may also test the sensitivity of the real system to the corresponding policy change. A sensitive policy parameter may well indicate just what the policy analyst seeks: a leverage point in the real system.

A number of examples of policy parameters have appeared in previous sections. The length of time a firm sets as its desired inventory coverage DIC can be interpreted as a policy parameter.[1] Changing DIC from, say, 8 weeks to 4 weeks represents a management policy to hold less inventory on the average, presumably to reduce inventory-holding costs. In the tiny epidemic model illustrated in section 3.4, the table function for susceptibles contacted per infectious person per day can be thought of as set of parameters partly within the control of actors in the system. Simulating a quarantine policy, for example, would involve changing somewhat the numbers in this table to represent decreasing the likelihood of contacts between susceptible and infected people.

The project model contains a number of potentially interesting policy parameters. Table 6.1 lists some of them, together with possible parameter changes and their interpretations as policy changes in the management of an R & D project.

For a detailed example, consider the constant ASMT, representing assimilation time, the time it takes for a person new to the project to become fully integrated into the project team and hence fully productive. A variety of policies might speed up the process of assimilation. Institute an explicit training program; set up a crash program for getting new people acquainted with the project team and the details of the project; pair people up in a buddy system in which each new person is assigned to an experienced person for supervision and guidance. Whatever the mechanism in the real system, it

---

[1] See the simple inventory control model used as an example in section 3.6. See also Lyneis (1980), p. 61.

Parameter Changes as Policy Alternatives 323

Table 6.1: Several policy parameters in the project model with possible interpretations of parameter changes

Parameter	Change	Policy Interpretation
Time to perceive progress	6 to 3 (months)	Speed reporting of progress; quicker response to changes in productivity
Workforce adjustment time	3 to 2 3 to 1 (months)	More aggressive hiring. More reliance on transfers within the firm than hires.
Schedule adjustment time	6 to 4 (months)	Less delay in responding to pressures to change the scheduled completion date
Assimilation time	6 to 3 (months)	Speed assimilation of new workers into the project team; a training program; more reliance on transfers within the firm than new hires

is simple and potentially instructive to simulate the model with a reduction in ASMT.

Figure 6.1 shows the results of a policy run in which ASMT is reduced from 6 to 3 months. The following table shows the printed summary information. The dynamics of the policy run are identical to the base run. Productivity is a bit higher because the detrimental effects of inexperience are not as great in the policy run, but otherwise the plots show few differences. Comparison of the printed summaries shows that the policy run takes 1 month less time and hires about 36 people instead of 38. As a result the policy run requires about 129 fewer man-months and costs about $386,000 dollars less than the base run. The schedule overrun is still awesome, but the run suggests that policies that can reduce the time to assimilate new workers have the potential to improve the behavior of the system

	TIME E+00	WF E+00	SCD E+00	CUMEFF E+00	COST E+03
Initial values	0.00	3.0	40.0	0.0	0.0
Base run	61.50	38.0	59.0	2082.5	6247.4
Policy run	60.25	36.4	57.9	1953.8	5861.4

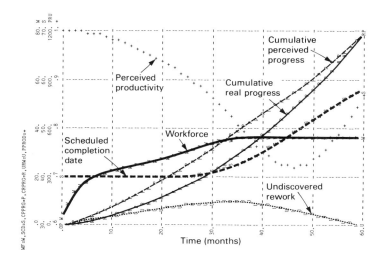

Figure 6.1: Simulation of the revised project model with assimilation time ASMT reduced from 6 to 3 months.

Uncovering the Reasons for the Effects of a Policy

It is not enough to know that a particular policy improves model behavior. The critical question is why. Because the model is not the real system, no one will implement a policy just because it made the model behave better. What is needed is fundamental understanding of why a particular policy improves model behavior. That understanding is then compared to what is known or believed about the real system. Only then does a model-based policy analysis have a chance to lead eventually to decisions about policy implementation in a real system.

If past experiences with the model do not clearly identify the reasons behind the success or failure of a policy simulation, model tests to locate critical structure are necessary. The techniques involved in the process of understanding model behavior have been described in detail in section 5.1. Here we illustrate the process briefly by locating the aspects of the project model that are responsible for the slight improvements observed in the policy run given here involving ASMT.

Parameter Changes as Policy Alternatives

Suppose we hypothesize that the reason for the improvement is the difference in productivity of new workers and experienced workers. The hypothesis suggests that, if we eliminate that difference and run the model with and without the policy parameter change, the runs should be identical. Therefore we change the table for the effect of experience on gross productivity EEXPGP to a string of one's, thereby incorporating the assumption that new workers and experienced workers are equally productive, and run the model with and without the policy change reducing assimilation time to three months. The printed summary results are the following.

The revised project model with TEEXPG=1/1/1/1/1/1:

TIME	WF	SCD	CUMEFF	COST
E+00	E+00	E+00	E+00	E+03
0.00	3.0	40.0	0.0	0.0
61.00	35.9	58.5	1955.0	5867.1

The revised project model with TEEXP=1/1/1/1/1/1 and ASMT=3:

TIME	WF	SCD	CUMEFF	COST
E 00	E 00	E 00	E 00	E 03
0.00	3.0	40.0	0.0	0.0
60.00	35.4	57.6	1892.0	5675.9

Since the results are not the same, our hypothesis about the structure responsible for improvement in the original model when ASMT is reduced must be faulty. In a complex model a lot of thought and a number of simulations may be necessary to produce a complete understanding, but here the reason our hypothesis failed is rather obvious: the model includes *two* kinds of effects of experience, and we eliminated only one of them. If we also wipe out the effect of experience on fraction satisfactory EEXPFS, the the model behaves the same for any value of ASMT:

The revised project model with TEEXPG=1/1/1/1/1/1 and TEEXPF=1/1/1/1/1/1, with ASMT=6, or with ASMT=3:

TIME	WF	SCD	CUMEFF	COST
E 00	E 00	E 00	E 00	E 03
0.00	3.0	40.0	0.0	0.0
59.25	34.8	56.9	1835.8	5507.4

We conclude in this simple example that the reason policies that reduce ASMT work in the model is the different characteristics the model assumes for NEWWF and EXPWF, the new and experienced workforces. To the extent those different characteristics exist among the personnel in real projects, the model suggests that efforts to reduce the time it takes to assimilate new personnel may bear a little bit of fruit.

The method used here to understand the reasons behind a policy improvement involves deactivating pieces of model structure until the policy run behaves no differently from the base run with the same pieces deactivated. Refer back to section 5.1 for more on understanding model behavior.

Caution

There is an important condition attached to potential conclusions from policy simulations. The policy investigated here assumed that only ASMT changed and that there were no detrimental side effects of the policy change. What else might change in the system when the policy to speed assimilation is implemented? Inferences from the policy run are not really justifiable until the question is fully explored.

There are two reasons for caution. First, on a conceptual level there is the need to be sure that the essence of the real-world policy has been accurately interpreted in terms of the model, to the extent that model aggregation allows. Second, on a more technical level, the fact that ASMT was modeled as a constant means that it is virtually certain the model contains no feedback effects from ASMT to other parts of the model. Both observations suggest that *the modeler must to some extent become part of the system to simulate accurately a policy change.* The modeler acts as the missing feedback links, changing other parameters manually, as it were, to simulate whatever far-reaching effects the policy change involves. The thinking is of the same conceptual sort that produced the initial feedback view of the system.

As an example, in the project model a more complete vision of a policy to speed assimilation of new workers would involve considerations of productivity. It is unlikely that ASMT can be reduced without some cost in lowered productivity of experienced workers. On the other hand, more attention paid to new workers should probably *increase* their effectiveness even before they pass into the experienced category. These observations suggest that in addition to changing ASMT to simulate more rapid assimilation of new people, we should lower somewhat the values of EEXPGP, the effect of experience on gross productivity, and perhaps raise those of EEXPFS, the effect of experience on fraction satisfactory.

Parameter Changes as Policy Alternatives 327

We leave it to the reader to speculate on the likely results of such a policy simulation, or to run the model to determine them.

These cautions apply also to policy simulations involving alterations in table functions.

Table Functions as Policy Parameters

Table functions frequently represent assumptions about existing policies in the real system. Table 6.2 gives some examples from the project model, together with possible interpretations of potential changes.

As an illustration, consider changes in WCWF, the willingness to change workforce, shown in the table function in figure 6.2. The variable represents the assumption that project managers become progressively more unwilling to bring new people into the project as it appears to near completion. Believing that it takes at least six months to bring new people up to speed, managers in our project model (curve A) begin to reduce significantly their tendency to bring on new people as early as twelve months before the end of the project (at which time WCWF equals 0.3).

A reasonable policy to try to reduce the schedule overrun would be to keep bringing more workers into the project even though it appears they can not be fully assimilated into the project team by the time the project will end. The alternative curves B and C shown in figure 6.2 represent such policies. The simulation run shown in figure 6.3 shows the behavior of the model substituting curve B for WCWF (obtained by setting TWCWF=0/0/.1/.9/1/1/1/1 in rerun mode). The printed summary information follows:

	TIME E+00	WF E+00	SCD E+00	CUMEFF E+00	COST E+03
Initial values	0.00	3.0	40.0	0.0	0.0
Base run	61.50	38.0	59.0	2082.5	6247.4
Policy Run	55.75	50.7	52.6	2169.2	6507.6

The policy reduces the schedule overrun substantially, dropping it from 21.5 months in the base run to 15.75 months in the policy run. The five-and-a-quarter month improvement is still just a drop in the bucket, however, and comes with a price tag of about $260,200, as the COST figure rises from $6.25 million in the base run to over $6.5 million in the policy run. The reason for the speedier progress and the higher cost are obvious. Many more people are brought into the project midway through.

Table 6.2 Several table functions from the project model interpreted as policy parameters

Change in table function	Policy interpretation
Normal fraction satisfactory TNFSAT=.5/.55/.63/.75/.9/1 changed to TNFSAT=.6/.63/.7/.8/.92/1	Greater quality control; closer allignment of project personnel to project purpose
Time to detect rework TTDRW=12/12/12/10/5/.5 changed to TTDRW=6/6/6/5//3/.4	Greater quality control, more rapid testing of prototypes
Current project definition TCPD=800/830/900/1000/1140/1200 changed to TCPD=1000/1000/1000/1000/1140/1200	Intentional overestimation of job
or to TCPD=800/830/900/1000/1140/1000	Redefinition of the project downward at the end
Willingness to change workforce TWCWF=0/0/0/.1/.3/.7/.9/1 changed to TWCWF=0/0/.1/.9/1/1/1/1	More willingness to change the workforce late in the project

They perform more tasks sooner, but not enough time is saved to offset the added wages required.

On the theory that, if a little change helps somewhat, a lot of change will help a lot more, we might try alternative C in figure 6.2 (TWCWF=0/.1/.9/1/1/1/1/1). The results would be mixed, however, as the following printed summary shows.

	TIME	WF	SCD	CUMEFF	COST
	E 00	E 00	E 00	E 00	E 03
Initial values	0.00	3.0	40.0	0.0	0.0
Base run	61.50	38.0	59.0	2082.5	6247.4
Policy Run	51.50	75.6	47.9	2308.1	6924.3

With policy C we drop the schedule overrun to just 11.5 months, an improvement over alternative B of 4.25 months but at the cost of an additional $416,700. The improvement of policy B over the base run was 5.5 months at a cost of $260,200; we are well past the point of diminishing returns.

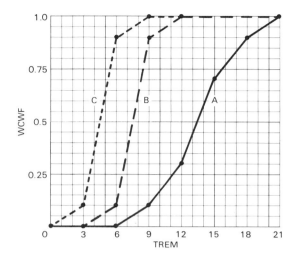

Figure 6.2: Table function for WCWF, willingness to change workforce: curve A is the original policy in the revised project model; curves B and C are alternative policies

Trying to solve the problem of overruns by bringing more and more people into the project right up to the end has its limitations. A glance at figure 6.4 shows why. The figure shows that in the simulation of policy C the tremendous push to bring in new personnel midway through the project reduces the experienced fraction of the workforce considerably. The less experienced workforce is less productive and performs a larger fraction of work requiring rework.

Limits on Policy Parameters

In the excitement of the chase for the perfect policy, the enthusiastic modeler may find a mix of policy parameter changes that work beautifully in a simulation but would be impossible to implement in the real system. It is easy to type a new number in a rerun but quite another thing to make the corresponding change in the real system. The sad fact is that the physics of the system delimits the range of possible values of any parameter in the model, including policy parameters. Furthermore it is likely that the more extreme the parameter change is, the more difficult the corresponding pol-

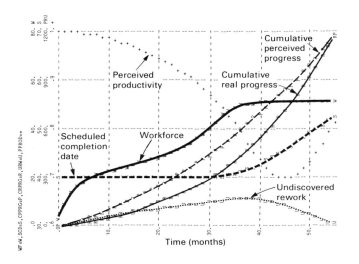

Figure 6.3: Simulation of the revised project model with greater willingness to change the workforce late in the project: TWCWF=0/0/.1/.9/1/1/1/1

icy is to implement in the real system. Consequently, the modeler must be careful to keep policy parameter changes within clearly achievable bounds (except when testing extreme conditions to gain insight into the behavior of the model). Bear in mind that the model does not test what can be implemented, merely what is likely to happen if it were.

Finding Sensitive Policy Parameters

It would be nice to have a set of rules of thumb for locating sensitive and beneficial policy parameters in a system dynamics model. In fact, given the apparent technological sophistication of a quantitative computer model, it is reasonable to expect such rules. Unfortunately, they do not yet exist in a generally applicable form. The model brings the real system into the laboratory so that it can be experimented with, but the direction of the experiments is guided only by the intuition of the modeler. Experience with dynamic models helps, of course, as does great familiarity with the particular problem being modeled, but we are very far from reducing

Parameter Changes as Policy Alternatives

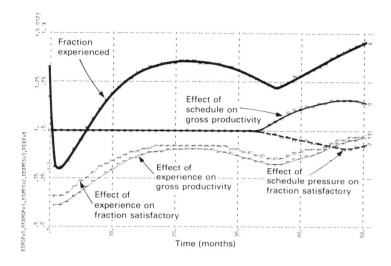

Figure 6.4: Simulation of the revised project model with still greater willingness to change workforce near the end of the project: TWCWF=0/.1/.9/1/1/1/1/1

policy analysis to a science. Here we can make only a weak stab at suggestions.

The first requirement for locating sensitive policy parameters in a model is intimate familiarity with the behavior of the model under a variety of scenarios. There are no substitutes for building and understanding the model in stages and for running lots of simulations. Exercising that understanding in verbal descriptions of runs may help to identify potentially sensitive policy parameters (for example, the cause-and-effect story for a run of the simple project model in section 5.1). Second, locating the feedback loops which dominate the behavior at various times is useful. Model behavior is not likely to be sensitive to a parameter that has no influence on any dominant loops. Do not seek the cure for urban decay solely among the feedback loops that dominated during the growth phase of the city; look particularly among the loops that conspired to halt growth. Third, identifying parameters that influence positive loops might be helpful. Since positive loops tend to be destabilizing, a parameter that affects the strength of a positive loop may well have an observable effect on model behavior. One can dampen the oscillations of a predator-prey model, for example, by slowing the tendency of the prey to multiply or increasing that of the predator. While vasectomies for rabbits (or aphrodisiacs for coyotes) are hardly a feasible policy to stabilize such a system, the importance of parameters affecting positive feedback loops can apply to other oscillating systems.

We have observed that feedback systems tend to compensate internally for parameter changes -- that model behavior tends to be rather insensitive to changes in parameters. That property means that policies represented as parameter changes frequently tend not to be very effective in system dynamics analyses. Small improvements, sometimes even significant ones, can be obtained from judicious adjustments in policy parameters, but dramatic results are more commonly obtained with changes in the policy structure, the subject of the next section.

6.2 STRUCTURAL CHANGES AS POLICY ALTERNATIVES

The structure of a feedback system--the patterns of its flows of material and information--tends to be the strongest single determinant of its behavior over time. Consequently, policy improvement in system dynamics studies frequently involves the addition of new feedback links that represent new, beneficial ways of manipulating available information in the system. Perhaps a sluggish information stream can be bypassed or shortcircuited; a different way of projecting needed resources might bring them more in phase with demand; perhaps basing desired output on the average order rate rather than on average production would have a desirably stabilizing influence; and so on. The system dynamics focus on information paths and loops is the basis of its value as a policy analysis tool.

This section discusses the addition of model structure to represent policy alternatives. It begins with the inclusion of simple constants, CLIPs, and SWITCHes to simulate policy changes. While not representing changes in feedback structure, such additions require changing the format of some equations after policy intervention points have been hypothesized. The section's primary focus, however, is on the conceptualization of additional model equations to represent new policy options that alter or extend the feedback structure of the system.

Structural Changes as Policy Alternatives

Adding Policy Parameters

The sort of R&D project our model represents gets into trouble in spite of managerial attempts to perceive productivity and progress accurately. A natural response to several such experiences is intentionally to overestimate the work to be done, the effort perceived remaining, or the manpower required. In our formulation of the project model, overestimating the size of the project is easy to simulate, merely by changing the values in the table function for the current project definition CPD. But, as it is formulated, the model can not overestimate, say, the desired size of the workforce. Instead, it tries to make a rational estimate of the number of people necessary to finish on time.

So as to allow the possibility of a policy that deliberately overestimates the workforce needed, we must add an overestimation parameter OE to the computation for the indicated workforce:

```
A     IWF.K=OE*(EPREM.K/TREM.K)
NOTE        INDICATED WORKFORCE (PEOPLE)
C     OE=1
NOTE        OVERESTIMATION FACTOR
NOTE           (DIMENSIONLESS)
```

(The original formulation for IWF is shown in the parentheses.) In the base case OE would be 1 and would have no effect on the model. To simulate a deliberate over estimation of say, 25 percent, OE would be raised to 1.25 in a rerun.

The addition of such policy parameters to a model poses no technical problems. The only potential difficulties are conceptual. Note, for example, that in the project model it makes a difference where the overestimation factor is added. Here, only the hiring policy of the project is affected (through IWF, to WFS, and finally to HR). If OE were inserted instead in the equation for EPREM, then the scheduling policy would be changed as well. Overestimating only in hiring produces the intriguing behavior shown in figure 6.5, where OE was set to overestimate by 15 percent the size of the workforce needed.

In this simulation there is an argument among the project managers. It is as if the person in change of the hiring policy insists on bringing in more people, in spite of the equally insistent director of scheduling who keeps claiming the project is well ahead of schedule. The hiring manager stops (after 24 months) only when everyone thinks the project is

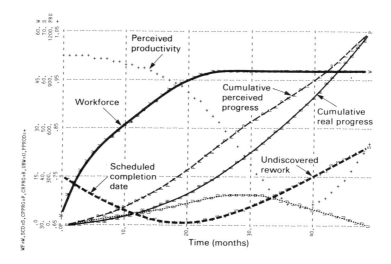

Figure 6.5: Simulation of a policy of deliberately overestimating by 15 percent the size of the indicated work force in the revised project model (OE=1.15)

within 6 months of completion, but then the project meanders on for another two years.

If the overestimation factor OE were placed in the equation for effort perceived remaining, EPREM, then the argument would disappear. Both managers would agree to the same distortion in the interests of reducing overruns.
Remarkably, however, a much *greater* overrun would result! The project would take about 58 months when EPREM is intentionally overestimated by 15 percent, whereas in figure 6.5 it took only 48 months.

The reason for the difference is vitally important for policy implementation. A careful analysis of the model with OE in the equation for IWF reveals that if OE is anything other than 1 the scheduling and hiring activities can never both be in equilibrium at the same time. If scheduling is in equilibrium, then SCD = ICD and TREM = TREQ = EPREM/WF, so the following must hold:

```
IWF.K    = OE*EPREM.K/TREM.K
         = OE*EPREM.K/TREQ.K
         = OE*EPREM.K/(EPREM.K/WF.K)
         = OE*WF.K
```

Structural Changes as Policy Alternatives

Thus, if OE ≠ 1, equilibrium in scheduling means disequilibrium in hiring. The converse can be shown in similar fashion. The point of the analysis is that overestimating in just IWF throws the model's rational hiring and scheduling behavior completely out of kilter. The model would, in fact, continue to hire people *ad infinitum* if it were not for the constraint imposed as the project comes within six months of completion.

Placing OE in EPREM does not have the same irrational character. The effort perceived remaining will be overestimated throughout, but as the project nears completion EPREM will head toward zero, as it should. The policy of placing OE in IWF is not a rational, implementable policy. Overestimating in the effort perceived remaining, however, makes a certain amount of sense.

In a fashion similar to the inclusion of OE, one could add simple constants to the model to test various other policies involving deliberately biased perceptions or estimates. For example, in the computation of effort perceived remaining, cumulative perceived progress could be multiplied by a fraction guessed satisfactory, in a deliberate attempt to manage the project better with a fudge factor.

Changing a Policy Parameter in Mid-Run

Frequently, the modeler or the clients are interested in the effect of a *change* in policy in the course of a simulation. "People will produce what's necessary," a worldly wise project manager might say, "if they were just given enough incentives. So show me the effect of introducing bonuses for bringing the project in on schedule." The real effects of bonuses on productivity could be debated endlessly. Some have even argued that they can have negative effects. But suppose we wanted to use our project model to test how effective a successful incentive plan might be. We would be interested in the dynamic consequences of a sudden, permanent increase in worker productivity partway through the project. The policy could be represented in a simplified way by a CLIP (or STEP) function raising the parameter normal gross productivity NGPROD to a new value. Changing the equation for gross productivity

    A   GPROD.K=NGPROD*EEXPGP.K

to

    A   GPROD.K=CLIP(NGPROD,GPI,PST,TIME.K)*EEXPGP.K

sets up the model to simulate a change in normal worker productivity from NGPROD to GPI when TIME exceeds PST. The following constants complete the formulation:

```
C     NGPROD=1
NOTE       NORMAL GROSS PRODUCTIVITY
NOTE        (TASKS/PERSON/MONTH)
C     GPI=1.2
NOTE       GROSS PRODUCTIVITY AFTER
NOTE        INCENTIVES
C     PST=24
NOTE       POLICY STARTING TIME (MONTH)
```

The CLIP (or FIFGE) function in this formulation takes the value NGPROD as long as TIME ≤ PST, after which it takes the value GPI. The values chosen above represent a 20 percent increase in worker productivity attributable to a (remarkably successful) incentive policiy.

Costs would no longer be simply proportional to the final value of cumulative effort; partway through the project new costs per man-month would rise as the incentive program is put in place. So we would have to reformulate cumulative cost in a level equation analogous to the equation for the man-months of cumulative effort CUMEFF. The jump in new costs per man-month could again be captured in a CLIP:

```
L     COST.K=COST.J+DT*(WF.J*ESPGP.J*CPMM.J)
N     COST=0
NOTE       CUMULATIVE COST (DOLLARS)
A     CPMM.K=CLIP(CPMMN,CPMMI,PST,TIME.K)
NOTE       COST PER MAN-MONTH
NOTE        (DOLLARS/PERSON/MONTH))
C     CPMMN=3000
NOTE       COST PER MAN-MONTH NORMAL
C     CPMMI=3200
NOTE       COST PER MAN-MONTH UNDER INCENTIVES
```

The value of CPMMI might represent the assumptions that $2,000 of the $3,000 cost-per-man-month is salary and the 20 percent rise in productivity is obtained by increasing salaries 10 percent.

The behavior of the model simulated with this policy looks much the same as the base run, with the slight improvement in cost and schedule shown in the summary data that follows:

	TIME E+00	WF E+00	SCD E+00	CUMEFF E+00	COST E+03
Initial values	0.00	3.0	40.0	0.0	0.0

Base run	61.50	38.0	59.0	2082.5	6247.4
Incentives	56.50	36.6	54.3	1820.0	5716.3

While our purpose here is to illustrate the use of the CLIP (or FIFGE) function in policy tests, we should note in passing that the remarkably successful incentive policy, raising productivity 20 percent for more than two years of the project, saved only 5 months and dropped the basic cost only $423,400. Why does a 20 percent rise in productivity for more than half the project reap only 7 percent and 8 percent reductions, respectively, in cost and duration?

A clue to the answer lies in the observation that the workforce hired is actually *less* in the policy run than in the base run. While using incentives to speed progress, these project managers enrolled fewer workers! The higher worker productivity the policy intended to produce is actually the cause of the underhiring. Managers in this run came to believe that with the higher productivity they eventually perceived, the project could be completed on time with fewer people. Silly project managers, you might say, yet they were trying to do their rational best, given the information they had. In fact, if they were quicker to perceive productivity, or adopted the incentive policy earlier in the project, they would have done even worse.

The structure responsible for the offsetting effects observable in this simulation is referred to as compensating feedback. Figure 6.6 shows a causal-loop diagram of the parts of the system that are involved. GPROD eventually effects the effort perceived remaining through two information paths. In both paths an increase in GPROD tends to lower EPREM. But then the goal-seeking negative feedback loop controlling the size of the workforce adjusts to the lower EPREM by hiring fewer people. Thus, the rise in GPROD boosts the apparent progress rate, only to have it pulled back down by the resulting smaller workforce. The system still tries to match perceived progress to the work schedule, and in the process tends to counteract the incentive policy.

Although the effect here is relatively minor, compensating feedback effects are a common and profoundly significant part of the structure and behavior of complex systems. They are a fundamental cause of policy resistance, the tendency of real systems to be insensitive to policy initiatives.

Adding Policy Feedback Loops

We turn now from the manipulation of policy parameters to the design of policy feedback structures. The modeler seeks

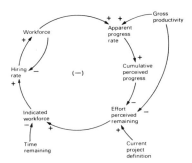

Figure 6.6: Feedback structure in the project model which compensates for changes in productivity and thereby weakens the incentive policy

additions to model structure that improve the behavior of the system by creating new loops of action and information. The key question is what links and loops to add. The number of alternatives is practically infinite compared to the number of policy parameters in a model. The modeler will have to be more selective in testing structural political alternatives. As with policy parameters, the primary guidance for the discovery of potentially powerful structural changes comes from familiarity with the real system and the model. But we have in addition the guidance that comes from our understandings of the behavioral tendencies of positive and negative loops. The addition of a link that creates a positive loop may have a significantly destabilizing influence on model behavior, while a new link that creates a negative loop has the potential to add stability. Unfortunately, things are seldom what they seem, particularly in feedback systems, so even these generalities must be tempered by specific circumstances.

Adding a Minor Negative Loop

To illustrate a situation in which structural principles can guide policy analyses, let us consider a simple system designed to adjust a workforce to meet the demand for its services. Figure 6.7 shows a flow diagram containing a level for the workforce WF, a level for the backlog B of orders, and a hiring policy striving to adjust WF to keep B at the level of desired backlog DB. The policy simply says desired output DOUT is based upon the average order rate AOR for services, plus a correction term for the backlog:

```
A    DOUT.K=AOR.K+CB.K  DESIRED OUTPUT
NOTE      (UNITS/WEEK)
```

Structural Changes as Policy Alternatives

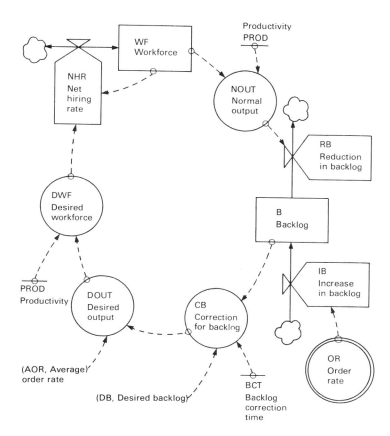

Figure 6.7: Flow diagram of a simple workforce-backlog system that exhibits oscillations in response to disturbances in the order rate OR

```
A     CB.K=(B.K-DB.K)/TCB  CORRECTION FOR
NOTE       BACKLOG (UNITS/WK)
C     TCB=2  TIME TO CORRECT BACKLOG (WEEKS)
```

(Average order rate AOR is not of interest here, but for the sake of specificity it could be a smooth of the order rate OR.) The policy then determines the desired workforce DWF, the number of people required to produce DOUT. The model assumes that productivity PROD is fixed and well-known, so all that is required is a simple division:

```
A    DWF.K=DOUT.K/PROD   DESIRED WORKFORCE
NOTE      (PEOPLE)
```

Hiring or firing then brings the actual workforce to equal the desired workforce after an adjustment delay, the time to correct workforce:

```
R    NHR.KL=(DWF.K-WF.K)/TCWF
NOTE      NET HIRING RATE (PEOPLE/WEEK)
C    TCWF=4
NOTE      TIME TO CORRECT WORKFORCE (WEEKS)
```

Normal worker output, NOUT, then determines the rate at which the backlog of orders for services is reduced:

```
A    NOUT.K=WF.K*PROD
NOTE      NORMAL OUTPUT (UNITS/WEEK)
R    RB.KL=NOUT.K
NOTE      REDUCTION IN BACKLOG (UNITS/WK)
```

Such a simple system oscillates in response to disturbances in the exogenous order rate OR. To make the claim reasonable without simulation (and to lead eventually to a nice structural insight), note that the structure in figure 6.7 is very similar to the mass on a spring structure diagramed in figure 6.8. Both are negative, second-order feedback loops. Their slight differences are largely cosmetic; only the extra minor link in figure 6.8 has any dynamic effect (and we shall soon see what that effect is). The point is that we know a mass on a spring oscillates, so intuitively we should expect the workforce-backlog system to oscillate, too.

Given such a workforce system, it would be desirable to find a policy change that acts to dampen the system's tendency to oscillate. After all, such instabilities in personnel and services would have nasty effects on service reputation and worker morale. Changes in policy parameters, such as TCWF or TCB, have to be extreme before damping is significantly affected. To discover a potentially beneficial structural change, we ask what a manager in such a system would actually do? It is likely that he or she would not simply hire and fire people to adjust the output of the workforce--people would first be placed on undertime or overtime to try to bring the backlog of orders to equal its desired value. So we are led naturally to consider the addition of an effect of undertime/overtime EUTOT which could be formulated as described in section 4.3:

```
R    RB.KL=NOUT.K*EUTOT.K
```

Structural Changes as Policy Alternatives

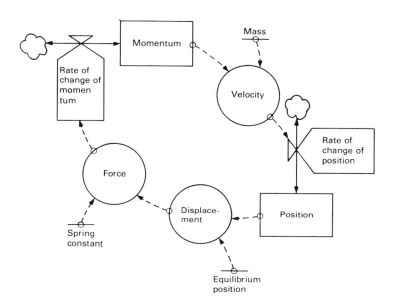

Figure 6.8: Flow diagram of a model of a mass on a spring

```
NOTE    REDUCTION IN BACKLOG (UNITS/WEEK)
A    EUTOT.K=TABHL(TUTOT,DOUT.K/NOUT.K,.7,1.3,.1)
NOTE    EFFECT OF UNDERTIME/OVERTIME
T    TUTOT=.85/.87/.91/1/1.09/1.13/1.15
NOTE    TABLE FOR EUTOT
```

Figure 6.9 shows the new structure. Not surprisingly, the addition of EUTOT dramatically damps the system, even for modest assumptions about the extent to which productivity can be increased or decreased by undertime and overtime.[2] (Here we have assumed workers vary their hours at most plus- or-minus 15 percent.)

The damping effect of this new link can be generalized. The new link creates a minor negative loop--a negative feedback loop containing just one level. It is a principle of feedback system behavior that *the addition of such a minor*

---

[2] The reader should try to verify this claim. We strongly urge that simulation experiments with the workforce/backlog system be performed, with and without the undertime/overtime policy. Oscillating systems abound; the more experience the modeler has with them, the better.

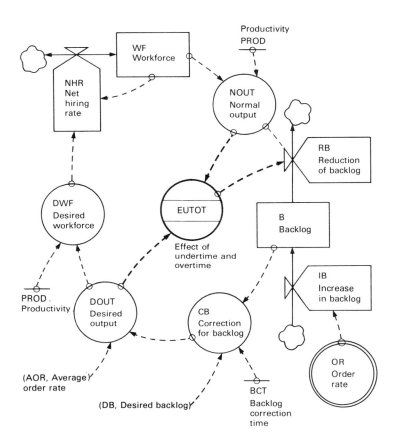

Figure 6.9: Flow diagram of the workforce-backlog system with the addition of a variable representing undertime and overtime, creating a minor negative feedback loop around the backlog level

negative loop in an oscillatory structure has a damping effect. Presumably, our intuitions about what undertime and overtime would do make the principle reasonable in the workforce-backlog example. The mass on a spring system provides further confirmation. Note that the system in figure 6.8 lacks any minor negative loops--it is, in fact, a pure oscillator that exhibits no damping. Putting the bouncing apparatus in a bottle of shampoo, say, would tend to damp out the oscillations, as the speed of the object creates fluid

Structural Changes as Policy Alternatives 343

friction that dissipates some of the force generated by the spring. Structurally, the fluid friction effect would be captured in a new feedback link from velocity directly back to force; the resulting new minor negative feedback loop would add damping to the system.

So if instability is a problem in your system, one way to attack it is to search for a policy that amounts to the addition of one or more minor negative feedback loops.

The principle of the minor negative feedback loop is a simple, well-known example of a growing set of principles relating feedback structure to dynamic behavior. The early work of Jay W. Forrester (1968b) and the more recent contributions of Alan Graham (1977) have extended and codified our understandings of these matters significantly. Their works are essential reading for practicing system dynamicists. Yet principles relating structure and behavior in nonlinear systems will probably always be too specialized to provide universally applicable policy guidelines. They can enhance, but not replace, the familiarity with the system that comes from the modeling process and many simulation experiments. For examples of the latter, we turn again to the project model.

Structural Policy Alternatives in the Project Model

The project model contains two primary causes of overruns in schedule and manpower: the assumptions that undiscovered rework will exist and that the initial project definition understates the size of the project. A more complex vision of project dynamics will no doubt contain more hypotheses capable of generating overruns. We know that we can cure the overruns in the model simply by raising FSAT to 1 and setting the table function for CPD to a constant, namely, 1,200 tasks. However, we must work with the assumption that such changes in the nature of a real project are simply impossible. We seek a policy, or a mix of policies, that can improve project performance in spite of the certain existence of undiscovered rework and underestimates in the scope of the project.

The project model has the peculiarity that a major component of its dynamic hypothesis--the level of undiscovered rework--must be assumed to be unknowable. It would be nice to suggest a variable called perceived undiscovered rework, obtained, say, as a SMOOTH of URW, and to subtract it from cumulative perceived progress CPPRG, but such a structural change makes no sense in the real system. People can't perceive what is inherently undiscovered. The structural change we will illustrate has the same basic purpose as this impossible policy but has a chance of being implementable in some real projects.

The information about undiscovered rework which is available to project managers is the rate at which URW is being detected. ("Thorndike, that's the 7th time in the last 10 months you've come in here to report a major boo- boo!") In our model, however, no use is made of that information. The structural policy change we shall formulate involves using the average rate of *detecting* rework to make a guess at the amount of undetected rework present in the current cumulative perceived progress.

The technique for using an average outflow rate to estimate a level is based upon the equilibrium interpretation of a time constant as an average dwell time (see section 4.1). Simple algebra tells us if

OUTFLO = LEVEL/TCONST,
then
LEVEL = OUTFLO*TCONST.

Hence *the size of a level is likely to be close to its average outflow rate multiplied by its (average) time constant.*[3] For example, if people are entering and leaving a supermarket at the rate of 33 every 15 minutes, and we guess the average shopping spree lasts, say, 45 minutes, then we can estimate there are about 100 shoppers inside, since

OUTFLO*TCONST = (33/15)*(45) = 99.

Suppose that the managers in our project model make a similar guess at the level of undiscovered rework. They would compute

A   AURW.K=ADURW.K*ATDRW.K
where
    AURW   = assumed undetected rework (tasks),
    ADURW   = average rate of detecting undiscovered rework
                   (tasks/month),
    ATDRW   = assumed time to detect rework (months).

They would then be able to combine AURW with their current perception of cumulative progress, CPPRG, to make a somewhat better guess at cumulative progress:

---

[3] Such an estimate is exact in equilibrium. If the level is rising, it *underestimates* the level; if the level is falling, it *overestimates* it.)

Structural Changes as Policy Alternatives

```
A    ACPRG.K=CPPRG.K-AURW.K
```

Assumed cumulative progress, ACPRG, would then be used instead of CPPRG to compute the effort perceived remaining:

```
A    EPREM.K=(CPD.K-ACPRG.K)/PPROD.K
```
where
- EPREM = effort perceived remaining (man*months),
- CPD = current project definition (tasks),
- ACPRG = assumed cumulative progress (tasks),
- PPROD = perceived productivity (tasks/person/month).

To carry out such a policy, project managers would have to estimate the average rate of detecting undiscovered rework, ADURW, and make some kind of guess at an assumed time to detect rework, ATDRW. For the former, we (and the project managers) could use a simple SMOOTH:

```
A      ADURW.K=SMOOTH(DURW.JK,TADURW)
NOTE       AVERAGE DETECTION OF UNDISCOVERED
NOTE       REWORK (TASKS/MONTHS)
C      TADURW=8
NOTE       TIME TO AVERAGE DURW (MONTHS)
```
where
- DURW = detection of undiscovered rework (tasks/month).

But project managers have practically no information on which to base ATDRW. The most that can be said is that the time to detect undiscovered rework is bound to be bigger near the beginning of a project than at the end. That suggests we formulate ATDRW as a table function of of FPCOMP, the fraction perceived completed. But what values should we assume in the table? Optimists might guess detection times falling from, say, 8 to 2 months over the course of the project, while pessimists could easily propose values at the beginning as high as 18 or 24 months. We'll start with an optimist's guess:

```
A    ATDRW.K=TABLE(TATDRW,FPCOMP.K,0,1,.2)
T    TATDRW=8/8/7/5/3/1.5
```

We can simulate the optimist, the pessimist, and even the perfect guesser in various scenarios to see who would most frequently fair the best in completing the project on time.

Figure 6.10 shows a flow diagram of the new policy structure and its connections with the project model. Figure 6.11 shows a run of the model with this estimation policy. The

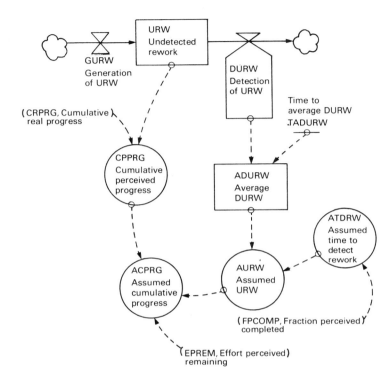

Figure 6.10: Flow diagram of the structural changes involved in the policy to estimate a quantity of assumed undetected rework

summary data that follows indicates that the policy has a small effect in reducing overruns.

	TIME E 00	WF E 00	SCD E 00	CUMEFF E 00	COST E 03
Initial values	0.00	3.0	40.0	0.0	0.0
Base run	61.50	38.0	59.0	2082.5	6247.4
Low ATDRW	58.50	41.8	57.2	2106.6	6319.8

Taking the values for the assumed time to detect rework given by the table TATDRW=18/18/16/12/6/3, the pessimist brings the project in two months earlier than the optimist, but at additional cost, as shown in the table below and the run in figure 6.12.

Structural Changes as Policy Alternatives

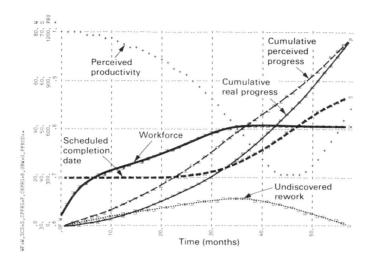

Figure 6.11: Simulation of the project model with the policy of assumed undetected rework, with an optimist's assumption of the assumed time to detect rework, given by TATDRW=8/8/7/5/3/1.5

	TIME	WF	SCD	CUMEFF	COST
	E+00	E+00	E+00	E+00	E+03
Initial values	0.00	3.0	40.0	0.0	0.0
Base run	61.50	38.0	59.0	2082.5	6247.4
High ATDRW	56.25	47.3	55.0	2145.4	6436.3

Surprisingly, the *assumed* undiscovered rework is almost the same as the actual undiscovered rework throughout the run, even though the true time to discover rework is about two-thirds as large as the time the pessimist assumes. That suggests that the perfect guesser, the one whose assumed values for ATDRW turn out to the the true values for TDRW, will not estimate as accurate a value for assumed undiscovered rework as the pessimist. Part of the reason for this anomaly is the unavoidable delay involved in averaging the rate of detection of undiscovered rework, and part is the fact that the LEVEL ≃ OUTFLO*TCONST approximation always *underestimates* when the level is rising, as it is for close to two-thirds the project. The overestimation of the pessimist compensates for the inherent underestimation of the policy.

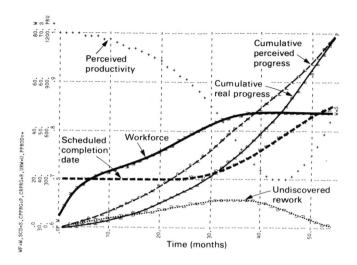

Figure 6.12: Simulation of the project model with the policy of assumed undetected rework, with a pessimist's assumption of the assumed time to detect rework, given by TATDRW=18/18/16/12/6/3

The applicability of this policy is limited. To produce reasonable estimates of AURW, there must be a moderately smooth character to the rate of detecting undiscovered rework, DURW. The policy seeks a pattern or trend in DURW, and that would be hard to see if DURW jumped around a lot. Because large numbers tend to smooth trends, a detectable pattern is more likely in a project with many tasks. The policy could not be implemented in projects with just a few tasks.

In summary the policy of assuming and estimating undetected rework is capable of reducing the overrun in the schedule in these runs, but it does not cure it. Rather it exacerbates the overrun in manpower and total project cost. Some savings would no doubt be generated by finishing earlier, getting the product out sooner, and being able to start another project quicker. But the policy does not address half of the problem, namely, the failure to estimate accurately the size of the project until near the end.

A good policy will be a mix of policies that work together, counteracting all the tendencies in the system to misbehave. It will be enlightened by intimate knowledge of the system being studied, by insights gleaned from the system dynamics perspective and the modeling effort itself, and by considerations of what is possible to implement in the real system. The next section illustrates one such mix of policies, and discusses the ever thorny question of the reliability of model-based policy recommendations.

6.3 POLICY RECOMMENDATIONS AND THE QUESTION OF VALIDITY

The final goal of a modeling study is the application of insights to a real world problem. One hopes that the model, or the model-building process, will help people manage some activity better in the future. The two characteristics of policy studies--the effort to be applicable, the inherent uncertainty of prediction--force the modeler to address two sets of questions in preparing model-based policy recommendations. One set of questions investigates the validity of the recommendations. The other set centers on the extent to which the recommendations are implementable. Before addressing these questions we note an inherent (and healthy) limitation of formal models.

What is Improved Behavior?

Throughout we have assumed that the characteristics of improved behavior in the system are well known to the modeler and the client. Presumably an inventory policy that tended to damp oscillatory tendencies would be a good thing, as would a mix of urban policies that economically revive a stagnating inner city. Unfortunately, people are not always in agreement about what constitutes improvement.

It is in the nature of feedback systems that *trade-offs* must be made. Policies aiming toward urban renewal, for example, must contend with the unavoidable competition for scarce urban land. A built-up city is constantly in the process of trading housing and open space off against businesses and factories. Policies that favor economic development tend to crowd living space, while an overemphasis on housing construction can inhibit the ability of the city to be economically self-sustaining.[4] Urban policy-makers are far from agreeing on what looks best for a city. Until they do, an urban policy consensus is impossible.

The project model has provided us first-hand with examples of trade-offs. In most of our policy tests, if we managed to reduce the schedule overrun, we increased the project

---

[4] See Forrester (1969) and Alfeld and Graham (1976).

cost. With the policy of assumed undetected rework, for example, using the pessimist's guess at the assumed time to detect rework, the project took 5 fewer months to complete but cost $368,100 more. Months are traded off against dollars. Whether or not the policy is an improvement depends on how one weighs the additional cost against the savings in time.

A system dynamics model does not set or evaluate the criteria for improved system behavior. It may vividly illustrate trade-offs, or force the conclusion that the short-term effects of some policy are opposite to its long-term effects, but it cannot determine what is a desirable scenario. People make the value judgements. A quantitative model merely adds to their information in weighing alternatives.[5]

Policy Validity

Section 5.4 focused on the question of model validity, translating the concept into the more tractable notions of model suitability and consistency. Here we shall assume that the issues emphasized in that section have already been dealt with successfully. The model passes a host of tests and is judged very suitable for its purposes and highly consistent with the reality on which it focuses. It is then used as a laboratory tool to test different management policies, and policy recommendations result. The question still remains, how much confidence should be placed in those recommendations? How valid are they?

Once again, a translation of the concept of validity helps to address the questions it poses. Given policy analyses based upon a suitable, consistent model, we prefer to ask, How robust are the resulting policy recommendations? and, Is the model adequate to support them?

Robust Policies

Robustness refers to the extent to which the real system can deviate from the assumptions of the model without invalidating policy recommendations based upon it. In the system dynamics perspective a policy recommendation is robust only if it remains a good choice in spite of variations in model parameters, different external (exogenous) conditions, and

---

[5] It is possible to formulate quantitative performance indices and to use model-generated data to help evaluate different policies. See Lyneis (1980), pp. 98-102, 109.

Policy Recommendations and the Question of Validity 351

reasonable alternatives in model formulation. A policy that is sensitive to such variations is suspect.

Policy robustness is vital because no model, mental or quantitative, is identical to the real system. The real problem will always have aspects that are not captured by a model. Consequently, to have effects in the real system similar to its effects in a model, a policy must, by definition, be robust. The judgment is necessarily relative. The best we can do is to seek policies that tend to be more robust, less sensitive, than others.

Testing for Policy Sensitivity and Robustness

The worst sin for a model-based policy recommendation is sensitivity to parameter changes. Few clients would place confidence in a policy that flip-flops as a parameter varies over a reasonable range.

Overestimation policies from the project model have the potential to be parameter sensitive. Suppose one proposed intentionally overestimating the job at the beginning by, say, 50 percent. Recall that the current project definition CPD assumed in the project model begins the project with 800 tasks and moves eventually to the true value of 1,200 by the end. An overestimate of 50 percent would start the project, by accident, at precisely the 1,200 tasks assumed to be the true requirement. But, if the initial estimate had been correct, rather than understated, the overestimation policy would *increase* the cost of the project.

The policy involving assumed undetected rework, AURW, is much more robust. It is neutral, neither helping nor hindering when CPD is varied, and it always helps whenever there is a tendency assumed in the model to generate undiscovered rework. For any values of the normal fraction satisfactory between zero and one, the policy reduces the schedule overrun. When NFSAT = 1, and very little undiscovered rework is generated, the policy always estimates a very small or zero value for assumed undiscovered rework and appropriately makes little or no correction to cumulative perceived progress. As the simulations of the optimist and pessimist in section 6.2 show, there is also a wide latitude for values of the assumed time to detect rework, ATDRW.

The modeler uncovers these facts by experimenting extensively with the model. Suffice to say, no policy recommendation should escape to the client or audience of a study without first passing an arduous set of parameter sensitivity tests.

A robust policy also ought to hold up in the face of reasonable alternatives in equation formulation. Perhaps an additive formulation and a multiplicative one for some variable appear equally reasonable. If policy recommendations based upon the model were different with each alternative, then the modeler has a problem. He cannot proceed without obtaining more information about the real system or reformulating the rest of the model in a way that removes the sensitivity.

The project model contains several potential spots for alternative formulations. In certain places in the model, for example, the reader may have questioned whether the fraction completed, FCOMP, or the fraction perceived completed, FPCOMP, should be used. In each case the modeler should be able to choose, based upon careful analyses of what is operating in the real system. But, if doubts remain, then proposed policies would have to be simulated with both FCOMP and FPCOMP in the questionable locations. If the choice makes a difference in the value of a policy, then there is a clear need for more information and/or reformulation of the model.

Finally, the success of a particular policy should also be tested under a variety of likely external circumstances. Can your new inventory control policy lower costs even in the face of steadily rising or declining orders, an oscillating order stream, or random shocks? How well does a particular agricultural land management policy stand up to vicissitudes in the weather? If money matters in your model, will your favorite policy have beneficial results even in the face of a business cycle downturn or sustained inflation? The external environment of a system is represented in a model in some of the parameters assumed and in explicit exogenous influences. The modeler must vary them in some systematic way to try to find the extremes under which a proposed policy breaks down.

In short, to build a robust set of policies for a system, one must spend considerable time and effort trying to prove they are not robust.

Policy Recommendations and the Suitability of the Model

Robustness is one aspect of policy validity. A second is the *adequacy of the model to support the recommendations based upon it.* A classic historical example helps to clarify the intent of the notion. In his *First Essay on Population* (1798) Parson Thomas Malthus described a (mental) model of population growth which we can picture in feedback terms as in figure 6.13.

Malthus argued that no "spot of earth," in particular, his own England, could support an arbitrarily large population. The diminishing availability of food would eventually cause

Policy Recommendations and the Question of Validity 353

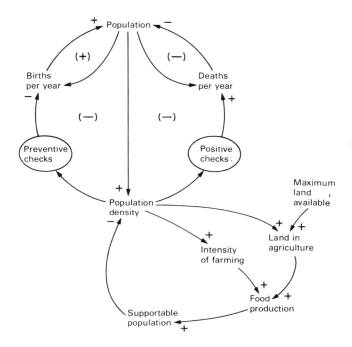

Figure 6.13: Malthusian view of population.

disease, starvation, and social conflict, all of which would eventually boost the death rate until it halted the growth of population. Malthus called these spectors "positive checks" on population growth. At the same time Malthus suggested that foresight could act as a "preventative" check. If society could perceive the inevitability of the positive checks, perhaps society could adopt traditions which lower its birth rate. Marry later in life; abstain. He was a realist, however, and dispaired that preventative checks would always be weaker than positive ones. "Passion between the sexes," he said, "is necessary, and will remain nearly in its present state."

Malthus's writings can be interpreted as advocating a population control policy. Without preventative checks, he predicted that England would outstrip its ability to feed itself within a century. History, fortunately, proved him wrong. Many point out that his model failed to take account of the coming industrial revolution and the increases in agricultural productivity and emigration that came with it.

From the perspective of this chapter, it is interesting to consider what errors, if any, he made in formulating and recommending the policy of preventative checks. While most people today would claim Malthus was in error, they would acknowledge that his model of population growth is essentially correct. Births and population do form a positive feedback loop that is the engine of exponential population growth. Natural constraints will conspire to halt population growth, at which time deaths per year will equal or even exceed births. Knowing that the crude birth rate for a population in equilibrium equals the reciprocal of the average lifespan, we can compute that a birth rate of 40 per 1,000 means an average lifespan of the population of merely 25 years. Yet humans could adopt measures to prevent high birth rates and lead to what most would claim to be more desirable equilibrium conditions. It really is not a bad model, highly aggregated but powerful in its simplicity, and probably correct in the long run.

While there are numerous ways to criticize Malthus, perhaps the most instructive focuses on the inability of his model to support the policy recommendations we presume he intended to make. Without more detail in the model, it would have been inappropriate for Malthus to recommend that England in his century adopt a policy of preventative checks to hold down the birth rate. The model is appropriate only for long term dynamic patterns. It can support only long term, global policy conclusions. It ignores significant shorter term effects on population, such as migration and productivity. It can not be used to argue for short-term policy, except to the extent that the short term is the first step in the long term.

The adequacy of a model to support policy recommendations is a consideration that brings together judgements of model and policy validity. It acts as a check (a preventative check) on the results of a modeling study. Long before policies are analyzed, the modeler tries to set the boundary of model appropriately for the problem being addressed and the questions being asked. Now, near the end of the study, the modeler must again address the same set of boundary questions. With policy recommendations clearly in focus, the modeler must ask if factors initially left outside the model boundary could invalidate the model-based analyses. We are brought back to the questions of model suitability outlined in section 5.4, but with a clearer view of what we require of the model.

Implementable Policies

No one would knowingly make a set of policy recommendations that could not be implemented in the real system. Yet

Policy Recommendations and the Question of Validity 355

there are two considerations related to implementation that a modeler might be tempted to overlook. First, can those responsible for policy in the real system be convinced of the value of model-based policy recommendations? Second, how is the real system likely to respond to the *process* of implementation?

People responsible for managing complex human systems base their policies on their own mental models. To be sure, a regression study may show that some price elasticity of demand is less than one in absolute value and that may add to a corporate officer's feelings that price should be raised, but the final decision to raise or lower price will be based fundamentally on what the corporate officer *believes.* His mental picture of the system will be far richer in some respects than any quantitative model of it. It will be his own mental model that he trusts.

If one accepts the premise that decisions are based on mental models, then there is a clear implication for the policy recommendations from a system dynamics study. To be implementable, they must be formulated and argued so as to fit into the mental models, even the intuition, of those to whom they are addressed. It is unlikely that model structure and a host of simulations, by themselves, will be convincing. The reasons behind the behavior of the model must be made clear. They must eventually be just plain obvious to the client. In a turn of events that seems to subvert the value of quantitative modeling, they must eventually be understandable completely on their own, even to those ignorant of the quantitative model that spawned them.

The need to insinuate model-based policy conclusions in the intuition of the client suggests the client be an integral part of the modeling process. Insights, when they come, are more likely to come out of the process rather than the final products of a modeling study. When clients cannot be involved in the modeling process, the task of the modeler is harder, and the likelihood of the results being implemented is diminished.

The second question raised above about implementation refers to the consequences of trying to bring about change in the system. A modeling study usually focuses on what policies will help, not on how those policies ought to be introduced into the system. Yet successful implementation depends upon not only on good policies and managerial acceptance but also on a sensitively planned transition from the old policy set to the new one. There is a vast literature on planned change. Suffice to say here, the modeler should be aware that the process of implementing policy change is a feedback problem in and of itself. In any given situation it may be as interesting as the original problem that stimulated the modeling study.

Policy Conclusions from the Project Model

The literature on the management of R&D is vast, and our little model has only scratched the surface of the dynamic hypotheses and management policies relating to overruns. Yet the reader who has made it this far may feel that after all the effort we ought to say something substantive about managing large projects. Therefore, as illustrations of the range of potential policy conclusions from a system dynamics study, we offer the following observations that stem from the project models we have developed.

- The hypotheses of undiscovered rework and underestimation of the size of the job are both capable, independently or in concert, of generating observed patterns of schedule and cost overruns.

- Rework can dramatically increase the number of tasks that have to be completed. With 1,200 tasks to accomplish and some fraction being done initially unsatisfactorily, the project can easily require the completion of more than 2000 tasks.

Undiscovered rework increases the total number of tasks completed (satisfactorily and unsatisfactorily) in the course of a project. In contrast, underestimating the size of the project does not directly increase the number of tasks eventually performed. It leads instead to the assembling of too small a project team.

- The sooner in a project that an accurate estimate of its size can be made, the less will be the tendency to overrun the schedule.

- Continuing to bring on new people throughout the project is more costly than assembling the required team to start with.

- An unwillingness to increase the size of the project team as the project nears completion can severely exacerbate a schedule overrun caused by underestimating the size of the project.

Underestimating the size of the project can lead to the perception that there is only six months of work left when in fact

Policy Recommendations and the Question of Validity   357

there may be two years or more. A willingness to increase the size of the project team, even when the project is perceived to be near completion, can prevent rambling on with a workforce that is far too small to get the job done on time.

- If underestimates are a persistent problem, overestimating the size of the project remaining may help, but the outcome is sensitive to where in the system the overestimates are used.

- Overestimating is essentially a nonrational policy. (Something outside the scope of our model must guide it.)

- Computing the effort perceived remaining with the best available information and *then* multiplying by an educated guess is potentially a reasonable overestimation procedure.

Overestimating the workforce needed, without a corresponding overestimation in scheduling, is a significantly unstable policy. Multiplying the man-months of effort perceived remaining by an overestimation factor has the advantage that the overestimate heads toward zero as the the project nears completion.

- Trying to estimate the current size of undiscovered rework by keeping track of the past rate of detecting it can reduce schedule overruns.

- Cost overruns caused at least in part by rework can not be significantly reduced without reducing the tendency to perform tasks unsatisfactorily.

Various estimation and planning procedures can help bring the project in on time, but rework is inherently costly. Completing, on the average, 30 percent of the project tasks unsatisfactorily can almost double the total number of tasks eventually completed.

A Mix of Implementable Policies in the Project Model

The list of insights the model-building process helped to generate could go on. Instead we shall leave the project model at this point, after showing one final policy run. The reader interested in depth in system dynamics modeling of the

---
[6] Roberts (1964). See also articles in Roberts (1980).

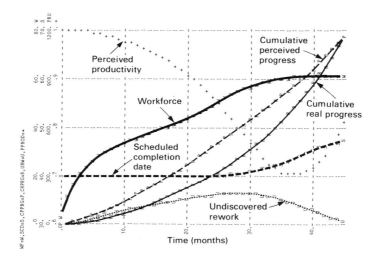

Figure 6.14: Simulation of the revised project model combining an overestimation policy with assumed undetected rework (OE = 1.5)

R&D process should consult the writings of Edward B. Roberts.[6]

The simulation of the project model shown in figure 6.14 represents two managerial attempts to combat the tendencies of a large project to always be larger than anyone anticipated and to involve rework. The run was performed with an overestimation parameter OE in effort perceived remaining and with our policy of assumed undetected rework. In this run OE was set at 1.5, and the values taken for the assumed time to detect rework were the pessimist's TATDRW=18/18/16/12/6/3:

	TIME E 00	WF E 00	SCD E 00	CUMEFF E 00	COST E 03
Initial values	0.00	3.0	40.0	0.0	0.0
Base run	61.50	38.0	59.0	2082.5	6247.4
Policy run	45.75	60.9	47.4	2187.9	6563.8

Policy Recommendations and the Question of Validity 359

As the graph and the summary data show, the combined policy manages to shorten the overrun dramatically, but at an extra cost of $316,400. Added to this might be costs involved in implementing the policies.

In spite of the observed improvement in the schedule overrun, the run could not be used to justify particular parameters in these policies. A more proper policy analysis procedure at this point would be to make numerous runs varying OE, for example, to check to see just how much trouble a project manager could get into by making such overestimates. Runs of the model, which the reader is welcome to make, demonstrate that in this model, with our particular assumptions about the current project definition and the fraction satisfactory, OE can be chosen as high as 4
5 before costs increase substantially.

These observations must be tempered, however, by the fact that our model nowhere assumes a limitation to the size of an effective project team. On another pass we would wish to remove the assumption hidden in the project model which amounts to saying that if one woman can have a baby in nine months, nine women can have one in one month. A larger workforce improves the rate of progress only up to a point.

Finally as we have noted, costs will only be significantly reduced if rework is lessened. Large projects, like lots of other undertakings, would be vastly improved if people did not make mistakes.

7 ADVANCED TOPICS IN DYNAMO USE

In the modeling efforts described in the first six chapters we confined ourselves to features of DYNAMO that are available in all versions of the language. In this chapter we shall explore two features that are available only in some DYNAMO's -- user defined MACROs (supported by DYNAMO II and III) and arrays (supported by DYNAMO III only). Both features are particularly useful for constructing large models. They simplify the writing of equations in situations where patterns of DYNAMO equations are *repeated* a number of time throughout a model.

User-defined macros enable the modeler to repeat any sequence of DYNAMO equations simply by invoking a single function defined by the user. Arrays allow a complex model that is composed of a number of structurally identical sectors to be written very compactly using subscripts. Because macros and arrays offer a logical way to simplify the writing of complex models, they help to clarify model conceptulization and formulation and also improve the modeler's ability to communicate model structure.

7.1 USER DEFINED MACROS

A macro is a collection of DYNAMO equations that are given a single name and invoked together as a function. We have seen several examples: the SMOOTH function, DELAY3, and TABLE (among others) are macros already built into DYNAMO. DYNAMO II and III allow the modeler to invent his own macros. Two types of situations suggest their use. The most common occurs when a piece of model structure is repeated a number of times throughout the model, perhaps with different parameters or inputs. Writing a single function for those equations vastly simplifies the model listing,

User Defined Macros

making it easier to formulate and understand. A second usage emphasizes ease of communication. The modeler might wish to formulate a macro for forecasting, for example. Even if used only once in the model, a macro named TREND or FRCST may help to make the model more readable, allowing a reader to pass over the details of the forecasting procedure on the first reading. Whatever the intent of its use, a macro allows one function to stand for as many equations as the modeler desires.

The SLOPE Macro

Suppose, for example, the modeler wishes to compute the current growth rates of one or more variables in a model. We wish to write a macro which will compute the change in a variable per unit time, for any variable in the model. We can select a time interval DEL and compute the change in a variable X.K over the time interval DEL as X.K - SMOOTH(X.K,DEL). The SMOOTH represents the approximate value of X DEL time units ago. If we give the name SLOPE to our macro for the growth rate, we would write the following statements near the beginning of the model listing:[1]

```
MACRO SLOPE(X,DEL)
A    SLOPE.K=(X.K-SMOOTH(X.K,DEL))/DEL
MEND
```

The SLOPE macro contains the minimum elements of a macro definition --

- a MACRO statement with the name of the macro expressed as a function,

- an equation defining the function named in the MACRO statement,

- a MEND statement (as in MACRO END) to signal the end of the macro definition.

The symbols X and DEL in the first line of the macro are so-called dummy arguments. They tell DYNAMO to expect

---

[1] Mathematically inclined readers may want to observe that the SLOPE macro is exactly equivalent to smoothing the derivative of X. The result is obtained, however, not by differentiating but by integrating (in the SMOOTH) appropriately.

two inputs whenever SLOPE is used and to substitute those inputs for X and DEL wherever they appear within the macro definition. They are written without timescripts because there is no telling what sort of variable in the model will be substituted for them. A dummy argument must be a quantity name; it can not be a number. It is advisable to identify dummy arguments with quantity names that are not used elsewhere in the model.

The equation that finally computes the value named in the MACRO statement can be any type of DYNAMO equation except C. If defined by an L equation, an N equation is, of course, required.

To use the SLOPE macro to compute the growth rate of population over the past two years, we could write

    A   POPGR.K=SLOPE(POP.K,2)

where
    POPGR   = population growth rate (people/year),
    POP   = population (people),

and the 2 represents the past two years over which the average growth rate is being assessed. POP and "2" are the real arguments which DYNAMO substitutes for the dummy arguments X and DEL, respectively. Note that their *location* in the expression SLOPE(__,__) determines which is which. In general, real arguments may be almost anything, including expressions involving arithmetic operations. (The averaging time here of two years could have been given a quantity name and been specified in a C equation.)[2]

To illustrate some of the possibilities of macros and some of the programming requirements that go with them, we shall describe a number of simple variants on the SLOPE macro. Some might feel that the SLOPE macro would be easier to read if we gave a variable name to the quantity SMOOTH(X.K,DEL) that appears in it. To do so we must define a variable that is *internal* to the macro and identify it as such with an INTRN statement, as follows:

---

[2] This macro has one hazard associated with it -- if the quantity whose slope is being sensed is a rate, we have a rate on the right of the auxiliary equation for SLOPE. The results are slightly distorted because the time interval between X and DELX is reduced by one DT as RATE1.KL is computed as a function of RATE2.JK. If that inaccuracy bothers you, then you must either avoid SLOPEing rates or, if SLOPEing only rates, replace DEL in the SMOOTH function with (DEL-DT).

User Defined Macros

```
MACRO SLOPE(X,DEL)
INTRN DELX
A    DELX.K=SMOOTH(X.K,DEL)
A    SLOPE.K=(X.K-DELX.K)/DEL
MEND
```

The quantity DELX stands for "delayed X." If we use the macro in several equations we want one DELX per SLOPE usage and no DELX to be confused with any other (or with a DELX introduced somewhere else in the model). By identifying DELX as a quantity internal to the macro, we signal DYNAMO to create a separate such quantity for each usage of the macro in the model.

As it stands, the SLOPE macro always computes the initial value of any growth rate to be zero. The following equivalent formulation shows why:

```
MACRO SLOPE(X,DEL)
INTRN DELX
L    DELX.K=DELX.J+(DT/DEL)(X.K-DELX.K)
N    DELX=X
A    SLOPE.K=(X.K-DELX.K)/DEL
MEND
```

The L and N equations here are equivalent to those DYNAMO uses for the SMOOTH macro. Since X = DELX initially, the value of SLOPE is always initially zero. Suppose we wish to be more general, to able to specify the initial value of SLOPE to be any particular number ISLOPE. We would add a new dummy argument ISLOPE to the MACRO statement and formulate the entire macro as follows:

```
MACRO SLOPE(X,DEL,ISLOPE)
INTRN DELX
L    DELX.K=DELX.J+(DT/DEL)(X.K-DELX.K)
N    DELX=X-ISLOPE*DEL
A    SLOPE.K=(X.K-DELX.K)/DEL
MEND
```

Substituting the new initial value of DELX (from the new N equation) into the equation for SLOPE shows that the initial value of SLOPE here is ISLOPE, as desired. Note that if ISLOPE = 0 this more general formulation reduces to the previous special case in which the initial slope is zero.

Output Arguments

The several variations of the SLOPE macro have illustrated the use of dummy arguments as placeholders for quantities specified *outside* the macro and used in its computations. Such quantities are input variables. It is also possible and often desirable to use a dummy argument to specify an output variable, a quantity computed *inside* the macro for use *outside*. The pipeline delay provides a common example. In section 3.5 we noted that DYNAMO has a version of the third-order material delay named DELAYP, which is written in the form

```
    R    OUT.KL=DELAYP(IN.JK,DEL,PIPE)
```
where
  OUT   = outflow rate from delay,
  IN   = inflow rate to delay,
  DEL   = delay time,
  PIPE = quantity in the pipeline.

PIPE is the sum of the three levels contained within this third-order delay. It is an output variable, a piece of information available for use elsewhere in the model about the quantity currently moving through the delay. A typical use is the following:

```
    R    RCPTS.KL=DELAYP(ORDRS.JK,DD,UOO.K)
```
where
  RCPTS = orders received (units/week)
  ORDRS = orders placed (units/week)
  DD   = delivery delay (weeks)
  UOO   = units on order (units)

Here UOO is an output variable from the macro, an important piece of information computed within DELAYP, which could be used in a supply-line correction term to ensure against overordering.

As an example of a complete macro with a dummy output variable, we present the following second order exponential material delay with pipeline term PPL:

```
    NOTE
    NOTE    SECOND ORDER MATERIAL DELAY WITH
    NOTE      PIPELINE
    NOTE
    MACRO   DELY2P(IN,DEL,PPL)
    INTRN   LV1,RT1,LV2,RT2
    L    LV1.K=LV1.J+DT*(IN.JK-RT1.JK)
```

User Defined Macros

```
N    LV1=IN*DEL/2
R    RT1.KL=LV1.K/(DEL/2)
L    LV2.K=LV2.J+DT*(RT1.JK-DELY2P.J)
N    LV2=LV1
A    DELY2P.K=LV2.K/(DEL/2)
A    PPL.K=LV1.K+LV2.K
MEND
```

(DELY2P is computed in an auxiliary equation to avoid the possibility of a rate on the right of a rate equation.)

Model Variables Inside Macros

Any variable which appears in a macro and which is neither an input variable nor an internally defined (INTRN) variable is assumed to be defined outside the macro. Its value in the macro is whatever it is computed to be in the model. Hence a quantity like DT need not (should not) be defined again in a macro because its value there is just what was assigned in the SPEC statement.

Summary of Rules for Macros

1. Start the definition with a MACRO card containing the name of the macro, a left parenthesis, and any number of dummy arguments separated by commas and followed by a right parenthesis.

2. Include one or more standard DYNAMO equations defining the name of the macro, and perhaps one or more arguments or internal variables. On the right of these equations may appear dummy arguments, internal variables, the macro variable itself, other macros that were defined earlier or built in, and quantities defined elsewhere.

3. Be sure that *dummy* arguments are simple quantity names, not algebraic expressions or numbers. It is also wise not to repeat these names elsewhere in the model.

4. End the definition with a MEND card.

5. Position the macro formulation ahead of the first usage.

6. Be sure the number and position of the real arguments specified in each usage of a macro correspond exactly to the dummy arguments used to define the macro.

7. Real *input* arguments may be algebraic expressions involving several model variables. Each real *output* variable, however, must be identified by a single variable name that is not defined anywhere else in the model.

Once defined, a macro can be used on the right of any equation in the model (except, of course, C equations), wherever in the structure it represents makes sense.

Errors in Macros

DYNAMO applies essentially the same effort to check for errors in macros as it does for standard DYNAMO equations. The notation used in reporting errors in macros is somewhat different, however. Suppose, for example, the SLOPE macro has been written into a model with a minor typographical error, as follows:

```
MACRO SLOPE(X,DEL)
INTRN DELX
A    DELX.K=SMOOTH(X.K,DEL)
A    SLOPE.K=X.K-DELX.K)/DEL   (NOTE "(" ERROR.)
MEND
```

A left parenthesis has been omitted here in the auxiliary equation for SLOPE. DYNAMO would not report the error in the macro directly, rather reporting the error in each instance of the use of the macro. Say, for example, it was used in the model to compute a population growth rate over an averaging time of AT:

```
A    POPGR.K=SLOPE(POP.K,AT)
```

DYNAMO would print the following sort of error message:

```
TOO MANY )
IN SLOPE - POPGR#7=POP-POPGR#7$1)</AT
```

From this message we easily get the general idea that parentheses are a problem here, but a full understanding of the message requires knowing what DYNAMO means by #, $, and < in this expression.

The simplest to understand is <. For errors in macros DYNAMO uses < to point to the error being described. This usage differs slightly from standard DYNAMO error messages in which the location of an error is signaled by a V placed over the spot where DYNAMO recognizes the error.

User Defined Macros

Understanding the significance of # and $ in this error message requires knowing how DYNAMO actually processes a macro. After DYNAMO reads the entire model, it expands the macros by inventing names for the internal variables and substituting the real arguments for all the dummy arguments. For example, DYNAMO would replace the equation for POPGR with

A POPGR.K=POPGR#7.K

and add two equations to the model:

A POPGR#7$1.K=SMOOTH(POP.K,AT)
A POPGR#7.K=POP.K-POPGR#7$1.K)/AT

(Note the parenthesis error is still present.) DYNAMO invented the name POPGR#7 for the variable SLOPE and POPGR#7$1 for DELX. Each invented name contains the name of the variable on the left of the equation using the macro, followed by a mysterious #7. DYNAMO appends such numbers after its invented variable names to ensure that each name is unique, not only different from other names it invents but also different from variable names chosen by the modeler. The particular number assigned has no significance for the modeler.[3]

But in the case of the internal variable POPGR#7$1, the number following $ is significant; it identifies for the modeler the particular internal variable the invented name represents. Dollar sign numbers are assigned in the order that the internal variables appear in the INTRN statement. Here POPGR#7$1 represents DELX, the first (and only) quantity listed in our INTRN statement. In the case of a large macro with many internal variables (and a few bugs), the dollar sign numbers printed out in error messages would be crucial to understanding and correcting the flaws.

The naming conventions described here should be enough to enable a modeler to decipher any error message about a macro, but the reader should realize that the names can get rather complicated. One of the nice characteristics of DYNAMO is that it places essentially no limit on the number of macros that can be embedded within a user-defined macro. A potentially nasty consequence of that flexibility is that

---

[3] FORTRAN does not allow # to be used in a variable name, so DYNAMO/F which is based on FORTRAN would write ZZ0700 and ZZ0701 instead of POPGR#7 AND POPGR#7$1. Again, the choice of names is to insure uniqueness.

DYNAMO may have to invent a variable name involving more than one # and/or $. In the population growth rate example, the real variable corresponding to DELX (the SMOOTHed version of X) is

    A   POPGR#7$1.K=SMOOTH(POP.K,AT)

But since the SMOOTH is also a macro involving a level variable for which a variable name must be invented, DYNAMO will create something like POPGR#7$1#10 and compute POPGR#7$1 as follows:

    A   POPGR#7$1.K=POPGR#7$1#10.K
    L   POPGR#7$1#10.K=POPGR#7$1#10.J+(DT/AT)*
    X   (POP.J-POPGR#7$1#10.J)
    N   POPGR#7$1#10=POP

The L and N equations compute the SMOOTH; the A equation assigns the result to POPGR#7$1.

Fortunately, we usually do not have to see these invented names. The point of this extended example is to note that it is conceivable you could encounter an error message about a variable named, say, SALES#4$3#12. If you do, interpret it to mean that DYNAMO had problems with a macro used to compute the *third* variable ($3) listed on the INTRN statement of a macro involving SALES, and hunt for the error accordingly.

7.2 EXAMPLES OF MACROS

To illustrate further the principles of formulating macros, we shall present several useful examples, including macros for a delay initialized out of equilibrium, a trend, pink (autocorrelated) noise, and summary statistics for a variable.

DELAY1 Initialized out of Equilibrium

The DELAYs and SMOOTH functions built into DYNAMO initialize themselves in steady state. The DELAY1 macro built into DYNAMO is the following:

    MACRO DELAY1(IN,DEL)
    INTRN LEV
    L   LEV.K=LEV.J+DT*(IN.JK-DELAY1.J)
    N   LEV=IN*DEL
    A   DELAY1.K=LEV.K/DEL
    MEND

Examples of Macros

Since the initial value of LEV is IN*DEL, the initial value of output of DELAY1 is

LEV/DEL = (IN*DEL)/DEL = IN,

so the output from the delay initially equals its input. To change this initial equilibrium assumption, the modeler would have to write a similar macro with a different N equation for LEV.

Suppose we wish to set the initial conditions so that the initial net change in LEV is given by an initial constant growth rate IGR to be specified as an input to the macro. We would require

IGR = (inflow - outflow).

Since the inflow to the delay is IN and its outflow is LEV/DEL, this initial condition becomes

IGR = IN - LEV/DEL,
which rearranges to
LEV = IN - IGR*DEL.

Thus the following macro, named DEL1G, is equivalent to DELAY1 initialized for linear growth equal to IGR:

```
MACRO DEL1G(IN,DEL,IGR)
INTRN LEV
L    LEV.K=LEV.J+DT*(IN.JK-DEL1LG.J)
N    LEV=IN-IGR*DEL
A    DEL1LG.K=LEV.K/DEL
MEND
```

If the modeler wished instead to specify an initial nonzero *fractional* growth rate IFGR, the initial condition would be, by definition,

IFGR = (inflow - outflow)/LEV.

In this case the IFGR would replace the third dummy argument in the MACRO statement, and the initial value of LEV would be given by

N   LEV=(IN*DEL)/(1+IFGR*DEL),

as the reader should check. Note that in either the linear or fractional growth formulations a value of zero for the initial growth constant results in an initial value of LEV equal to IN*DEL, just as in DELAY1.

The TREND Macro

To assess the general trend of growth or decline in a variable, the modeler might turn to the SLOPE macro used in section 7.1 to introduce the formulation of macros. But that macro has somewhat limited applicability because it assumes that the current value of the input X is instantaneously knowable. As a consequence it may respond more rapidly to changes in X than is reasonable in many real systems. It is likely that the true value of X in most applications is perceivable by actors in the system only after some delay. To handle these criticisms, the modeler can first smooth the input X and then determine the slope. One could formulate

```
MACRO TRND(IN,TPPC,TERC)
NOTE    IN = INPUT
NOTE    TPPC = TIME TO PERCEIVE PRESENT
NOTE       CONDITION
NOTE    TERC = TIME TO ESTABLISH REFERENCE
NOTE       CONDITION
A   TRND.K=SLOPE(SMOOTH(IN,TPPC),TERC)
MEND
```

provided, of course, that the model already contains a formulation of the SLOPE macro. TPPC is the delay in perceiving or acting on the value of the input. TERC is the denominator of the slope expression, the "delta t" in "(delta y)/(delta t)."[4] Because of the embedding of macros here, TRND looks quite simple, but when all the macros are expanded it actually involves seven equations (and some pretty messy looking variables).

Frequently, one needs a more general formulation allowing a nonzero initial trend to be specified. The L and N equations implicit in the SMOOTHs must be written out explicitly so that the N equations in the macro can specify the initial trend. Having formulated a version of the SLOPE macro that equals ISLOPE in initial conditions (see section 7.1), we have done all the conceptual work required to formulate a TREND macro that can be set initially to equal ITREND. The

---

[4] While TPPC and TERC may represent conceptually different delay times in the real system, as a practical matter there is little reason to select different values for them.

Examples of Macros

following listing should look familiar in its parts, even if the whole is new:

```
MACRO TREND(IN,TPPC,TERC,ITREND)
INTRN PPC,RC
L    PPC.K=PPC.J+(DT/TPPC)*(IN.J-PPC.J)
N    PPC=IN-ITREND*TPPC
NOTE     PERCEIVED PRESENT CONDITION
L    RC.K=RC.J+(DT/TERC)*(PPC.J-RC.J)
N    RC=PPC-ITREND*TERC
NOTE     REFERENCE CONDITION
A    TREND.K=(PPC.K-RC.K)/TERC
MEND
```

where
- IN = input
- TPPC = time to perceive present condition
- TERC = time to establish reference condition
- ITREND = initial trend (slope)

The TREND macro is built into Mini-DYNAMO.

A likely use of TREND is to compute a forecast based on the extrapolation a recent trend. To forecast a sales rate over a forecast period FP, for example, one could write

```
A    FSALES.K=AVSALE.K+(FP+TAS)*TREND(SALES.JK,TAS,
X    TETS,ITS)
```

where
- FSALES = forecasted sales rate (units/month)
- AVSALE = average sales rate (units/month)
- SALES = current sales rate (units/month)
- FP = forecast period (months)
- TAS = time to average sales (months)
- TETS = time to establish trend in sales (months)
- ITS = initial trend in sales (units/month)

Pink Noise

In Section 3.6 we described the random number generator in DYNAMO called NOISE and used it to display the natural modes of behavior of a simple inventory model. At the time we noted a number of theoretical problems with the simple-minded usage of NOISE illustrated there. The NOISE function generates random numbers that fit a uniform or rectangular distribution. We might prefer a pattern of disturbances more like the bell-shaped normal distribution. Furthermore, since the NOISE function spits out a pseudo-random number every DT, for example, decreasing DT in a run inadvertantly puts more random shocks into the system. We would prefer a formulation for random disturbances that is not so sensitive to DT.

A more serious concern is the complete lack of autocorrelation in the numbers spewed out by NOISE. It is as if the amount of rainfall in one hour or one month has absolutely nothing to do with the amount in the previous hour or month. Real random disturbances from a single source are much more likely to show some correlation over time, so we desire a formulation that allows the value of a random shock in one DT to be more or less correlated with the one in the next DT.

An even more theoretical concern, but one with practical implications for dynamic simulation models, involves the noise frequencies present in the random disturbance. Noise can be thought of as the sound produced by an infinitely large choir in which each voice hums a different pitch. The random numbers that DYNAMO's NOISE function emits act something like the sound from a choir with more tenors than basses, more altos than tenors, more sopranos than altos, and so on to higher and higher voices. It tends to contain more power in the higher frequencies. It is akin to the engineer's construct called white noise (named after white light which contains all the frequencies of the visible spectrum). A random disturbance that acts like white noise tends to excite high frequency responses more than low frequency responses. We would like a noise formulation which tends to excite more equally the frequency responses present in a model.

What we want is called pink noise. The following macro produces it by first generating pseudo-white noise and then smoothing the result:

```
MACRO PKNS(MEAN,STDV,TC)
NOTE    MEAN = ARITHMETIC MEAN OF PKNS
NOTE    STDV = STANDARD DEVIATION OF PKNS
NOTE    TC = CORRELATION TIME CONSTANT FOR PKNS
INTRN WTNS
L   PKNS.K=PKNS.J+(DT/TC)*(MEAN+WTNS.J-PKNS.J)
N   PKNS=MEAN
A   WTNS.K=STDV*SQRT(24*TC/DT)*NOISE()
MEND
```

The smoothing embodied in the level equation for PKNS tends to generate some autocorrelation in the noise over time. The larger TC is, the more autocorrelation is present. The smoothing also tends to tone down the high frequency components of the noise; there are fewer sopranos in PKNS. Finally, smoothing is a kind of averaging process so PKNS tends to be normally distributed. The expression for

Examples of Macros   373

pseudo-white noise WTNS embodies a number of technical subtleties that we will not discuss.[5] The reader may wish to note, however, that DT enters in two places in the macro in such a way that changes in DT have somewhat offsetting effects on the value of PKNS. While the details are no doubt fuzzy, it should be clear that some of our concerns about NOISE() are being addressed by the formulation of PKNS.

To use pink noise select a mean and standard deviation for the random disturbance that are appropriate for the application. Roughly two-thirds of the time the value of PKNS will fall within plus-or-minus one standard deviation of the mean. Only rarely will its value be more than three standard deviations from the mean. The value of the correlation time constant TC should theoretically be related to what engineers call the bandwidth of the system. Select a value larger than the shortest time constant in the model. Experiment to find a value that seems appropriate for the extent of autocorrelation present in the real random disturbances the PKNS application is intended to approximate.

Macros for Summary Statistics

Statistical summaries of quantities in model runs are sometimes useful in comparing different runs with each other or with real data. We shall illustrate the general ideas with a macro that computes the mean, variance, and standard deviation of a variable with respect to time. To compute the mean with respect to time we simply accumulate the values of the variable in a level over the course of a run and divide by the elapsed time of the simulation. The variance of a quantity with respect to time is defined as the average squared deviation of the quantity from its mean. The computation for the variance given here in the STAT macro is equivalent to the definition but has the virtue that the mean does not have to be obtained first in a separate run before the variance can be computed. The standard deviation is simply the square root of the variance.

```
MACRO STAT(X,MEAN,VAR,STDV)
INTRN CUMX,CUMXSQ,ITIME,ELPST
L     CUMX.K=CUMX.J+DT*X.J
N     CUMX=0
NOTE        CUMULATIVE X
A     MEAN.K=CUMX.K/ELPST.K
```

---
[5] For a sightseer's trip through some of the mathematical details see Britting (1973).

```
A    ELPST.K=TIME.K-ITIME+1E-30
NOTE    ELAPSED TIME
N    ITIME=TIME
NOTE    INITIAL TIME
L    CUMXSQ.K=CUMXSQ.J+DT*(X.J*X.J)
N    CUMXSQ=0
NOTE    CUMULATIVE X-SQUARED
A    VAR.K=(CUMXSQ.K/ELPST.K)-(MEAN.K*MEAN.K)
A    STDV.K=SQRT(VAR.K)
MEND
```

In the STAT macro X is the only dummy input argument; MEAN, VAR, and STDV are dummy output arguments. The tiny number 1E-30 prevents division by zero during the first DT without affecting subsequent results.

The real argument corresponding to the dummy input argument X can be any variable or combination of variables from the model. The STAT macro could be used, for example, to compute summary statistics for the difference between a model variable and a historical time series the variable is supposed to match. The time series would be placed in the model as the y-values of a table function of TIME.K. The variance of such a difference is the mean-squared-error for the two time series.

7.3 ARRAYS IN DYNAMO III

DYNAMO III provides the modeler with the capability to write variable names with *subscripts.* Instead of simply writing WF.K to represent a workforce, for example, one can write WF.K(S,T), where S could represent, say, the skill level of the workforce and T the task to which it is assigned. S and T are called subscripts. Although the notation may initially appear cumbersome, there is a dramatic advantage to being able to write variable names using subscripts. It means that one variable name can represent more than one quantity, and one equation can represent more than one element of model structure. The capability to write variables with subscripts enables the modeler to duplicate an entire sector of a model, with different parameters, any number of times without writing more equations. It allows complex, disaggregated models to built up from small, conceptually aggregate sectors.

A subscripted variable such as WF.K(S,T) actually represents an array of numbers. Suppose there are three tasks to which our workers can be assigned, and suppose for each task workers are classified into two skill levels. The number of workers in our workforce can be displayed in a

Arrays in DYNAMO III

Table 7.1: An array showing a hypothetical workforce of 90 people arranged according to skill level and task asigned.

		Task number assigned:		
		1	2	3
Skill Level:	A	10	13	21
	B	17	11	18

rectangular array like the one shown in table 7.1. Because of the common usage of the word "array" for an arrangement of numbers like that in table 7.1, we shall call any set of numbers represented by a subscripted variable an *array*. DYNAMO III allows variables to have one, two, or three subscripts, so we can have, in effect, one-, two-, and three-dimensional arrays.[6]

Disaggregating a Simple Epidemic Model

To illustrate the kind of situation in which subscripted variables are helpful and to introduce the basic requirements of a DYNAMO III model written with subscripts, we return to the very simple model of an influenza epidemic listed in section 3.4. Suppose we wished to track the spread of the disease in different age groups in the population. We might be interested in various quarantine or innoculation policies focusing particularly on school-age children or the elderly, for example. If we assume that the structure underlying the spread of the disease is essentially the same in all age groups, we are led to build a model that replicates the simple structure of the chapter 3 model for several different age groups, as shown in figure 7.1.

The structure shown in figure 7.1 assumes that children, adults, and the elderly contract the disease by coming in contact with someone who has it (CNTCTS). In each age group there is a certain probability (FRSICK) of catching the disease once contact between a susceptible and an infectious person has been made. If a person in any of the groups becomes ill, the disease runs its course (with no deaths assumed) in some number of days (DUR). Disaggregating the population allows us to select different parameters for these

---

[6] Variables can also, of course, be written in DYNAMO III without subscripts, as in the other DYNAMO's.

376 Advanced Topics in DYNAMO Use

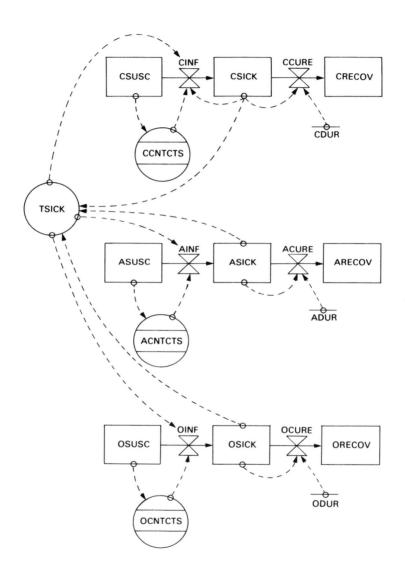

Figure 7.1: Structure of a simple model of an epidemic (from figure 3.6), disaggregating the population into three age groups.

Arrays in DYNAMO III

processes in the different age groups, obtaining a richer structure and perhaps more intricate dynamic interactions, without changing the assumed underlying mechanism generating the epidemic.

We could write the equations for this disaggregate epidemic model by linking together three versions of the simple epidemic model listed in figure 3.7. The result would be a model with 19 active (L, R, and A) equations--the six active equations from the aggregate epidemic model repeated three times, plus an equation for TSICK, the total number of sick people in the population. The alternative in DYNAMO III is to assemble the whole thing in seven active equations, simply by using a subscript to represent the three different age groups.

Let A represent the age groups, defining it as follows:

A = 1 corresponds to the subpopulation of children;
A = 2 corresponds to the suspopulation of adults;
A = 3 corresponds to the subpopulation of elderly.

We then write the equations of the simple epidemic model, appending the subscript A to each variable that stands for a quantity in each of the subpopulations. The following single equation, for example, computes the number of sick people in each age group:

L   SICK.K(A)=SICK.J(A)+DT*(INF.JK(A)-CURE.JK(A))

We would like to instruct DYNAMO III to compute this equation for A = 1, 2, and 3. Therefore, before the first use of the subscript A in the model we include the following DYNAMO III statement:

FOR A=1,3

The FOR statement instructs DYNAMO III to compute the subscripted equations for all interger values of A from 1 to 3, that is, A = 1, 2, and 3. If we would like to remind ourselves of the meanings of these values, we can write

FOR A=1,3=CHLDRN,ADLTS,ELDRLY

giving acceptable DYNAMO names to the numbers 1, 2, and 3. Doing so enables DYNAMO III to recognize SICK.K(CHLDRN) and SICK.K(1) as identically the same quantity.

Writing the subscript A in parentheses in each of the active equations handles the repeated structure in the disaggregated model. The next task is to assign values to the various parameters in the model. For the sake of illustration, let us suppose that the probability of a susceptible person contracting the disease after contact with an infected person differs among the three age groups, as follows:

```
C   FRSICK(CHLDRN)=0.05
C   FRSICK(ADLTS)=0.01
C   FRSICK(ELDRLY)=0.03
```

We could type these three constant equations into the model, or we could replace the names CHLDRN, ADLTS, and ELDRLY, respectively with 1, 2, and 3, or we can type the entire assignment of values as one T statement, as follows:

```
T   FRSICK(*)=0.05/0.01/0.03
```

DYNAMO III interprets the * as a command to substitute, in turn, each of the values of A, assigning 0.05 to FRSICK(1), 0.01 to FRSICK(2), and 0.03 to FRSICK(3), as desired.

Initial values for the nine levels in the disaggregate model can be assigned similarly. The initial values for the three SICK subpopulations can be written:

```
N   SICK(CHLDRN)=10
N   SICK(ADLTS)=5
N   SICK(ELDRLY)=5
```

Or, more compactly and with the capability to change initial values in rereun mode, we can write

```
N   SICK(A)=ISICK(A)
T   ISICK(*)=10/5/5
```

Figure 7.2 shows a DYNAMO III equation listing of the disaggregate epidemic model we have been describing. By referring back to the simple listing in section 3.4 and the conventions given here for subscripts, the reader should be able to decipher almost all of the listing at this point. Only the formulation of TSICK, the total sick population, and the use of TABCON(*,A) remain to be explained.

TSICK is written as

```
A   TSICK.K=SUM(SICK.K)
```

The equation introduces the DYNAMO III function SUM which operates on arrays. SUM instructs DYNAMO III to add all the values in the array named SICK. Thus, in our model

```
* EPIDEMIC MODEL WITH THREE AGE GROUPS
NOTE
FOR  A=1,3=CHLDRN,ADLTS,ELDRLY
NOTE
L    SUSC.K(A)=SUSC.J(A)+DT*(-INF.JK(A))
NOTE   SUSCEPTIBLE POPULATIONS (PEOPLE)
N    SUSC(A)=ISUSC(A)
T    ISUSC(*)=1980/980/695
NOTE   INITIAL VALUES OF SUSC
R    INF.KL(A)=TSICK.K*CNTCTS.K(A)*FRSICK(A)
NOTE   INFECTION RATES (PEOPLE/DAY)
NOTE
A    TSICK.K=SUM(SICK.K)
NOTE   TOTAL SICK (PEOPLE)
L    SICK.K(A)=SICK.J(A)+DT*(INF.JK(A)-CURE.JK(A))
NOTE   SICK POPULATIONS (PEOPLE)
N    SICK(A)=ISICK(A)
T    ISICK(*)=10/5/5
NOTE   INITIAL VALUES OF SICK
T    FRSICK(*)=0.05/0.01/0.03
NOTE   FRACTION SICK (DIMENSIONLESS)
A    CNTCTS.K(A)=TABLE(TABCON(*,A),SUSC.K(A)/TOTAL(A),0,1,.2)
NOTE   SUSCEPTIBLES CONTACTED PER INFECTED PERSON
NOTE   PER DAY (PEOPLE/PERSON/DAY)
T    TABCON(*,CHLDRN)=0/2/3/3.5/3.8/4
T    TABCON(*,ADLTS)=0/1.5/2.3/2.6/2.85/3
T    TABCON(*,ELDRLY)=0/1/1.5/1.8/1.93/2
NOTE   TABLES FOR CNTCTS
N    TOTAL(A)=SUSC(A)+SICK(A)+RECOV(A)
NOTE   TOTAL POPULATIONS IN EACH AGE GROUP (PEOPLE)
NOTE
R    CURE.KL(A)=SICK.K(A)/DUR(A)
NOTE   CURE RATES (PEOPLE/DAY)
T    DUR(*)=7/8/10
NOTE   DURATION OF DISEASE (DAYS)
L    RECOV.K(A)=RECOV.J(A)+DT*CURE.JK(A)
NOTE   RECOVERED POPULATIONS (PEOPLE)
N    RECOV(A)=IRECOV(A)
T    IRECOV(*)=10/15/0
NOTE   INITIAL VALUES OF RECOV
NOTE
NOTE   CONTROL STATEMENTS
NOTE
SPEC DT=0.5/LENGTH=50/PRTPER=0/PLTPER=1
PLOT SICK(CHLDRN)=C,SICK(ADLTS)=A,SICK(ELDRLY)=O
PLOT SUSC(CHLDRN)=C,SUSC(ADLTS)=A,SUSC(ELDRLY)=O
```

Figure 7.2: DYNAMO III equation listing of a simple epidemic model disaggregated into three age groups.

SUM(SICK.K) = SICK.K(1) + SICK.K(2) + SICK.K(3)

If SICK had been formulated as a two- or three-dimensional array, SUM(SICK.K) would have added up every number in the array, not just those in one row. Our epidemic model assumes that the infection rates in each subpopulation depend on the total number of sick people--in our model children are just as likely to catch the disease from the elderly as they are from other children.

The notation for the table functions in the contacts equation CNTCTS is similar to FRSICK(*). The model contains the following table equations:

```
T    TABCON(*,CHLDRN)=0/2/3/3.5/3.8/4
T    TABCON(*,ADLTS)=0/1.5/2.3/2.6/2.85/3
T    TABCON(*,ELDRLY)=0/1/1.5/1.8/1.93/2
```

The * in these statements takes the place of an index (or subscript) for the table entry. DYNAMO III stores this table data as a large array, with three rows each of which contains six numbers, corresponding to the min-x, max-x, and x-increment in the auxiliary equation for CNTCTS. The * in TABCON tells DYNAMO III to assign the numbers listed in one T statement to an entire row of the table data--the implied index for the row ranges over all its possible values (here, 1 to 6). An * in a T statement can always be interpreted as an index ranging over all of its possible values in the model.

Note the extension of the T statement in DYNAMO III. The statement does more in DYNAMO III than supply values for TABLE functions; it can be used to specify any vector of numbers. We have used it above to set initial values (ISICK), assign values to constants (FRSICK), and determine tables (TABCON). To specify the values of a table which has more than one dimension, the name of the table is followed by two or three subscripts, the first of which is always an asterisk.

With these understandings and the addition of familiar simulation control statements, the model listing in figure 7.2 is complete. The model is simulated by typing

   DYNAMO3 EM3

where EM3 happened to be the filename given to this particular model. The results are shown in figures 7.3 and 7.4.

Arrays in DYNAMO III

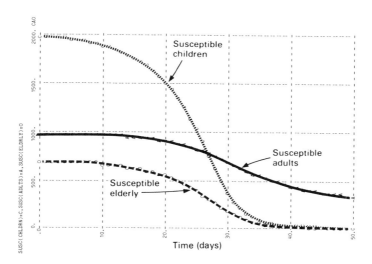

Figure 7.3: Simulation of the disaggregate epidemic model, showing the behavior of the sick populations over the course of the epidemic.

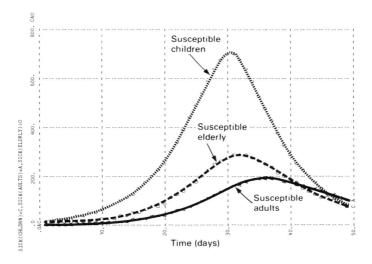

Figure 7.4: Plot of the three susceptible populations from the disaggregate epidemic model.

An Aging Chain

A second example, using subscripted variables to simplify the writing of equations in an aging or vintaging structure, further illustrates the potentials of subscripts. An aging chain monitors the flow of people or objects as they move from age group to age group. Figure 7.5 shows a four-level aging chain for a population: POP1 could represent children, POP2 young adults, POP3 middle-aged people, and POP4 the elderly. Note the contrast with the disaggregated epidemic model. Although children, adults, and the elderly were explictly represented, the epidemic model did not include the flow of children into the adult subpopulation, or adults into the elderly subpopulation. (The time constants associated with those flows (20 to 45 years) are so much greater than the time horizon of the epidemic (50 days) that the aging rates properly have no place within the model boundary.)

Consider the problem of writing equations for the flow diagram in figure 7.5. The figure and the variable names selected suggest that with only a few minor exceptions the structure surrounding each level in the chain is the same. The level and rate equations for POP2 are quite representative of the rest of the structure:

```
L    POP2.K=POP2.J+DT*(MOUT1.JK-MOUT2.JK-DR2.JK)
N    POP2= ?
R    MOUT1.KL=POP1.K/MT1
R    MOUT2.KL=POP2.K/MT2
R    DR2.KL=DRF2*POP2.K
```
where
```
POP2     = population in level two,
MOUT1    = maturation rate out of level one,
MT1      = maturation time for level one,
MOUT2    = maturation rate out of level two,
MT2      = maturation time for level two,
DR2      = death rate out of level two.
```

Seeing the similarity between these equations and those we would write for the other levels in the chain, we are led to try to formulate the entire structure with a subscript representing the number of the level in the chain, something like the following:

```
L    POP.K(AGE)=POP.J(AGE)+DT*
X    (MOUT.JK(AGE-1)-MOUT.JK(AGE)-DR.JK(AGE)
N    POP(AGE)=IPOP(AGE)
T    IPOP(*)= ?
R    MOUT.KL(AGE)=POP.K(AGE)/MT(AGE)
```

Arrays in DYNAMO III

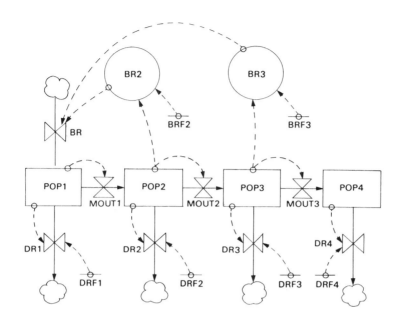

Figure 7.5: Flow diagram of an aging chain

R   DR.KL(AGE)=DRF(AGE)*POP.K(AGE)

where AGE is intended to be a FOR variable ranging from 1 to 4. Unfortunately, the structure for POP1 involves a birth rate BR and POP4 has no maturation out rate, so some special adjustments must be made.

Before making those special adjustments, we note the use of AGE and AGE-1 as subscripts in the level equation for POP.K(AGE). When AGE = 2, for example, the equation states that MOUT(1), the maturation rate out of level one, *adds to* POP(2), while MOUT(2) *subtracts* from it. The use of AGE and AGE-1 here accurately captures the sequential flows *out of* one level in the aging structure and *into* the next.

There are a number of ways of curing the special problems posed by the rates involving POP1 and POP4. The lack of a maturation rate out of POP4 is perhaps the simplest to handle. We could fit POP4 into the pattern of the subscripted equations by pretending that MOUT(4) exists and arranging it so that value that would be computed is always essentially zero. That is, we could select a vector like the following for the maturation time times MT:

T   MT(*)=20/20/20/1E30

This T equation guarantees that the average time an individual stays in each of the first three levels is 20 years (ignoring the outflows due to deaths) and that the maturation rate out of POP4 is

MOUT(4) = POP(4)/1E30 ≈ 0.

This trick allows the general subscripted equations for POP(AGE) to apply to POP(4). There is a clear tradeoff here. In achieving a concise model listing that exposes the common structure among these four levels, we must distort a bit those elements of structure that do not quite fit. In the case of MOUT, the tradeoff results in a net gain in the transparency of the model's structure.

The rate structure surrounding POP1 is another matter. The birth rate BR affects only POP1. It would be too much of a distortion to try to fit POP1 into the pattern of the other levels by inventing fake zero "birth rates" flowing into the other levels or trying to redefine a spurious MOUT(0) as the birth rate BR. We are led to write a separate level equation for POP(AGE) when AGE = 1, as follows:

L   POP.K(1)=POP.J(1)+DT*(BR.JK-MOUT.JK(1)-DR.JK(1))

The equations given previously for MOUT(AGE) and DR(AGE) will work here as well, so our task now focuses on an equation for BR.

The flow chart in figure 7.5 shows that the model generates births from the two middle population levels, POP2 and POP3, presumably corresponding to young adults and middle aged people. The intent of the flow diagram is a computation like the following:

BR = BRF2*POP2 + BRF3*POP3

where BRF2 and BRF3 represent potentially different birth rate factors for POP2 and POP3. In our subscripted version of the model we would want to assign values to BRF(AGE) for AGE = 2 and 3 and then compute in the pattern shown, preferably with some sort of subscripted shorthand.

The computation is in the form of the *scalar product of two vectors*.[7] DYNAMO III has a built-in function SCLPRD that computes such a pattern for any two vector quantities. The form of the usage is almost SCLPRD(BRF(*),POP.K(*)), but the SCLPRD function applies in more general situations. As a result it requires specifying the *first* and *last* elements of BRF(*) and the *first* element of POP(*) that are involved in the computation. For the computation BRF(1)*POP(2) + BRF(2)*POP(3) we would write

R   BR.KL=SCLPRD(BRF(*),1,2,POP.K(*),2)

The usage is always in the form

SCLPRD(A, a_{first}, a_{last}, B, b_{first})

(without spaces, or course) where A and B are vectors (strings of numbers) and a_{first}, a_{last}, and b_{first} are the indices of the *first* and *last* elements of A and B appearing in the computation.

In our computation for BR, for example, we will assign birth rate factors as

T   BRF(*)=0.1/0.05

Hence we want to start our scalar product computation SCLPRD with the *first* element of BRF and end with the *second* element of BRF, and we want to multiply those numbers with the elements of POP beginning with POP(2), the *second* element of POP. Thus the numbers 1, 2, and 2 in the above rate equation for BR signal that the sum begins with the term BRF(1)*POP(2) and ends with BRF(2)*POP(3). We do not specify the last element of POP in the computation because it is forced by the choice of the first and last elements of BRF.

The equation listing in figure 7.6 draws together these ideas. All of it should look familiar except the output statements, which will be discussed in section 7.5, the use of NAGES and the two different FOR variables. NAGES represents the number of age groups in the model. AGE1 and AGE2 are used as subscripts and take on slightly different values: AGE1 ranges from 1 to 4, while AGE2 ranges from 2 to 4. AGE1 is used in most of the subscripted equations of the model, whenever the computation applies to all four age

---

[7] A vector can be thought of here simply as a string of numbers. If vector X = (x_1, x_2, x_3, x_4) and vector Y = (y_1, y_2, y_3, y_4), then the scalar product XY is the single number $x_1*y_1 + x_2*y_2 + x_3*y_3 + x_4*y_4$.

```
* POPULATION AGING CHAIN
NOTE
I     NAGES=4
FOR AGE1=1,NAGES=CHLDRN,YADLT,MDLAG,ELDRLY
FOR AGE2=2,NAGES
NOTE
A     TPOP.K=SUM(POP.K)
NOTE
L     POP.K(CHLDRN)=POP.J(CHLDRN)+DT*
X     (BR.JK-MOUT.JK(CHLDRN)-DR.JK(CHLDRN))
L     POP.K(AGE2)=POP.J(AGE2)+DT*
X     (MOUT.JK(AGE2-1)-MOUT.JK(AGE2)-DR.JK(AGE2))
N     POP(AGE1)=IPOP(AGE1)
T     IPOP(*)=1000/950/900/400
NOTE
R     BR.KL=SCLPRD(BRF(*),1,2,POP.K(*),2)
T     BRF=0.1/0.05
NOTE
R     MOUT.KL(AGE1)=POP.K(AGE1)/MT(AGE1)
T     MT=20/20/20/1E30
NOTE
R     DR.KL(AGE1)=DRF(AGE1)*POP.K(AGE1)
T     DRF=.002/.005/.006/.06
NOTE
SPEC DT=1/LENGTH=50/PRTPER=25/PLTPER=1
PRINT 1)TPOP,2)POP,5)BR,7)DR
PLOT POP
```

Figure 7.6: Equation listing of a population aging chain

groups in the model. AGE2 enters in the level equation that computes POP.K(2), POP.K(3), and POP.K(4). The FOR variable in that equation can not include AGE = 1 because POP.K(1) is computed in a separate level equation. This use of different FOR variables for the same subscript allows the modeler to control completely the way different pieces of model structure are replicated.

The number of age groups in the model is specified in figure 7.6 in an *index* statement:

I   NAGES=4

The statement could have been omitted and the FOR variables AGE1 and AGE2 specified as

FOR AGE1=1,4
FOR AGE2=2,4

However, the modeler would then have to remember that the 4 in these two statements represents exactly the same quantity. If one value is changed to investigate an aging chain with

more levels, then the modeler must remember to change the
other accordingly. In such a situation it is better to give the
memory work to the computer and write the FOR statements
as shown in figure 7.6.

To focus on the essential elements of an aging chain, the
model in figure 7.6 contains no constraints on the growth of
the population. It exhibits unlimited exponential growth in
all age groups. A valuable exercise at this point would be to
add effects of crowding or carrying capacity to the model,
raising the death rates of the various age groups (or lowering the birth rate) as TPOP grows larger and larger. The
additional structure, formulated as table function multipliers
on the death rates (or the birth rate), should change the
basic behavior of the model from exponential to S-shaped
growth and may produce some interesting dynamic interactions as limits are reached.

Summary of Rules for Subscripted Variables

The following conventions on the use of subscripts in
DYNAMO III are largely necessary consequences of the definitions implicit in the preceding examples.

1. Arrays in DYNAMO III are represented by variables written with subscripts in parentheses. Like any other
quantity in DYNAMO, a subscript may be named with
from one to six characters. A subscript that takes on a
range of values specified in a FOR statement is called a
FOR variable. FOR variables may be used only as subscripts.[8]

2. The FOR statement specifies the range of integer values
assumed by a subscript in a DYNAMO III model. The
endpoints of the range may be given as numbers or as
named quantities. If names are used, they must be specified in index statements signified by the letter I in the
leftmost column. (See NAGES in the population model.)
Indices for the range of a FOR variable may be used
elsewhere in the model, but they may not be changed in
rerun mode.

---

[8] The user can work around this restriction by defining a
counting vector such as

    T   NUM=1/2/3/4/5/6/7

and then substituting NUM(I) wherever it is desired to use a
FOR variable as something other than a subscript.

3. The only FOR variables permitted on the right of an equation are those that also are used on the left. DYNAMO is happy with A(I)=B but produces an error message for something in the form A=B(I) or A(I)=B(J) because conceptually they make no sense. The purpose of the subscripts is to allow repeated structure to be represented in one equation. An equation with a FOR variable on the left that does not appear on the right would mean that the equation does not represent a piece of repeated structure.

4. The integers represented by a FOR variable may be given names in the FOR statement (such as CHLDRN and ADLTS in the epidemic model). A subscripted quantity can then be referred to using either the numerical subscript or its name: SICK(ADLTS), for example, is identically the same as SICK(2) in the epidemic model.

5. A quantity may have as many as three subscripts. For example, a workforce might be disaggregated according to skill level, task, and location. Except where noted later, all references to the array must include that number of subscripts, in the proper order. If you wish to push the language to its limits you will find that the maximum number of elements in any one dimension (the maximum number of skill levels, for example) is 999.

6. A FOR variable plus or minus an integer permits the easy specification of flow between elements of an array. (See the use of AGE2-1 in the population model.)

7. The use of two or more FOR variables with different ranges for the same dimension is a general technique for handling similar classes that have significant exceptions. (See AGE1 and AGE2 for examples.)

8. The values assigned to elements of arrays must be well-defined. That is, every element must be assigned a value in the model, and no element may be computed redundantly (in more than one active equation). As illustrated by the array POP in the population model, several equations may be used to compute the elements of an array, but for any given array all the computations must be of the same equation type, and there must be no overlap.[9]

9. Elements of an array may not be computed in terms of other elements of the same array. DYNAMO will reject

such computations, even those that the user knows are valid, because of the chance of creating simultaneous equations.

7.4 FUNCTIONS ON ARRAYS

DYNAMO III provides a number of special functions that handle arrays. In the preceding examples we have discussed two of them: the SUM function and the scalar product SCLPRD. SUM(ARRAY) adds all the numbers in an array of any number of dimensions. SCLPRD generates in the inner or scalar product of two vectors. In addition we have illustrated the use of the TABLE function with different sets of y-values specified for different values of a subscript (see the computation for CNTCTS in the disaggregated epidemic model).[10]

Multiplying Arrays

Although none of the powerful equations of matrix algebra are supported by DYNAMO, equivalent equations can be written. The product of two arrays can be constructed by using the scalar product function SCLPRD. Suppose, for example, that A is a matrix (a two-dimensional array) with M rows and K columns and B is a similar matrix with K rows and N columns. Then the matrix product C = AB is a matrix with M rows and N columns, and C(I,J), the element of C in the I-th row and J-th column is

   A(I,1)*B(1,J) + A(I,2)*B(2,J) + . . . + A(I,K)*B(K,J).

Whether or not this fact looks familiar, the reader should note that this expression is the sort of thing SCLPRD computes. Therefore,

   C(I,J)=SCLPRD(A(I,*),1,K,B(*,J),1).

---

[9] If the entire array is initialized, the active equations need not compute every element. Initial values of elements in an array may be assigned in a mixture of N and active active equations, as long as each element eventually receives an initial value.

[10] TABHL and the other DYNAMO table functions are used in the same way with subscripted variables.

To compute the entire matrix C = AB, the model must contain two FOR statements prescribing that the FOR variables I and J range, respectively, from 1 to M and 1 to N.

The SUMV Function

The SUM function described earlier indiscriminantly adds up all the elements in an array, no matter how large. A more flexible summing function that requires three arguments is SUMV, a sum of elements along one dimension of an array. The form of the function is

SUMV(VECTOR,FIRST,LAST)

where
VECTOR = a one-dimensional array,
FIRST   = first element to be included in the sum,
LAST   = last element to be included in the sum.

If, for example, VECTOR represents the string of numbers

VECTOR = 8,3,-4,12,-1,-10

then

SUMV(VECTOR,2,5) = 3 + (-4) + 12 + (-1) = 10

since SUMV begins here with the second element (3) in the string and ends with the fifth (-1). Although SUMV operates only on a single string of numbers, it may be used to sum along a row or column of a two- or three-dimensional array by fixing the subscripts of all but one of the dimensions and letting the remaining subscript vary over the prescribed range by substituting an asterisk for it. For example, if a workforce is disaggregated by skill and location so that we would write WF.K(SKL,LCTN), then

SUMV(WK.K(*,LCTN),1,MAXSKL)

would compute the sum of employees of all skill levels at various locations LCTN.

SUMV(WK.K(SKL,*),LCTNA,LCTNB)

would compute the total number of workers with particular skills in the range of locations from LCTNA to LCTNB. For and additional example, note that in the disaggregated epidemic model we could have written

Functions on Arrays

A TSICK.K=SUMV(SICK.K(*),CHLDRN,ELDRLY)

In this particular case SICK is a one-dimensional array so the result is the same as summing over the whole array using SUM, as we did. Since they amount to the same thing, we used the simpler formulation involving SUM. Had we wanted to sum only the child and adult subpopulations, however, we would have had to use SUMV, replacing ELDRLY with ADLTS.

Wherever no confusion will result, DYNAMO is comfortable with the omission of the (*) notation in vector functions. The above equation for TSICK, for example, could have been written

SUMV(SICK.K,CHLDRN,ELDRLY) or
SUM(SICK.K)

because DYNAMO knows (and assumes you know) that the first entry in each of these functions is a vector.

The PRDV Function

PRDV is analogous to SUMV, but it multiplies rather than adds the elements of a vector. The form of usage is the same as SUMV:

PRDV(VECTOR,FIRST,LAST)

This product function could be used to formulate concisely a rate equation involving a number of dimensionless multipliers:

R RATE.KL=NORM.K*PRDV(MULT.K(*),1,NMULT)
where
NORM = normal or reference rate,
MULT = multipliers from various sectors,
NMULT = number of multipliers.

As noted before, the (*) notation could be omitted here because DYNAMO does not have to be reminded that the first entry in PRDV is a vector.

The SHIFTL and SHIFTC functions

The exponential delay and smoothing functions provided by all DYNAMO's are usually sufficient to capture the dynamics of any real delays. Occasionally, however, it is useful to be able to store a quantity and to recall its exact historical value some time later in a simulation. What, for example, was the average monthly sales rate one year ago? The exponential delay formulations will not do the job because current values affect the smoothed or delayed values to some extent. The *linear shift* function SHIFTL available in DYNAMO III provides the machinery necessary to hold onto historical values.

The SHIFTL function permits the user to assign a value to the first element of a vector, shift that value periodically to higher and higher positions in the vector, and finally, to obtain its value when it reaches the last position. (The cyclic shift function SHIFTC operates exactly the same way, but returns the value from the last position to the first position again.)

Suppose a model computes a sales rate SALES, measured in units per day, and we wish to know the average sales rate over a 30-day period exactly twelve months (of model time) later. We accumulate SALES in the first of a string of thirteen levels:

L   CUMSAL.K(1)=CUMSAL.J(1)+DT*SALES.JK

Then we invoke the shift function SHIFTL to move the contents of CUMSAL(1) to CUMSAL(2), CUMSAL(2) to CUMSAL(3), and so on, every 30 days. CUMSAL(13), the last element in the string, 12 months removed from CUMSAL(1), is given by

SHIFTL(CUMSAL.K,SI),

where SI is the *shift interval,* here equal to 30 days. To compute the average monthly sales rate a year ago, we must divide that month's cumulative sales by the length of time of the accumulation, namely SI = 30 days. Hence we need the following equations:

FOR M=1,13
   A   AMSY.K=SHIFTL(CUMSAL.K,SI)/SI
where
   M   = FOR variable indicating months
   AMSY   = average monthly sales one year ago
   SHIFTL   = linear shift function
   CUMSAL   = cumulative sales over time interval SI,
   SI   = shift interval (days).

The complete set of equations is the following:

FOR   M=1,13
   L   CUMSAL.K(1)=CUMSAL.J(1)+DT*SALES.JK
   N   CUMSAL(M)=ICS(M)

Functions on Arrays 393

```
T   ICS=0/?/?/?/?/?/?/?/?/?/?/?/?
A   AMSY.K=SHIFTL(CUMSAL.K,SI)/SI
C   SI=30 (DAYS)
```

The initial values of CUMSAL(M), for M = 1 to 13, would be specified in the T statement for ICS. To accumulate accurately the values of SALES, the initial value of CUMSAL(1) must be zero. Note that CUMSAL.K in the expression for AMSY does not have a subscript; the output of the shift function SHIFTL is always the quantity in the final element in the string. The graph of the average monthly sales a year ago AMSY will hold constant for thirty days, then move to perhaps a new value and hold for the next thirty days, and so on, moving in steps every thirty days.

The cyclic shift function, SHIFTC, is used similarly, but has the property that the content of the last element in the sequence is returned to the first element, as if the elements were boxes arranged in a circle. A likely use of SHIFTC is to provide seasonal information. For details see Pugh (1976).

7.5 PRINTING AND PLOTTING ARRAYS

The printing and plotting features of DYNAMO have been extended in DYNAMO III to permit the user to PRINT or PLOT whole arrays simply by mentioning their name. In the disaggregated epidemic model, for example, we could have obtained the graphs shown previously in figures 7.3 and 7.4 by typing

```
PLOT SICK
PLOT SUSC
```

In these plots DYNAMO would identify the points on the curves by numbers: 1's would be typed for points on the graph of SICK(1), 2's would be typed for SICK(2), and so on. If names (such as CHLDRN, ADLTS, and ELDRLY) have been assigned the FOR variables involved in such plot statements, DYNAMO will head the plot with a key to the graphs, such as

SICK(CHLDRN)=1, SICK(ADLTS)=2, SICK(ELDRLY)=3

To obtain such labels for a PLOT statement entered in rerun mode, the user must type the subscript in the PLOT statement, as in the following

```
PLOT SICK(A)
```

If one plot showing all sick and susceptible populations is desired, the model could contain the single statement

    PLOT SICK=A,SUSC=1,

Here, the graph of SICK(CHLDRN) would be indicated with A's, the graph of SICK(ADLTS) with B's, and SICK(ELDRLY) with C's. SUSC(CHLDRN), SUSC(ADLTS), and SUSC(ELDRLY) would be marked with 1's, 2', and 3's, respectively.

The commas in the previous PLOT statement tell DYNAMO to place *all* of the immediately preceding variables on the *same* scale. Frequently, it is preferable to graph, say, the SICK subpopulations on the same scale, with the SUSC subpopulations on a different scale. To accomplish that, type

    PLOT SICK=A,/SUSC=1,

The first comma tells DYNAMO to plot the three SICK subpopulations on the same scale, while the slash indicates that the scale chosen for the three SUSC subpopulations may be different. The final comma calls for one scale for all of the SUSC subpopulations.

In these PLOT statements DYNAMO selects the scales itself. To specify particular scales include the ranges in parentheses in the PLOT statement, as described in section 3.5.

The PLOTS Statement

DYNAMO III provides a plot statement called PLOTS that is particularly suited to subscripted variables. To generate three separate plots for various variables associated with chldren, adults, and the elderly, all that is required is a single PLOTS statement:

    PLOTS (A),SUSC(A)=S,SICK(A)=I,RECOV(A)=R/
    X   INF(A)=I,CURE(A)=C

The FOR variable A represents the age groups in the population and was specified earlier in the model listing. Here the single PLOTS statement produces three graphs, one for each of the values of the FOR variable A; on each graph there will be five curves. If not all the subpopulations are desired, the modeler could define a new FOR variable, say, A1, write the PLOTS statement with A1 in place of A, preceded somewhere in the model listing by a FOR statement for A1 specifying the desired range.

Printing and Plotting Arrays

The PLOTS statement is particularly powerful if more than one subscript is involved. Suppose, for example, we had disaggregated the population aging chain model by *region* as well as by age. Each variable in the model would have two subscripts, say, AGE1 and RGN. If three regions and four age groups are involved, POP(AGE1,RGN) is an array containing twelve quantities. The following PLOTS statement would therefore produce twelve graphs, each containing three curves:

PLOTS (AGE1,RGN),POP(AGE1,RGN)/MOUT(AGE1,RGN),
X   DR(AGE1,RGN)

Alternatively, the modeler can request plots for just one of the regions:

PLOTS (AGE1),POP(AGE1,2)/MOUT(AGE1,2),DR(AGE1,2)

This statement would produce four plots, one for each age group in region number two.

Printing

The form of the statement to print the values in an array over time is analogous to the simple statement for plotting one. The statement PRINT POP or PRINT POP(AGE1) in the disaggregated population model would produce printed output arranged in columns headed by the numbers 1, 2, 3, and 4, representing children, young adults, the middle aged, and the elderly. (The range of the FOR variable AGE1 and the names for its successive values were specified in the FOR statement for AGE1 given at the beginning of that model.)

As in other DYNAMO's, the modeler can control some aspects of the format of printed, tabular output. The columns in which printed output appears are numbered by DYNAMO from left to right. The user can request output in particular columns by preceding the quantity name by the number of the column, as illustrated by the statement

PRINT 1)TPOP,2)POP,5)BR,7)DR

that appeared in figure 7.6. This PRINT statement puts the output in the first, second, fifth, and seventh columns across the page. The four different subpopulations will be printed one under the other in column two, with each value appropriately labeled on the left with CHLDRN, YADLTS, MDLAGE, or ELDRLY.

7.6 DEBUGGING DYNAMO III MODELS

DYNAMO III checks whether all array elements are properly computed before it runs the model. If it encounters a subscript error, it tries to describe the problem accurately in readable English, but sometimes seemingly absurd error messages occur. These messages are almost invariably related to the presumptions that DYNAMO must make when encountering a subscripted quantity. When a name is first seen, DYNAMO carefully notes how many subscripts are associated with it. If the name is followed by a left parenthesis (with no intervening asterisk) DYNAMO assumes that at least one subscript is involved. Any subsequent occurance of the name must have the *same number* of subscripts.[11] Once DYNAMO's mind is made up about a quantity it is very insistent that all subsequent uses conform. If DYNAMO III tells you something about an array that just does not seem to make sense, check to be sure that all references to the array have the same number of subscripts.

Another disconcerting type of error message occurs when the model either fails to compute some element of an array or computes some elements more than once. Either error, if left undetected, would cause much grief when the model is run. If the user believes that all the guidelines in preceding sections have been followed and still fails to see the source of an error, he should review every usage of the array on the right of equations. Some usage probably includes the wrong FOR variable (or the wrong order).

Erroneous subscripts can sometimes slip by DYNAMO's error checking. For example, if two FOR variables have the same range and the modeler accidentally uses them in the wrong order somewhere, DYNAMO will not detect the error. Printing every DT, as one does for other mysterious simulation errors, should expose the error, provided the user is sensitive to the possibility of reversed subscripts.

PRINT and PLOT statements can produce a related set of error messages. If a quantity name represents an array and is followed by a left parenthesis, DYNAMO assumes that a subscript follows. But specifying scales in a PLOT statement also requires parentheses. If the user wishes to supply a scale range on a PLOT statement, he must first satisfy

---

[11] The only exception is an one dimensional array that appears as an argument of an array processing function (or any array used as an argument of the SUM function); these do not require subscripts. An example is SUM(POP.K) rather than SUM(POP.K(*)).

DYNAMO's appetite for subscripts, enclosing them in parentheses, and then introduce the scale range in parentheses. Leaving out one or more or the parentheses can produce strange results.

SCALING LETTERS FOR DYNAMO OUTPUT

Letter	Decimal Equivalent*	Mnemonic**
A	E-03	
B	E+09	Billion
C	E+27	oCtillion
D	E+33	Decillion
E	E-06	
F	E-09	
G	E-12	
H	E-15	
J	E-18	
K	less than E-30	
L	E-21	
M	E+06	Million
N	E+30	Nonillion
P	E+24	sePtillion
Q	E+15	Quadrillion
R	E+12	tRillion
S	E+21	Sextillion
T	E+03	Thousand
U	E-24	
V	E+18	
W	E-27	
X	E+00	
Y	E-30	
Z	greater than E+33	

* The E notation is DYNAMO's shorthand for multiplication by powers of 10. E-03 means mulitplication by 10^{-3}.

** In American usage, names for large numbers correspond to the number of commas that would be used to separate groups of three digits. Nonillion, from the Latin root for nine, suggests nine commas and hence ten groups of three digits, which would be represented in scientific notation as 10 to the 30th power.

ACKNOWLEDGEMENTS FOR REPRINTED MATERIAL

Figures

2.1
Schroeder, Walter E. and John E. Strongman, 1974. Adapting Urban Dynamics to Lowell, in *Readings in Urban Dynamics,* vol. 1, Nathaniel J. Mass, ed. Cambridge, Ma.: The MIT Press. 197-223.

2.3
Naill, Roger F., 1973. The Discovery Life Cycle of a Finite Resource: A Case Study of U.S. Natural Gas, in *Toward Global Equilibrium,* Dennis L. Meadows and Donella H. Meadows, eds. Cambridge, Ma.: The MIT Press. 213-256.

2.4
Reprinted by permission of the Harvard Business Review. Exhibit from "Advertising: A Problem in Industrial Dynamics" by Jay W. Forrester (March-April 1959). Copyright c 1959 by the President and Fellows of Harvard College; all rights reserved.

2.11
Reprinted with permission from *The Persistent Poppy,* copyright 1975, Ballinger Publishing Company.

2.12
Randers, Jorgen and Dennis L. Meadows, 1973. The Dynamics of Solid Waste Generation, in *Toward a Global Equilibrium,* Dennis L. Meadows and Donella H. Meadows, eds. Cambridge, Ma.: The MIT Press. 165-211.

2.14
Reprinted with permission from *The Persistent Poppy,* copyright 1975, Ballinger Publishing Company.

2.15
Meadows, Dennis L., William H. Behrens III, Donella H. Meadows, Roger F. Naill, Jorgen Randers, and Erich K.O. Zahn, 1974. *Dynamics of Growth in a Finite World.* Cambridge, Ma.: The MIT Press.

2.16
Reprinted from "Market Growth as Influenced by Capital Investment," by Jay W. Forrester, *Sloan Management Review,* Vol 9, No. 2, pp. 83-105, by permission of the publisher.

Acknowledgements

Copyright c 1968 by the Sloan Management Review Association. All rights reserved.

4.33
Redrawn from Kormondy, Edward J., 1969. *Concepts of Ecology.* Englewood Cliffs, N.J.: Prentice-Hall.

Peanuts, by Charles Shultz. c 1977 by the United Feature Syndicate, Inc. Reprinted by permission.

Excerpts

Models in the Policy Process: Public Decision making in the Computer Era by Martin Greenberger, Matthew A. Crenson and Brian L. Crissey. c 1976 Russell Sage Foundation. Reprinted by permission.

"Strategies for Effective Implementation of Complex Corporate Models" by Edward B. Roberts, *Interfaces,* Vol. 8, No. 1, Part 1, November 1977, c 1977, The Institute of Management Sciences. Reprinted by permission.

Industrial Dynamics by Jay W. Forrester. c The MIT Press. Reprinted by permission.

"Market Growth as Influenced by Capital Investment," by Jay W. Forrester, *Sloan Management Review,* Vol 9, No. 2, pp. 83-105, by permission of the publisher. Copyright c 1968 by the Sloan Management Review Association. All rights reserved.

The Persistent Poppy by Gilbert Levin, Edward B. Roberts, and Garry Hirsch. Copyright 1975, Ballinger Publishing Company. Reprinted with permission.

REFERENCES

Alfeld, Louis and Alan K. Graham, 1976. *Introduction to Urban Dynamics*. Cambridge, Ma.: The MIT Press.

Britting, Kenneth R., 1973. Correlated Noise Generation Using DYNAMO. System Dynamics Group working paper D-1908, Alfred P. Sloan School of Management, M.I.T., Cambridge, MA. 02139.

Cooper, Kenneth G., 1980. Naval Ship Production: A Claim Settled and a Framework Built. *Interfaces,* 10,6(1980):20-36.

Forrester, Jay W., 1959. Advertising: a Problem in Industrial Dynamics. *Harvard Business Review*, March-April, 1959.

Forrester, Jay W., 1961. *Industrial Dynamics*. Cambridge, Ma.: The MIT Press.

Forrester, Jay W., 1968a. Market Growth as Influenced by Capital Investment. *Industrial Management Review* (now the *Sloan Management Review*), 9(1968):83-105.

Forrester, Jay W., 1968b. *Principles of Systems*. Cambridge, Ma.: The MIT Press.

Forrester, Jay W., 1969. *Urban Dynamics*. Cambridge, Ma.: The MIT Press.

Forrester, Jay W., 1971. Counterintuitive Behavior of Social Systems. *Technology Review* 73(1971):52-68.

Forrester, Jay W., 1973. *World Dynamics*. Cambridge, Ma.: The MIT Press.

Forrester, Jay W., 1975. *Collected Papers of Jay W. Forrester*. Cambridge, Ma.: The MIT Press.

Forrester, Jay W., Gilbert W. Low, and Nathaniel J. Mass. The Debate on *World Dynamics:* A Response to Nordhaus. *Policy Sciences,* 5(1974):169-190.

Forrester, Jay W. and Peter M. Senge, 1980. Tests for Building Confidence in System Dynamics Models. *System Dynamics. TIMS Studies in Management Sciences* 14(1980):209-228.

Garn, Harvey A. and Robert H. Wilson, 1972. A Look at Urban Dynamics: The Forrester Model and Public Policy. *IEEE Transactions on Systems, Man, and Cybernetics,* SMC-2(1972):150-155.

Goodman, Michael R., 1974. *Study Notes in System Dynamics.* Cambridge, Ma.: The MIT Press.

Graham, Alan K., 1977. *Principles on the Relationship Between Structure and Behavior of Feedback Systems.* Ph.D. dissertation. Department of Electrical Engineering and Computer Science, Massachusetts Institute of Technology, Cambridge, Ma., 02139.

Graham, Alan K., 1980. Parameter Estimation in System Dynamics Modeling, in *Elements of the System Dynamics Method,* Jorgen Randers, ed. Cambridge, Ma.: The MIT Press, 143-161.

Greenberger, Martin, Mathew A. Crensen, and Brian L. Crissy, 1976. *Models in the Policy Process.* New York: Russell Sage Foundation.

Hardin, Garret, 1972. *Exploring New Ethics for Survival; the voyage of the spaceship Beagle.* New York: Viking Press.

Kormondy, Edward J., 1969. *Concepts of Ecology.* Englewood Cliffs, N.J.: Prentice-Hall.

Levin, Gilbert, Edward B. Roberts, and Garry Hirsch, 1975. *The Persistent Poppy.* Cambridge, Ma.: Ballinger.

Lyneis, James M., 1980. *Corporate Planning and Policy Design.* Cambridge, Ma.: The MIT Press.

Malthus, Thomas, 1798. *First Essay on Population,* 1966 edition. New York: MacMillan, Inc.

Mass, Nathaniel J., 1973. *Readings in Urban Dynamics,* vol. 1. Cambridge, Ma.: The MIT Press.

Mass, Nathaniel J. and Peter M. Senge, 1978. Alternative Tests for the Selection of Model Variables. *IEEE Transactions on Systems, Man, and Cybernetics,* July, 1978.

Meadows, Dennis L., 1970. *Dynamics of Commodity Cycles.* Cambridge, Ma.: The MIT Press.

Meadows, Dennis L. and Donella H. Meadows, eds., 1973. *Toward Global Equilibrium.* Cambridge, Ma.: The MIT Press.

Meadows, Dennis L., William H. Behrens III, Donella H. Meadows, Roger F. Naill, Jorgen Randers, and Erich K.O. Zahn, 1974. *Dynamics of Growth in a Finite World.* Cambridge, Ma.: The MIT Press.

Meadows, Donella H., 1980. The Unavoidable A Priori, in *Elements of the System Dynamics Method,* Jorgen Randers, ed. Cambridge, Ma.: The MIT Press. 23-57.

Meadows, Donella H., Dennis L. Meadows, Jorgen Randers, and William W. Behrens III, 1972. *The Limits to Growth.* New York: Universe Books, A Potomac Associates Book.

Naill, Roger F., 1973. The Discovery Life Cycle of a Finite Resource: A Case Study of U.S. Natural Gas, in *Toward Global Equilibrium,* Dennis L. Meadows and Donella H. Meadows, eds. Cambridge, Ma.: The MIT Press. 213-256.

Nordhaus, William D. World Dynamics: Measurement Without Data. *The Economic Journal,* 83(1973):1156-1183.

Peterson, David W., 1980. Statistical Tools for System Dynamics, in *Elements of the System Dynamics Method,* Jorgen Randers, ed. Cambridge, Ma.: The MIT Press. 224-245.

Pindyck, Robert S. and Daniel L. Rubenfeld, 1976. *Econometric Models and Econometric Forecasts.* New York: McGraw-Hill.

Pugh, Alexander L., III, 1976. *DYNAMO Users' Manual,* 5th ed. Cambridge, Ma.: The MIT Press.

Randers, Jorgen., 1980. *Elements of the System Dynamics Method.* Cambridge, Ma.: The MIT Press.

Randers, Jorgen and Dennis L. Meadows, 1973. The Dynamics of Solid Waste Generation, in *Toward a Global Equilibrium,* Dennis L. Meadows and Donella H. Meadows, eds. Cambridge, Ma.: The MIT Press. 165-211.

Richardson, George P., 1981. Statistical Estimation of Parameters in a Predator-Prey Model: an Exploration using Synthetic Data. System Dynamics Group working paper D-3314-1, Alfred P. Sloan School of Management, M.I.T., Cambridge, MA. 02139.

Roberts, Edward B., 1964. *The Dynamics of Research and Development.* Cambridge, Ma.: The MIT Press.

Roberts, Edward B., 1980. *Managerial Applications of System Dynamics.* Cambridge, Ma.: The MIT Press.

Roberts, Edward B., 1977. Strategies for Effective Implementation of Complex Corporate Models. *Interfaces,* 8,1(1977)Part 1.

Schroeder, Walter E. and John E. Strongman, 1974. Adapting Urban Dynamics to Lowell, in *Readings in Urban Dynamics,* vol. 1, Nathaniel J. Mass, ed. Cambridge, Ma.: The MIT Press. 197-223.

Senge, Peter M., 1977. Statistical Estimation of Feedback Models. *Simulation,* 28(1977):177-184.

Shaffer, William, 1976. *Court Management and the Massachusetts Criminal Justice System.* Ph.D. dissertation, Alfred P. Sloan School of Management, Massachusetts Institute of Technology, Cambridge, Ma. 02139.

Shaffer, William, 1978. *Mini-DYNAMO Users' Guide.* Pugh-Roberts Associates, 5 Lee Street, Cambridge, Ma. 02139.

Wiener, Norbert., 1961. *Cybernetics: or control and communication in the animal and the machine.* Cambridge, Ma.: The MIT Press.

INDEX

A equation. *See* auxiliary
Accumulations, 30-31, 75,
 176-178, 186-188, 262. *See
 also* Level
Aging chain
 insect population, 138
 with subscripts, 382-386
Aggregation. *See also* Disaggregation
 in project model, 51
Area between rate graphs, 183n
Argument, in macro
 dummy, 361-362, 365
 input, 362, 364, 366, 374
 output, 364, 374
 real, 362, 366
Arms race, 29
Array. *See also* Subscripted
 variable
 definition of, 374-375
 dimension of, 375
 use of, 374, 375
Array functions
 PRDV, 391
 SCLPRD, 384-385
 SHIFTC, 392
 SHIFTL, 391-392
 SUM, 378, 380, 391
 SUMV, 390-391
Asterisk
 as multiplication, 78
 in T statement, 378, 380
 in array functions, 380, 391
Asterisk statement, 92
 and DOCUMENTOR, 217
Auxiliary equations, 80-82,
 159-176
 computation of, 82
Auxiliary variable, 80

as information, 81, 159
vs. level, 178-180, 186-188
dimensions of, 159
symbol for, 82, 89
as table function, 82
units of, 160
AUX.K*LEVEL.K, 146
Averaging. *See* SMOOTH;
 Smoothing Information
Bank balance, 135
Boundary. *See* System bounda-
 ry
C statement, 75
 in rerun, 99-100
Causal link, 25-26. *See also*
 Feedback link
 negative, 26
 polarity of, 26
 positive, 26
Causal-loop. *See* Feedback loop
Causal-loop diagrams, 25-30
 guidelines for, 28-29
 hybrid, 34
 limitations of, 33-34
 practice with, 29-30
Causality, and feedback, 268
Cause-effect, 268
Chickens and eggs, 268
CLIP function, 119-120
 in LENGTH, 209
 in policy test, 335-336
Coffee-cooling, 69-72
Computation interval, 76. *See
 also* DT
Conceptualization. *See also*
 Causal-loop diagrams; Reference mode
 guidelines for, 18-19
 example of, 45-61

without reference mode, 25
summary, 61-63
tools for, 19-38
Conserved
  flow, 262
  subsystem, 262
CONST*LEVEL.K, 134
Constants, in DYNAMO, 75
  computed in N statements, 87
  symbol for, 89
Contact area, 6, 162
Cost
  in project model, 294
  with incentives, 336
CP statement, 99
Criminal justice, 13, 27, 183
Crowding, 161
Cycles, 30
Data, 49-50. *See also* Parameter estimation
Decisions, representation of, 265-266
Debugging
  definition of, 247
  DYNAMO III models, 396-397
  methods for, 247-261
  by printing every DT, 256-257
Deer growth factor, 235-237
DELAY1, 104. *See also* Delays
  initialized for growth, 368-370
DELAY3, 105. *See also* Delays
DELAYP, 106. *See also* Delays
Delays
  as auxiliaries, 159n
  initialization of, 105, 114n, 368-370
  information, 113-115
  material, 103-108
  order of, 108-109
  pipeline, 106
  symbols for, 114, 116
  using SHIFTL, 392-393
Delivery delay, 139
  and sales, 82-83
Desired vs. actual
  in production rate, 158
  in decisions, 265
Dimensional analysis
  examples of, 53-54, 139, 161, 199
  in conceptualization, 62, 175
  in principles of model building, 264
Disaggregation, 28, 31-33
  for client, 296n

for model behavior, 295
for policy testing, 295
in project model, 296-298
and subscripts, 374, 375
Disequilibrium, 185
DLINF3, 113. *See also* Delays
Doctor's workload, 30
Documentation *See also* DOCUMENTOR
  in model listing, 92-94, 213, 218-219
DOCUMENTOR, 214-229
  model definitions for, 215-216
  options, 217-218
  output, for project model, 220-229
  preparing model for, 215-217
  reformatting with, 218
Doubling time, 142
  and parameter estimation, 237
DT, 68
  effect of size of, 73-75
  in project model, 209
  PRTPER equal to, 256-257
  selection of, 112-113
  in SPEC statement, 88
Dwell time, average. *See* Time constant
Dynamic hypothesis, 55-56, 63
  enrichment of, 300-301
Dynamic thinking, 19-25
DYNAMO, 67
  arithmetic in, 78, 137n
  computation scheme, 68, 73, 82
  equations, 72
  equation types, 75-77, 79-82, 86-87
  error messages, 248-256
  format rules, 78
  hand-computed example, 69-72
  order of statements, 94
  output, 88, 90-92, 96, 97
  quantity names, 77
DYNAMO III, 360, 374-397
E, in DYNAMO numbers. *See* Scientific notation
Effect, 160. *See also* Multipliers; NORM.K + EFFECT.K; NORM.K*EFFECT.K
Effort perceived remaining, 190, 198-199
Electric blanket, 11-12
Endogenous
  variable, 63

Index

point of view, 63-66
Environmental damage, 5-7
Epidemic model, 96
 and subscripts, 375-380
Equilibrium, 185
 initialization in, 241-245
Errors. See also Debugging;
 Simultaneous equations
 common, 260
 with subscripts, 396-397
 types of, 248-249
Exogenous variable, 63
 symbol for, 89
EXP function, 119
Exponential decay
 structure for, 139
 properties of, 140-142
Exponential growth
 structure for, 135
 properties of, 142
External view, 63. See also
 Internal; Exogenous variable
Feedback, 2
 definition of, 3
 view of R&D project
 management, 52-61
Feedback link 6. See also Causal-link
 negative, 26
 polarity of, 26
 positive, 26
Feedback loop
 compensating, 285, 337
 definition of, 4
 dominant, 285
 and internal view, 65-66
 negative, 7-8, 11-12
 polarity of, 27-28
 positive, 7, 9-12
Feedback system, 2-3
FIFGE function. See CLIP
FIFZE function. See SWITCH
Flow diagrams, 31-36
 information links in, 33, 89
 material flows in, 33, 89
 symbols for, 89
FOR statement, 377, 386
FOR variable
 definition of, 387
 examples of, 377, 383
 rules for, 387-388
Forecast. See TREND macro
Fraction satisfactory, 194

Functions in DYNAMO. See also
 Array functions; Delays; Macros; Table function
 mathematical, 115-116
 logic, 117-121
 test, 121-132
Gain, 164n
Gasoline reserves, 187-189
(GOAL-LEVEL.K)/ADJTM,
 143-145
Goal-seeking loop, 7, 143-145.
 See also Feedback loop, negative
Gross productivity, in project
 model, 194
Half-life, 141, 235
 of DDT, 146-147
Hiring in the project model, 50,
 52, 55, 190, 202-203
Heroin
 addiction, 34
 and crime, 29
I statement, 386
Implementation
 and mental models, 355
 and planned change, 355
 strategies for, 39
Incentives, 335
Index statement, 386
Industrial dynamics, 1
Information
 auxiliary variable as, 81, 159
 availability of, 58, 63, 262-263
 not conserved, 262
Information link, 33
 in flow diagram, 89
Initialization, 240-247
 for equilibrium, 241-245
 for growth, 245-247, 368-370
Initial value
 changing, in rerun, 101, 286
 equations, 86
 estimation of, 240-247
Insensitivity to parameter
 changes, 281-286. See also
 Sensitivity
 examples of, 282-284, 286-289
 reasons for, 284-286
Interest rate, 135
Internal variable. See INTRN
 statement
Internal view, 15-16, 65. See
 also Endogenous
INTRN statement, 362
 examples, 363

Inventory
  control system, 30
  as decoupler of rates, 180
  and workforce, 36
J. See Timescript
K. See Timescript
L. See Timescript
L equation. See Level equation
Length of statement
  in DYNAMO, 78
LENGTH
  parameter in DYNAMO, 88
  in project model, 209
Level equation, 76-77
Level variable, 31-33, 176-186, 189. See also Principles of model structure
  vs. auxiliary, 178-180, 186-188
  characterization of, 189-190
  as decoupler of rates, 180-183
  initial values for, 240-249
  in project model, 192-193
  snapshot test for, 176-177
  symbol for, 32, 89
  and time horizon, 177
  units of, 177
LEVEL.K/AUX.K, 146-147
LEVEL.K/LIFE, 137
LOGN function, 117
MACRO statement, 361, 365
Macros, 361-374
  examples of, 361, 363, 364-365, 368, 370-374
  errors in, 366-368
  purpose of, 360-361
  rules for, 361, 365-366
MAX function, 117
  and absolute value, 118
  and division by zero, 118, 204n
  cautions about, 118-119
Mean
  defined, 373
  of NOISE, 131
  in PKNS macro, 372
  in STAT macro, 373
Mean square error, 374
MEND statement, 361, 365
Migration, in city, 152
MIN function, 117, 250
  cautions about, 118-119
Model, 2
  formal, 2-3
  mental, 2
  as a laboratory tool, 2, 276
Model testing
  by deactivating loops, 268-274, 325
  using extreme conditions, 268-269, 317
  using test functions, 121-132, 275
Modeling
  as an iterative process, 268, 293
Multipliers, dimensionless, 152-154
  caution with, in net rates, 155
N statement, 86-87
  for TIME, 88
Natural gas production, 22
Newton's law of cooling, 69
NOISE function, 130-132. See also Pink noise
Normal point, in table function, 166-167
NORM.K + EFFECT.K, 148-151
NORM.K*EFFECT.K, 152-155
NOTE statement, 93-94
Open-loop thinking, 4
  and feedback view, 4-6
Overestimation
  of workforce needed, 333-335
  of effort required, 335, 357, 358
Overruns in R&D project, 46-49
Overtime, 170-173
Parameter
  definition of, 230
  and reality, 232
  types, 231
Parameter estimation
  accuracy of, 230-232, 237-240
  using behavior, 235, 237-238
  using bounds, 234
  using process, 235-237
  using statistics, 238-240
Perceptions, 52, 58, 63
  in R&D project, 48-49, 58-59
Pink noise, 133, 371-373
  equations for, 372
  meaning of, 371-372
  use of, 373
PLOT statement, 90-92, 304
  with comma, 91, 394
  in rerun, 98
  with slash, 91, 394
  and subscripts, 393-394
PLOTS statement, 394-395
PLTPER, 90
  in project model, 209

Index

PKNS macro, 372
Polarity
  of causal-link, 26
  of feedback loop, 27-28
Policy
  criteria, 294, 349-350
  insight, 324-327
  lever, 62
  limits, 329
  parameters, 322-323, 333
  recommendations, 352-356
  robustness, 349-352 (see also Sensitivity)
  structure, 341-345
Policy simulation, 321-348, 357-359
  caution in, 326
  using CLIPs, 335-336
  using constants, 323, 328
  using tables, 328
  using new structure, 338-346
Pollution, in *World Dynamics*, 147
Population
  aging chain, 382-385
  positive loop, 38
  and Malthus, 352-354
PRDV, 391
Principles
  of model building, 264-266
  of model structure, 341-343, 262-264,
PRINT statement, 90-91, 304
  in rerun, 100
  and subscripts, 395
Prison population model, 181-185
Problem definition, 18-24
  examples of, 40-42
  reference modes in, 19-24
  summary, 62-63
Problem-solving, 5
Productivity
  in R&D project, 52
  perceived, 58-59, 200-201
  real, 201
  of new and experienced workers, 298-300
Production sector, 14
Progress, in R&D project, 51
  real, 52, 190, 193
  apparent, 56, 193
  cumulative, 52, 194, 198
  perceived, 190, 198
Project model, simple
  base run behavior, 210
  description of behavior of, 276-277
  DOCUMENTOR output for, 220-229
  DYNAMO listing of, 211-212
  problem definition for, 46-50
  reference modes for, 46-49
Project model, revised
  base run behavior of, 306
  DYNAMO listing of, 305-309
PRTPER, 90
  in project model, 209
  equal to DT, 257-258
  as an auxiliary, 259, 304
PULSE function, 127-129
Purpose of model, 18, 38-44, 62
  examples of, 39-42
  of R&D project, 50
  and conceptualization, 38-44
  and evaluation, 310
R&D project, 46
  flawless, 47
  description of, 276-277
RAMP function, 125-128
Random number generator. *See* NOISE function
Rate equations, 79-80, 134-158
Rate variable, 31-33. *See also* Principles of model structure
  symbol for, 32, 80, 89
Reference mode, 21-22, 62
  inferred vs. observed, 48
  for project model, 47-49
  practice, 23-24
Reference line, in table, 171
Reference point, in table, 166-167
Reformatting model, 218
Rerun mode, in DYNAMO, 99
  cautions about, 275
Rework, 51, 55-56
  assumed, 343-344
  undiscovered, 57-58
RUN statement, 94
Salesmen loop, 10, 163-164
Saturation, 171, 172, 178n
Scalar product, 382
Scaling letters, 94-95, 399
Schedule in R&D project
  adjustment of, 54
  equations for, 205-206
  overruns in, 46-49
  and pressure, 302
Scientific notation, 94, 399
SCLPRD function, 384-385

Sensitivity. *See also* Insensitivity
 behavioral, 278, 281
 numerical, 278, 281
 to parameter changes, 281-286, 331-332
 policy, 278, 322
 responses to, 291-292
 to structural changes, 279-281
Sensitivity testing, 281-292
Shape, of table function, 167
SHIFTC function, 392
SHIFTL function, 391-392
Simulation, 67, 133, 209
Simultaneous equations, 249-256
 in active equations, 253-256
 in initial values, 250-252
 in new structure, 301
SIN function, 115, 128-130
Sine function. *See* SIN function
Sink, 31
 symbol for, 89
SLOPE macro, 361, 363
Slope, of table function, 167
SMOOTH, 109-111. *See also* Delays
 and smoothing time, 112
Smoothing information, 109
Snapshot test, for levels, 176-177
 in project model, 192
Solid waste, 35
Solution interval, 76. *See also* DT
Source, 31
 symbol for, 89
SPEC statement, 88
 examples of, 90
SQRT function, 115
Standard deviation, 373
Star statement. *See* Asterisk statement
STAT macro, 373-374
STEP function, 122-124
Structure
 and behavior, 15
 and parameter values, 268-269
Subscripted variable. *See also* Array; Array functions; FOR variable
 in aging chain, 382-383
 definition of, 374
 in epidemic model, 375-379
 examples of, 374, 377-379, 382-384

 in T statement, 378
 in table function, 379-380
 in PLOT statement, 393-394
 in PLOTS statement, 394-395
 in PRINT statement, 395
 purpose of, 374
 rules for, 387-389
SUM function, 378, 380, 391
SUMV function, 390-391
SWITCH function, 120-121
System boundary, 42-45, 62-63, 382
 examples of, 43-45
 for project model, 50-52
System dynamics
 history, ix, x, 1
 focus, 1-2
 goal, 6
 philosophy, 15-16, 63-66
Systems approach, 1
T statement, 84. *See also* Table function
 in rerun, 100
 with subscripts, 378, 380, 384
Table function, 82-85
 aggregation in, 172
 alternatives, 288-291
 with arrays, 379, 380
 beyond range of, 84-85, 172-173
 cautions about, 175-176
 corners in, 172
 example in project model, 196-198
 formulation of, 164-174
 guidelines for, 173-174
 normal point in, 171-172
 normalizing, 170
 reference points in, 167
 reference lines in, 171-172, 298-300
 representing two effects, 169
 symbol for, 85, 89
 subtleties in, 298-300
TABHL, 85. *See also* Table function
TABLE, 83-84. *See also* Table function
TABPL, 85n
TABXT, 85n
Tasks, in project model, 52
Time
 in DYNAMO, 68-69, 88, 206
 remaining in project, 56
 required for project, 56

Index

Time constant, 140-143
 and time horizon, 188, 382
Time horizon, 21
 for project model, 49
Time to detect rework, 196
Timescripts, 68-69
TP statement, 100
TREND macro, 370-371
Underestimation of project,
 300-301, 356
Undertime/overtime, 170-172
Undiscovered rework, 57-58,
 190, 194-195
 generation of, 195
 detection of, 195
Units. *See* Dimensional analysis
Urban model
 construction in, 166-169,
 289-290
 migration equations in, 152
 reference modes for, 20
User-defined functions. *See*
 Macros
Validation, 310-320

Validity
 and model purpose, 310
 and policy recommendations,
 352-354
 as suitability and consistency,
 311-312
 and subjectivity, 312
 and model utility, 312
 of project models, 319
 tests, 313-319
Variable, in DYNAMO, 75
Variance, 373
Vector, 382, 390
Weighted average, 112, 200
Where used list, 215
White noise, 372
WHR-CMP (where computed), 215
Willingness to change workforce,
 204
Wine, 27
Workforce
 and inventory (*see* Inventory)
 in R&D project, 52, 296

> "Surrender to what is. Let go of what was. Have faith in what will be." – Sonia Ricotti

This book is dedicated to the one person that saw something in Cole and wouldn't let me let it go until the rest of the world was able to see it too. Chauna, this one's for you, doll. Love your face!

Prologue

Cole

There has got to be some kind of rule against doing this.

Relationship to the family, being too close to the situation, something. Anything I can use to get myself out of the mess I've found myself in.

"We know this is a trying time for you, Cole. We appreciate you making the time to see us."

The trying time he's talking about is my Uncle John's death. The uncle I've seen maybe six times over the last fifteen years, but the same one that saw fit to have it written into his will that I take over the family business.

Private Investigation.

What the Alexander men do best.

It started with my grandfather, passed down to my father, and when he passed on after a heart attack at the age of fifty-five, had fallen to John until he took the red eye to the sky a few months ago.

Leaving all of his open cases piled in my lap, of which Xander Grayson is one.

The reason for that being that seventeen years ago, when I was eight and she was eighteen months old, my best friend Luke's sister was taken and hasn't been seen since.

At first, the family left it in the hands of the police department, but after years of that wielding less than stellar results, they'd taken it to my dad. When his relentless searching seemed to get nowhere, he passed it on to John.

From that point on, John Alexander made it his life's mission to find this little girl. And now, in the midst of my supposed grief, because he came up empty, it's up to me to do the same.

Another thing he made sure to take care of in his will.

I had to bring Sunday Grayson home.

How the hell I'm supposed to do what he spent the better part of a decade trying and failing at is beyond me, but now that I'm here and as loyal as always, I've got to at least try.

"It's okay, Xander. I'm just not sure what else I can tell you that you haven't already gone over with John before he died."

"The reason we were so adamant that we speak to you is because Allison thinks she may have found something."

"What kind of something?"

Allison Grayson chooses that moment to enter the room. Her hands tightly gripping a silver tray, complete with three mugs of what I hope is some strong coffee. After the way last night ended, nothing else will suffice.

Sitting in a dark apartment with only a forty of Jack Daniels to keep you company while you attempt to drink away memories, well, it never ends well.

Lesson learned.

Placing the tray down onto the table, she slips into the spot beside her husband on the sofa and with a quick glance between them, turns her attention to me.

Waiting for her to speak, I grab one of the mugs from the tray and adding three spoons of sugar, lean back and drain the mug dry as she speaks.

"Being so close to us over the years, I'm sure you're already aware, but when Sunday was born we took precautions. Our insurance company at the time was working hand in hand with the police on a program designed to help parents keep their kids safe. We received a package in the mail in which we could take a lock of her hair, a prick of her blood for typing and all of her personal details at the time. Sunday being so young, we knew that a lot of the details would change over time, but the DNA from her hair and her blood wouldn't."

As thankful as I am for the explanation of how proactive they were in terms of their children, she's right. This is all information I've heard before. Having stayed in contact with Luke all these years, he'd been the one to tell me all of this right before my dad passed away. What I need now is less story and more details I can actually work with.

"What does that have to do with what you found?"

"Despite the case remaining open, the police in recent years haven't been as vigilant as we would like them to be. They've apparently lost hope along with a lot of the general populace. Except for one officer. It was in speaking to him yesterday that we obtained new information."

Allison is right. As the years passed with not so much as a bread crumb to follow that may have led to Sunday being alive, the police department had all but given up. It comes with the territory. But her admission that there was one officer still actively involved is new. The news that he may have stumbled across new information even more so.

"What new information, Allie?"

"According to Officer Bradley, there was an admission to a hospital in Alliston that matches Sunday's aged description and whose blood work appears to be a match."

Now this is something I can work with.

"If you've been in contact with the police why bring me in?"

"I assumed by now that would be obvious, Cole." Xander speaks up, placing his hand tenderly on his wife's knee in an effort to take back control of the conversation. "Sunday has been missing for seventeen years. Not once in that time were the police able to do anything or get us closer to an end result. The same could be said for Samuel and then John, but at least with them, they were family. We trusted them. The same way we trust you."

When Samuel Alexander plucked me out of the foster home I was wasting away in and brought me back here, Luke Grayson was the first person I'd met. With a mutual love of hockey, it wasn't long before we were practically living at each other's houses. Xander calling us family is accurate. Luke was the brother I was never lucky enough to have. His family became almost as important as my own.

I still remember the day they brought Sunday home from the hospital and how insane Luke was about showing her off. Giving me countless warnings before I was allowed to touch her. Treating her like some precious piece of glass that would shatter the second I made contact.

I can also still remember the way she looked at me that day. A look that in the seventeen years since still haunts me.

"What is it you want me to do?"

"We want you to do what you've always done. Nose around and find out who this girl is. Get close and find out if it's our Sunday."

My uncanny ability to be a nosy asshole aside, I'm not sure how comfortable I feel stepping on local PD's toes. This is still considered an open missing persons investigation even though based on protocol it should have been closed years ago. Strolling in and forcing their hand is only going to cause shit that if we want to find Sunday, we don't need.

"Not to be a buzzkill, but do you really think it's smart for me to throw myself into an ongoing investigation?"

"It's what Sam and John would have done." Xander responds and there's no argument I can mount because he's right.

It's the code of the PI. They can get the shit done that the police can't because doors that normally won't be cracked for a uniform, open easily for a guy in a pair of jeans and a black leather jacket. A guy looking for information about a girl that once upon a time he cared pretty deeply about.

"We just want to know if there's anything to this, Cole." Allison explains, leaning across the table that separates us until she's placing her hand on mine. "This is our little girl. Our Sunday."

God damnit.

Why did John have to leave me the business when he died? Why didn't he reach out to his own son and made him take this on?

I never wanted this. If anything I've spent the last seventeen years trying to escape it. It might have been a losing battle considering I was helping them on cases when they were both alive, but it's still not what I wanted for my life.

Except you did want to bring Sunday home.

Shit. I've got no argument I can give here. I did want that. It's all I've ever wanted. For myself and for Luke and his family.

I may be a son of a bitch to the rest of the world, but when it comes to family, I'm as loyal as a Saint Bernard. They're always going to come first and with the way I feel about this family, it's pretty obvious what way this is gonna go too.

"I'll look into it, but guys; I can't promise anything. This could end up being another false lead. I don't want you getting your hopes up."

"We won't. At least no more than we already have. Like Allison said, we just want answers. And if there's anyone besides John that can get them for us, it's you, Cole."

Their faith in my ability is overwhelming. It's almost impossible to live up to. Right now with what they're asking me to do, it's like I'm their personal savior and I'm weighed down by the pressure that being that brings.

I don't want to be the one that comes back here in a few days and destroys the one bit of hope they've had in the last seventeen years. I can't lose the only family I've got left.

"Are you going to be able to handle it if this turns out to be nothing and we're right back where we started?"

"Yes." They both answer simultaneously and with a curt nod of my head, placing the mug back on the tray, I lift myself from the chair. Waiting as both of them follow suit and Xander holds his hand out between us for me to take.

Gripping his hand tight we shake and the request from my Uncle John's will is brought to fruition. What I've been putting off since he died, I now have to face head on.

It's time to find Sunday Grayson and bring her back where she belongs.

Even if it turns out to be the last thing she wants.

Chapter One

Reagan

I can't believe I did it again!

Throwing back the comforter, I slip out of the bed and seeing the time flashing its bright red warning at me—the result of hitting snooze seven consecutive times in a row and not once getting up the way I was supposed to—I make a beeline for the door and the solace a hot shower will give me.

Slamming straight into the brick wall known as my brother Jake.

"Where's the fire?" he jokes as I push my way past him, cursing under my breath before diving into the shared bathroom. "Nice talking to you too, Reagan!"

This is the second time this week I've done this. Spent half the night tossing and turning. Deciding at the last minute to read until I pass out and then sleeping past the alarm when I finally do reach the bliss that comes from sleeping soundly.

Making quick work of the shorts and baby tee that qualify as my pajamas, I turn the water on and step in, welcoming the warmth the second it hits my body and the relaxation that follows.

It's only when I've successfully washed off the night before and am sliding back the curtain in order to step out that I feel the draft from the door as it opens and my mother steps in. A knowing smile crawling across her face as I grab a towel, wrap it around myself and step out. After a few seconds of silence between us as we both share the mirror, she breaks the silence by laughing.

"Let me guess. You're gonna need a ride to work?"

"If you could that would be great."

"Meet me downstairs in five and you've got yourself a deal." She says, placing a soft kiss to the top of my dripping hair and making her way back out.

It's only after she's left and I've turned back to the mirror, going through the motions of brushing my teeth and running the brush rough and quick through my hair that I'm struck again by just how different I am from her.

Taking how different we look entirely out of it, my mother hasn't been late for a job or appointment in her life. If anything, she's the one that's got everything planned down to the letter and is ten minutes early. Proving what an asset she is. Where I'm leaning more toward the liability end. The woman that even knowing she's got to drive to the opposite side of town to drop her daughter off at work, still walks around with a smile.

Nothing fazes Katherine Carter.

Not even me.

It's just further proof that when the stork was dropping babies off the night I was born, he dropped me at the wrong house.

Securing the towel around me, I open the door and slip out, padding my still wet feet across the carpet as quickly as I can manage until all that's left behind me as I shut my door is the footprints left behind.

Grabbing my work clothes off the hanger in my closet, I make quick work of getting dressed and with one final brush through my unruly mane of brown hair, I head out, hitting the front door at the exact moment my mom does.

"Ready?"

"As I'll ever be."

"How much time do we have this time?"

Looking down at my arm and seeing it bare, I grab my purse off the hook at the door and slip my phone out, turning it toward her to show her the time.

"Five minutes and counting."

"You're gonna be late."

"I know. I'd rather it be a few minutes late than the hour I would be if I took the bus, though. So let's go."

Following her out the door and along the gravel path that leads to where her car is parked in the driveway, I use the time to take her in.

Short blonde hair in the form of a pixie cut, make up covering her face done immaculately as always. Her blue eyes the complete polar opposite of my brown ones, but the perfect match for my dad's light green ones. Her six foot frame giant compared to mine.

Reminding me again of just how different I am.

Hearing the click from the automatic lock, I slide myself into the passenger seat as she makes her way around to the back, throwing in her briefcase before sliding in beside me.

"What made you oversleep this time?" she asks as she starts the car and proceeds to pull out onto our street. "It wasn't Walker was it?"

Eww. The last thing I want to talk about before I've had my morning coffee is anything to do with Walker Matthews.

The guy that thought it was completely okay to screw around with other girls while his clueless girlfriend sat at home believing she was his one and only.

Walker is *the last* person that would be keeping me awake.

"No, but it was a guy."

Even from her position staring out the windshield as she's driving, I can see her eyes raise and a smile cross her face.

"Is this a guy I know or someone you have yet to tell me about?"

"You don't know him. He's actually dating this other girl right now. It's only a matter of time before he dumps her for me, though."

This gets her attention as her head does an Exorcist style swivel around to me. Seeing the way her eyes bug out, I stop trying to contain myself and laugh.

"It's a book, Mom. I fell asleep with a book boyfriend."

"As long as it's not Fabio, I think I can handle a book boyfriend or two."

"Fabi-who?"

"Never mind." She laughs as her eyes fall to the rearview mirror before hitting her blinker and pulling into the right lane.

Driving the rest of the way in silence, she speaks again when she's pulled into the bank parking lot and pulled the car to a stop, my seatbelt already off and ready to make what I hope isn't my last walk through the bank.

"Have a good day at work, honey. I love you."

I haven't wanted to say anything before, but with it happening again now, it's becoming impossible to shove down and ignore completely.

Ever since I fell getting off the bus last month and spent the night in the hospital because of suspected damage to my head, she's given me the same strange look when she tells me she loves me.

Her eyes glaze over first and then the tears appear.

She's never been closed off or unaffectionate before, so I'm sure I'm reading into it, but I can't really figure out a reason for her to be acting like this. There was no permanent damage from the fall. She can clearly see that, so this makes no sense.

"Mom, is everything alright?"

"Of course it is. Why do you ask?"

"I don't know. You just look sad. Is everything alright with you and Dad?"

"Your father and I are as happy as we were the day we got married. Maybe even as happy as we were the day we had you and Jake."

"Okay well, if it's not dad, is it something else?"

She smiles softly before reaching out and running her fingers through my hair, her eyes locked on it, almost as if all the answers to my questions can be found in the fly away strands now wrapped in her hand.

"Everything is fine, Reagan. Sometimes I just forget that you're not my little girl anymore and when I open my mouth to speak, the emotion takes over."

I'm pretty sure there's more to it than that, but since it's as good an excuse as I can allow with how late I am, it's gonna have to do.

Leaning over and giving her a quick hug, I flash her a smile, saying my final piece before getting out of the car and heading for the bank.

"I'm always gonna be your little girl, Mom."

Cole

"Come on, Frank. How many times have I done things for you without asking for anything in return?"

I knew it was going to be a long shot, but before I went behind their back, I thought it only fair I do my best trying to their face first. Helping my dad and John over the years when they weren't able to get around, I'd managed to earn a few contacts within the police department. Frank Greavey being one of them and my best bet at getting the information I'm after.

"More times than I can shake a stick at, Cole, but if anyone gets wind of me helping you, it's not your ass on the line."

"All I want is an address. One simple street name and number and I'm out of your hair. Never to bother you again."

"Why don't I believe that?"

Maybe it's got something to do with the fact that I'm lying through my teeth. First chance I get, I'll be back here bugging the piss out of him for information on an unrelated case and he knows it as well as I do.

I can deny the business all I want, but there's no denying it's what I do best.

"No one needs to know where I got the information. Come on, man. You know you wanna do it."

"Fine," he concedes as he grabs the pad and pen. Scribbling out the address for Reagan Carter, the girl Allison and Xander believe is Sunday. "But you didn't get it from me, you hear?"

"Didn't get what from who?" I ask and when Frank finally settles enough to laugh, I smile. "Thanks, Frank. You're a lifesaver."

Slipping the paper from under his hand as he passes it over, I pocket it and make my way out of the station. Only stopping to inspect it when I'm safely tucked away in my car and away from prying eyes.

Where I expect to find the girl's home address, I'm surprised to see Frank went ahead and did me one better. Not only can I do a drive by her house, but I can also do the same at her place of employment.

Looks like it's time for the part of the job I hate the most.

Surveillance.

Sitting around and waiting for the target to leave whatever place it is they have themselves stashed away in so you can catch a glimpse. Attempting to interact with them when you're finally given the opportunity. Drinking disgusting amounts of coffee and trying to pretend that I'm not a stalker.

It's worse this time around though.

When I pull out of here, it's not just some random person I'm going to head off in search of. This isn't a Joe Schmo that took off and is trying to keep his identity a secret.

This is my best friend's sister.

If there's one thing I learned growing up with Samuel Alexander; it's that blood doesn't lie. And with what I learned from the hospital earlier after slinging them a bullshit story I'm hoping won't come back to bite me later, I'm in possession of a report that backs up that lesson.

A smoking gun.

Reagan Carter based on the blood work is Sunday Grayson, and once I get visual confirmation, I'm officially going to be the bastard that blows her world apart.

Reagan

When you're as solitary as I am, you'll have these lulls in your day where no one is at your counter, and you can just watch the people moving about around you.

When I first got the job as a teller with FNB—First National Bank—I did a lot of people watching. And by a lot, I mean, I sat around when I wasn't serving a customer and made up mini stories in my head of the lives they would lead when they walked out and back into what I deemed the real world.

It's been awhile since I've done it. The majority of my days lately so busy because of the end of the month that I can barely get a second alone to use the bathroom—let alone a break to watch people. With the lack of people in line now though, it's exactly what I'm doing. The object of my storytelling abilities this time being a guy that looks as though he stepped right out of the book I fell asleep reading last night.

Day old stubble lines his face, along with matching dark shadows under his eyes, speaking to an extreme level of insomnia that only an all-nighter with a woman or an evening long drinking binge can create. It's not those things that jumps out at me most though. His eyes do that.

Light blue orbs that the lights above his head only seem to accentuate. Adding that to the shortness of his jagged cut brown hair—looking a lot like someone who just rolled out of bed, and it's a miracle my jaw isn't needing to be picked off the floor.

Almost two years of working here and not once in that time have I come across someone that looks quite like him.

I might be pathetically single and destined to live my life as a spinster because of my lack of game when it comes to going out and picking up men, but it doesn't mean I can't appreciate a good looking guy. Even if he does look like he's in desperate need of sleep and judging by the scowl on his face wants to be anywhere but where he is right now.

Trying to gauge his height based on the way he's lounging and putting him at around six feet, I don't even realize I'm not alone until the person beside me clears their throat.

Startled away from my dissection, I turn and am face to face with my best friend Bethany. The only other person on the planet that looks as mesmerized in the moment as I do.

"I see you've noticed Mister Hottie."

Laughing at the absurdity of the name, but unable to disagree with the assessment, I just shake my head and pretend to busy myself with the papers in front of me. Maybe if I fake it long enough, Bethany will actually take the hint and get back to work before our boss notices our ogling and decides to call us on it.

"He made that much of an impression that he earned a name?"

"They all earn names, Reagan. I'm surprised hanging out with me as long as you have, you're just now figuring it out."

Two years we've been best friends and despite knowing the way she is, I still hold out hope she'll do things differently. Read my mind and just do what I need her to do.

So much for pretending the guy doesn't exist and focusing on work.

"Do my eyes deceive me or does Reagan actually seem interested in a guy?"

"Don't be ridiculous. The only thing I'm interested in is making sure everything balances out."

"You keep telling yourself that. I saw the way you were drinking him in. Someone finally overheated your lady parts."

"Enough Beth." I bark out, equal parts appalled at how loud she's saying it and annoyed at how right she is. It might not be as dramatic as she's making it seem, but there's no denying that with the way I was just staring at him, he did peak my interest.

"Touchy, touchy." She giggles and before I can respond, she's leaning over into my personal space, her face completely lit up in a grin that normally signals all kinds of trouble. "Hottie's all yours. Your lady parts can thank me later."

"My lady parts are just fine thanks. Besides, judging from the way he looks, he's more your speed than mine."

"Which you're only saying because that asshole Walker did the mambo on your heart."

Any reminder of my past with Walker is enough to turn a day from bright to sour instantly and no one knows this better than Bethany.

"Can we not bring Walker into every conversation?"

"Fine, but my offer still stands. You say the word and I'll make sure his pleasure stick never pleasures another person again. Himself included."

There's my best friend.

The one that when she heard about what Walker did at the party had gotten in her car and gone there, armed and ready to take him out before three of his buddies yanked her off. She might be slightly man crazy, but at the end of the day, I know where her loyalties lie.

Where they always have.

It doesn't change the fact that what I said before is still true. No matter how affected I am by the guy across the room, he's definitely been around. Which means he's not my type.

Besides, a guy that looks as good as he does is probably taken anyway. A problem I don't need. I won't flip sides from scorned girlfriend to the person doing the scorning.

The last thing I need with the way I've been feeling lately is another guy coming into my life and turning it upside down.

I've had more than enough of that already.

Chapter Two

Cole

It's confirmed.

Reagan Carter is Sunday Grayson.

The second I laid eyes on her, or felt hers on me anyway, I knew it as easily as I knew my own name.

She doesn't look exactly like Luke, but there's enough of a resemblance that even with my only memory of her being that of a drooling baby, there can be no doubt about her parentage.

Where Luke has his mother's blue eyes, Sunday has her father's brown ones. Add that to the way her lips curve downward when she scowls at the redhead standing to her right and it's all the confirmation I need. She looks just like her brother.

When the police are actively involved in missing person's cases, especially ones that have lasted as long as Sunday's, they will use software that takes the missing person's features and age them until they appear as close to the real person as you can get.

My mind is working a lot like that software now. Maybe even better because I can take note of things that a computer program wouldn't pick up. Mannerisms, facial tics and expressions are all fair game, and the more she moves around and interacts with her co-workers, the more apparent it becomes just who she is.

There's just one problem with realizing all of this. Now that I know the truth, I have no clue what the hell to do with it.

Casting a momentary glance down toward my phone, I resist the urge to call Luke. It would be so easy to dial the familiar number. Tell him that I'm sitting not four feet away from his missing sister and that if he wants the chance to reconnect, he needs to get his ass down here, but there's something stronger stopping me from going through with it.

Her.

It wasn't part of the plan to make my way in here, but after twenty minutes in the car bored out of my mind, I'd sauntered in and thrown myself down in the chair to watch and wait her out.

The first thing I noticed was the way she interacted with the bank clients. How even before they stepped forward to do business, she smiled a welcoming smile to put them at ease.

A smile that completely gutted me on sight.

What I wouldn't give for a woman to look at me that way. I can't even remember the last time anyone of consequence smiled at me period, much less the innocent way she does with people she doesn't even know.

From there it became the twinkle I caught in her eye when one of the older customers would attempt to flirt with her. It reminded me of the times I came to the bank with my dad as a kid. It was embarrassing at the time, but seeing it play out in front of me now, it's entertaining. And the way she reaches over the counter to them, her smile changing, but not completely going away, speaks to the level of patience and respect she has for them.

Chestnut brown hair with eyes to match, she's a Grayson through and through.

Watching her now though, that's not what I'm being driven by. I'm not looking at her like she's my best friends' sister. I'm not a private dick sent here to tell her the truth and get her home to her family.

I'm just a guy sitting and staring at the most beautiful girl I think I've ever seen. She's fucking gorgeous and it's become so damn obvious to me how affected I am by her that I'm having to swallow down the urge to jump the ropes in order to get closer.

"Can I help you with something?"

I wondered how long it was gonna take the stuffed shirt to finally get up the nerve to talk to me. I'm sure with the way I look right now, I'm the last person he wants to deal with. To be honest, I'm not all that excited to be dealing with him either.

"Who do I see about opening up an account?"

"That would be one of our account managers, sir. Let me see if one of them is available." He turns to go but thinking better of it, or forgetting some pertinent piece of information I'm sure he thinks I can't live without, he faces me down again. "I'll only be a minute."

Before he can get more than a foot or two away, realizing how fast this whole plan of mine can fall to shit, I call out to stop him. "Are the tellers capable of doing it?"

"They can. I suppose if you're pressed for time you can go through them."

If he's attempting to appear anything but self-centered and snarky, he's failing. If I wasn't here with a specific purpose in mind, I'd have no problem putting the jerk in his place. If there was a jerkoff that needed it in the moment, it's definitely this guy, but one look to the counter and the person that's getting dangerously close to having face time with my girl, I swallow it and jump the ropes until I'm directly in front of her counter.

Her head looking down and completely unaware of my presence, it takes the redhead from earlier clearing her throat for those eyes I caught earlier—the ones that look so much like Xander's— to finally look up and land on me before her mouth falls open in a perfectly shaped O.

"I'm so sorry! I got so caught up with this I didn't even realize anyone was in line."

"Don't worry about it. I was enjoying the view."

She offers up a shy smile as her cheeks flush and despite knowing who this is and what my job is supposed to be, I respond with a smile of my own. A natural one that takes next to no effort to make happen. The complete opposite of the way I normally am.

Take a step back, Cole. I chide myself. *This isn't some random girl you need to flirt with in order to get information. This is Luke's sister.*

"So umm," she stammers, the blush deepening until its spread over her entire face. "How can I help you?"

"The guy that was stupid enough to wear that ugly ass plaid shirt said I could see you about opening an account."

Starting to laugh before catching herself and bringing her hand up to curb it, I allow myself the luxury of smiling again.

This is crazy. I've been in front of her for less than five minutes and already I'm swept up in how natural—how real— her responses are.

The complete opposite of the way it is with Luke.

"You can do that." She nods. "What kind of account were you interested in opening? Personal or business? Checking or Savings?"

Since the last damn thing I want to do is open another bank account when I can barely get a handle on the one I've already got, I take things in a different direction.

"What kind of account gets me coffee with the teller opening it?"

Reagan

I've gotta hand it to him. Hottie works fast.

I should have known from watching him that if he ended up in front of me, the first thing he would do is flirt. He looks like the shameless type. Willing to do whatever it takes to get the girl alone and preferably in his bed by the end of the night.

The last guy I'd ever be interested in.

Which is exactly why you're single and having to look to book boyfriends for pleasure.

"I'm sorry, what?" I respond, pretending not to hear him while at the same time hoping that what I did hear was wrong and just my hormones playing tricks on me.

"Shit. That was stupid of me. Forget I said anything."

He's giving me an out and even though I know I should take it, there's something about how quickly he's backing off that makes me rethink my earlier assessment of him. If he was well versed in picking up women, the last thing he'd be doing is backtracking. So what exactly is he after?

"You don't really want to open an account, do you?"

His eyes shift around as a miniscule half smile appears and it's obvious I've caught him. What should offend me, him wasting the time I could be using with actual customers, doesn't. Now I just want to know what it is about me that made him have to tell a lie in order to talk to me.

"Is it that obvious?"

"Considering the last account I opened for someone I didn't get asked out for coffee, I'm gonna go with yes."

"So you did hear me."

"Yeah, I did, but I'm having a hard time believing it."

Eyebrows lifting, he just stares. "Not to be a total dick, but I'm having a hard time believing you don't get hit on by every male in the place the second they come through the door."

The lack of control I have over my body's response is starting to annoy me. No sooner do I seem to have overcome my earlier blush than another one threatens to rise up and replace it. The way I'm acting right now, you'd think I'd never received a compliment in my life, which is total crap. With the way Jake's friends drool over me, it's a daily occurrence, even if right now the situations are reversed and it's not a sixteen year old doing the complimenting.

"If you count sixty year old widowers, I'd have no choice but to agree with you."

"I do count them, but the old guy that was up here a little bit ago was wearing a ring. So it's more than just the widowers."

When he made his way in earlier, I thought Bethany and I were the only ones doing the watching. The majority of our clients being business men and women and seniors, he was like a breath of fresh air for us. A real person in a sea of my own make believe. Hearing him now, and realizing that he'd caught sight of the sweet elderly man that was at my counter about thirty minutes before, means we weren't the only ones watching.

"You saw that?"

"I told you. I was enjoying the view." He flirts again and just like the last time, I swallow down the urge to blush. "With the smile he had when he left, I assumed he'd gotten whatever it was he was after."

Remembering Mr. Wilson, my two time a week regular, I smile. He had left happy. Spending the last year tending to his banking needs, listening as he talked about the love he had for the wife he referred to as his rose, seeing him smile for the first time since she died, well, it had done good things for my heart.

"Don't tell me the old guy got to you before I could."

"If Mr. Wilson was fifty years younger, he just might have." I joke and the guy smiles, which seeing it up close, sends a jolt straight to my heart. After watching the scowl he'd been wearing since the moment he walked in, this is a nice change.

"So now that I know the old man didn't steal you completely away, will you let me buy you a cup of coffee?"

I've always been a big believer in keeping my job separate from my personal life. Putting it out there for the entire bank to hear when James—the guy in the plaid—asked me out a few months before. Having this guy ask me out, even if he isn't an account holder or client, makes me feel like I'm treading close to a line I said I'd never cross.

My body may want to betray me and say yes, but the levelheaded part won't let me.

Parting my lips and reading myself to give him the polite version of thanks but no thanks, Bethany interrupts and answers for me.

"She'd love to. She goes on break in fifteen."

Shooting her what I hope is my *'I'm going to kill you for this later'* look, she laughs before walking away and leaving me completely alone with my embarrassment.

"She do that a lot?" Hottie asks once Beth is out of earshot.

"You have no idea."

"That's where you'd be wrong. I've got this buddy that does the same thing every time we go out. It's like if I don't have a girl hanging off me, something's not right."

Laughing at his explanation, thankful for the distraction as the embarrassment begins to fade, I take a risk and give in to his request.

"If you want me to say yes to coffee, you've gotta do something for me first."

"I don't really want to open the account."

"I know. It's not about that."

"Hit me with your best shot, teller girl."

"I have a rule about going out with strangers. So if you really want to grab coffee, I'm gonna need a name."

"I'll tell you mine if you tell me yours. Fair's fair."

"Reagan." I blurt out quickly.

"Nice to meet you, Reagan. I'm Cole." He responds with a smile. Resting his hand on top of mine, the spark that shoots up my hand on contact quickly fades until I can feel is heat. Smiling again when I chance another look at him, he turns and saunters back to his seat on the other side of the bank. Only calling out as his body falls into the chair and he's leveled his piercing blue irises back on me.

"See you in fifteen."

Chapter Three

Cole

I'm crossing so many fucking lines right now and it's only a matter of time before it comes back to bite me.

Nothing that ever comes as easily as the conversation I just had with Reagan ends without some sort of price being paid. In this case, it's gonna be her kicking the shit out of me once she realizes that I'm not just some random guy hitting on her, but her biological brother's best friend.

The right thing to do would be to leave before she goes on her break. Go back to the Grayson's and tell them that the girl I saw isn't their daughter and I'm sorry I couldn't be of more help. Keeping Reagan and her life intact.

But there's only a handful of times in my life where I've done the right thing and when it does happen, it's almost always about family. Xander, Allison and Luke; they're family. Which means I can't walk away the way I probably should.

Attraction to this girl aside, she's as good as family too. So no matter how big a mess this is going to be in the end, I've gotta see it all the way through.

Asking her out for coffee though, especially when it's the first time in a year that I've done anything like it, is definitely not about doing the right thing by family.

It's me being selfish and doing something for me. Which is crazy really, since I swore I was never gonna get caught up in shit like this again.

I don't date. I don't ask girls for coffee like I did with Reagan. I don't anything. No walks on the beach, no dinners over candlelight or cuddling while watching movies. I don't do romance. I learned my lesson about doing stupid shit like that after what went down with Miranda.

Other than a couple of casual hookups I had with random women shortly after she traded me in for an older and richer model, I've stayed true to my word. I learned my lesson and distanced myself as far away from this crap as I could.

Dating and becoming a couple, offering commitment and the promise of a long term arrangement can't happen. Not with my job and definitely not with my relationship track record.

The last damn thing I want is another woman coming along and doing what Miranda did. Once has got to be enough. Fool me once, shame on you. Fool me twice, shame on me and I refuse to be taken for a chump again.

Aware of the person standing to the side, but too pissed at myself for not making good on leaving when I had the chance, I make no motion to look up and acknowledge their existence. At least I don't until a throat clears and I feel a hand—Reagan's—come to rest on my shoulder. The electric feeling from earlier alerting me to exactly who it is. Shaking the effect of it off, I grow a set and meet her eyes.

"You ready to grab that coffee now?"

The way she asks, her voice starting off strong and then fading off, shows me she's as nervous and unsure about this as I am. Our reasons for it might be completely different, but there's no denying the comfort I get knowing that in the moment I'm not so alone.

"Absolutely. Seeing as I'm a little out of my element here though, you might need to lead the way."

"Sure thing, but before we go, I feel like I should warn you about something." Trying to get a read on her, figure out what it could be that she feels the need to warn me about, she starts talking again and fills in the blanks. "Once you've had Mama D's coffee, you're not gonna be able to go back to whatever it was you were drinking before."

Considering the stuff I make for myself is basically the equivalent of pouring motor oil in a mug and adding sugar to it, I'm not about to argue. It's bad enough that whenever Luke and I get together he asks me what road construction site I grabbed the tar from. At this point anything is an improvement.

Especially when it brings about the flash of excitement I see dancing in her eyes. If I'm not completely reading into the way she's reacting, it's almost as if there's something more hidden underneath her warning.

Like maybe I'll like the coffee so much it'll have me coming back again for more.

She doesn't have the first clue that in just the short amount of time I've been watching and then speaking with her, she's already made me want to come back for more. Indulging until I've had my fill. Something with the way my heart starts racing as she slips her hand in mine, I don't see happening anytime soon.

I can't see anyone ever getting their fill of this girl.

Following along blindly until she comes to a stop about thirty feet from the bank, I look up and take in where we are. It's obviously a coffee shop, what with the giant mug present on the sign, but it's nothing like the bigger chains. It's quieter with a lot less foot traffic judging by the few people I can see moving around inside.

I'm definitely not in Toronto anymore.

Feeling her hand slip from mine, but so swept up in my thoughts that I don't register that she's moved to the door, it takes her clearing her throat to pull me back from whatever trance I found myself in.

"Is everything alright, Cole?"

"Yeah, um, everything's great."

Since when do I stammer?

Since you asked a girl out for coffee under false pretenses and have been unable to stop staring at her body since she leaned it up against the door waiting for you to go in.

Knowing who she is, what I'm here to accomplish and where I'm gonna head the second she goes back to work, doesn't stop my reaction to her even though it should.

How I can't seem to tear my eyes away from her body, is actually the last thing I'm interested in when it comes to her. It's her eyes and the way her entire face seems to transform when she smiles that I'm caught up in most.

It's also those things that if I don't get a grip on this soon, will be what destroys everything.

For the first time in a year, I've got to tame the beast inside. The one actually interested in a woman.

"Are you having second thoughts about grabbing coffee?"

"No." I respond quickly, which judging from the way her eyes seem to light up, is the first thing in the last few minutes I've been able to do right. "Just got sidetracked taking everything in."

"You weren't lying about not being from around here."

"Nope. I'm a fish out of water."

"Where are you from?" she asks, again making her way over to the door and this time when she leans her body against it in an effort to usher me inside, I do it.

"Orangeville originally, but living in Toronto since I went away to school."

"Ah, okay. That explains it. You're a city boy."

"I haven't been a boy for a long ass time."

"So how old are you if you're not a boy anymore?"

"Twenty-four."

"Hmm." She ponders before waving to the barista behind the counter. Slipping her hand easily into mine and pulling me with her, she weaves in and around a bunch of tables before stopping at one directly in the back.

"Don't we need to order?" I ask when she motions to the table for me to sit.

"I always order the same thing when I'm here, so when Sara brings it over she can take your order."

Having just met this girl, the apprehension I expect with her not knowing if I'm a standup guy or not, is nonexistent. She's speaking to me with an ease that is so fluid I'm jealous of it. Even knowing that I'm sitting here with Sunday and for a year and a half of her life I'd been around her, I'm still shaken by how she acts as though we've known each other for years.

Is the reason this seems so easy because we already have a connection?

"So I have to ask."

"That sounds ominous."

She laughs and the sound along with the slight shake of her body as she does it brings the urge to touch her back full throttle. What it is about this girl that makes me lose what's left of my self-control? The way I respond to her, it's almost like I'm back in junior high again.

Which is not exactly a time I'm itching to repeat.

"Do you normally come into small towns and pick up random tellers?"

Filling the space around us with a hearty laugh of my own, I shake my head. She needs to know this is as much a first for me as it is for her.

"No. I don't make a habit of picking up women at all. Let alone the way I did with you."

"How did you end up at the bank?"

"I came to town on business. It brought me to the bank. Turns out the lead I was following was a dead end."

Her smile, which up until I spoke had been vibrant and bright, lowers until it's dangerously close to a frown.

"I'm sorry to hear that."

"You know, I'm not." I admit and the second our eyes connect, her earlier disappointment recedes.

"Here ya go, Reagan. Scalding hot. Just the way you like it." The barista interrupts the moment as she slides the cup across the table. "Now handsome, what can I get ya?"

"I'll have what she's having."

She laughs the minute I reply and it's not long before I realize what I've done and am flushing a darker shade than Reagan.

"I didn't picture you for much of a *When Harry Met Sally* fan, but alright. One insanely hot Reagan special coming right up."

Oh sweetheart, you have no idea how much I wish that were true.

As she steps away from the table and we're alone again, I chance a look her way, only to find hers already on me and a crooked grin on her face.

"What?"

"Do you have any idea how I take my coffee?"

"Not a clue, but it can't be any worse than the crap I normally drink."

"Which is?"

"Motor oil, or according to my best friend, the thickest black tar off the street."

"No offense, but that sounds kind of gross."

"None taken." I laugh. "It's definitely an acquired taste. So how do you take yours?"

"A full cup of sugar with a pinch of milk and coffee thrown in."

Someone as sweet as you shouldn't need to drown themselves in that much sugar. I think and thankfully don't say. The many ways I'm completely blowing this job to shit right now is enough on its own. Speaking my thoughts out loud would only make it worse.

"I guess I'll find out soon enough if I can take it."

"Something tells me you can." She smiles before sitting up straighter in the chair. "So who were you looking for at the bank?"

Her question is the exact reason I never should have walked in there to begin with. I can't give her the truth, but I also can't leave her hanging. For whatever reason, my body has decided that Reagan Carter is what it wants and there's no swaying it once it's locked and loaded.

And I'm not even sure I'd want to even if I could.

"Just a guy that skipped out on his family."

"Are you a bounty hunter?"

"No." I laugh again, not at all surprised that with the information I'd given her, that's where her mind went. "Private Investigator."

Sensing the atmosphere around us changing, I look up just as Sara returns and places my cup down in front of me. Waiting until she's done, I flash her one of the fakest smile I can manage before leaning across the table and lifting my cup to Reagans.

"So, it's my turn to ask. Do you normally say yes to city boys who walk in off the street and ask you out for coffee?"

"Can't say that I have. This would be the first."

I get the feeling as I watch her tip her cup back and drink, that there's a lot of things I could say and do with her that would be firsts. And despite knowing where this is going to end up, I can't help the road my thoughts take the more I think about it.

How I really wouldn't mind being her first.

Shit. No. I can't think about it like that. This is Sunday.

"Well then, I guess that's as good a thing to drink to as any. To firsts." I say lifting my cup and tipping it toward hers again, only this time, her meeting me in the middle.

"To firsts." She toasts, but before our cups can knock off each other, she speaks again. Making my body, which I'd finally managed to get under control, come completely undone. "But also to guys who take off and can't be found. Your loss seems to be my gain."

Reagan

I'm not the girl that does things like this.

You know the girl that you pass in the hall at school or the one you see at the mall with her earphones stuck in her ears, focused strictly on getting where she needs to be and not on anything going on around her?

That's me. I'm that girl.

I'm outgoing enough when I want and need to be, but most of the time am content to have my nose stuck in the latest romance or lost in the music. Getting so lost in it sometimes that it's amazing I haven't walked into walls or even numerous people while going about the task of living my daily life.

Sometimes I wonder how I even ended up dating Walker at all with the way I am.

Though I'm pretty sure if I actually allow myself to think about it, a lot of that came about because of Bethany and Melina. If I hadn't been friends with the both of them, I'm pretty sure I would have remained single all throughout high school.

I'm a slightly popular nerd. That's my story and I'm sticking to it.

Making me the complete opposite of my brother Jake. The guy that brings a new girl through the house every other day and claims them as his girl. Has tons of friends and keeps the phone tied up for hours at a time, making it impossible for anyone else in the house to get a word in edgewise.

Being the complete opposite of him is why Walker was the only person I dated, other than Colin in fourth grade, but he doesn't count. I swear that was a love hate relationship. He loved to pull my hair and I hated every second of it. Well, the times I wasn't kissing him anyway. That was about all I liked.

Sitting here with Cole and admitting that not finding the guy he's after is my gain, I don't do that. The girls in the books I read do that, but I definitely don't. Whenever I would try and flirt with Walker—rev his engine or whatever, it always came out garbled and awkward.

The complete opposite of what's happening now.

It's easy to flirt with Cole even though it shouldn't be. Even easier to sit and stare at him while he speaks to me. I have no idea what it is about them, but from what I've been able to tell so far, they seem honest.

Getting picked up at work and going with him to get coffee goes against every warning my dad ever gave me, but because of those damn eyes and how sincere and honest they appear with everything he says, I'm ignoring it all. I feel safe. Or that's not it at all and it's all because of what Bethany said earlier.

My lady parts really have taken over my brain.

Feeling the table shake as Cole shifts his legs under it, I focus my attention back on my cup and what I can now feel is those eyes of his on me again.

"This is probably gonna earn me a slap, but how old are you, Reagan?"

"Eighteen. Well no, that's not true. According to my brother Jake, I'm six months away from legal drinking age."

"Older or younger?"

"Younger."

"That explains why he's focused on the drinking age. He's jealous."

"If I was choosing things about someone to be jealous of, that would be the last thing on my list."

"Not much of a drinker?"

"Not a drinker at all. A fact my best friend never lets me live down."

"The redhead at the bank?"

"One and the same." I proudly admit. Bethany might come across as too much sometimes, but I'm not sure I would have survived the last few months without her. "Can I be honest?"

"I'd appreciate it if you were."

"I figured if you were gonna ask anyone out for coffee, it would have been her."

"And why's that?"

The way his brow furrows and his head dips to the side proves my earlier assessment of him. Even now he's being honest with his response. It genuinely confuses him as to why I would say what I did.

"She seems more your speed."

"I didn't think I had a speed, but since it appears I do, care to share what you think it is?" he asks before lifting the cup to his lips and taking the world's most long and drawn out swallow, my eyes focused not on his words, but the way his Adams apple bobs as he does it.

The guy even looks sexy drinking coffee. *Damn.*

"Fine, but answer something for me first?"

"Shoot."

"What's your idea of a good time?"

"Is this where you wait for me to say something disgusting like 'anytime I get to have a girl like you under me'?"

Am I that transparent? It wasn't exactly what I expected him to say, but close enough. It's what Walker would have said. Probably still would say too.

"Is that your final answer, Cole?" I laugh, doing my best to swallow the embarrassed blush I know is going to appear any second now that I've been found out.

"Hell no. My idea of a good time is any time I get to myself."

"Does your work keep you that busy?"

"Not exactly, no. I'm just not a fan of people."

"Yet you're sitting in a random coffee shop in Alliston with a person."

"Present company excluded." He smirks, causing me to laugh.

"Nice save."

"If only I could make saves like that on the ice, I might have had a whole different life right now." He quips. "Now that I've answered your question though, it's time for you to answer mine."

Right. Why I think Bethany is more his speed than I am. *Great.* Now that I know what his answer is to my question, I'm not even sure I feel the same.

"Bethany has always been the more outgoing one. She says what she thinks all the time and doesn't give a crap who knows or hears. It's so bad that sometimes I wonder how she even has as a job. I guess it seems like that might be more your style than the quiet girl whose idea of a good time is a good book and hot chocolate."

"I happen to *love* hot chocolate."

Changing the subject, not wanting to give in to the flutter taking place in my chest with what he said, I change the subject and attempt to learn more about what he does for a living.

"You not being able to find the guy you were looking for, what happens now? Do you go back to the city or do you have plans to wait him out?"

"Honestly, I should probably head back and report in. Let them know what I found or in this case didn't find, but not sure I'm ready to do that yet."

"Why's that?"

"Despite being a so-called city boy, I'm enjoying the pace here. A lot more than I thought I would. There seems to be a bit more to it though."

"What else could there be?"

"Can I be honest with you?"

"I'd prefer it if you were." I repeat his earlier words back and his entire face lifts into a smile, which the second I catch it sets off the butterfly effect again.

"There's this girl I'm intrigued by and if I leave, I'll never get to ask her out for hot chocolate."

Chapter Four

Cole

"Hey man," *I found your sister and she's pretty amazing.* "Give me a call when you get this. I need to talk to you."

Walking Reagan back to the bank and waiting by the door while she made her way through the throng of people now standing impatiently in line waiting to be seen and catching death rays from Mr. Plaid, I leave and of course my first call is to Luke.

It really should be Xander and Allison, but with how eager they are for any scrap of information I can give them, and not being at all ready to drop this on them yet, he's the next logical choice.

What his parents don't realize, but that I've known for years because of our friendship, is that Luke while not giving up entirely, washed his hands of his sister's disappearance a long time ago. It became too much for him, watching his parents suffer, and he had enough.

I also know that walking away for Luke isn't completely giving up and if anyone is going to help me navigate this with his parents, it's him. Plus, if I've gotta be the one to drop the bomb on Reagan, I'd rather not do it alone. Having Luke there not only as her brother, but my best friend, might just be the one bright spot in a really shitty situation.

It also doesn't help that after only fifteen minutes with Reagan, I'm invested and it's not just about reuniting their family anymore. It's about keeping her safe while it happens.

She didn't ask for this and if there's anyone that gets that, it's me.

It's not the same situation, but being plucked from the foster care system and dragged into the Alexander home as a kid, I get what it's like to be two people. There's the kid I was before

Samuel Alexander changed my life, and the guy I became after it. Sometimes, I'm a combination of both.

Whether she likes it or not, Reagan is like me. She's two people. She's the girl she became that day seventeen years ago when her babysitter stripped her away from the only family she'd known, and she's Sunday—the baby girl that hadn't gotten a chance to enjoy the truly amazing family she'd been given when she was born.

Despite needing to stay separated from this, one coffee, a few smiles and my body's visceral reaction to this girl makes staying away from it and remaining impartial, impossible.

Like it or not, I'm in this now and I'm in it for long haul.

With nothing better to do and a whole lot of time to kill now that Reagan is back at work and I'm waiting on a call, I make a plan to follow through with what I'd told her not fifteen minutes before.

Finding a place to stay.

Unlocking the door, I throw my body down into the seat, tossing my phone down across from me. Casting a final look back to the bank and exhaling deeply, again being reminded of the load of shit I'm creating for myself getting involved with this girl, I turn the key in the ignition and gun the engine until I'm pulling out. Destination unknown.

Driving along, I think about everything I know about the case. What both my dad and uncle drilled into my head before both of them had to go and leave me with the fallout.

The day that started out like every other one before it. Allison getting Luke up in the morning and preparing to take him to school, getting breakfast ready and making sure lunches were packed. Xander leaving for work early with the promise of taking his wife out for their annual weekly date night before heading off.

Everything following the same routine that had been happening from the day they brought Sunday home a year and a half before. Allie had kept herself busy, feeding and playing with her daughter, even taking her to the park in the early afternoon, both of them not having a care in the world.

Being there to meet Xander when he got home and after a time of them both getting cleaned up and changed, calling the babysitter and proceeding to go out on their date.

Never for a second thinking that their lives were about to change forever a few hours later.

Hallie Francis—the babysitter—was a girl from the neighborhood. We all knew her even though she was ten years older than me and Luke at the time. She seemed like a pretty dependable girl. She'd even looked after me a few times when I first got to town and Sam was busy on a case. There was no way with what we knew about her, she could have been involved in this.

A few weeks after Sunday went missing, Hallie was found dead near Monora Park. Another innocent victim. It didn't take long after that for the entire investigation to go cold. With Hallie dead and her case as cold as Sunday's, it started looking like she was either dead like our old babysitter, or at the very least out of the province. Maybe even the country.

Leads came in, of course, but none of them panned out and over the years, I remember my dad getting so pissed off with all of the false ones that he swore he wasn't going to look into them anymore until there was something solid attached.

DNA would have been solid, but there hadn't been a hint of it. It was like Sunday was taken from her house that day and just vanished into thin air.

Something about this popping up now and not when John was looking into it doesn't sit right with me. If there was anyone that could give my father a run for his money in terms of the family business, it was John. He was the best of the best. Knowing the way things are now, well baby checkups and what takes place at one of them, I find it highly unlikely that this hadn't been picked up on before. Which means somewhere along the way, John dropped the ball.

If I want to get to the truth, go back to Xander and Allison and give them answers, I need to go back to John's notes and familiarize myself with them. Maybe even look into whatever my dad kept from his time on the case.

Noticing a Super 8 on the right side of the road, I quickly signal to change lanes, merging easily and pulling off until I'm driving into parking spot closest to the door. Now that I've found a place that will give me ample time to wait out the call from Luke and keep me close to Reagan in order to be able to see her again, there's only one thing left to do.

Find out why John didn't bring Sunday home sooner.

Reagan

"Please tell me that you got to see what he's hiding under those jeans."

No sooner do I put my purse down, pull the next business day sign down from the top of the counter and make quick work of helping the line of people then Bethany is slipping her own sign up and making her way over nosing for details.

As much as I wish she had something better to occupy her on her own break, it looks like I'm the excitement for the day.

"We went for coffee, Beth."

"I've gone for *coffee* plenty of times. I know what it means." She winks and I just shake my head. I want to ask when the definition of coffee changed, but knowing Bethany the way I do, I'm not entirely sure I'm going to enjoy the answer. "With the way the two of you were undressing each other earlier, I figured you'd at least find out what the guy's packing."

Remembering what Cole told me about his job, now curious to know if he does indeed carry a weapon, I just smile and when Bethany catches it, she wastes no time unloading her thoughts on my coffee date.

"You sly bitch. You did do something with him! Come on, Reagan! Inquiring minds wanna know."

"There's nothing to tell. Like I said, we had coffee. You know, the brown liquid that people pour into their cups every day and are extremely addicted to?"

"You're no fun."

"I didn't realize it was my job to be your entertainment."

"A guy looking the way he does walks in here, gets you in his sights until he's jumping the ropes in order to get up here to ask you out and you don't expect me to want to know everything?"

"There's seriously nothing to tell, Bethy. We just had coffee and talked. It was nice."

"Did you at least pretend to trip so he could catch you with those tight arms of his?"

Shaking my head again, this time laughing, amused with how deeply involved she seems to be in my lack of a love life, I attempt to focus my attention on the work in front of me and she sighs.

"This isn't a book, Bethany. I'm starting to think you're the one spending too much time reading. You're confusing real life and fiction again."

"Does Hottie have a name?"

"Cole." I reply, overjoyed that there's finally something I can answer.

"And when are we seeing Cole again?"

"I don't think *we're* going to be seeing him at all, but as for when I'm going to see him again, I suppose if and when he comes back to the bank."

Something about the way she shrieks, grabbing and yanking me to her, causing the eyes of a few of the managers to look away from their work and focus their attention on the sideshow now taking place across from them, makes me reach out and smack her.

"You're telling me you didn't secure a second date?"

"It wasn't a date, Beth." I sigh. "It was coffee, and no I didn't."

"Well, if you're not gonna make a play for the guy, do you mind if I do?"

Something about the way she's asking to go after Cole makes my blood pressure rise. We only went for coffee so I shouldn't be reacting, but the idea of Bethany making a move upsets me.

It's no secret how Bethany spends her time. How she's content remaining single and just having fun with a guy before discarding them and moving on to the next one.

I can't let that happen with Cole.

Even if nothing ever comes of what he said and I never see him again, there's no denying that he made an impression during our short time together. An impression I want to give attention to. It's the first time since Walker cheated and my heart got smashed in pieces that I'm feeling something other than awkward and broken.

"Judging by the scowl on your face, I've hit a nerve. Good. It means I was right. You do like him and whatever happened between the two of you on break was a whole lot more than coffee."

"You did that on purpose didn't you?"

"Whatever do you mean?" She laughs, batting her eyes at me innocently but looking anything but.

"You know exactly what I mean Bethany Thompson, so don't even play that game with me."

"You're right, I did and I'm not sorry. Rea, I've been telling you the same shit for months now, but because no one has ever come along that's sparked your interest enough, it's been in one ear and out the other. What Walker did was wrong, and if I ever see his stupid ugly face I will pound it into the ground, but you can't let his mistake ruin the rest of your life. You need to move on and maybe Cole's the one to help you do that."

Going for coffee with Cole, enjoying myself and coming back to work happier than I've been since I started working here, I wanted to enjoy it a little bit longer. But of course, once the conversation comes around to Walker, everything goes to shit.

Bethany's right. I can't keep hiding myself away because one guy wasn't who he said he was. It's just hard to get up and go when that same guy took everything you had with him when he left.

I've spent so long on the outside of things and feeling different, both at home with my family and at school with the people I called friends that when Walker and I got together, I'd thrown everything I had into it. Given it my all and now, even though we've been broken up for months and he's moved on, it's like I'm stuck.

"One coffee with a guy doesn't make him the answer to my romantic prayers, Bethy."

"Maybe not, but if you're too afraid to try you're never going to know."

"You've been working on this speech the entire time I was gone, haven't you?"

"I admit nothing." She grins. "Just do me a favor, okay?"

The last favor I did for Bethany had me dressed up like a naughty nurse at a college party for a fraternity neither one of us was pledging. The last thing I want to do is have a repeat of that because it's engrained in me to promise my best friend the world.

"Depends. Does it require fishnet?"

"No. It just requires you to take a step out of your comfy little box and live a little."

"Fine. What am I promising?"

"If Cole does come back and he wants more coffee," she says, her voice dripping in innuendo with the way she drags out coffee. "Don't question it or try to find ways to compare him to Walker. Just say yes."

"So that's all, huh? Just say yes?"

"Yeah, that's it. I saw the way he looked at you today before he even got up the nerve to come talk to you. He's interested and with the goofy half smile you've been wearing since you got back, I think it's safe to say you're interested too."

"For the sake of argument, let's say I am. He's from Toronto. Hours away from here. What could we even have with that kind of distance?"

She laughs softly and when I throw her a look of utter confusion, her cackling only gets louder.

"You're being all logical about it. You can't do that with dating and you damn sure can't with love."

"Why'd you have to go and bring the L word into it?"

"Because that's what makes the world go round."

Shaking my head, I attempt to steer the conversation away from the road its going so that she can take off back to her corner and I can finally focus on what's important. Work. The one aspect of my life where thinking logically actually comes in handy.

"If I'm overthinking things, you feel like telling me how you would handle this?"

"That's easy, Rea. Save the horse. Ride a city boy."

Chapter Five

Cole

"Alright bro. You got my attention. What the hell is so important that you needed to actually pick up the damn phone when you know I'm coming down this weekend?"

Gee, Luke. I don't know. Maybe I had to pick up a damn phone and call because I found the sister that you've believed is dead, and she's alive, well and sexy as hell.

"You talked to your parents lately?"

"As a matter of fact, no. The last time I talked to my mom she brought up Sunday so I hung up on her. You know how I am about that shit. I can't go there with her. It's been seventeen years for Christ's sakes."

I do know how he is about it, but I would have thought that with them reaching out to me, they would have done the same with him. Apparently staying out of it really did mean he was completely out of the loop.

"Remember what I told you about John's will?"

"Yeah. He made you take over the family business and you're pissed as hell about it. That about cover it?"

"In a nutshell. Look, your parents got in touch with me. They want me to pick up where he left off."

"Jesus Christ." Luke curses and I can't help seconding his sentiment. "Please tell me you told them to go screw themselves."

"You know damn well I didn't. They're family, man."

"So they've got you in on this pointless shit now? That's just fucking great. Is this what we're gonna be talking about this weekend when I come down? My parent's inability to let go?"

He has no clue how wrong he is and how their inability to let go is the reason I was able to stand in front of his sister less than six hours ago.

"Are you done getting pissed at me for doing the right thing? Can I get to the point of the call now?"

Another string of curses falls before he finally clears his throat and barks out his agreement.

"Fine. What's up?"

"Luke," I pause, debating again whether or not this is the right move. "Man, I don't know how the hell to even say this."

"You open your mouth, form the words and say them. Seems pretty easy to me. What the hell is going on, Cole?"

"I found her, Luke. I found Sunday."

Patiently waiting as the line goes silent, I wait him out until a few seconds later I hear him again. Only this time his voice is quieter. More subdued. The shock of what I've just laid on him obviously taking its toll.

"You what?"

"I found her. It's a long story and when we're both back in the city I'll explain it, but it's definitely her. She's got Grayson written all over her."

"She's alive." He repeats slowly.

"Yeah bro. She's alive."

"You're gonna need to give me a sec here. I need to process this shit."

"Take as much time as you need. Call me back when you're ready. I just figured that if anyone needed to know about this, it was you."

"Yeah, okay. I'm gonna go. I'll hit you back in a bit."

I hear the click before I can respond and tossing the phone back onto the bed, I lean back into the bedpost and sigh. Calling Luke and letting him know about his sister was the right call, deep down I know it was, but I still feel twisted up inside and it's easy to see why.

I watched Luke spend the last seventeen years going through the motions. Telling me time and again about how his parents needed to move on. That she was probably as dead as Hallie and that dragging this search out was only going to cause more pain. The big speech I knew deep down he didn't really believe in, but said and lived because it was easier than the alternative.

It's half the reason we managed to stay so close when our lives went in different directions a few years back. We were doing the same thing. Our versions of living were forever altered when that girl was stripped out of our lives.

From the second he first laid eyes on Sunday, Luke had fallen in love. After the time I spent with her today, it's easy to see why. There's a magnetism whenever you're around her and it all centers around that damn smile of hers. It pulls you in until you're powerless to fight against it and you just want to spend as much time in its presence as you can.

He doted on her when we were kids. Whenever she would cry, Luke was always the first responder. The one scooping her up in his puny little arms and holding her while he fed her, burping her even though it always ended up in him being splattered in baby puke, and even being the one to hold her tiny little hand for the first time while she toddled along the floor attempting to walk.

Luke made Sunday his everything and when he lost her, he'd lost a good chunk of himself, becoming the guy he is now. A great friend to be sure, but a lot of the time jaded and cold. Closed off.

Believing Sunday dead gave him the right to die right along with her and with the bombshell I've laid on him, he's going to have to struggle with being brought back to life.

Being able to do this for him, bring her back into his life again, it should fill me with a sense of happiness. Doing the right thing by the people you love is supposed to do that, but all it's really doing is leaving me as cold as he is.

No matter what pretty picture I want to paint for how this plays out in the future, it's never going to compare to the reality. This is going to be hard and not just for Reagan, Luke and the rest of the Grayson family.

For me too.

Luke

When Sunday went missing, our family shattered. Dad made sure he was working around the clock so he could be anywhere but at home where he was needed. Mom checked out completely, some days it being a miracle she even got out of bed at all, and me?

I stopped talking.

In a way I guess I was shutting down the same as my mom, but I got up every day and did the shit that was expected of me still. I just did it all silently.

This didn't sit right with my teachers and it wasn't long after they called home and talked to my parents that I was sitting in a therapist's chair. Two different therapist's trying to break me wide open in order to put me back together the way they thought I should be.

When they tell you that time heals all wounds, they're blowing smoke up your ass. I didn't know it before, but with what Cole just told me, I'm positive of it now.

Seventeen years went by and not once in that time have I ever had a night where I wasn't haunted by what happened. What I should have been home to prevent, but was too busy playing street hockey a few streets over to care about.

I can't heal because it's my fault.

That's what the nightmares are about. The memories I bring to life so vividly in my mind when I'm alone. What I need to abuse alcohol in order to escape. What I want to give into so badly now.

If I had just been there the way I promised her every night when I kissed her chubby little cheeks before bed, none of the agony that my family went through would have happened at all.

This isn't Hallie's fault for being a part of a plan to take her, if she even was. It's not even on the people that did take her and hid her away this long.

It's all on me.

I wasn't a good enough brother. I couldn't be what Sunday needed then and even with the knowledge that she's out there somewhere, alive and well, I don't think I can be what she needs now either.

All because of the way I handled everything from the start.

The day my mom sat me down and tried explaining to me how our family was going to change, I'd been such a little jerkoff. I'd spent the last five years being the only kid and I liked it. No way was some other little brat going to come in and steal my thunder. I treasured the attention I got from my mom and don't even get me started on the way it felt having my dad's. I won the damn lottery every time he would pick up the hockey stick and meet out in the road.

I spent nine months hating her and she wasn't even here. Steeling my scrawny seven year old body the day my aunt came over to stay with me while my parents went to the hospital. I had it all planned out in my head. I was going to be filled with so much venom that I would get her before she got me.

God. Even thinking about this now turns my stomach. It's no wonder my dad enjoyed smacking me upside the head whenever I would talk about how much I hated them for making everything change.

I was a moron.

My mom walked in with this pink bundle and at the time, I thought she'd just brought me home a present and hadn't taken the time to wrap it properly. I had no idea at the time that the bundle in her arms was this baby I had just spent the last few months loathing.

As it turns out, when she did finally unwrap that blanket and those tiny eyes looked up at me through sleep ridden lids, I really had gotten the best present ever.

My Sunday.

The baby sister I didn't think I'd ever see again.

Sliding off my bed and making my way over to the phone, I punch in the familiar number and wait him out as the rings go in. The longer it takes, the more my impatience starts to build so when he finally answers, I snap out the question I'd planned on being relaxed before asking.

"Where did you find her?"

"Alliston."

"Are you there now?"

"Yeah. Well, on the outskirts anyway. You know how small this town is. You drive for five minutes and you've seen all you need to."

"Did you talk to her?"

"I did."

"So she knows about us?"

"Luke," he sighs and I swear I can feel my chest actually freezing over and seizing. "I didn't get that far with her."

"Cole, how the fuck is this even possible?"

"You need to call your parents. They can fill you in."

"I'm not calling my parents for information I can get from you right now!" I yell and even though it's low, I hear Cole curse and I immediately feel guilty. He's only doing his job, being a friend and here I am being a total ass. "Sorry."

"All good, Luke. If the roles were reversed, you'd be getting a whole lot worse, trust me."

If there's anyone that gets the way I feel right now, it's definitely Cole Alexander. When we were sixteen and his search for Sunday wasn't getting anywhere, he had the bright idea to go looking for his bio parents. After running into one brick wall after another, he finally found them. Everything he found out that day had messed with him for a long time after. The same way this is all messing with me now.

"There was a DNA match. That's how I found her."

I've got so many questions rolling around in my head right now that I can't even think straight. I have no idea where to even start. Pinching my arm and sucking in my breath at how hard I actually did it, Cole laughs softly and brings me back from the mental cliff I find myself on the edge of.

"You just pinched yourself didn't you?"

"That obvious, huh?"

He laughs again before exhaling a breath of his own and turning things serious. "I know it seems unreal, but Luke, I spoke to her today. She exists. This is as real as it gets."

"Is she okay?"

"She's perfect."

There's something about the way his voice softens that makes me wonder just how up and close and personal he got with her, but throwing it to the back of my mind, I fill him in on what has to happen now.

"Tell me where you are."

"Why? What are you thinking?"

"I'm not thinking anymore. I'm acting. Tell me where you are, Cole."

"Super 8 motel on the way into Alliston, but Luke, I got this. You don't need to do anything. We'll talk about it this weekend."

Not happening, and if Cole considers himself a friend at all, more importantly my best friend, then he damn well knows it.

"I'm on my way. I'll see you in a few hours."

Chapter Six

Reagan

"Come on! I never ask you for anything!"

I'd been home exactly five minutes before Jake started in on me.

A little over a year ago, him and a few guys from school decided it would be cool to start a band. Withering Embers was born that day and they haven't looked back since.

What started as four guys in the garage of our house taking popular pop songs and giving them a rock edge, is now them playing in all age's clubs all over Alliston and the surrounding areas.

Gone are the days of covers. Now they're writing and performing their own material, and even though I give him a hard time a lot—mainly because his friends are total perverts—I've never been prouder of him.

Jake doesn't realize this, but he didn't have to hound me the second I stepped in the door to go see him play tonight. I was planning on it anyway, but watching him sweat out my response is too funny to pass up.

"Did you or did you not ask to borrow my iPod a few months ago and then return it cracked? I'm pretty sure that's you asking me for something and making me regret saying yes later."

"You're never gonna stop torturing me with that are you?"

"Nope." I laugh as I turn back to the fridge, grabbing a can of soda out before kicking it closed and heading for the kitchen table. If he's going to continue going at me about this, I'm going to be comfortable for it.

"Besides the iPod incident, you know I never ask you for anything. So pretty please with a huge ass cheesecake on top, will you come see us play tonight?"

"I don't know, Jakey. I've got a pretty hot date lined up for later."

Bursting out laughing before catching his mistake and shutting his mouth, the caught laugh now coming out sounding a whole lot like a snort, he slides down into the chair across from me.

"The only hot date you have is with the book I caught on your nightstand. So cut the bullshit and just say yes."

"I'm pretty sure if you want me to say yes, making fun of the way I spend my nights is not the way to do it."

"Someone's gotta call you on your shit."

Done playing around, I finally give in and give him what he wants.

"Fine. I'll go, but since my car is still in the shop, you're gonna need to take me with you."

"So done!" he yells excitedly, jumping from his chair and making his way around until he's grabbing me up in the tightest hug he can.

The way Jake's acting now makes all the annoyances worth it. We may have two years between us and be nothing alike, but there's no denying that when he's happy about something, I'm even happier. For an annoying little brother, he's the best. I'm thankful every day that out of all the people I could have been stuck with, it was him that I got.

When he finally releases his hold and I'm able to breathe again, I take a long swallow of my drink before asking the question I know is bound to piss him off the most.

"Are you gonna be alright if Beth meets us there?"

"Reagan, no! You don't need her. You've got me."

"The guy that's going to spend the night on stage singing and having girls throw their panties at him?"

He shakes his head obviously wanting to argue, but with the crooked grin creeping across his face, aware there's no argument he can make. Jake may be young and stuck playing small venues, but in the short time he's been doing it, he's got his share of crazy fan girls. With the way he looks, it's not exactly hard to see why that is.

Jake is a combination of the best parts of our parents. He's got my dad's light brown hair color and my mom's blue eyes. Add the sound of his voice and a guitar and he's pretty much any teenage girl's fantasy. Jake Carter is definitely the Alex Gaskarth of his time.

"It's not even like that, Rea."

"It's totally like that, Jake, and I get it. It's cool. If I invite Beth along, at least you can take off and do whatever you want when the show's over and not have to worry about your older sister cramping your style."

"Cramping my style?" he repeats with a raise of his eyebrow. "Did you get stuck in the nineties again?"

Reaching across and smacking him and him answering back with a tap of his own, he breaks and gives in exactly the way I want him to.

"Fine. She can meet us there, but Reagan, I swear. If she goes after Tyler like she did the last time, I'm not gonna hold back. Keep her on a leash."

Tyler Wilson is the drummer for Withering Embers. He's sixteen like Jake and one of many guys Bethany's gotten in her sights and been determined to have a little fun with. Jake noticed it first and let me know about it, but by the time I'd gotten to her, she'd moved on to someone else and the subject was moot.

If Bethany's been going after him again its news to me.

"Did something happen?"

"No, but between Tyler being the biggest horn dog in the band and Bethany screwing around with anything that moves, it's only a matter of time before it does."

"She's not that bad," I argue, knowing the way my best friend really is and hating the light Jake is painting her in. "But fine, I promise to keep in line."

"Good. I get the feeling that tonight will be the night and I don't want anything screwing it up."

"Does that mean what I think it means?"

"Yeah. Rick said that his uncle came through and there's gonna be a rep from Divinity there tonight."

Divinity Records came out of nowhere about three years ago. They built their label from the ground up and signed some pretty good bands, sometimes outselling other, better known labels with their brand of indie rock. It's also the one place Jake wants to get signed. If Rick's uncle really did come through then tonight very well could be the night that my brother's dream comes true.

Making it even more important that I show up to give him my support.

"What time do you have to be there?"

"The gig is from nine to midnight, but I wanna be there at eight to set up."

Looking over to the time flashing in green LED on the microwave, I take another long swallow of my soda before slipping out of my seat and pouring the rest down the drain. Making sure before I head out to toss the can in the recycling bin under the counter.

"I'm gonna get ready. I'll meet you at the car in an hour." I say and starting to make my exit, I pause, turning back to him with a smile. "And Jake. If I've gotta work on Bethany, you've got to do the same with Ty. Make sure he keeps it in his pants."

"I'd do that if I could be sure he would actually show up wearing pants."

Not sure what to say back but unable to hold back the laughter, I let it fall with a shake of my head and head for the stairs.

I might not say it to him often, but what Jake's doing with his life—living his dream— all while maintaining the same level of focus on school and his extracurricular, I'm proud of him. Maybe even a little envious too.

If I can't manage to get mine off the ground, I'm glad that he can, and tonight when we go to the bar, I'm determined to make sure the world knows how proud I am.

He deserves it.

Cole

No sooner do I get back to the room with my dinner then there's a knock on the door and without even so much as a hello when I open it, Luke shoves his way in and throws his body down onto the bed.

"Hello to you too. By all means, come in and make yourself comfortable. It wasn't like I had a bunch of work lying around or anything."

"Don't mind if I do." He smirks before his face turns serious again. "You didn't even wanna do this shit anyway, so who gives a shit about your work?"

Until I found Sunday earlier, I agreed with him. Now that I've done what it seems like my uncle couldn't do though, everything is different. Besides, the shit he's now sitting on is all of the journals and notes that John kept and they're important if I want answers.

"This isn't just work, Luke. It's shit from John." I say in way of explanation as I make my way over the bed and start packing up the paperwork.

Ignoring me completely, he leans his back up against the headboard and shoots me a look I'm familiar with. Luke wants answers and he's not going to waste time getting them.

"Tell me everything."

"I told you everything I know on the phone."

"No you didn't. You told me that you found my missing and presumed dead sister. There's more to it than that. I could hear it in your voice. So spill it."

"Remember the shit you told me about a few months back? The stuff your mom had done for you and Sunday when she was born?"

"Yeah. What about it?"

"They called me in to look into it because they got a call from one of the detectives they kept in touch with about a possible bloodwork match. I went to the hospital first and confirmed what I already knew and then hit up Frank at the precinct for an address and details about the girl that was admitted."

"So I lied, Cole. I don't really give a shit about any of that. Tell me about *her*."

I knew it was only a matter of time once I started giving the details that he'd zone out and cut to the chase. It's another one of the ways we're alike. Neither one of us wants to wade through a sea of bullshit to get answers.

"You're gonna need to be more specific, bro."

"What's her name? How does she look? Does she go to school? Work? Does she live with anyone?"

"Her name's Reagan Carter, and like I told you on the phone, she looks like you. Long brown hair that's kind of wavy and your dad's brown eyes. She's about your mom's height. She looks healthy and well kept. Happy." Pausing to give him a minute to let everything I've said sink in, I give him the rest. "I have no idea if she goes to school or not, I didn't get that close to her, but she works at a local bank. She's a teller. She mentioned having a brother, and with the information I pulled from the DMV, it looks like she still lives at home with her parents."

There's no mistaking the hurt in his eyes the second I mention the family. Especially what I said about her having a brother. If our roles were reversed, I can't say I wouldn't be feeling the same ache knowing you're not the only one in her life.

"You talked to her?"

"Yeah. I wasn't going to. The plan was to check her out and report back to your parents about what I found, but it didn't go as planned."

"Nothing with you ever does."

"Ain't that the truth?"

This is brutal. I want to be able to say more. Tell him every single thing that happened earlier with Reagan, but I can't and it blows. I've never had to have a filter with Luke before, but there's no way I can explain the way it felt seeing his sister. The way that even talking about her now, I can still see that damn smile of hers so clearly. How it awakens sensations in me that I thought were long dead.

"She has no idea who you are, right?"

"No. I told her that I was there investigating a husband that skipped out on his family. Probably not the smartest move, but if you'd seen her, you wouldn't have wanted to drop this either."

"I want to see her."

"Really? With the way you booked it here from the city, I had no idea that's the way you were leaning."

"Smartass."

Luke wanting to see his sister isn't surprising, I expected as much when I called him about it. I'm just not sure him wanting to get right on it is the right call.

Reagan still has no idea about who she really is and the last thing I want to do is bring her face to face with Luke and slam it home.

"We'll get there. I want you to be able to see her, I do. I just don't think forcing yourself on her is the way to go here."

There must have been something in what I said that tipped him off because after a few seconds of silence where I think he's mulling over what I've said, he turns and studies me, something obviously going through his head that I wish he'd just spit out so we can get on with it.

"Is there something you're not telling me?"

"No."

"Hmm, alright. You just seem pretty invested in how she's going to handle all of this after one conversation."

"Tell me that again when you're the one around her. I guarantee your answer is gonna change. She's definitely a Grayson."

It hasn't happened a lot in our decade's long friendship, but when it does it always has the ability to twist me up. Lying to Luke, even when it's by omission the way it is now, is not okay. It's not the kind of friendship either of us wants, but I can't dump the way I'm affected by Reagan on him.

Being interested in someone I'm being paid to find—Luke's sister no less, is like tossing a live grenade into a group of innocent people. It's not right and until I can get a handle on it, it's gotta remain my secret.

"Is it wrong that I want to be around her this bad, Cole?"

"Nah, man. All things considered, I think it'd be weird if you didn't."

"Then what the hell are we doing?"

"Come again?"

"I'm sitting here with a private investigator, even if it's the last thing he wants to be called. A person that knows where my sister is and can take me to her. End this shit once and for all. Why the hell are we sitting here talking when we could be out doing something about it?"

"Luke—"

"No, Cole. I'm here. She's here. I'm not gonna be able to settle this shit until I see her. I need to see Sunday."

Except she's not Sunday anymore.

"I'm right there with you, bro. I know how you feel, but do you really want her first impression of you to be a forceful one?"

It's a long shot, but I'm hoping that being the voice of reason will get through to him. The last thing Reagan or Luke needs is for this to happen when emotions are running high.

"No, but I need to do something. Can we at least drive by her place?"

Still against doing this, but knowing this isn't a request I can argue my way out of, I concede, which the second it happens makes Luke genuinely smile.

A smile seventeen years in the making and one that now that it's back, I hope I never see him lose.

Luke

I've known Cole a long ass time, but it's been forever since I've seen him like this.

I was there when he was the emotional little kid that went around our neighborhood rescuing stray dogs and cats and bringing them back to their owners. Telling them no when they wanted to give him a reward for doing the right thing. The same way he was with my family when Sunday was first taken and the way he's been with us ever since.

Cole was the kid that would give you the shirt off his back if it meant that you'd be alright. He had the biggest heart of anyone I've ever known, both then and now. Though he'd also be the first one to call you a liar when you called him on it.

He was too damn conditioned by the time with his mom to believe he could be something other than the junkie's kid.

Flash forward to high school, the search for his biological parents and the period where he finally made good on wanting to be someone different. Gone was the good guy and in his place was a first class jerk. One with a chip on his shoulder so huge it's amazing he didn't crumble from the pressure of holding onto it as tightly as he was. It became his mission in life to rebel against every adult that tried to get through to him and attempt to pave out his own path, separate from the one Sam Alexander wanted.

He didn't want to be that other guy.

The soft one that cared too much for his own good.

I lost track of the amount of times he called me and my dad to come bail him out when he got himself caught up in shit he couldn't get out of. It's only when he met Miranda in our senior year that he seemed to even out and turn back into the kid I knew growing up. The one that kept me sane when I was losing hope over ever finding Sunday again.

Other than the odd time with me, no one seemed to be able to crack that wall he built around himself the way Miranda could. There was just a look in his eyes any time he got within a foot of her that spoke to how deep he was in it with her. How determined he was to make her his everything.

When she got up that day a year ago and disappeared, taking his money, pride and heart with her, it was like we were thrown right back in time again. It was him against the world. He was pissed off, tortured and for a really long time, numb from all of the alcohol he was drinking in order to forget.

I'd become so familiar with his attitude that when I called him back earlier and heard the way he was tripping over telling me about Sunday, I was thrown for a loop. I wasn't even sure I was talking to Cole, that's how out of left field it was.

How it still is.

He can tell me all the wants that there's nothing going on, but the look I saw in his eyes when we were kids, is there again now. For the first time in a year Cole's alive and if there's anyone who knows what a miracle that is, it's me.

I've been living like I'm dying for years.

I'm just not sure why he's trying to hide it from me. With as long as we've been friends, he knows he can talk to me about shit and I'm not gonna rag on him. All I can figure is my sister is the cause and that's where the idea to drive by her place comes into play.

If a few minutes in her presence can soften Cole after a year of him being ice cold and shut off from the rest of the human race, it's amazing to think what she'd be able to do for me.

Maybe she'll be able to bring back the boy I was when I lost her.

"You wanna remind me why we're doing this again?"

"Because I would have haunted your ass if you didn't?"

"That can't be it. You've been haunting my ass for eighteen years already."

"Good point. Alright, let me try again. You're doing this because it's my sister on the other side of that door and you wouldn't be able to live with yourself if you didn't at least try to do the right thing by us."

"Pretty sure that's wrong too. I'm not the Saint you're making me out to be."

"Maybe not, but you and me, we're family. I've never seen you give up on family yet."

"Point taken."

"Cole," I start but am forced to stop as his hand snakes out in front of me, pointing toward the front door where a guy and a girl are making their way out.

"That's her."

From his position parked a few feet away from the house, I take in the girl that's now shoving her body into the guy walking beside her. The smile on her face matching his until they make it to the SUV and it disappears as they slip inside.

It's hard to tell from this distance whether or not the girl is Sunday, but judging by how revved up and tense Cole is, there's no denying that he can, which only confuses me more.

"Okay, man, what gives?"

Averting his eyes away from the SUV long enough to throw me a glance before whipping his head back, he clears his throat and brings his hands down hard on the steering wheel.

"I never asked for this shit, okay? That's what gives. I don't know what John was thinking throwing this on me! Sitting outside someone's house and watching them like some kind of stalker, even if it is for you and Sunday, doesn't sit right with me."

"There's more to it, man. Deny it all you want, but it's been written all over your face since I got here. It's not like you weren't helping them out with cases the last couple of years. So this is about more than the job."

"You don't know what you're talking about."

"Stop it, Cole. If this was a case you weren't personally attached to, you'd be able to sit here all night. It's because of who you're watching that's got your losing your shit."

Catching movement out of the corner of my eye, I turn toward it and Cole quickly follows. The SUV is backing out. Despite wanting to force my friends hand and get him to admit what's really going on, I see this moment for what it is.

It's decision time. I can either do what Cole said earlier and wait this out, or do what my heart wants and get close enough to Reagan to see for myself if she's Sunday or not.

Twisting the key in the ignition before I can sort out what way I want this to play out, Cole brings the car to life and pulls away from the curb, making the decision for me.

He follows them.

Chapter Seven

Reagan

"Since when does Jake sing like that?" Bethany leans over and yells, trying to be heard over the band that's now three songs into their set. "Wasn't it like last week we were waiting for his voice to drop?"

Tearing my eyes away from my brother as he works the stage, I laugh and put my attention back on my friend. The same friend that wasted no time the second she met me here to start drinking.

Making the promise I made to Jake about keeping her in line harder to accomplish.

"He hasn't been that kid for a long time, Bethy."

"No kidding. Looks like he finally grew into those baggy jeans."

Attempting to keep a straight face and failing, I stick my fingers down my throat and gag, only mildly kidding about the way she's talking about Jake. A move that once Bethany catches just makes her laugh before turning her attention to the band. More specifically, Tyler.

Done with him my ass.

Thankful for the silence, I turn to the bar and flash the bartender Angel the sweetest smile I've got. Returning the smile with one of his own and making his way over, he slides the tall glass out from under my hand and shakes it.

"Another water on the rocks, Rea?"

"You know it."

Turning his back and busying himself with my water, I keep my eyes trained on him the entire time. My father's repeated warnings about bars and men who buy and fill the drinks in them on repeat. Comfortable in the knowledge that Angel won't be drugging me this evening, I allow my mind to wander to the person that even now hours later is still front and center in my thoughts.

What would Cole think about all of this? Would it be something he would be into doing or is this even less his speed than it is mine? What would he drink if he was here? Beer like Bethany or something harder?

"Earth to Reagan! There's a hot bartender holding a tall glass of water that's sweating the way most of the women in the bar do whenever they're near him standing in front of you!"

Blushing at Angel before taking the glass and bringing it to my lips, I polish off about half of it before placing it back down on the bar. Earning a laugh from him and wide eyes from Bethany.

I'll definitely take wide eyes over an open mouth any day.

"Mmmm...refreshing." I moan, earning a smirk from Angel before he heads to the other end of the bar.

"I know that look Reagan Carter."

"And what look would that be exactly?"

"The over the moon dreamy one you had on your face before Angel came over. You were thinking about Hottie again."

"I was not." I lie and she just smirks. Even when I'm doing my best attempt to hide the truth, she still sees past my bullshit.

"What did I tell you about securing a second date? If you'd just listened to me, it could have been him here with you tonight."

"But if I was here with him, I wouldn't have all the perverted comments and we all know that's no way to live."

"Ain't that the truth? Whoever said that is brilliant."

"Oh yeah," I laugh before bringing the water back to my lips and finishing it off. "She's a real scholar that one."

Turning in her seat, she bends over in a half bow and I burst out laughing again, this time making sure to spray her with some of the water I hadn't gotten around to swallowing.

"You're such a bitch!" she shrieks, jumping off the stool and frantically wiping at her newly sprayed clothes.

"Takes one to know one."

"So this is what tellers do when they're not stuck behind counters all day." A voice interrupts and looking at Bethany before turning to acknowledge the intruder, I watch as the smile wipes completely off her face and her eyes go wide.

"You know, when you came into the bank earlier, I never figured you as a stalker. It's a shame too. Stalking is wasted on someone who looks like you."

"Bethany!" I exclaim before turning around and finally facing down the person she just attacked. "Cole?"

Just as he parts his lips to speak, another body comes out and inserts himself in between us.

"I'm Luke."

"You sure about that?"

This stops the guy cold as he actually pauses before taking a slight step back, his gaze flittering between me and Bethany before landing on Cole.

"You'll have to excuse my friend. He's only allowed out in public a couple times a month so he hasn't figured out how to interact with people yet."

"Screw off, Cole."

"Luke, buddy, we've talked about this. When you screw on you get better results."

If I didn't know for certain it was two guys talking, I'd think I went back in time and was having this conversation with Bethany. No sooner does Cole respond to his comment then Luke turns around and sticks his tongue out at him.

"Like I was saying," Luke turns back flashing me a bright smile. "I'm Luke Grayson."

"Has anyone ever told you that you're kind of rude, Luke?"

"Before he lies," Cole cuts in, elbowing Luke in the side and causing him to move back so I can have a clearer view. "He gets that a lot and not just from me."

"I figured as much." I laugh and his eyes crinkle as he smiles. A move that has the temperature in the room raising at least another twenty or so degrees with how affected I am by the sight of it.

He's got a really great smile.

"Since she's too polite to ask, I guess I'm going to have to. What are you doing here? Better yet, how about you tell us how you knew we were gonna be here?"

Cole's response is immediate as he slips his gaze from mine to Bethany.

"We're here because I promised someone I would stick around town. Luke was crawling the walls of the hotel we're staying in so I agreed to going out. I had no idea the two of you would be here, but now that I see you are, I can't say I'm all that upset about letting this jackass drag me out."

"Way to throw me under the bus, pal." Luke grumbles and Bethany visibly thaws as she laughs.

Turning his attention to my best friend, he moves around me until he's standing beside her, bending down and whispering something in her ear that again makes her laugh, leaving me completely open to give Cole my full attention.

"I thought you didn't drink?" He asks with a quick glance to my glass on the bar.

"I don't, unless you're counting water. Come to think of it, though, I have had one too many of these tonight. I'm thinking I may need to be cut off soon."

Shaking my head in embarrassment as he slides his body up on the stool, I attempt to look anywhere but back at him.

One of the drawbacks of being with someone for years is that once you've snagged them and you're comfortable, the need for things like flirting and mastering the art of conversation completely falls out the window. A fact that's never been more apparent than it is right now.

"I'll agree to a cut off, but only after you share one more glass with me."

Tapping the bar, Angel turns and with a look between me and Cole, he makes his way over. "What can I get ya, man?"

"I'll have what she's having." Cole says and for the second time today I'm having to resist the urge to bury my face in my hands from the affect those words are having on me.

Get a grip, Reagan.

"Be careful with this one, man. She'll drink ya under the table."

"It's about time I met someone who could." Cole jokes back before Angel grabs my glass and busies himself getting us our drinks.

"Can I ask you something, Reagan?" he asks once we're alone.

"Sure."

"Does your friend always move that fast?"

Twisting my body in the direction of where I left Bethany with Luke, I see that they're now standing closer to the band. The two of them looking the opposite of two people who just met with how closely their bodies are pressed to one another.

"Truthfully? She normally moves a lot faster. Luke's gonna have his hands full."

"Somehow, I think he can handle it." He chuckles softly, taking in our friends one final time before turning to me. "So I told you what I'm doing here. Care to share what brings you here tonight?"

Looking from Luke and Bethany and back to Cole as I hear the opening chords to the only ballad Jake said they would ever play, I motion to the stage and smile proudly.

"Is this your way of saying you're with the band?"

"In a way, yeah."

A flash of something resembling jealousy passes through his eyes and against my better judgment, I reach out and rest my hand on his arm.

"The lead singer is my brother Jake."

Without realizing he's done it, I feel his body visibly relax with my explanation.

"Your brother. Right."

Bothered by the way the conversation is going, I attempt to pull my hand away, but he grips tighter to stop me.

"I'm sorry. I thought…" Pausing, he laughs nervously. "It doesn't matter what I thought. It was stupid."

"I'm sure it's not." I respond softly and after he drags a hand through his hair and releases a heavy sigh, he looks up again. This time, I see the smile there matches the lightness in his eyes.

Everything is alright again. Mini crisis averted.

"So after I dropped you off earlier, I realized I made a mistake."

"I'm not sure what you mean."

"I told this girl I met that I wanted to take her out for hot chocolate, but didn't get her number in order to set it up."

"If it helps, the girl didn't exactly ask for your number either."

"Does she regret not getting it?"

Liking the way it feels talking about our time together this way, I smile and keep the game going.

"Hmm, I'm not sure. I guess that depends on whether or not the guy regretted not getting her number before he disappeared."

"He regrets it. He actually feels pretty stupid for not doing it and it took everything in me not to beat his ass earlier for being such a bonehead."

"Does he want it now?"

"If she wants to give it to him, which if she doesn't he would totally understand."

Holding out my hand and motioning for him to hand his phone over, I watch as he slips his hand off mine and into the pocket of his jacket, bringing it out and handing it over.

As my fingers manipulate his lock screen and head into his contacts in order to add my number, I feel his eyes on me and it's so intense that I sneak a quick glance up at him, both of us caught in our staring contest and laughing.

"Sorry." He says, breaking the spell I seem to be under long enough for me to finish adding in my number and hand the phone back.

"You do that a lot." I observe and when his head dips to the side, I explain. "Apologize. Seriously, it's okay."

Drumming his free hand on his leg once he's placed the phone back in his pocket, and shaking his head the way I've been known to do when I'm having an argument in my head, I slip my hand off his arm and rest it on top of the moving one. A jolt or a spark of some kind hitting the second our hands make contact, causing me to jump back in surprise.

"Stop me if this is too forward, but with Luke occupied by your friend and a bar the last place on earth I want to be, I was thinking that we could have our hot chocolate date now."

I want to say yes, but all it takes is one look toward the stage to remind me of the reason I'm here. Jake hasn't so much as looked in my direction since he got here, so I really don't think he'll notice me gone, but there's something about just up and bailing that doesn't feel right.

I've never bailed on my brother before and even though spending more time alone with Cole is causing me to have serious second thoughts about sticking around, I don't intend to start now.

"I promised my brother I'd be here for him. I don't feel right leaving."

"You know, I think I can work with that."

"How?"

"Your brother's gotta come off stage sometime, right? So we wait until he does, then let him know what's going on."

This has got to be some kind of dream. There's no way he's actually willing to wait around for Jake to come off stage so we can go out. I dated Walker for years and not once in that time did he ever do anything remotely thoughtful like this. How is it possible that a guy I've known for less than twenty-four hours would be willing to?

"You'd really do that for me?"

"I'm pretty positive with the way I'm feeling right now, I'd do a whole lot more than that Reagan, but yeah. I can wait for your brother to finish his set before stealing you away."

Wanting to get this out before the traces of hesitation that were present before decide to make another appearance, I squeeze his hand and give him the brightest smile I can summon up in the moment.

"It's a date."

Cole

I gave Luke every warning I could think of before we entered the bar and what did he do the second I caught sight of Reagan and her friend? Forced my hand until he was standing in front of her, about to blow everything out of the water.

Before we even left the motel I knew he wasn't going to keep himself in check. Luke operates on one setting. He sees what he wants and he goes after it. He can't be bothered factoring things in and giving it the time and thought it might need. It's full speed ahead and if I hadn't stepped in and turned it around, there's no doubt I wouldn't be sitting where I am now.

I tried fighting this shit, I really did, but the way I react whenever I'm near her isn't the worth the struggle.

I'm ignited and alive with every sense heightened, feeling better than I have in years and that's just what happens when I'm near her. When she speaks, smiles or blushes, all bets are off. My temperature rises, my body is revved and ready and all I want to do is get her alone, grab her face and force my lips down until I'm drowning in her taste.

Having a female blush shouldn't be a hot button for me, but with how often she does it and the way not only my head reacts, but the rest of my body too, it's pretty damn obvious it is. The only reason we're not off in some secluded area of this dive right now is because the second I get swept up in all things Reagan, the reality gets slammed home that she's Sunday and I'm acting like a complete pervert.

I'm in the perfect place to drown myself, but I don't even attempt it. I drink the water the bartender brought me and watch her do the same. Enjoying every damn second of it even though the idea of drinking until I'm completely numb to my reaction to her holds a very strong appeal.

"How long has he been doing this?" I inquire motioning to where Jake is now holding out his hand to one of the girls in front of him in an attempt to pull her on stage.

"Since he was thirteen, but he's only gotten serious about it over the last year or so."

Despite being caught up in Reagan, I've been paying attention to the music and he's not bad. He's got one of those voices that would work perfectly for radio, but it's obvious from the stuff he's playing and the way the rest of the band is, that the last thing he cares about is fame. He's in it for the love of the music.

"He's good."

"I know." Cocking an eye up at her confidant response, she laughs. "I know how that sounds, but I've been a part of him doing this for years. He wasn't so good in the beginning, but he's dedicated and there's no doubt, at least for me, how good he is."

"Is he aware you're his biggest fan?"

"You know," she laughs. "I think he might be. It's half the reason I'm here tonight."

"What's the other half?"

"It's been brought to my attention that my idea of a good time is actually pretty boring. That at my age I should be out having the time of my life. Doing this is my way of getting them to back off."

"So you're proving a point?"

"Most definitely."

The more she opens up about herself, the more I hang on what she's saying and want to push for more. I can claim that learning about her is about the job all I want, but it's not. There's something about her that I'm intrigued by and it's got nothing to do with her being Sunday.

"I take it staying home and reading isn't high on the list of things your friend wants for you?"

"No, but this time it wasn't her. It was Jake and then before we left, my parents. Mostly my mom."

"She'd rather you be out partying?"

"No, but she's made no secret of her dislike for how secluded I've become since I broke up with my ex."

If I'm honest, I wondered about her relationships. It was obvious fairly quickly that she wasn't dating anyone or the coffee with me would have been a no go, but that didn't mean there wasn't a long line of guys in her past. Ones that might still mean something. With my track record with Miranda, who she was before we got together and who she'd been with, I've got no doubt a girl that acts and looks like Reagan has probably left a ton of broken hearts behind.

"What happened?"

Her face scrunches up, her eyes going dim as her body stiffens and hating the reaction that my question caused, I immediately do damage control.

"You don't have to answer that. Forget I even asked. It's none of my business."

"He cheated on me." She quietly answers, ignoring my request.

"So you dated an idiot?"

"Yeah, I dated a *really big* idiot."

"How long were you together?"

"A couple of years. He was my high school sweetheart."

The way she refers to her ex as her high school sweetheart tells me that it's another way we're alike. Her utter disdain for it makes her sound makes me think I'm looking in a mirror. I'm not the only one that had the carpet of their life ripped right out from under their feet.

Reagan has too.

While I try to formulate a response, debating whether or not to give her my own horror story, she settles it for me.

"Have you ever put your all into something? Gone all in without reservation and then woken up one day and been completely blindsided? Broken down until you're not even sure where you end and the pain begins?"

There's zero hesitation once she's finished asking the question.

"You're damn right I have."

"So you dated an idiot too?"

"Something like that."

"Then you might just be the only person in the world that gets what I went through with Walker."

You're damn right I know what she went through with Walker. I also know who I want to go find when I finally walk away from her and pound into the ground for hurting her.

Back off, Cole.

Instead of saying what I'm thinking, letting Reagan know beyond a shadow of a doubt what a douchebag I think her ex is, I turn it around until she's not the only admitting personal things about herself.

"Her name was Miranda. For six years I thought she was the love of my life. There wasn't another person on the planet that understood my complicated shit the way she did. I was determined to be with her forever. Three bedroom house, white picket fence, two point five kids, the whole bit. I worshipped her."

"What happened?" she questions softly, and before I can answer, her hand is slipping through mine, giving me a reassuring squeeze. Infusing me with strength I'm not even sure she's aware she has to give and making the response I've spent the day trying to tame rise up again, until all I can focus on is how right this feels.

Something so wrong feeling this right is a joke. It has to be.

I don't do right. At least not since my two reasons for living were buried six feet underground and the other went missing.

"She traded me in for a newer model. Or I guess, an older model considering the guy was in his thirties."

Talking about Miranda makes me sick and brings up memories I've been doing everything in my power to bury. The fake tears in her eyes when I found out about the other guy and called her on it. Her empty pleas about there being nothing going on and how determined she was to keep us together.

Right along with the morning after when I woke up to go to work only to find the closet stripped of all of her things and my bank account as empty as my closet and eventually my heart.

I really fucking hate thinking about this.

"I'm sorry, Cole."

"It's all good. It's been a year, I'm over it."

"Why don't I believe you?"

"Maybe because you can see through my bullshit the same way Luke can?"

"Well, as long as I'm not the only one." She laughs and even though the subject turns my stomach, I join her.

"Okay, enough talking about our idiotic exes. Tonight is supposed to be about fun. They've taken enough already. They don't get to take this too."

Her outlook and how easily she switches gears, I'm envious of it. It's easy to see from the vacant look in her eye when she mentioned her ex-boyfriend that it's still a raw subject for her, but being able to see past it, change things up and get us back on a better track, it makes the bond already starting to develop that much stronger.

Maybe if I stick around for a little while longer she can teach me how to do the same thing.

"What exactly is this, Reagan?"

"A few different things I guess. If I had to choose, though, I guess I'd say its two people enjoying each other's company."

Sensing movement, I look up and see Luke and Bethany making their way toward us, but more than that, I catch Jake laying his guitar down and jumping down off the stage. His destination clear when he finally looks up and locks eyes with me.

Knowing how Reagan feels about him, I expect him to be pretty pissed with how close I am to her, especially with the way her hand is still on mine. Shooting me a look of death right now would be a sensible reaction, but it's not what I'm getting.

He doesn't look happy, but it's not the tension that gets me. It's the surprise. For whatever reason, my contact with Reagan shocks him.

Strange.

Looking from him to Luke, I see my best friend noticing the same thing, which only makes his own instincts kick in as he moves away from Bethany and closer to Reagan. Reaching out once he's close enough and resting his hand on her arm protectively.

Spinning around and locking eyes with him, I see her face go from the happy demeanor of a few seconds ago to the complete opposite. Slipping off the stool in an effort to head this off at the pass, Jake gets there first. His eyes hard and filled with anger all I'm able to catch before all hell breaks loose.

Jake pulling on Luke, his strength enough to twist my best friend toward him, and leveling him with a look that I know all too well. Protective rage. After all the fights I was in at his age, I know what's going to happen now, and despite knowing that it could possibly just make things worse, I do my best to stop it.

"What the hell you think you're doing grabbing her like that?"

"None of your business, kid. Why don't you run along and play with your friends and leave the adults alone."

"Luke," I call out, stepping around Reagan until I'm standing between both guys. "Back off."

"Yeah, Luke. Why don't you do what he says and back off." Jake smugly responds and reaching out before he can do what I know he wants to, I grab him and pull him back. The last thing I need happening is for Luke to lose his shit and get into a fight with the only family she's got.

"Let go of her, man." I say, motioning to the grip he's still maintaining on her arm and willing him to release it. After looking between me and her, he gives me what I'm after and he releases the hold.

Reagan demanding answers almost the second she's done tending to the sting Luke's hand must have left behind.

"Why did you grab me like that?"

"I didn't like the way he was looking at you."

"The way who was looking at me?"

"Him." Luke nods toward Jake. "It didn't feel right."

"You got a damsel in distress radar or something?" Bethany interrupts and despite how tense the moment is, I laugh. This girl has no idea how right she is.

"Something like that. I didn't trust the look I saw."

"I'm her brother, jackass. I looked at her the way I did because I saw her holding some random guys hand. Last I checked it was my job to protect her, not yours, so how about you back the hell off?"

Before I can speak up in order to settle Jake down, Reagan beats me to it, obviously on the same page as me and wanting to diffuse the situation.

"Jakey, it's okay. I'm alright."

The same way she seems to do with me, she does with Jake as he visibly starts to relax when she talks to him, though still making sure his intention is clear as he moves in as close as he can to where she's seated. Doing what I know has to be eating Luke alive.

Guarding his sister.

"Are you sure?"

"Yes, I'm sure. Jake, this is Cole. Cole," she says turning toward me with an apologetic smile. "Meet my brother, Jake."

"Not exactly the way I wanted this to go, but nice to meet you. Sorry about Luke. He seems to have slipped his leash tonight."

Knowing I'm gonna pay for throwing Luke under the bus later and not much caring as long as this misunderstanding can get cleared quickly, I smirk when Jake laughs and holds his hand out for me to shake.

"It's cool, man. I guess we both misread what was going on. Nice to meet you."

"Now that we've got that settled, there's something I want to talk to you about little brother." Reagan says, slipping herself down off the stool, sliding her arm through his and dragging him away.

When they've walked far enough away for us not to be able to hear their conversation, Bethany reminds us of her presence as she whistles quietly.

"I think I like it when you let him slip his leash. It's been awhile since I've seen Jake get his back up over anything, especially where Rea is concerned."

"What do you mean? I got the impression they were close."

"That's because you got that impression from Reagan. She adores Jake and doesn't bother hiding it. I'm just not sure sometimes that it's returned."

Well, this is news. I think at the same time Luke chooses to speak.

"He better start giving a shit. I'd kill to have that kind of relationship with my sister. And you can be damn sure if she cared about me half as much as she seems to about that mouthy little brat, I'd worship the ground she walked on."

"You have a sister?" Bethany catches and I nudge him before he lets anything else spill out. The last thing I want happening is for Reagan's true paternity to get dropped on her by her best friend.

"I did. Past tense." Luke responds and I breathe a sigh of relief. At least for the time being, he seems willing to play along.

"I'm sorry."

"Yeah, me too."

Sensing the struggle Luke's having pretending to be someone that's not completely invested in the girl now making her way back over to us, I change the subject and inform him of my plans for the night.

"You think you can get back to the hotel on your own?"

"Yeah. Why?"

"I made plans with Reagan."

His eyes lift and sensing the million questions he's just dying to ask, I nod, silently letting him know that I'll answer all of them later.

What I don't expect is him smirking at me as Reagan and Jake make their way back over and her hand slips its way down into mine.

"I've got his blessing. You ready to go?" she whispers and I nod.

I'm definitely ready to go, especially when my best friend's grin is growing significantly larger with each passing second we stand here motionless.

Moving toward the door after Reagan gives Bethany a small hug goodbye, Jake's voice calls out before we can make our way through, pausing us both in our tracks.

"Don't make me regret being okay with this! If you hurt her, I will end you."

Chapter Eight
1997

Cole

"Cole, you remember Mr. Samuel, don't you?" Anita, the lady that's paid to watch over us asks as she steps into the room. The giant of a man with a gentle smile following closely behind.

Samuel Alexander.

Of course I remember him. He's been 'dropping by' as he calls it for weeks now, but where his first visit had been to see everyone, every other one since he's only been staying to visit with me.

Ever since I got here, they've been trying to get me to open up. Tell them about my life before they ended up my doorstep. They said all the questions were so they could come up with a way to help me better, but I know different.

Even a two year old could see through it.

They want me to talk because they want to make my mom pay for what she was doing to me. They're not gonna get what they want. I might have hated my life, but not enough to turn on anyone.

"Hello, Cole."

"Hi, Mr. Alexander."

"Now what did I say about that?"

He hates it when I call him Mr. Alexander. He told everyone that when he first showed up here and it's been the one thing he's said that's stuck every visit since. Doesn't mean I'm gonna call him by his first name though. It feels weird. He's an adult and we're not friends.

"You want me to call you Sam."

"That's right. So how about from now on," he proposes sitting down on the edge of the bed. "We stick to first names."

"Okay."

"The reason I'm here today, instead of later in the week the way we planned is because I have a proposition for you."

I have no idea what proposition means, but there's something about the way he says it that makes me wanna buy whatever he's going to sell.

"Okay." I repeat.

"During our previous visits, you've spoken very openly about your hatred of the group home. That you miss home, even though you don't miss what took place there. What if I said I had a solution for you?"

See what I mean? I want to buy into what he's selling.

"What solution?"

"I want you to come live with me. Call it a trial run, but after spending these last few weeks getting to know you, I think I can offer you the solution you're looking for. A place to be that's not littered in violence. With someone who cares and would love you the way you deserve."

Littered in violence. That's an interesting way to put the way I was living before Children's Aid removed me.

A mother so addicted to drinking, drugs and sex that most days it was a miracle she even woke up at all. Leaving the tiny apartment we lived in to rot. Mold gathering in every room, bottles littering the place, a lot of them cracked and broken because when she got too drunk, she got angry and took it out on me.

Boyfriends, a different one every other week, looking for a warm place to shove their dick while they fed her addictions. Sometimes even starting stuff with me, creeping into my room at night until I'd finally gotten the courage to knock the one guy in his nonexistent balls to stop it. They stuck to beating the crap out of my mom and me after that.

The last time, the one that someone finally summoned up the courage to call about being when my mom beat me so bad I blacked out and woke up hours later in the hospital.

With bruises and a box cutter sized slice in my stomach as my parting gifts.

Littered in violence might be right after all.

"Why?"

Seems like a good question to ask. I mean, he knows what I came from and what I'm probably gonna end up being like because of it, so why take the chance?

"My wife and I tried for years to have a child. It was to no avail. People would tell us that it just wasn't in God's plan for us and we were meant for more. I always considered that a bunch of bullshit quite frankly, but after spending time with you, I'm starting to think maybe they were right."

"I don't understand."

"There's a reason you stood out to me, son. Why I've made time each week to come back and sit with you. Why of all the people that you could choose to open up to, you've done it with me. The truth is, Cole, I see a little bit of myself in you and because of it want to help you."

He sees himself in me? So he was a midget with a bad attitude and nightmares too?

"Why are you telling me all this?" I ask instead. "You know how it works here. You could have just talked to the people running the place if you wanted to foster me."

"Because that's not how I do business, son."

I don't wanna admit it, but when he calls me son—which he's been doing a lot lately—my chest does this little flip before it settles and everything gets all warm.

There's something about it that I like.

Could it be because I want to be Samuel's kid?

"And how do you do business?"

"I could have easily gone through the proper legal channels and left you out of the decision making process altogether, but I would much rather talk to you about it. Get your opinion so I know where I stand."

"And if I said that I wanted to go with you?"

"Then I would ask your permission to put everything in motion so I can take you home as soon as possible. Is that what you're saying, Cole?"

Here it is. One of the defining moments the therapists keep rambling on about when they're in the middle of their allotted time with me. Samuel is leaving the decision about what happens next up to me, and even though it might turn out to be the worst decision I've ever made, I need to see it through.

I need to go home with Samuel.

"Yeah, Sam."

He releases a large exhale of breath as his features soften from the tight way they'd been, and there's something about the way it happens that tells me I made the right choice.

Samuel Alexander is going to be the father I never had.

*****

Holy cow.

Samuel didn't say anything about being loaded.

We've only been here ten minutes, but I only needed one of those to see that this isn't some busted up apartment in the city.

The house is three stories tall, windows with the curtains open letting the sun in and giving the place a glow that makes it appear like a figment of my imagination. Or maybe Heaven.

Yeah. Definitely Heaven.

What exactly is his job if he's got a place like this?

The only time I've ever seen grass this green was in a book and that's only because it was colored in. Pictures are always more colorful.

I've gotta be dreaming this.

"Welcome home, Cole." Samuel announces, moving behind me and bringing his large hands to rest on my small shoulders.

"You live here." I state in complete disbelief, still unable to wrap my mind around this now being my home.

"Yes, but now you live here as well."

"This place is like a museum."

"Believe it or not, I've heard it called far worse than that. Cole, I know it's a lot to take in, especially with what you're accustomed to, but I meant what I said. This place, my home, it's yours too. I want you to make yourself at home."

"What about your wife?"

"Rachel should be home shortly. With it just being the two of us, we were severely lacking in the food department, so she's making a quick run to the store. She's excited to meet you."

I've never doubted anything Sam has said before, but there's something about the way he looks away—the first time since we pulled up—that tells me his wife is not nearly as excited to meet me as he wants me to believe.

Didn't figure Sam for a liar, but here we are.

"Are you sure about that?"

"Of course. Why would you ask that?"

"Whenever my mom was hiding something from me, she would look away and get this look in her eye, like there was nothing there. You just did the same thing."

"That's very observant of you."

"Does that mean you'll tell me the truth now?"

"Rachel is happy to have you here, Cole. She wants this as much as I do. It's just a little bit more of an adjustment for her because even though we've been told that we can't have any children of our own, she's still holding on."

Well, that's better.

I hated thinking Samuel was lying to me. With as good as it's felt since he offered me a place to live, having it be any other way twists me up in a way I haven't felt since before I was taken from my mom.

"I'm sorry."

"You have nothing to be sorry for, son. This is the hand we've been dealt and over time, Rachel will come to love you as much as I do."

He loves me?

When's the last time anyone said that to me? Has anyone ever said it? I lived with my mom for six years, memorizing thousands of things she threw at me during one of her rages, but I don't think love was ever a part of it.

Can I even be loved?

Before I can ask him if he means it, there's yelling from across the street and looking toward it, I make out the shape of a guy about my size running toward us.

Coming to a stop, the boy huffing and puffing like he'd just run a marathon, he sticks his fist out and even though it's the strangest thing I've ever seen, Samuel does the same.

Samuel fist bumps people too?

Who is this guy?

"Lucas, it's nice to see you again."

"Who's that?" he points to me.

"This is Cole."

"Hi, Cole!" the guy named Lucas says, sticking his fist out.

What is with this guy and fist bumping?

Shoving my hand out, the fist I've made weak in comparison to the strength behind his when it connects, I shake my hand before shoving it back into the pocket of my shorts.

"Cole is going to be staying with us from now on." Samuel explains when Lucas turns his attention away from me. "I'm hoping that you'll make him feel at home here."

Nodding to Samuel, he turns back to me. "You like hockey?"

"Who doesn't?"

"Good answer." He grins and turning back to Samuel again leans forward. "Me and a couple guys from down the street are gonna set up a game. Is it okay if Cole comes?"

"As long as it's alright with him, I see no problem with that."

Great. I'm on the spot again. There's a part of me that wants to go with this Lucas guy because I want to at least try and fit in, but considering I haven't even seen my room or met Sam's wife yet, I don't feel right about it.

Maybe I should stick around for a bit.

"I don't know." I admit to which Samuel pats me on the shoulder.

"Go ahead, Cole. Everything will be here waiting for you when you get back. We've got all the time in the world."

If only I could believe that.

I might only be seven, but after living with my mom, I know the difference.

Time is fleeting.

"Are you sure? Shouldn't I stay until Rachel comes home?"

Somehow sensing my struggle, he motions to Lucas to give us a minute before pulling me a few steps away and leaning down to his knee, giving me his full attention.

"I know you think that if you leave all of this going to disappear, but I assure you that it won't. If it would make things easier though, I can come get you when Rachel is home and dinner is close to being done."

Samuel Alexander really does have an answer for everything.

"Okay. I guess that works."

"Great. Lucas is a good kid. You'll be alright with him."

Something tells me that Sam says that about a lot of people he interacts with. I mean, he knows all about my home situation and still thinks I'm the perfect kid to foster. I'm gonna need more than his endorsement on Lucas. And the only way to get that is to go with him now and decide for myself.

"You'll come get me later?"

"Absolutely. Now go." He points over to where Lucas is standing. When I start to do what he asks, I get about halfway to the guy when he calls out again. "And Cole?"

"Yeah?"

"Have fun."

*****

"Psst! Cole!" Luke calls as I do a quick check both ways and cross the street. "Hurry up. I gotta show you something!"

It's only been a couple months since Samuel brought me home, but in that time I've gotten used to the way Luke is. Whenever something pretty cool happens, he's practically burning a hole in the ass of his pants in excitement over it. Though, I'm pretty sure I know what it is this time that's causing it.

His mom—Mrs. Grayson, is pregnant, and the last time I was over, she'd been complaining that the baby was never going to come. At the time, Luke hadn't been all that happy about it, but with the look on his face as I step onto his porch, somethings obviously changed.

I always knew that if I had a brother or sister, I'd be happy about it. I always wanted one, so when he was acting like a brat to his parents, telling me that when his parents brought her home he was gonna run away, I wanted to hit him. He had no idea how cool it was having a baby sister.

Something tells me he gets it now.

Babies are cute. Well, they are when they're not throwing up all over you and smelling like a sewer.

"Okay I'm here. What's up?" I ask once I'm inside and he's shut the door.

Pointing to the family room but putting one finger up against his lips shushing me, he starts walking forward until we're standing just outside the living room.

On the sofa is Mrs. Grayson and in her hands is a pink bundle. Her eyes tired but a smile on her face. One that looking back, I always wished I'd seen my mom wear but was never lucky enough to see. Swallowing down the tight squeeze in my chest at the sight of her, I shake myself and look to my best friend for what we're supposed to do now.

"It's a girl, Cole. Her name's Sunday."

Yeah, he's changed his tune completely.

I knew he cared.

"Like the start of the week?"

"Yeah. I looked it up too. All I could find was that it meant the day of the week, but I think it means more."

"Like what?"

"She's like the sun when it hits your face. When my parents brought her home, she opened her eye just a little and suddenly my face was warm. Sunday, she's my sunshine."

"Boys," Luke's dad says as he makes his way over to us. "If you promise to be careful and keep the volume down, you can hold her."

The closest I've ever come to a baby was passing by one in the grocery store. I know next to nothing about holding them. Only that they're super tiny and squirmy. Do I really want to risk something happening by holding her?

Short answer is, yeah I do.

This is as close to a sister as I'm ever gonna get with the train wreck that my mom is, so I want to be a part of it. Especially since sometimes when I'm alone, I wish Luke's family was mine.

"For real, Dad?"

"Considering the way you've been all over your mom since we got home, something tells me if I don't let you do this, you'll just steal her out of her crib later. So yes, Lucas. As long as you're careful, you can hold her."

Going completely against the no loud noises rule, Luke starts bouncing up and down in place, screaming his excitement at the top of his lungs, and I just shake my head.

For all of the weeks he spent complaining to me and the other guys about how much he hated his parents having another kid, he's acting completely different now. It's hard to believe he's even the same guy.

"What do you say, Cole?" Xander turns and asks. "Would you like to hold Sunday?"

"Yes sir."

"Well, let's do it. Maybe while you're holding her, Luke can figure out a way to settle himself."

Smirking at his son and shaking his head like I did a few seconds before, he motions to the sofa and following all of his instructions to the letter, sitting down all the way back, and getting my arms resting the right way, it's only a couple of seconds before the bundle of pink I was viewing from the other side of the room is in my arms.

Looking down, finally able to see her face around the blanket, I'm reminded of what Luke said before. How true it is. There's no other reason for the warmth that's now flooding through me looking down at her half opened eyes.

Sunday Grayson really is the sunlight.

One I hope never goes away.

Chapter Nine

Reagan

"Can I ask you something?"

Something about the way his eyes shift away and he swallows hard, clues me in pretty quick to how he feels about random questions. I'd gotten to see shades of it when we were at the bar earlier, but here, away from all of the noise and the quiet, it's even more obvious.

Maybe I'm overthinking things. Maybe he's like this because of what he does for a living. I can't imagine a private investigator being the most forthcoming in his responses. It doesn't help me shake the feeling that it's more than that, though.

It's like he's hiding something.

"Sure. Ask away."

Here goes nothing.

"Your friend. Does he have any siblings?"

Damnit! There it is again. Him lowering his gaze away and looking at the mug of hot chocolate.

I knew I wasn't imagining it.

"Why do you want to know about Luke?" he asks, running his fingers in a circle around the edges of the mug, his eyes never wavering from their position on it. Making the need to get answers even stronger.

"When he got into it with Jake, I saw this look in his eyes. It didn't last very long, but it was there. I don't know how to explain it, but it made me wonder about him."

"You mean him acting like a rabid dog?"

Shaking my head, I try and explain. "It was something else. For this brief moment his eyes were hard. Protective. Like he really believed Jake was a threat he needed to eliminate. I've seen it before, so I was curious to see if it was the same thing."

"Seen it how?" he asks, finally breaking his gaze away from the mug and placing it back on me.

"My dad. Whenever he had to come to school for me and Jake. He would go to bat for us and whether we were at fault or not, he always had that look. At least he did until we got home and the truth came out."

"Has anyone ever told you that you're perceptive?"

Now it's my turn to look away. It might only be a question to anyone else, but for me it's a compliment. One that coming from him because of the intensity falling off him in waves as he says it, has the power to completely unravel me.

If reacting this way was confusing, what's even more so is why it's never happened until now.

It's my job to interact with people. Generate conversations. Never once in all that time, even at school, have I ever had a reaction like this. One so heavy that I can't even maintain eye contact for more than a few brief seconds.

"If by perceptive you mean nosy, then yes."

"Hey," he speaks softly, reaching across and brushing his hand over mine until I look back up. "I didn't mean nosy. I really did mean perceptive because that look you saw…you weren't wrong."

"So he does have a younger sibling?"

Shaking his head, he confuses me, but before I can ask, he gives me what I'm after.

"He had a sister."

"Had?"

As eager as I am to hear an explanation so I can understand, I'm not sure I'm ready for it. If Cole tells me that the reason he said it the way he did is because Luke's sister died somehow, I'm not sure I'll be able to keep it together.

I may not know much about them, but they're still human and I'm still me. If something happened to her, it's going to make my heart hurt.

"I don't know how comfortable I am talking about this."

Right. Of course. I can't believe I didn't think about that before I asked.

Cole barely knows me so there's no reason for him to tell me his best friends business.

"Forget I said anything. I was just curious and wanted to make conversation."

"No, Reagan. It's fine."

"It's obviously not fine if the second I ask, you avert your eyes."

Noticing the way he's been acting since we sat down was supposed to be for me only. It was never supposed to be something I blurted out, but now that I have and it's earned me what I'm after with his eyes back on me, there's no going back.

"See what I mean?" He sighs.

"Now I'm confused."

"I used the wrong word I think, but the idea is still the same. What I meant to say was you're observant."

This time, the warmth I experienced the last time doesn't happen. It's because this time, it's less compliment and more fact. I *am* observant and what I'm observing now tells me that this guy I agreed to go out with, the one that turns me to jelly when he looks at me, is hiding something.

Walker spending our entire relationship hiding the fact that he was seeing other girls behind my back has to be the last time it happens. I can't repeat the same mistake with someone else.

Even someone with eyes as soulful as Cole's.

"This was a mistake."

Pushing the chair back from the table and standing, his hand comes out again. This time wrapping itself tight around my wrist, locking me in place.

"You make me nervous." He admits and despite the reservations I have about him, I make no move to pull away or distance myself.

"Luke's sister was kidnapped seventeen years ago. Never heard from again. She was his entire world and he lost her. He hasn't been the same since. When he saw Jake with you earlier, it set him off."

My body reacting the way I expected about Luke's sister, my heart aching before dropping deep into my stomach, my next words should be of the sympathetic variety. What comes out is anything but.

"I make you nervous?"

Chuckling under his breath and running a hand through his hair nervously, he raises his eyes to meet mine and motions to the table.

"Yes Reagan, and if you sit back down I'll explain why."

"Every excruciating detail?"

"Every single one. Even some that are downright gruesome."

"Well in that case," Sliding myself back down once he releases the hold on my wrist, I bring the warm mug to my lips and take a long swallow, letting the warmth coat my throat before placing it back on the table. "I'm all ears."

"This is gonna sound crazy," he begins, again shoving another rough hand through his hair before looking from the table to the front of the coffee shop and back. "But being around you makes me nervous because I feel like I've known you a lot longer than a day. A few hours really. Reagan, I feel like we've known each other forever."

Just remember you asked for this. I admonish myself as soon as his admission falls. He's using lines, and judging by how quickly I sat back down, they're working.

He's wrong, though. There's nothing crazy about what he said. The only thing crazy here is how pathetic I am for falling into the trap to begin with. For that split second where his words actually penetrated my heart and I believed he meant them.

Cole is just like Walker.

"Say something, please?"

"Okay. How many times has that line actually worked?"

"Rea—" He starts, but before he can even get my name completely out I hold up my hand to stop him.

"No, Cole. I get it. For a second there you had me going. I thought there might be something different about you. Something that set you apart from my ex and everyone else. I was wrong. Everything you said at the bar and since we've been here, it's all been a game."

"I saw you in the bank and you looked like someone I knew a long time ago." He blurts and before I can stop him, he continues. "It was your eyes that did it first, but then when you were standing with the old guy, it was the expression on your face. Your eyes were just so damn expressive and full of feeling. I knew I had to approach you after that."

"Cole..."

"No, Reagan. I heard you out, now please hear me. This isn't a game. I can't explain why it feels like I've known you forever, I just know that I do. It makes me want to tell you things. Everything really. That's dangerous."

"Dangerous? Why?"

"Because there's a very real something buried deep inside my everything that has the power to destroy you."

"That's a tad bit overdramatic, don't you think?"

Reaching across the table, taking my hands in his and squeezing softly, I wait with baited breath for what he's about to say next. Knowing deep down that whatever it is will determine what happens next. Where we go from here. Tonight *and* in the future.

The truth of it is written in his eyes. Determined yet erratic.

"My life is built on secrets. Keeping and uncovering. It's only a matter of time before that twists itself around the innocence that is being here with you, and turns it bad."

"Cole, it's just hot chocolate."

"You're wrong and I can prove it."

Shoving the chair back from the table, he stands and before I've even had the chance to blink he's standing beside me, the scent of his cologne mixed with perspiration hovering and his hand gripping my face. His physical presence making me dizzy, I grip the arm of the chair tighter, sucking in a breath and releasing it.

The second it falls and I'm readying myself to do it again, I feel the brush of his nose against my skin as his lips, rough and ready, press down hard on mine.

My only thought as his hands snake around and pull me in closer, one of complete and total surrender, but also truth.

It's a whole lot more than hot chocolate.

Cole

Well, this just got a lot more complicated.

Who am I kidding? This has been complicated from the second I landed at the bank this morning. It's just gotten worse since I turned into a selfish bastard.

It's a damn good thing I left Luke back at the bar. No doubt with what just happened, I'd be a dead man if he was anywhere around.

Truth is, I'm already a dead man. It's just a matter of when.

When my lips touched hers I'd been hoping she wouldn't taste as sweet as she is, but just like I was wrong about this not being personal, I was wrong about that too.

Because god damnit. She's sweet. Sinfully sweet.

Lips like candy, skin so god damned soft, and those fucking eyes. One glance across the table between us and it was like I was seven all over again.

She's still the sunlight and right now, with the hunger I have for her, she's threatening to burn me alive with her rays.

Stepping back from the kiss, removing myself from whatever it is about her that seems to put me under a spell, I attempt to find my breath. The breath she'd stolen the second she kissed me back.

Damn straight this wasn't just hot chocolate. I knew it. No one that can nip someone's lip the way she did mine is in it for something as innocent as hot chocolate.

Reagan is reacting the same way I am and it's so damn wrong that it's right.

"So…I'm thinking I was wrong." She admits, her cheeks now that I'm able to look at her again without combusting, darker than they'd been when I grabbed her face and marked her lips as mine.

"It's a little bit more than hot chocolate is what you're saying."

Yeah, I was right. She's flushing completely red now. Except I'm not sure if it's from the embarrassment or from the way we seem to react to each other. My vote wants to be the latter.

Because fuck, she makes me hot.

"I probably didn't need to be kissed in order to admit it," she continues. "But yes, it's something more."

God, I was right earlier. Whatever is driving me here, what connects us—other than the obvious—it's making me want to tell her everything. Say screw it to playing this one close to the vest and just lay it all at her feet.

Even if doing it that way means she walks away.

I don't know how much longer I can play this like I'm some private dick here after a deadbeat dad. A lie here or there when I was getting a feel for her over coffee is one thing. Keeping it up for days afterward is another.

What the fuck was John thinking throwing me this case? Did he somehow know that when I saw Sunday again it would do this? Turn me into a sweaty palmed jerkoff that with the way her lips are pursed now, makes me want to shut the rest of the world away and have my way with her?

Chill Cole. What would Sam do in this situation?

My dad. Another set of memories I'd been flooded with on the way here from the bar. How we came to be the way we are. The way he plucked me from that group home like it was nothing and gave me the one thing that for seven years leading into it, I'd never had.

A family.

Sam Alexander gave me everything, and with the wayward road my thoughts are taking with Reagan, wanting nothing more than to strip her bare and taste every part of her, I'm shitting all over it and him.

Some son I am.

Maybe now that my heads off what just happened with her, I can channel him and save this moment and the memory he left when he passed. I owe him that.

Stepping away despite the urge that still runs thick in my veins to move closer, I slide my body back down into the chair and take a long swallow of the now lukewarm hot chocolate.

Time to think with your other head, Alexander.

"So now that we've established there's more going on here, where do we go now?"

"I'm not sure. Where can we go? Once you're done with your case you're going back to the city."

She's got a point. My cover story would kick me back home. Little does she know she's the case and with the way I'm feeling after that kiss, I'm even more invested than I was.

It's not all that crazy to be willing to relocate for a girl I've only really known a day, right?

"That wasn't supposed to upset you."

"Huh?"

"You're scowling."

Maybe her being so observant isn't a good thing after all. She shouldn't be able to see what happens when I'm in the middle of mentally kicking my own ass.

"It wasn't because of what you said. At least not entirely."

"Well, I've been told I'm a fantastic listener. I mean, I know we don't exactly know each other, but if there's something on your mind that's making you scowl, I'm willing to listen if you want to get it out."

Cue me spilling my guts in 3-2-1.

"I was thinking that it's not such a bad thing that I've gotta head back."

Great. Now I'm not the only one scowling.

"Shit Reagan, I'm sorry. That came out wrong."

"Are you sure about that? Seemed pretty honest to me."

"Do you remember what I said earlier? You don't want to get into this with me. My life is complicated. Just in the couple of hours I've been around you, I can see that you're the least complicated person I've ever met. You have a brother that loves you, a crazy best friend that seems like fun and a stable job. You don't hunt down cheating spouses, bail jumpers and criminals."

"So I'm too boring for you?"

Fuck. This is not going the way I want it to. I can't deny that her taking me the wrong way would make things a lot easier, but I'm not exactly trying to hurt her.

"No, okay? It's not that. If you were boring, I never would have asked you to do this with me. I just don't want to complicate your life."

For the first time since I sat down with her, there's not even the hint of a lie laced into my response. I'm being completely honest with her, and judging by the way her face seems to soften, the rift I'd created fading away, she knows it.

"My life isn't as black and white as you make it sound. It's just as complicated as yours. Try living in a house where you don't look or even act like anyone else there. Being misunderstood at every turn. I'm not musical like Jake, I'm not Super Woman like my mom and I don't have a mathematical mind like my dad. Hell, I'm not even outgoing and bouncy like Beth. I'm a loner who enjoys books and quiet."

She has no idea how enticing that is.

After years spent with Miranda. A woman that for all intent and purposes was a social butterfly. Letting her treat me like her little project, parading me around and attempting to get me to fit into situations I wanted no part of, Reagan is a breath of fresh air.

The speed she operates at is one I can definitely keep up with.

"Don't even get me started on Walker. Nothing in my life was ever as complicated as my time with him. I'm sure he would say differently, but being in a relationship with a guy and being one of many was definitely complicated."

Any mention of the ex-boyfriend makes my blood boil. I don't know if it's the attraction to her or my loyalty to who she really is that's causing it, but just like at the bar earlier, I feel the need to find this guy and end him.

"What if I said that anything you start with me would turn out worse than what you had with Walker?"

"Correct me if I'm wrong, but earlier at the bar you had multiple chances to hook up with any girl there. One of them being my best friend. You only had eyes for me."

She's got me there.

"You're right."

"Then you're already better than Walker. Maybe I'm naïve and too trusting, but I believe you when you say you're not playing a game, Cole. I might be wrong and live to regret it, but for once in my life I'd like to take my chances with whatever this is."

"Even if it turns ugly in the end?"

"In my experience, it's only ugly if you let it be."

After I broke up with Miranda, women were the last thing on my mind. I'd been put through the ringer with her both emotionally and financially. The last thing I wanted to do was set myself up for more of the same treatment by getting into something like that with someone new.

If I was going to do anything, I was gonna maintain control at all times and keep it as casual as possible. No actual dates and overnight stays. There was nothing that could be considered the beginning of a long term arrangement.

Something tells me Reagan isn't the casual type, though. If I agree to explore whatever this is between us, I know for a fact that when the truth does come out or I have to head back to the city, it's going to leave her affected in a long term way.

I'll be the asshole that caused her heart to break and that's something I don't feel right about. Don't even get me started on the ramifications for Luke. I will have fucked over his sister. I can't see our friendship surviving long after that.

The argument I'm staging in my head is perfect, but when I do finally speak again, it's not at all what I say.

There's a special place in hell for guys that act on their impulses the way I am now and I've officially got a place with my name on it reserved.

"I'm not ready to walk away, so you win, Reagan. Let's see where this goes."

Chapter Ten

Reagan

After the way the night ended with Cole, the way he brushed his lips across mine in a goodnight kiss with a promise to call me when he woke in the morning, the last thing I expect to deal with when I do turn over in bed in order to slam my fist down on the alarm clock is the smiling face of my mother.

"Mom?" I hoarsely whisper, rubbing at my eyes before turning over in the bed and stretching and sitting up. "What you are doing in here?"

"Well, I brought you breakfast." She says, picking a plate up off my nightstand and handing it over. "But that was only to butter you up."

"Butter me up for what?"

"Last night, I could have sworn I saw you wearing something you haven't since before you got with Walker."

Oh no. I knew she was awake when I got in last night, I mean, I'd said goodnight before heading up the stairs, but I had no idea I'd stood around long enough for her to get anything from me, least of all me wearing what I know she means is a smile.

Sure, I was smiling when I got in. After waving to Cole before stepping inside, I'd been grinning like an idiot, but the sliding doors to the living room weren't even open all the way. How could she have caught it?

"Do you want to tell me who's got you smiling like that?"

Not really. "It wasn't anyone specific, Mom. You know I went to watch Jake play. It was just a good night and I was happy when I got home."

"You know, it wasn't all that long ago I was your age and had young men interested in me. Some of them even making me smile the same way you did last night. Ones that even made my cheeks flush red the way yours are now. Are you sure it wasn't anyone specific?"

Damnit. Why am I such an open book?

"His name's Cole!" Jake calls through the half open door and my stomach rolls over. So much for attempting to lead her off the scent.

"Jerk!" I yell back and his laughter in response just annoys me even more.

"That's Jake!"

"Alright you two." Mom laughs despite her attempt to admonish us. "That's enough."

"Yeah sure. Tell him that." I grumble, for the first time in forever wishing my brother would just leave us alone.

"So, Cole huh?" She asks and I nod as my cheeks overheat.

"Where did you meet this Cole?"

Okay, this I can work with. Maybe I can give her enough of the truth that she'll accept it and move on. "Work."

"Customer or new employee?"

"Customer."

"If you met him at work, how does Jake know about him?"

If she just thought about it, she'd realize how Jake knows, but in the interest of getting this over with quickly, just in case Cole does make good on his parting words last night and calls, I spell it out.

"His friend came into town and they ended up at the bar last night. We watched Jake play and then went for hot chocolate."

"Wait. You left with him?"

"Yeah, Mom. I did what you've been telling me to do since Walker. I accepted a guy's invitation to go out for hot chocolate. Jake met him before we left, I had my phone on me and he was a perfect gentleman, so nothing to worry about. Hence the whole smile thing you're in here trying to dig up dirt on."

What I think but don't say is that not every second of our time together was innocent and he wasn't always the perfect gentleman. His rough lips pressed against mine when he was trying to prove his point was anything but gentlemanly.

But I sure did enjoy it.

Hearing the familiar squeal of my door when any weight is laid on it, I avert my eyes from my mom and ready to tell Jake off again for not giving us privacy, turn to meet the sobering eyes of my dad.

James, or as my mom calls him, Jimmy Carter. Named after the 39th President of the United States and using it to his advantage every day since in the business world. Also the man I love more than life itself.

"What's this I hear about a perfect gentleman? And what exactly does it have to do with the daughter that shouldn't be dating until she's forty?"

"Dad, please." I roll my eyes before turning back and silently pleading with my mom to end this. I might not have wanted to give her anything when I woke up, but when there's a choice between her and my dad, I'll gladly spill it all if she makes him go away.

This is the guy that the night of my first date with Walker, pulled out his 9 mm handgun in an attempt to scare him off. There's no way I'm telling him about Cole.

It's just too bad that my mom doesn't feel the same way.

"It appears as though our daughter met a young man at work yesterday and he took her out last night for hot chocolate."

"Does this 'nice young man' have a name?"

In other words, he's asking if Cole has a living will and insurance policy because if he even so much as looks at me and my dad finds out about it, he's going to be buried six feet under. I've been down this road before. For all of the things I adore about my dad, his overprotectiveness is a bit much.

"Cole. Though we were just getting around to a last name."

No way! She's selling me out. They're both staring at me now. Waiting patiently for me to give a last name so the second they leave the room they can check him out.

You'd think with the way they act sometimes that my dad actually was the President. I mean, I gotta figure with the way they background check everyone that comes in and out of our lives, we're far more important than we actually are.

"Alexander." I give up once I've had enough of their prying eyes. "His name is Cole Alexander and before you press me for more, he's not from around here. He's from the city."

"What city?" My dad probes and I growl under my breath. Here we go. The inquisition has begun.

"Toronto, but that's all you're getting. He's a really nice guy and I'd like to see him again. Which if I tell you anything else, won't happen."

"I have no idea what you mean, sweetheart."

Sure he doesn't. *I call bullshit.*

"So you didn't background check Adam Ramirez before he took me to the park when I was eight? You didn't do the same thing to Walker when you found out he was sniffing around?"

This gets him, but not in the way I was expecting. Where I thought he would look at least a little guilty at being caught, he's grinning. Like background checking someone at eight years old is the most natural thing in the world.

My dad is nuts.

"You are the love of my life apart from your mother. The day you came into my life everything changed. So if being a little proactive in terms of your safety is an issue, I'm sorry but I'm not changing. Your well-being will always be my top priority. It's what you do when you love someone."

"Sure it is."

"When you have a daughter of your own, I want you to watch the way her father reacts and come back and tell me if you feel the same way."

Okay. He's overbearing and kind of a tyrant when it comes to safety, but he's right. I don't even need to be settled with a kid to know that when I do have children, I'm going to be concerned about their safety too. Paranoid even.

I'm just not sure I'm going to be the handgun on the kitchen table level of paranoid.

That's all for James Carter.

With a look over to my mom, his face a mask, not giving away much about the way he feels minus the smile that doesn't quite meet his eyes, he sighs before focusing his attention back on me.

"When are you seeing this Cole Alexander from Toronto again?"

It's there and gone in an instant, but for a split second I swear his features give away his feelings on the subject. His lips pressed tightly together, the cold flash of what looks like anger in his eyes as he crosses his arms across his chest awaiting my answer.

That's weird.

Why would he be acting like that when my mom is wearing the same smile she walked in with? Usually when they're against something, they're on the same page. That's not what's happening here.

Why is my dad upset about this?

"He's supposed to call me later. So I guess I'll let you know when I know." I offer up while studying his facial tics, which I can now see in the grinding of his teeth causing his jaw to shift uncomfortably.

There is definitely something more going on here, but after the way last night went, there's no way I'm gonna sit here and try and figure it out. Not if it's something that in the end could ruin the buzz I went to bed with. Nothing is going to ruin the way I feel about Cole and what we shared and agreed to last night.

Not even my dad.

James

How did this happen?

I figured after coming across Samuel and John Alexander in the past and ridding myself and my family of their nosy and incessant digging into our lives, it would have been over and done with.

To find out now that there is yet another Alexander that has inserted himself into our lives, this time getting closer than either of his two family members by latching onto Reagan directly, blows the peaceful tranquility that has been our lives, apart.

Why wasn't I told about this? And just how long has this weasel been in our lives?

Picking up the phone and pounding my fingers down hard on the numbers, desperate for answers that damn sure better be at the ready, I wait as patiently as I can as the rings go in and a voice finally answers.

"James. What do I owe the pleasure of this call?"

"You screwed up Martin, and you have exactly two minutes to explain to me how that is, given the assurances you made me the last time we spoke."

Martin Francis is a gun for fire. A one-time junkie turned paid killer, who along with his daughter had been able to give us the gift we had spent ten years fighting to have.

Our Reagan.

What the rest of the world—including my family—doesn't know, but I have been living with every day since, is that our baby girl wasn't always Reagan.

Before she came to stay with us, she was known by another name. A name that first with Samuel and then with John Alexander, has consistently haunted us every step of the way.

Reagan was once Sunday Grayson.

The Grayson family hadn't been well to do, but they did have the love, respect and adoration of those in our small neighborhood. They were beloved. A true portrait of the picture perfect family. A son, Lucas, who was almost eight years Reagan's senior, a rock solid marriage, and then the little bundle known as Sunday to complete the angelic package.

Years we lived on that godforsaken street watching as everything was given to them without question, while Katherine and I struggled year after year to make ends meet. Fighting tooth and nail before my ventures took off to have the family that just never seemed to materialize.

It was never my intention to kidnap Sunday and make her ours, but as time wore on and the struggle to conceive a child began to take its toll on both our marriage and our finances, something died inside me and the plan formed.

Night after night I spent pacing the halls of our hollow house until I finally reached out to various business contacts. Ones that didn't exactly play above board, finalizing what would eventually take place exactly one year later.

Hallie Francis was a blessing in disguise. When attempting to figure out the best way to reach Sunday when the time came for us to take her, Hallie had been building up quite the following babysitting. Once we learned of that, she quickly became our greatest asset. It also helped that she happened to be the daughter of the crooked bastard I am now on the phone with.

"What the hell are you talking about, James? Did I not do as you requested and make his death look like an accident?"

John Alexander. The private investigator that came across us most recently, and the one that once we dug a little deeper and learned had been diagnosed with colon cancer, we'd had driven off the road and killed. Setting it all up to appear as though it happened naturally.

A family that for all of his touted accomplishments, Martin hadn't researched as deeply as he made it seem.

What other explanation can there be for Cole's emergence now?

"You handled John Alexander the way we discussed, but what you didn't inform me of was the other male in the family. What do you have to say for yourself, Martin?"

"Cole? He's some young punk Sam adopted. Dad was a convicted rapist and mom was a junkie. He was well on his way to playing hockey in the NHL the last time we looked into him so there was no need to inform you. Are you saying that something's changed?"

"You're damn right something's changed. He's here, Martin. He's been spending time with Reagan. So whatever or wherever you got your information, you were told wrong."

"Watch your tone. Just because you've used me in the past doesn't mean I have to sit here and take your abuse now. You want someone to be pissed at, look in the mirror. All the information I obtained back then I got from your people."

Grunting over the line, knowing he speaks the truth but not wanting to admit that it was my careless mistake that may have caused what is happening now, I huff out a large exhale and forge ahead.

I've gotten the answers I'm after. The threat to my family is real. Now all that needs to be done is that threat to be eradicated. And there is no one better to accomplish that then Martin Francis.

"I need you again."

"I was wondering when you were going to get around to that." He smugly answers. Judging from the thickness of it in his tone, I can almost hear the smirk on his face along with it.

"If you make it quick, I'll pay you double what I did for the previous job."

"After everything we've done, what makes you think I would help you again?"

"Because if this comes out, it's not only my ass on the line, Martin. I will name names and I will bring all of you down with me. Even your daughter, whose name has been clean as a whistle for seventeen years despite being just as dirty as her fathers."

"You watch your fucking tone, asshole. No one speaks about Hallie that way."

"Then get over whatever reservations you have and do your damn job, Martin. We both know you want to. There is no one in the world that garners enjoyment from the chase and kill like you."

I didn't go into this thing blind when it came to him. Not only do I have him for everything he did with the Alexander family and the abduction of Sunday Grayson, but I also know what he does to the bodies of those he kills after he's done it.

When one says Martin Francis is a perverse killer, it's not just a label. It's a fact. It's the only way he's been able to get a sexual release in years. It's also what he will burn for long after he pays for the crimes he's helped me with.

I may be a sadistic son of a bitch with what I put in motion all those years ago, but Martin is definitely the monster hiding under your bed.

"Triple the last job?" he asks and I smile. I knew it. If there's one thing that always gets through to assholes like this, its money.

"Yes. I want proof of death and a promise that Reagan will never know about this. In fact, I want her as far away from this as possible. If his death blows back on her at all, you will pay."

Not waiting for a response, I slam the phone down and pull up the file in front of me. The one that details not only the Grayson family, but any and all other families attached to them. Most notably, the Alexander's.

When all of this started, it was driven by the need to fulfill Katherine's dream of having a family. It was never supposed to be about murder, but here we are.

As much as I wish I could regret the decision I made back then, I can't. Getting to wake up each day and be met with the smiling face of my daughter has demolished all of the reservations I may have had.

No matter how much blood is spilled in the process, Reagan will never find out she's Sunday Grayson. Even if it means killing the first person to make her smile in months.

I will never let him take her away from me.

Chapter Eleven

Cole

No. This can't be right.

I'm tired. When I read this in the morning, I'll see how wrong I am and think I'm the world's biggest idiot.

Sitting in front of me clear as day in one of John's old journals is an entry related to Sunday Grayson. At least who Sunday became.

Reagan Carter.

He found her.

John knew she was alive for five years and never did a damn thing about it.

This should put to rest all the questions I had when I arrived in Alliston about why, with as good a PI as he'd always been, Sunday never made it home. I could make excuses for my father because towards the end he was losing significant chunks of time and had never been one to write all of his thoughts down, but there's no excuse for John. Not one thing I can use to justify this in my mind.

For five years, the one person on the planet that Xander and Allison put their faith in to bring their missing daughter home had known about her and kept it under wraps.

You're no better. You've known who she is and done nothing with it other than making sure Luke knew.

No way. I'm not like him. I will head back to Orangeville and tell her parents, but not until I get the chance to make sure Reagan knows and has had time to adjust first.

Who are you kidding? You're not keeping this secret to protect her. You're doing it to protect you.

"Ugh. Just shut the fuck up, would you?"

Shaking off the crazy, thankful no one's around to hear me as I begin my slow descent into madness, I'm thrown back when Luke responds.

"Nice to see you too, sweetheart. Why am I shutting the fuck up again?"

"Not you. Me."

A look on my face, the truth buried in my eyes, or something about the rigidness of my body must give me away because in a matter of seconds he's throwing his body down onto the tattered bed I attempted to sleep in the night before and glancing over all of John's journals I've got scattered everywhere.

"Okay, you wanna tell me what's here that's got you rattled?"

"Five years, Luke. John knew about Sunday for *five fucking years* and never said a word. Not to me, you or from what I can tell, the police."

"Wait," Luke leans over and begins really scanning what's in front of him. "Repeat that again. I'm not sure I heard you right. Are you saying that he knew the entire time he was on the case?"

As much as it pains me to admit it, knowing that even though we're not blood, he was still family, I nod before tossing the book over.

"That's exactly what I'm saying and just so I know I'm not going crazy, it's all detailed in there. Read it and tell me I'm wrong."

Where I expect his eyes to move over the pages quickly, he picks up the book and begins reading aloud, only confirming as he does, exactly what I'd been hoping had been a fabrication.

"After speaking with Detective Travers, I made my way to Alliston and kept watch over the Carter home. Catching sight of Sunday Grayson, renamed Reagan Carter, as she played happily with a young sandy haired boy on the lawn of what appears to be her familial home."

Slamming the book closed, we sit in silence, drowning in our disbelief until after a few minutes he breaks it.

"Holy fuck, Cole. He knew. He knew the entire time."

"At least the entire time he was contracted to find her, yeah."

"This doesn't make any sense. I remember the way he was when he would meet with my dad. All business. He was determined to find her and bring her home. Why would he find her and keep it a secret?"

"I don't know, but I'm half tempted to dig him up and beat the shit out of him until I get answers."

"Dead men tell no tales, Cole."

"Dead men haven't met me yet."

Grimacing, he looks down to the book in his lap, studying it before speaking again. "There's something that doesn't add up."

I'm not sure how I know what Luke's about to say next, call it best friend ESP, or me questioning the same thing before I even realized he was back in the room but before he can even bring it up, I throw it back at him.

"Sam didn't know about this. The old man was a lot of things, but a willing participant in a kidnapping wasn't one of them."

"Agreed, but come on, Cole. Do you really think John orchestrated Sunday's kidnapping on his own?"

No actually, I don't believe that. It wasn't like I spent a lot of time around my uncle growing up, but the one thing I did catch onto quickly is that while a fantastic investigator when he needed to be, he wasn't the brightest bulb away from it.

What happened to Sunday seventeen years ago, it would have needed meticulous planning and that couldn't have come from John. Which means there are more pieces that still need to be uncovered before we can get the answers we need.

"John was an asshole, but he doesn't have the skill needed to plan and pull of something like this and keep it under wraps this long. He was only a small cog in a bigger machine."

"So where do we go from here?"

Until now, I didn't mind Luke being a part of this. In fact, knowing how losing his sister had changed him, I'd been alright with it because I knew it was helping. Now though, with everything I've come across in the last few minutes, it changes everything.

Luke can't be a part of this anymore. It's up to me alone to sort through the mess that for whatever reason my uncle found himself a part of. A fact that as soon as I open my mouth to speak and he begins shaking his head—another instance of our closeness at play—I know he doesn't like.

"We don't go anywhere, man. You need to go home and I need to dig deeper into this."

"You tell me my sister is alive after all these years, and that your family played a part in her being gone so long and you expect me to just walk away? You're kidding, right?"

"Not this time. Luke. I have no idea what's going on here, how deep this runs, or what I'm going to find when it's over. I'm not going to let you walk in blind."

"Good thing I'm not giving you a choice then, huh? This isn't some random chick I want you to get the goods on like when we were kids, Cole. It's Sunday. I'm not going to walk away. Not again."

"Luke—"

"No. Save it. It's a waste of breath. I've made up my mind."

"I can go over you here."

"Then do it, bro, because not only am I not walking away from Sunday again, I'm not letting you go it alone either. Family remember?"

As well as I know Luke, how often he's used that same argument against me when we were kids in order to settle me down or talk me down from the edge, I should have seen this coming. The one word that I'm powerless to fight against.

Family. And not just any family, but the only one I've got left.

"Okay renegade, but you follow my lead." Pausing the second he rolls his eyes and shaking my head before glancing back down toward the journals on the bed, I wait until he follows and lay out the only way this is going to go. "You want to get to the bottom of this, then we do it by my rules or we don't do it all. Deal?"

After a few seconds of silence where he mulls over what I've said, he nods. "We've got a deal, but when we find the person behind this, I want you to give me five minutes with him before we bring him in."

"Fine, but only if there's anything left when I'm through with them."

Chapter Twelve
1998

Cole

"Get in here, quick!" Luke hisses, grabbing and yanking me inside.

Pulling to a stop once we're in with the door shut firmly behind us, I do a quick scan of the house. He's been like this a lot lately, but there's something about the way he pulls me into the house, practically dragging me to the living room that feels different than other times.

Standing at the table in the middle of the room is Sunday, drool slipping out of the corners of her mouth as she flashes her gummy grin at us. Well, at Luke anyway. She always looks at him that way.

"Dude, do you see it?"

"See what? All I see is you about to jump out of your skin and Sunday standing at the table laughing at how stupid you look."

That's when it hits me. It's Sunday. What he's so eager to show me, it's this.

Sunday is standing.

"Holy shit."

"Right? We were just sitting here tossing blocks around and she stood up!"

I don't know a whole lot about babies, but this seems pretty huge. She's not even a year old yet. "Did you tell your Mom?"

Shaking his head, he motions to the stairs and explains why he hasn't filled Allison in on the very obvious milestone her baby reached.

"She was tired so I told her to go sleep for a bit."

"This is like her crawling again isn't it?"

Luke shakes his head but its total crap. When Sunday started crawling across the floor at four months, Luke had been the only one in the room to see it and instead of finding his parents and filling them in, he kept it secret until about a month later when she did it again for the entire family.

Her standing is a big deal for the family, but Luke and the way he is with her, it's like he wants to keep it just his.

With how rough he is when we're playing hockey with the guys or out screwing around at the park wrestling or fighting, I never expected to see him like this.

My best friend is a gigantic softie for his sister.

"I want to try something. Will you help?" he interrupts my thoughts and I nod. I'm not sure what the hell it is he wants to do, but since I've never not gone along with something he wanted to do, I'm not about to start now.

"Stand behind her. I'm gonna sit on the sofa and see if she'll walk to me. If it looks like she's gonna fall, you can catch her."

I don't know how I feel about this. It's one thing to not fill Luke's parents in on her standing up. She'll probably do that again for them real soon, but to try this? What if something goes wrong and even with me behind her, she falls anyway? I would never be able to live with myself if anything happened to her. Especially if it's something I could have prevented.

She's family. Same as Luke.

"I'm not sure."

"Come on, Cole! Don't you wanna see if she'll walk?"

Not really, but I don't tell him that. He seems so excited.

"Why is this so important anyway? Isn't it enough that she's standing?"

"It's important because I think she wants to do it. Look at her. She likes making her big brother lose his mind."

I can't deny that. With the grin she's had on her face since I showed up, she definitely seems to feed off Luke's excitement. I'm finding it hard not to feed off it myself with as eager as he seems to be.

"Fine, but if she doesn't seem like she wants to, we stop."

"I'm not gonna force her, Cole."

Positioning myself behind her, my eyes never once wavering off her tiny form as Luke gets comfortable on the sofa, I listen as he smacks his lap and calls out to her. When her head lifts and turns toward the sound, I catch her lips part more, her mouth—complete with drool—opening wider as she grins at him.

Believing myself ready for whatever happens next, I'm blown away when smacking his lap again and calling her name, one of her chubby little legs starts shifting.

Holy shit! She's actually doing it.

Wobbly on her feet but determined, her foot moves, followed up by another and moving with her, it isn't long before she's bumping her body against the sofa and resting her head in Luke's lap while he yells at the top of his lungs.

"She did it! She freaking did it! Cole! Did you see that?"

Before I have a chance to respond, the sound of heavy footsteps on the stairs filters in and Allison Grayson comes running into the room, her hair dishevelled, looking like she's been pulled straight out of a dead sleep.

"What's going on? Is Sunday okay? Luke, if you were having issues why didn't you wake me? Why did I even sleep to begin with?"

"Sunday walked, Mom!" Luke exclaims in way of explanation, which once done, causes Allison to be able to breathe again. "Cole, call to her. I want to see if she'll do it again."

With Allison looking over us, I'm even more nervous about doing this now, but there's something in the way her eyes soften and she motions toward me with a slight smile that puts me at ease. If she's okay with this then maybe I don't have to worry about it after all.

Doing what Luke said, I call out to Sunday, and even though it takes a few seconds for her to respond, she finally lifts her head from her brother's lap and looks toward me. Mirroring the smile she's now throwing my way, I follow Luke's earlier motion and tap against my legs with my hands, willing her to do it again so that Allison can see.

Turning her body toward the coffee table, she slaps her hand down hard before stepping around the rim of it, stepping out at the exact moment I step forward to catch her in case she falls, our movements synchronized so well that her hand immediately slaps down into my bigger one.

Picking one of my fingers, she wraps her hand tightly around it until she's taking the two tentative and wobbly steps needed to rest safely against my legs.

Once her body has connected to mine, I lift her in my arms and just like Luke before me, start spinning her around as she laughs, whispering in her ear as I do.

"You're amazing, Ray."

Shortly after that first day when I held her, fueled by what Luke told me about the meaning of Sunday, I started calling her Ray. It's stuck ever since. And with the way she's smiling up at me now that I've stopped moving, her eyes lit up and dancing, she's owning the name the only way she can. Warming me the same way she did the first time.

Sunday is my ray and with the way it feels in my chest being a part of what she did today, she always will be.

*****

"Earth to Cole!"

Shaking off the memory and looking up to my best friends questioning gaze, his kid features now morphing into his fully grown ones, I smile weakly.

Luke isn't the only one that my zoning out affected. I'm getting the same concerned look from Bethany and Reagan too.

Shit.

"You alright, buddy?"

"Yeah, sorry."

Acutely aware of how close Reagan is to me as her body turns into mine and she leans in to speak, I do my best to swallow down the image of her holding onto my hand when we were kids and focus on the woman she's become.

The woman she has no idea she really is.

"Are you thinking about your case again?"

"Not this time, no." I admit easily. "I just remembered something that happened a long time ago and I guess I got lost in it for a bit. I'm sorry."

Slipping her hand under the table until it's resting on mine, she squeezes softly and smiles. She's subtle about it, but she's letting me know she understands. Something that judging from the look I get from Luke when I finally pull my attention away from her, I see I'm not going to get from him.

"Where the hell did you go just now? You went catatonic."

"Maybe if you weren't so damn boring that wouldn't have happened." I attempt to redirect, which only earns me a raised eyebrow from both Luke and Bethany.

"Considering the glossed over look in your eye that appeared after Reagan spoke, I'm pretty sure you just called her the boring one."

"Never." I hurriedly respond as I squeeze her hand back. For some reason, I can't let her think that what just happened has anything to do with the bullshit line I used on Luke.

"Well, now that you seem to be among the living again, I was just telling the girls that the movies about to start. So if we wanna get decent seats, we should probably get up there."

Wanting to spend time with her, claiming it's for work purposes but knowing that it's just about needing to be around her, a couple of days after the hot chocolate date we'd come up with the movie idea. No pressure for anything past the four of us sitting together in a darkened theatre and watching something that it seems none of us has had the time to see until now.

It was the perfect way to get to know the woman she's become. At least until she started talking about something she went through as a child with Jake and my mind instantly went back to 1998.

"Yeah, okay. Just give me a sec. I'm gonna hit the head first."

Slipping my hand out from under hers and hating the separation the second the contact is broken, I break off and head off in search of the washroom.

I don't even have to go, but considering what just happened, I need the peace and quiet to pull myself together before I put on another performance for everyone.

If the memories are gonna haunt me when I'm with her, the same way they do when we're apart, it's not going to be much longer before everything comes spilling out. That can't happen. Not until I've got more answers than questions. When I do sit her down and blow her world wide open, I want to be sure I have as much information as possible and right now that's not the case.

Pushing my way into a stall and throwing my body down onto the seat I close my eyes and exhale.

It takes a few minutes, but knowing he wouldn't be able to leave it alone, I hear the door open and footsteps as they make their way down the line of stalls, the sound ceasing as he pulls to a stop right outside my hiding spot.

"What the fuck is going on, Cole?" Luke's voice booms through the door and I debate whether to answer. I knew that shit I pulled out there wasn't gonna fly with him, but I still held out hope that he'd at least let it go until we were alone later. I'm already having a hard enough time dealing with the influx of feeling that came from remembering Sunday. I don't need Luke's pressure on top of it.

Flipping the lock on the door, I stand and slip my way out heading for the sink, completely ignoring him altogether until I've cupped a handful of water in my hands and splashed it over my face.

"I remembered something when we were out there. She said something about her and Jake and it triggered a memory. Something I haven't thought about in years. I needed to get the fuck out of there before I said something I won't be able to take back."

"You mean like telling her the truth?"

"Yeah. Exactly like that."

"I get that this might not be the best place to do it, but would it be such a bad thing if you did tell her the truth?"

"No, but not being able to give her the answers I know she's gonna want? That's not good."

"What did you remember?" Luke changes the subject and after going through the motions showering my face with a handful of water again, I turn and lean back against the sink and let it all out.

"I remembered the day she walked."

"She was ten months old." Luke says as the smallest trace of a smile begins to form. "I was so blown away that she was standing, but when she walked, I'm pretty sure my head exploded."

"Yeah, it was pretty amazing."

"So what did she say that triggered it?"

"I don't even know, man. It was just, one minute she was talking and the next we were back in your house and she was standing and walking to us."

"Well this complicates shit."

"No kidding. I already knew this was going to be different than other cases I worked before, but I wasn't expecting to get hit with the memories again. I thought I drank them all away and replaced them with my own."

"You can't replace Sunday."

"I'm starting to see that."

Leaning back against the counter, he releases a heavy sigh of his own, his eyes crossing the room, seemingly lost in memories of his own as the room falls silent. So much for being able to keep this shit under control. With each passing day I spend here, especially with her, it seems like I'm fighting a losing battle with it. I can't control anything.

Not my body's response to her, not the memories, or even the time I need to put everything together before I tell her the truth. It's as if everything has a mind of its own and I'm just a player in the movie that's going to go on with or without my input.

"You gonna be okay to go back out there?"

"Yeah man, I got this. I just needed a second to set shit straight."

"And you're straight now?"

I'm as straight as I'm going to get. I know the minute I'm back out there with her again and she looks at me with those doe eyes of hers, every emotion she feels on display for the world to see, I'm going to be at risk for falling under her spell, but I can keep it together.

I have to. I don't have any other choice.

What I do have a choice about though, is how fast I get these answers and move on from this. With the way this night is already going, it's something I know I need to deal with the second we drop the girls back home after the movie. I already want Sunday more than I should. If I don't want to risk falling into something I can't come back from, solving this case has to take precedence.

"Yeah, I'm good. Let's go."

What I want to tell Luke and can't is that what I really want is out there waiting for me in the theatre and I don't think that even when the truth comes out, with the connection between us, I'll be able to walk away.

The imprint of Sunday is going to be with me forever.

Reagan

Its times like these that make me wonder why Bethany and I are such good friends.

How two people that are seemingly so different could be as close as we've become over the years. How I could consider this boisterous, bouncy, ready to party at a moment's notice girl, my best friend.

No sooner does the movie end and we all make our way out to the parking lot than she's suggesting the night doesn't have to end and we can continue it down the road at this club she's been dying to go to.

When for the majority of your life your idea of a good time is staying inside, the idea of going to a club is frightening. I'm not that girl. I tried to be a few times when I dated Walker, but it never worked. I can't be the party girl like Beth and I can't be the rocker like Jake. All I can be is me and well, I'm the one who runs on a much slower speed.

But we do agree on one thing and it's what inevitably gets me to agree to the club.

I don't want the night to end. At least not until I've gotten a chance to be alone with Cole and make sense of what happened earlier that seemed to change him.

I'd mentioned it at the table before we even entered the theatre, thinking that the way he zoned out on all of us was due to the reason he was in town to begin with. When he blew that off, saying that he got lost in the past when I was talking, I wanted to probe more, but he'd taken off before I could.

"Rea, I need a favor." Bethany interrupts leaning in, bumping her body against mine and keeping her voice low so the guys can't hear.

"I already said we could go check out the club. What more could you want?"

"For you to lighten up for one thing. You were the one that said after you broke up with Walker that you wanted to try new things. This is me giving you the chance to do that. Besides, with the way you two keep looking at each other," she grins slyly while motioning to Cole. "The dance floor seems like the best place in the world for you to release some of the sexual tension."

"You're delusional. We're not looking at each other like anything and there's definitely no tension. Sexual or otherwise."

"Who are you trying to sell that line of bullshit to? I've been watching the two of you all night. You talk a good game, and you even try your best not to make it obvious, but you can't fool me. You two were setting the theatre on fire."

You see? This is what I'm talking about. Because Bethany is so free and open with her sexuality, not shying away from attraction and the way it makes her feel physically like me, it's a miracle we're friends.

"I think you're reading too much into a few looks, Bethy."

"And I think you're in denial about what that man is doing to your nether regions."

"Nether regions? Really?"

"Tell me I'm wrong and I'll drop it."

Redirecting her, wanting to get off this topic before she gets what she's after and I'm admitting just how affected I am being around Cole, I try to put her back on her original topic. Which once I do, earns me a knowing smirk.

"What was the favor?"

"Since we both drove here separately, I was thinking you could drive over with Cole and I'd shack up with Luke in my car."

With the dance in her eyes and the grin on her face, I can tell that wanting me to drive over with Cole is less about wanting to help me out and more to do with the way her and Luke have been practically attached at the hip since we all got here.

Bethany isn't exactly known for sticking with one guy. Not knowing much about Luke means he could be just what she needs, but this is where our friendship comes in handy because while she's taking stock of the way I am with Cole, I've been doing the same with her all night.

She has this look in her eye when she interacts with Cole's friend and I can tell it's about more than just getting him into bed. Bethany is searching for something. Something that lasts more than one night and knowing what Cole is here to do and Luke by association, I'm worried she's only going to get hurt in the end.

The same way I will.

"Come on, Reagan. Be my wingman."

"Don't you mean wing woman?"

"Whatever. Please just say you'll go with Cole?"

"Fine, but Bethy. Be careful."

"I'm pretty sure I should be the one saying that to you, but if it means you not turning into my mom, fine. I'll be careful."

Quickly embracing me, I watch her bounce over to where the guys are and releasing a heavy sigh, gearing myself up for what's about to come next, I follow her over. Pausing beside Cole and attempting to explain the change in plans.

"Bethany wants me to ride over with you. Is that alright?"

"Best idea I've heard all night."

Not wanting to read too much into his words, but completely unable to ignore the flutter taking place in my chest, I smile softly as he motions across the parking lot where his car waits in the distance.

"Do you know where this club is?"

"Yeah."

"Have you been there before?"

"I drink water in bars and spend the majority of my nights falling asleep with a book on my face, Cole. Pretty safe bet I've never been."

Immediately regretting the raw edge to my voice along with the admission of just how boring and singular my life is, I clear my throat and attempt to rectify the damage.

"Sorry. I'm just not really into the whole club scene."

"Well that makes two of us."

Slipping his hand down and into mine, locking our fingers together, he lifts them in a wave to Luke and Bethany as we start crossing the parking lot silently.

Using the quiet to attempt to settle the quickened pace of my pulse when his hand made contact with mine and getting it under control, I pick up the conversation where we left off.

"So if neither of us is into this, why are we doing it again?"

"To placate our friends?"

He's got a point, but there's a little more to it for me and I don't waste any time filling him in. Something that with as unfamiliar as Cole is to me, is surprisingly easy. It's astounding how in such a short period of time I feel as though I could tell him everything about me and he'd just get it.

"When Walker cheated on me, one of the things he said to me after we broke up was that I was boring. That the reason he did it was because I wasn't into the same things as him. All I wanted to do was spend time together alone, away from the world. I didn't want to live. Bethany said that was shit when I told her, but I took it to heart."

"She said its shit because it is, Reagan. Were you someone different before you two got together?"

"No." I say as we reach the car and he slips the key into the lock on my door before making his way around to his own.

"Then he knew what he was getting into when he agreed to date you. He's full of shit. He just wanted to make you feel bad for him being a prick."

Watching as his body lowers and his head ducks down into the car, I slide the door open and slip inside, letting his words resonate with me as I do. The rightness of them, yet words that up until recently I didn't believe in.

Despite my best attempt at not letting Walker get to me or leave me with scars, he did.

Which only serves to make me believe that I was as weak as everyone claimed I was. If I was stronger, his words wouldn't have gotten to me and turned me into even more of a recluse.

Turning the key in the ignition and tapping his fingers on the dash a few times, his gaze lingering outside to the cars moving around us, he exhales before turning in his seat and leveling me with an apologetic look. .

"Reagan, that didn't come out right. There's no excuse for cheating. No reason anyone can give for doing it. I wasn't trying to say that if you were different before you two dated that he had the right. I'm sorry."

"I understood what you were saying, Cole. It's okay."

There's something about my acceptance that seems to give him what he needs in order to relax, his body settling comfortably into the seat and his eyes softening. Reacting to someone this way is so new, but I like the way he seems to respond to me. His eyes, the deep soul piercing blue that I've been haunted by every day since we met, are so much more moving when they're settled.

"Reagan," he speaks softly, his eyes never wavering or breaking away, studying mine. "Can I be brutally honest with you?"

"Of course."

"I don't want to go to this club tonight. I would much rather drive someplace quiet and maybe take a walk and talk. The idea of going there, even for Luke, makes me sick. It's not my scene. I've been to enough bars and clubs to know what happens at them and I'm over it."

He wants to take me someplace secluded so we can talk?

"Maybe we can reach a compromise."

"What do you suggest?"

"We go and do this for our friends, since I'm pretty sure you've caught on that tonight is more about them than us. We dance to a few songs and tell them we're taking off. Then we can do what you want."

"That night at the bar you said that your family wanted you to climb out of your shell more. Is that why you agreed to this club thing? Or are you really doing it for Beth and Luke?"

"A little of both. I told you I bought into what Walker said. I took it to heart and made myself a promise that I would step out of my comfort zone more in the future. Try new things and meet new people. It's not exactly easy and most times I don't want to do it, but I feel like for myself, I need to."

"Well in that case, dancing it is. I only have one request before I drive over."

"Name it."

"You save the last dance for me."

I wish I could work up the nerve to tell him that I'd already planned on fulfilling that request long before it was made. The gravitational pull I feel toward Cole is so strong that even if I was given other options, his arms around me and our bodies pressed together on the dance floor is the only one I'd want. The only one I'd choose.

I'm going to break out of my shell and do something different, be a little wild for a change, and Cole Alexander is exactly who I'm going to do it with.

"Deal."

Cole

If I didn't know that for the last twenty minutes she's been drinking water, I'd start questioning how many drinks her friend had been slipping her when I wasn't looking.

When we were in the car and she was leaning back in the seat with her eyes closed with a playful smile on her face, she looked at home. Serene. Peaceful, but most of all happy. When we entered the club, though, and were practically unable to make a move without bumping into a body covered in what looked like glitter and sweat, it was like she morphed into a completely different person.

If her words earlier mean anything, she's putting on this show for herself as much as she is her friend. That jerkoff she dated getting to her in a way I hate and making her think that the way she is wasn't good enough.

Bethany had dragged her onto the dance floor almost the second we broke away from the sea of bodies and instead of following her the way I expected him to, Luke made a beeline straight for the circular bar on the opposite side, slapping down a ten and having a beer passed over to him. Where he's been ever since.

"Looks like she's got a bit of Grayson in her after all." Luke acknowledges as I study the way she's dancing with Bethany.

"Are we talking about your mom? You forget I grew up with you, Luke. You hate dancing."

"I do, but at least I can move. Unlike you."

"Give me a good enough reason to move and I might actually do it."

"You're telling me that she's not a good enough reason?" He laughs, motioning out to where the girls are, his insinuation not lost on me.

As hard as I'm trying to not be completely fucking obvious with the way Reagan is affecting me, him commenting on it proves I'm not doing it well enough. It's not gonna be long before he ends up pulling me aside and kicking the shit out of me for looking at her like she's a Sunday freaking brunch.

Sunday.

Fuck.

All I wanted was a few hours where I didn't have to be reminded of who she is and what we're really here to do and of course, my mind can't co-operate.

Maybe she has the right idea after all. I'm sick of being the Cole that's gotta guard every damn thing he's doing. Maybe it's time I break out of my own shell and become someone else.

Become the guy that heads out onto that floor, pulls Reagan away from her friend and makes sure it's my body she's rubbing hers all over.

"I'll be right back." I say to Luke and when his only response is an eyebrow raise and a smirk, I turn and brave the crowd. My eyes never once leaving the girl on the dance floor that even now with the decision made, I'm a little too eager to get my hands on.

"Took you long enough!" Bethany yells over the music when I finally make my way through and directly behind Reagan, my hands immediately slipping around her waist, body moving and pressing into her once the shock of my touch settles and she's caught on to who's behind her.

Responding in exactly the way I want her to, Beth laughs before slipping away from us, interacting with the other bodies that under the lights are glistening in color.

As Reagan begins to move her hips, grinding them back into my crotch, my body temperature rises as I try to restrain its natural response. The jeans I'm wearing not feeling like such a good investment anymore, as what was first her hips moving into me is now the grind of her ass.

My hands never leaving her hips, I move with her, my reaction pushed to the back of my mind as I fall victim to the feel of her skin against mine. Lights beating down on us, bodies moving so closely they're boxing us in and our breath heavy and laden with sexual desire, I drown in the scent of her body and the sensual way she moves.

Focusing so intently on the way my body seems to be responding, I don't feel her body shift and turn until it's too late and I'm forced to react to her breasts pressing against my chest and her lips meeting mine.

The moment no longer a dance meant for the rest of the world but an explosion of what I know has been the fire building between us for days.

We're not Reagan and Cole anymore. We're what becomes of those two people when their backs are against the wall and they're forced to give into their most basic and carnal needs.

Hunger. Desire and an intense, soul crushing ache to connect.

Under the bass of the music it's hard to make out in sound, but with the vibration against my lips, there's no denying that she's moaning, which only serves to push me deeper. Crave the feel of her even more.

My dick pressing hard into my jeans, threatening any moment to break free, I tip the scales in my favor, pressing my lips back onto hers hard, gripping their softness with my teeth and tugging, feeding into her moan with a growl of my own. One she responds to as her hands fall from their place around my back, sliding seductively around until she's rubbing the beast demanding to be freed.

Alarm bells are going off in my head as we continue our delicious dance. The grip she has on my jeans making me come unglued, overpowering the need I have to put the brakes on before I take this further than either of us are ready for.

Sliding my hand up her back and behind her neck, overcome with the need to taste more of soft flesh with my tongue, I tip her back and give in. Sucking on her neck until the taste of her is all I'm able to experience.

If this has to end, and the rational part of my brain is fighting to make me believe it does, then before it happens I want to take something to remember it by.

Remember her by.

I want the taste of her with me forever.

"Cole…"

"Reagan," I growl as she presses closer. The contact threatening to unravel me. "You're so fucking sweet."

Running a hand over the mark I've managed to leave on her now sweat laden skin, I give into the need to taste her again by pressing my lips softly back over the spot. Tasting her again before righting us. Our eyes locking on one another once we're standing our want and needs no longer hidden, but on display for the other to see.

My want to take her right where we stand and her need to let me.

Gripping my hand on her ass, while slipping the other around her back, I lift her, taking the hand off her ass long enough to curl her legs securely around my back.

"I'm not done with you, but we *are* done giving a free show. Whatever happens next we need privacy for."

Resting her face into my neck, her hot breath on my skin driving me forward faster as the need to get her alone to finish what she started overpowers all of my normal defenses.

When I've been with women in the past, it was all mechanical. A need to release what my body naturally couldn't control. With Reagan it's going to be more than that. There is nothing at all mechanical about the way my body craves hers. Wants to be nestled in between her legs as I'm tasting what I just know is going to be Heaven.

Catching Bethany's eye once I finally make it out of the throng, thankful in the moment that it's her and not Luke, I nod toward the door of the club and she grins before motioning with her hands for me to go.

Her acceptance all I need to guide me out of the club and into the parking lot, where I take another taste of her lips before finally placing her down on the ground.

Our breath ragged, clothes drenched in the sweat created from our dance, and our flushed faces as we stare silently at one another, all we have in the moment.

"We're leaving." I decide and feeling my common sense beginning to kick in again, I give her time to tell me otherwise. When after a few seconds no response comes but her eyes becoming heavier, I slip my hand into hers and head for the car.

So much for enjoying a few dances together and going for a walk.

At the rate we're going, I'm not going to be able to hold myself back from getting her into the backseat, stripping her bare and filling her the way I wanted to on the dance floor.

And something tells me that if that happens, she won't be the only one destroyed when the truth comes out. Because there's no way with the way I feel that I'll be able to walk away unscathed.

In just a few days she's tattooed herself on my soul.

I'm never going to recover from this.

Stopping at the door and preparing to unlock it, she slips her hand from mine and swipes the keys out of my hands, grinning wickedly as she pockets them and blocking me as my hand immediately reaches out to take them back.

The control I'd managed to garner on the dance floor changing hands as she reaches out and runs her hand across my face, her smile still firmly in place before pushing me back against the passenger side door and kissing me again. This time using her tongue and running it across my lips before slipping it inside to taste mine.

"Take me home, Cole." She murmurs as she pulls away, her fingers lingering on her lips over the bare outline of where my stubble has left her raw and exposed.

"Are you sure?"

Great. Nice going, Cole. Now is the *perfect* time for your conscience to kick in.

"I'm sure. I'm done dancing and a certain someone promised me some quiet conversation."

Wait, what?

Giggling at my obvious confusion, she strokes my face with her hand again before placing the softest kiss I've ever received to the side of my lips.

"As hot as that was, and trust me, Cole, it was hot, I want what you were offering earlier more and I think I know the perfect place to go."

"Your house wasn't exactly what I had in mind."

"I said take me home. Not that we'd be going inside. Where I want to take you, it's someplace I've never taken anyone else. So I guess the question is, will you come with me?"

I'll go anywhere as long as you're there.

Despite knowing it's the wrong move and it's not something I can admit, it doesn't change the fact that it's the truth.

I would follow her anywhere. No questions asked.

"Lead the way."

Chapter Thirteen

Reagan

"Turn right onto Concession Road 7 when it breaks off and then follow the signs."

It's been awhile since I'd been there, but ever since my parents brought Jake and me when we were kids, I've made it a point to come back a few times a year on my own.

Because it's a provincial park, at any time I could be surrounded by a sea of people, but after spending so much time here I'd managed to find areas of seclusion. Ones where I could relax and block out the rest of the world.

I've always loved the water, and there's no better place to experience that then at Earl Rowe. I just hope that once Cole gets us there, he feels the same.

"Earl Rowe Provincial Park? That where you want me to go?"

"That's the place."

"I can't believe that after all the driving around I did when I came into town, I never realized that you live exactly seven minutes away from a park this size. Let alone the bed and breakfast just around the corner."

"Do I smell a hotel relocation in your future?"

Following the prompts of the signs on the road, he laughs as he pulls off and into the front of the park.

"You're damn right. Nothing against Super 8, but that bed and breakfast looked amazing from the outside."

"My parents are friends with the people that run it. It's even nicer on the inside too. I have no clue how you missed it."

Pulling the car into a spot and turning off the ignition, he turns and leans over in the seat, resting his forehead against mine, a playful smile on his lips as he speaks again. "I thought it was obvious that my attention had been stolen by something far more beautiful than a B and B."

"Come on, Casanova. I promised you a walk and that's what you're gonna get." I change the subject, too nervous after the show we put on back at the club to show him how much his words affect me.

Undoing my seatbelt and slipping out of the car, I turn to head over to his side just in time to catch him coming around to meet me, his hand out from his side and slipping easily into mine when he reaches me.

"I'm sorry if what I said made you uncomfortable."

"It wasn't. It was nice."

"Then why the quick escape?"

"Wanting to show you my special place wasn't a good enough reason?"

"It was. It just wasn't a believable one."

"You're not the only one who's nervous." I admit, reminded of his own admission over hot chocolate days before.

"I don't want to make you nervous."

"You know, I might not know you well, but I believe you mean that. It just can't be helped. I'm pretty inexperienced with everything that's happened since you walked into the bank."

Starting to walk before he has a chance to comment on what I've said, I'm not at all prepared to be pulled back and into his arms when he notices how quick I am to get away.

"Is this about what happened at the club?"

"No—yes." Chucking to myself, I lift my hand to hide the flush and meet his gaze. "Maybe a little."

"I wasn't myself back there, Reagan. You need to know that. I don't know if it was the lights, the heat or all the people, but I don't normally do things like that."

Blushing what I'm sure is a deep shade of magenta remembering the way we were on the dance floor, I bury my face into the space where his shoulder meets his neck.

"That makes two of us. I don't know what got into me."

"I'd say it was Bethany, but considering it wasn't her body that you had yours pressed into while you made love with your mouth, I'm pretty sure she wasn't the cause."

Did he really just say that I made love to his mouth when I kissed him?

Oh boy. I'm way out of my element here. I have no idea what to even do with that. Let alone what to say.

"Do you remember what you said the night we had hot chocolate?"

"I said a lot of stuff that night, Reagan. You wanna be more specific?"

"The feeling you have when you're around me."

"How it feels like I've known you a lot longer than a few days?"

"That's the one."

"What about it?" He asks but before I have the chance to explain, he leans in and brushing his lips against the bridge of my nose, smiles before pointing out in front of us. "Hold that thought. You've got a place to show me. Hopefully one that gives us a place to sit so we don't have to have this conversation in the middle of a parking lot."

"It does." I laugh softly and pointing over to where I can make out the beach in the distance, I let him guide the way silently, not even attempting to say another word until I feel our shoes begin to sink as we finally hit the sand.

Slipping his hand out of mine and bending over, giving me what might possibly be the best view of his lower back as it's flexed, along with his ass as it tightens, filling out his already tight jeans perfectly, I look away just as he slips his shoes and socks off.

As good as he looks in that position, my need not to treat him like a piece of meat wins out.

"It's been so long since I've been to a beach. I needed to feel the sand on my feet."

Of all the things I expected him to say when he finally spoke again, it definitely wasn't that. Something that connects us more with my favorite thing to do when I'm here walking the beach barefoot.

"Join me?" he asks motioning to my feet and with a nod I slide down to my knees. Feeling the intensity of his eyes on me, I attempt to undo the laces, my hands fumbling nervously until finally after struggling with them for a few minutes, I've gotten them off and I'm standing again.

"Ready to walk?" I ask, catching exactly where his eyes are before the sound of my voice pulls them away.

"Definitely."

Strolling slowly across the beach with the sound of the waves hitting the shore guiding us, I lift my hand and point out to the secluded rocks I can make out across the way. The trees taking up residence behind them hanging over like an umbrella, bringing my earlier words to life.

"You weren't kidding about knowing a spot."

"I never kid about hiding spots."

"Be careful, Reagan. I happen to be a master at hide and go seek."

When he sticks his tongue out, there's no stopping the laugh that escapes. Just like the moment we shared in the club was new to both of us, so is this one. For someone who's spent the majority of our time together on edge, it's a nice change seeing him act so childish.

"I'm gonna test that later."

"Promise?"

"Cross my heart." I joke tracing an X over my heart before stepping toward him and doing the same over his.

"Just don't get mad at me when you learn you can't hide from me."

"Never." I say as we reach the rocks. Tossing my shoes down onto the sand, I slide across and wait for him to do the same. Turning my attention out to the water to hide the reaction that comes when he finally lowers himself beside me and his leg brushes mine when he slides himself in closer.

When he kissed me, I expected to feel some kind of spark. I used to feel something similar when Walker and I first got together. With Cole, though, it happens more frequently. I can burst into flames from a look, feel like I'm being electrocuted from a touch and don't even get me started on what happens when we kiss.

I'm amazed I'm not cinders of ash by now.

"So what you said earlier—" he starts, and knowing where he's going with this, I cut him off as I lift my hand and speak.

"You're not the only one that feels that way. I know I said you were using a line on me, but at the movies earlier, in the car on the way to the club, hell, even when we had hot chocolate that night, I could feel it. I believed in it, even if I wasn't ready to admit it."

"Does that have something to do with what happened at the club?"

"I told you what that was. I wanted to be someone other than same old boring Reagan."

"There's nothing about you that's boring, but I am pretty sure you accomplished what you were after."

"Yeah, but what you don't know is that if it had been some other guy, it wouldn't have happened. I didn't feel like Reagan anymore, so that part would have been the same, but I never would have let Bethany walk away when she did."

"So what you're saying is that what happened between us was because you feel as connected to me as I do to you?"

"Pretty much, yeah. It scares me, though."

"I don't want you to be scared. Not of me or of this, but I do get why you would be."

Exhaling a heavy breath, he turns toward me, resting both of his legs cross legged on the rock, his gaze moving from my face to the trees behind us and out to the water before making their way back again.

Whatever he's about to say or do, it's pretty obvious he's unsure about it. Nervous.

"Can I ask you something personal? You can slap me, not answer or tell me to fuck off once you hear it. I'm good with whatever, but I have to ask."

Already knowing so much about me, especially when it comes to my insecurities, there's really nothing he could ask me now that would be too personal. My need to hide away and blend in has always been my best kept personal secret. Though, in a purely playful way, I wouldn't mind telling him to fuck off. I'm sure the response I'd get would be more than worth it.

"Sure."

"You and Walker, uh…how serious were the two of you?"

Wow. I was wrong. He can ask something pretty damn personal still. Crap. Do I really want to get into this with him?

Yeah, Reagan, you do. The same way you want to know it all about him.

"I'm not a virgin if that's what you're getting at."

"It was but wasn't at the same time."

"Explain?"

"I met Miranda in high school. I was pretty screwed up at the time, but she seemed to even me out. Yeah, we slept together. Any chance we could get really…" he pauses once he catches me averting my eyes. Something I do to protect myself from what he's going to say next, but that I can't do entirely because I can still hear it.

Knowing Cole has slept with someone shouldn't make me feel anything, but there's this knot in my chest that no amount of rubbing can make go away and I just know it's because the idea of him being with someone else, even someone from his past, hurts.

How ridiculous is that?

God. I need to chill.

"I was committed to her is what I'm getting at. I'm sorry, Reagan. I didn't mean to make you uncomfortable."

There he goes apologizing for nothing again.

"You don't have to apologize every time I react, Cole. It's okay. I guess I just don't understand why you want to know about Walker and me."

"The truth?"

"Sure, we can try that." I joke, but even to my own ears it falls flat. "Of course I want the truth."

"I knew when we had coffee that someone had hurt you. There's this guarded look in your eyes sometimes. Like when I said it felt like I knew you forever and you wanted to bolt. He hurt you so damn bad Reagan and he's still doing it."

I'm not known for being able to hide things, wearing my heart and every other emotional response on my sleeve for the world to see, but having someone that barely knows me pick me apart so easily is crazy.

Am I really that transparent?

"What does that have to do with what you asked?" I try again, hoping he'll get to the point so I can put this entire conversation out of my head. This wasn't what I had in mind when I wanted to bring him here.

"He cheated on you, and honestly, even though I don't know what this is between us, thinking about that makes me want to kill him. I guess I asked because I needed to know that when you were with him that it was something you wanted and not just you giving into the pressure of being his girlfriend."

"I thought I loved him, Cole. It happened because I wanted it to."

It's not entirely true. There was pressure there for a while, but a lot of it was what I put on myself. I didn't want to be the only virgin in the senior class, or be the ice queen that people believed me to be before I finally started dating Walker.

"Has there been anyone besides Walker?"

"No, and there won't be until I'm ready."

"The dance floor earlier…your hands said something a lot different than what those perfect lips are saying now."

Completely ignoring the comment about my lips and honing in on everything he said before, I swallow hard before letting him have it. "How many people have you slept with? Since we're taking inventory here, might as well take yours too."

If I had known it was going to turn into this, I never would have done any of it. Screw trying to branch out and be someone different. I was safer in my room with my damn books.

Where I expect him to say my name softly the way he has before right before he apologizes for the hundredth time, he surprises me when he answers the question instead.

"Four. I lost my virginity at a party while I was drunk. I don't even remember the girl's name, which works most days because I'd rather not remember the night at all. I wasn't in my right mind. Then Miranda came along and for years it was just her."

"And the other two?" I ask, curious despite myself.

"Rochelle I think, and Denise, but because those two happened while I was drunk, horny and in desperate need of sticking my dick in something that would help alleviate it, my memory is kind of hazy."

Oh my god.

Cole sleeping with other women isn't a big deal. I knew the moment I met him that he wasn't a virgin. He exudes sexuality. It's his admission they were one night stands—names he can't even remember, that twists the knot in my chest tighter. Making me realize that if I continued what we started at the club, I would have become another notch on his belt.

Another nameless one night stand he would forget about when he left and went back home.

I can't believe I was so stupid. That I saw something different in him.

He's just like all the rest.

"Well, this was very informative. Maybe it's time you took me home."

Scowling before sliding his legs out and over the rock, he sighs heavily as he stands. Obviously not pleased with the way things are taking place, but still doing right by me.

There's something about how quick he is to give into what I want that gives me pause. Makes me feel guilty. He just admitted how many people he's been with in his life and what did I do?

Freaked out.

He's not my boyfriend. I'm not even sure we can call what's going on here more than hanging out and having fun, so what right do I have to judge him on something that happened before I even met him?

This isn't me. I'm not like this. *Shit*. I need to turn these feelings off before I alienate the only guy since Walker to give me the time of day.

"When's the last time you had sex?" I ask, torturing myself more because knowing the answer to this will determine what I do next.

And based on the way I can feel the electricity between us, even in a tense moment like this, I'm hoping like hell he says it was right before he met me so I can walk away without feeling bad.

"Both happened right after Miranda left, within a couple of weeks of each other. So, a little less than a year."

Damnit. Of course it wasn't days before he met me.

Because Cole, I'm starting to see, is damn near perfect and I'm so messed up from Walker that I'm pushing away anything that might look remotely good so it doesn't happen again.

If Walker could cheat, what's stopping Cole from doing it?

"Would you have slept with me tonight if I wanted you to?"

"No."

Oh hell no. He's not going to lie like that. He wasn't stroking himself in the damn club. That was all me. I damn well know that had I come on a little stronger, he would have given in. I can still feel his hardness pressing against his jeans, craving my hand almost as much as my hand craved it.

"You're lying."

"No, Reagan, I'm not. Do I want you physically? You're damn right I do, but I don't want to fuck you."

"Then what do you want?"

Groaning loudly, he shakes his head before turning his back and walking away. Each step taking him farther out until he stops at the water. Giving him a few minutes, watching silently as he bends down letting the water run over his hands, I finally summon up the courage to call out.

I want answers, damnit. Why was that question so hard?

"If you can't give me an answer, I'm leaving!" I yell out. "Don't worry about taking me home! I've got no problem walking!"

Determined to stick to my words, I turn and grabbing my shoes, start stomping my way through the sand, hating the way it feels now as it filters through my toes and scratches my feet.

Screw Cole and screw this.

I don't know what I was thinking coming here. I'm obviously not in my right mind. Giving a shit about someone I've known less than a week and having a connection on top of it, is insane and I'm sick of it.

Making quick work of the beach, I'm halfway to the stairs when I feel rough grip around my wrist, spinning me around and straight into the crosshairs of a very pissed off Cole.

With barely enough time to prepare for another round of this tug of war, his lips crash down hard onto mine, stealing the breath straight from my lungs. The familiar burn building when his arms slide around my weakened form and grip me tighter to him as he pushes his tongue against my lips.

Parting them, desperate for air and a chance to think this through and change it, he slides his tongue in and teases mine. The contact releasing a moan from deep inside me. A call for more that he answers as he probes deeper, so connected are we in the moment that I'm struggling to find where I end and he begins.

Is this what kissing is supposed to feel like? Quickened heartbeat, body on fire and needing to strip completely bare just to try and combat the sensation of burning alive?

"Now," he murmurs against my lips. "If that doesn't tell you what I want from you, give me a minute to recover and I'll repeat it many times as needed until it does."

A minute to recover? How about a few hours? Days? Weeks? Can you really put a time limit on this? Because I don't really think I'm ever going to fully recover from that kiss.

He's wrong, though. All this tells me is that he desires me. I already knew that. I knew it the first night and tonight when he has his hands all over me at the club. Attraction and desire are not our problem.

Cole holding back is.

What other reason could there be for him turning his back on me and running off to the water?

"Because there's a very real something buried deep inside my everything that has the power to destroy you."

He's already told me that there's something he's holding on to that could potentially hurt me. He's been warning me right from the first day, and stupidly I ignored it because I wanted this. I wanted him.

I'm acting like a completely different person and it's gotten so bad that I don't even recognize the person I was before anymore.

"I know you're attracted to me, Cole."

"Good, but I don't care. I haven't tried to hide that. I'm more interested in what else you know."

"What else is there?"

"Four days, Reagan. I've been in town four days. Been around you two of those four and in that time, I've thought about murdering your ex, kissing you until the breath has been taken from both our bodies and spending endless days making love to you, just so I could hear the sound of my name falling while I worship you."

Whoa. Say what?

"But before you go and say that all of that is because I'm physically attracted to you, let me make something very fucking clear. There's something else that I've thought about more, and despite knowing it might make you run once I've said it, I'm having a hard time giving a shit because I need to get it out before it eats me alive."

"W-what are y-you talking about?" I stammer, the grittiness of his tone mixing with the soul crushing agony I see in his eyes from the need to get whatever it is he's holding back out, making me nervous.

"I told you!" He exclaims as he pulls back and releases me. "I said that being around you would make me wanna tell you everything. That this connection, this attraction with a twist that we have, would cripple me and make me want to be something I swore I never would be again."

Swiping a hand through his hair, grasping it and pulling, he huffs out a breath before doing it again, his body becoming even more rigid under the stress his words are causing him.

This is surreal. I want to say something to ease the struggle for him, but I don't have the first clue what he's even going to say so I'm stunned into silence. Witnessing the way he seems to be cracking under the pressure though seems to be creating a ripple effect as my stomach twists and I'm struck with my own wave of nausea.

"I'm not supposed to want you this bad! Do you get it yet, Reagan? You and I aren't supposed to be anything, but whenever I'm around you, I forget all of that because I all I want is for us to be everything!"

Cole

I'm losing my fucking mind.

It all started when I brought up her relationship with that douchebag she thought she loved. For purely selfish reasons, I needed to know if she'd given herself to him. Gone to the point of no return and given him the gift that someone like him doesn't even deserve to hear about, much less experience.

Then I screwed it up more by giving up my past with other women freely when she asked. I honestly thought that with everything that had already gone down with us, she deserved to know the truth. What an idiotic move that was. No sooner did it come out than I watched her face change from desire to disgust.

I became Walker in her mind. I saw it the exact second it happened and I couldn't let it go down like that.

I was a fucking disaster after Miranda left. Even worse than I was before she came into my life because of the feelings I'd given in and allowed myself to have, but god damnit, I had nothing else to live for. No one at home to give a shit about.

The only father I ever knew was dead. Luke was persona non grata because of what Sunday being taken did to him, and then there was her.

Sunday. The only other tangible hold I had on the world until she went missing. A little kid that made more of an impact on my life in the year and a half she was in it than those I was closest to, like Sam.

In my head, once Miranda walked, I'd lost everything. So every move I did make, especially sleeping with those two random women, was purely mechanical. I had urges I had to deal with. It wasn't like I ever planned on connecting to another woman ever again, so what the fuck was the problem?

Well, when you've been one of a dozen girls, it's a pretty big problem. She didn't even try to hide it. Which only made me hate myself more than I already did.

Asking me if I wanted to fuck her. Shit. I'm a guy for Christ's sakes. Of course I want to, but that's only when I focus on the need. If I put the attention on something other than the physical and I remember who she is and the impact she had on my life, it's about a whole lot more than needing to be buried inside her.

It's about me wanting to be with her. Period.

End of sentence.

Four days and I'm ready to go back on the pact I made never to let another woman get one over on me. Four days and I want to give this girl the world. I want to give her everything I have. Physically, emotionally and mentally. I want her to possess and own me the same way I want to do with her.

And I'm losing my fucking mind after devouring those perfectly sculpted lips because I'm about to make her run like a spooked rabbit when I tell her just how deep this runs.

Telling the girl that I feel as though I've known her forever and telling her that I want to be with her forever are two different things. Especially when I can't tell her why it is I want to be with her so bad.

What makes all of this worse is not being able to say for certain exactly who I want to be with.

Sunday or Reagan.

The memories I have of her are mixing with the desire I have to know the person she's become leaving everything muddled and confusing.

Am I falling for Reagan or am I falling for the idea of Sunday?

"Why can't this happen? What's so wrong about us that it's making you act like this?"

This can't happen because you're my best friend's sister and I'm here to figure out the truth of what happened so I can drop it on you. I'm not the one to heal the damage done to your heart. I'm just the asshole that's going to break it even more.

Shit. I can't do this. I'm not cut out for this shit. I'm not even sure knowing Sam the way I did, how he did it for so long. Blowing someone's life apart for a paycheck is bullshit. It's even worse when you know the person and feelings are involved.

"I'm never going to be able to be completely honest with you. There are things I'm going to have to keep secret, not only to protect you, but because my ability to keep my mouth shut is a pivotal part of doing what I do."

"You wanting to tell me everything is what's making you act like a crazy person?"

"Yes. You deserve honesty. I can't give it to you. At least not all the way. You deserve to have it be all the way, Reagan."

"I never asked you to!" she screams at me before stepping forward and slamming her tiny fists down on my chest. "I just want you to tell me what you're holding back right now!"

"You really wanna know?"

"Yes!"

"I agreed to have fun with you, see where this could go, but I can't do that anymore."

"Me either," She agrees, but before I can question what she means, she pulls the words straight out of my head. "Because I already know where this is gonna go."

"I'm going to hurt you, Reagan. I don't want to, but with everything that's happened in the past; already driving one woman into the arms of another guy, it's inevitable. I *will* hurt you."

"What did you call Walker earlier? A prick?"

"Yeah, so? What about it?"

"Well if Walker is a prick than your ex-girlfriend is whatever the female version is. You didn't drive her away, Cole. She was gone before you even got together."

No matter how much I want to buy into what she's saying it still doesn't change the truth. I know where I want this to go and I know where it's going to go and they're two different paths.

"I don't want you to hurt anymore."

"Well, you wanna know the easiest way to assure that I don't?"

Shrugging and shaking my head, she wastes no time enlightening me.

"Tell me what you want."

"You." I don't even hesitate admitting. "I just want you."

Lifting herself up on her toes and smiling, she presses her lips to my cheek softly. The innocent move melting the iceberg that for the past year my heart has become a little bit more.

She's doing it again. The same way she did before. She's warming me and melting away the layers of hard ice I'd built to block the rest of the world out.

"I'm here. You've got me."

"I thought you wanted to go home?"

"First rule of girls, Cole. Sometimes we change our minds." She grins, hitting me with another shot of the warmth that I'm starting to see is as natural to her as breathing. "If the drama of the last few minutes hasn't completely done you in, I was thinking we could go back and just enjoy the quiet a little longer."

Pulling her close to me, breathing her in and committing it to memory, I release her back to the sand below and slip my hand through hers before beginning to walk back. All the while thinking about the words I didn't say. Ones that I hope one day if she's able to forgive me for everything, I get the opportunity to finally speak out loud.

I'm falling for her.

Chapter Fourteen

James

"I've got it under control. Whatever he thinks he knows is all speculation at this point and we didn't leave a trail. Let him chase his leads. He'll just go in circles."

This is the third call like this so far this week. Each time, the voice on the other end of the line becoming more unhinged. What he doesn't seem to get is that by acting that way, he's giving away more than anything the private dick sniffing around might uncover.

If we go down for this, it's going to be because Cameron couldn't keep it together.

He's a lawyer and judging from the way he's becoming increasingly more unglued with each passing day, a pretty useless one. Where's the grit? The take no prisoner's attitude that wins him cases in the courtroom?

Does the way he's sniveling now really get him far?

"James, I don't think you understand the severity of our situation. Cole Alexander is good at what he does. He comes from a long line of men that will not rest until they get what they are after. Which right now, is your head on a platter."

Cole Alexander is a joke. I've looked into the boy. He's nothing but a junkie's kid who tried to turn good when some poor shmuck took pity on him. Nothing more.

The sooner Cameron gets with the program where this little punk is concerned, the better for everyone involved.

"Cole Alexander is a twenty something guy who thinks more with his dick than his brains. He has nothing to go on but some useless bread crumbs his father and uncle laid out before they met their untimely demise. Crumbs that amount to nothing."

Looking toward the door at the sound of the knock, I whisper Cam a warning before pulling the phone from my ear as Katherine enters and makes her way toward the desk.

"What's going on, Jimmy?"

My wife. So beautiful. So stunning. The woman of my dreams brought to life. Also the woman that seventeen years ago, I'd made the ultimate sacrifice for, even though she still has no idea just what that entails.

It's been hard keeping her away from this. I want to be able to tell her the lengths I've gone to over the years to give her the child she so desperately wanted. Well, at least she did before we got the surprise that is our son Jake two years later. It just can't happen. She must never find out what I've done in order to give her Reagan.

Especially with the nonsense she's been talking lately about telling our beloved daughter the truth.

She would never forgive me and who could blame her?

In the dead of the night, I signed my name in blood to an invisible contract with no expiration date. It would be with me forever, same as it would Cameron and all others involved.

Katherine and her big heart, the one thing that no matter how old we get will never fade, can't be anywhere near this. She must never know that our beautiful Reagan used to belong to someone else. Someone who to this very day, still searches for and loves her. .

"Business call. Couldn't be helped." I mouth with a motion of my head to the receiver. "Five minutes, I swear."

She's used to this. Calls at all hours. Doing business internationally means that sometimes I have to work with odd hours, but there's something about the way I'm acting or what I've said that makes her pause this time. Almost as if she knows there is more going on in the moment than I've admitted to.

"The bed is cold without you."

"Sweetheart, I'll be right there. I'm just finishing up."

She has to believe me. I can't finish this call with her standing in the room and with how on edge Cameron seems, I have to put a stop to this in the only way he's going to understand.

Threats.

"Five minutes, Jimmy."

"Yes honey."

Blowing her a kiss from my spot behind the oak desk, I smile softly when she as is her way, blows one back before making her way to the door and out of the office. Ignoring the feeling I always get whenever I'm reminded of the lies I'm telling by withholding information, I pick the phone back up and make my point loud and clear.

"Cameron, you need to get yourself in check. I will not take you calling here in the middle of the night again. The next time you get paranoid, go to church. I'm sure they'll be more than willing to listen to your delusions. If Cole becomes a problem, he will be dealt with. You can be assured of that. It's been handled. But if you call here again and it is about anything other than business, I will make sure that you experience the full weight of what I have in store for him."

Smiling at the intake of breath that comes sharply across the line, I'm warmed.

Message received loud and clear.

Cameron won't be a problem anymore. At least he won't if he wants to keep breathing.

"Are we clear?"

"Yes, James. Crystal."

"Glad we could clear that up. Goodnight, Cam…and remember what I said. Leave Cole Alexander to me."

Cole

If I want to know why John held back, rather than spend every waking moment going through his old journals, I've got to go back to the beginning. When Sunday was taken and when her babysitter was found dead.

The answers are hidden in there. I'm sure of it.

The death itself is public record, so going back to that time for information proves rather easy. Every news organization in Canada and the US reported on it. It's only when it comes to the lesser details that my fact finding mission proves more difficult.

I know how Hallie died, how she was found and what the official cause of death was, but no matter how much paperwork I sit on this grungy ass bed and go through, I can't find any of the police reports, notes or other information I would have expected John to have.

Could John have put them somewhere before he died?

The better question is, could he really have played a part in her disappearance? Could he be the one behind this so that sometime in the future, he could have used it to his advantage and gotten national recognition for it?

Is the John Alexander I remember really the callous bastard he appears to be?

"I'm gonna have to go home."

Looking up from one of the journals that I've had him scouring over since we learned John knew about Sunday, Luke grunts.

"For what?"

"What I expected to find here, things you would expect a private investigator to have, isn't. It's weird too because John was usually better about paperwork than Sam was."

Sam. Not Dad. I'm already pulling away from the Alexander name.

Damn John for making me doubt the only father I've ever known. If he wasn't already dead, I'll kill him again. My sentiments from the other day running wild again.

I'm still thinking that digging up his corpse for answers is a viable option.

Time for a break, Cole. You're losing it.

"These journals he kept, he glosses over Sunday and stuff that he's doing, but there's nothing here that could help us now. Maybe it's in another book."

"Or my dad kept things from him."

There you go. He's dad again.

"Cole, you know how much I respect your old man, but have you given any thought to what you're going to do if it comes back that he had something to do with keeping this quiet?"

"He was dead by the time John found her, Luke, but yeah I thought about that."

I don't want to think that Samuel Alexander was anything but the upstanding guy he always appeared to be, but the thought has crossed my mind that he might have gotten himself into something and not been able to get out of it before he died.

"So when are you heading back?"

"I was thinking of doing it in a couple of days. I need to give your parents a progress report. It's been too long already. I should have checked in the day I found her. I'm sure they're wondering what's up."

"Don't tell them."

Whoa. Now that wasn't expected. With as eager as my best friend has been to get the truth out there, wanting to tell Sunday himself the first night he saw her, I assumed he would have been all for me letting his parents in on what we've found.

"You're gonna have to explain that one."

"When you tell Sunday the truth, she's already going to hate us. I think deep down you know that. I mean, if this was happening to me, I'd hate you. How do you think she's going to feel about my parents? They're the ones that put all this shit in motion. Ruined her perfect life. I don't want them dealing with that."

"At some point they're going to have to. I've already kept this a secret longer than I intended when I took the damn job."

Luke laughs before lowering his head back to the journal in his lap.

"What are you laughing about?"

"You."

"Care to fill me in on what I said that was so funny?"

"Even with everything we've been through, you still take me for a moron."

"Get to the point, Luke."

"This was never just a job. Sure, it might have started that way because of what John put in his will. We've been over this. For a day or two after I got here and met her, I figured your reasoning for keeping all of this a secret was because you wanted to do right by my family, but it's so damn obvious that it's more that even a blind man can see it."

"So my wanting to limit the fallout here is amusing to you?"

"No. You having a thing for my sister is amusing to me."

Deny, deny, deny.

That's what I need to do, but there's something about the smirk on his face and the lift in his eyes that tells me that he's ready for me to do exactly that and it won't get me anywhere.

I know I haven't exactly been shy about the way I feel when I'm around Sunday, but damn. I didn't think it was that obvious that she'd gotten under my skin. I thought I could keep it under wraps a little longer. At least until I could sort out why I feel the way I do and rid myself of it once and for all.

Bro's before hoes and all that.

Except Sunday isn't a hoe.

"You thinking up a way to deny it?"

"No. I'm thinking of a way to change the subject."

"I loved her from the second my parents brought her in the door, you know. She was my sunshine. My entire fucking world. When I walked into that bar and saw her for the first time in seventeen years, it was like someone punched me in the gut. It took everything I had not to hug her and never let go."

"Does this story have a point?"

"There's only one other person on the planet I know that loved her like I did."

"And?"

"And he's standing right in front of me."

Oh, hell no. I'm not going there with Luke. I care about Sunday sure, but I am not in love with her. My feelings may be taking me down that road, but only time will tell. For the time being, we're nowhere near the same.

"You're messed up man. I cared about her because she was your sister, which meant by association she was my sister. That's it."

"Stop lying, Cole. You suck at it."

"I'm not lying. She's your sister and for a little over a year, she was this cute little baby that we used to like making laugh. I didn't love her, at least not the way you're describing. I was seven. I didn't even love myself. How could I love her?"

"How long have you been here?"

"Are you serious?" I ask and when he nods, I release an annoyed breath. "A week. But what does that have to do with anything?"

"You've had a week to tell her the truth. Show her what you know to be true and instead, you've been either out with her or holed up in here trying to get answers to what actually happened back then."

"I'm dangerously close to being over this conversation."

"Multiple chances to tell her that she's Sunday and yet nothing happens. It's because I wasn't the only one sucker punched in the gut when he met her. She nailed you with it too. Try and make me believe otherwise all you want, but it's written all over your face. You give a shit again."

He's right, but I'm not about to give him the satisfaction of telling him. Sunday gave me something to believe in again and despite my every attempt at backing away, ignoring it or trying to pretend it doesn't exist, nothing changes.

It's because you can't change what's written in stone. It's there forever.

Sunday is written on the stone that is me. Forever.

Its why other than phone calls, I haven't seen her since our night at the beach. If Luke is able to see the truth, she's going to be able to see it too and I'm not sure she's ready for that yet. Even if she wasn't Luke's sister, because of what that asshole she dated did to her spirit—the way she sees herself—it's going to take more than a week for her to feel the same way about me as I do her.

"Even if what you're saying is right and she did make me feel again, it doesn't matter. Nothing is gonna come of it."

"Why? Because she's my sister and you're like the brother I never had?"

"That, and I'm seven years older. Not to mention, I'm the heartless piece of shit that will eventually hurt her, which will in turn cause you to react and kill me to defend her. Can't forget that little tidbit."

"Not to get all chick flick on you here, bro, but age is nothing but a number, and the only one killing anyone around here is you killing yourself. I'm not seeing a real downside to this."

"You're tell me you'd be cool with me and Sunday?"

"Yeah. I mean, if that's what you both wanted to do, who the fuck am I to stop you? Besides, I know you. How many times growing up did you have my back even when you didn't know the score? You protected me on blind faith alone. If anyone can take care of Sunday and give her everything she deserves, it's you."

He's wrong. I'm no better than my mother. We're both lost causes. Victims to our vices. The Cole that Luke is fondly remembering right now is not the one I became when my dad died. It's the one I was before. The guy that I don't think I can ever be again. I will hurt her. It's only a matter of when.

This isn't getting us anywhere. I'm not here to create some perfect fucking love story. I'm here to bring a girl home that's been away far too long. Once that's done, back to my life I go.

Just me, a bottle of Jack and the stench of everything I've loved and lost that no matter what I do, I can't seem to wash off.

"Whatever. It doesn't matter because it's never gonna happen. So can we just focus on finding out what else John was hiding please?"

"Yeah, but Cole. If I'm even a little bit right about what's going on here, you need to get straight with it before you end up getting hurt. Or worse…killed."

Chapter Fifteen

Luke

"I've been doing some thinking."
Ever since I called him out about his feelings for Sunday, he's been relatively quiet. We've said maybe a total of twenty words to each other over the last hour, where he joined me pouring over the journals and coming up as empty as the beer bottles on the bedside table.
The ones I polished off well over a half hour ago and have been craving another one of ever since.
Why Cole allowed himself to be saddled with this is beyond me.
I could never sit pour over case files and basically stalk people twenty-four seven. It would bore the hell out of me. But I guess that's why I run my own construction company and he's the one contracted for stuff like this.
He may have bitten off a whole lot more than he can chew with this, though. Sure, he found Sunday, which back when we were kids and she went missing, I always knew he could do if he put his whole heart into it, but it's still too much for one person to have to deal with. This goes deeper than Sunday and her kidnapping.
Throw feelings into the mix and it won't be long before the explosion hits and the entire town of Alliston is leveled from the blast.
If there's one thing I kept that my parents taught me, it's that nothing worth doing ever comes without scars and we're all walking away with scars on this one, but no one more than Sunday.
"Scary thought."
"Very funny, dickhead. You want to hear what I'm thinking or not?"

"Lay it on me, brother. Anything that keeps me from having to read through another one of these boring ass journals, I'm all for it."

"What I said earlier about going back to the beginning. I don't need to go back to Orangeville for it."

"That's good, but what changed?"

"What changed is that I've got a line to the past sitting right beside me."

Me? He thinks I can offer up something? Since when? We've exhausted everything I know. It's half the reason I moved the hell away. I couldn't deal with going over it anymore. There's nothing I know that Cole doesn't.

I might not have been honest with everyone else in my life, especially when she went missing and I shut down, but I always have been with him. He's the only one that knows the ugly truth about what happened back then. How I caused it all by not being home where I should have been.

"What do you think I can give you that you don't already have?"

"A firsthand account of that night."

Nope. Not going there.

It took me years not to have that night on repeat in my dreams. I'm not fucking reliving it.

"You already know what happened. I also know that John has my statement somewhere in that mess over there. He wrote about it." I say, motioning to the book in front of me as if that's going to stop him.

"He does have your statement, but if I really want to do this right, I need to go the one source I have at my disposal. Unless you'd rather I talk to your parents about it. I mean I can get it just as easily from them."

Son of a bitch. He knows my parents are my hot button. I don't want them knowing about any of this yet, same as he doesn't want Sunday knowing. So it's either I torture myself by going back in time and risk the nightmares or I risk the wrath of my parents when they find out just what we know so far.

"You're an asshole, you know that?"

"Been called worse."

"Cole, you fucking know what that night did to me. How many years I spent fucking mute and drowning in memories. I can't believe out of all the people that could be asking me this, it's you. How can reliving something that happened seventeen years ago help you now?"

"You're older now."

"Yeah, well all that means is that my memory is probably shittier. You weren't the only one that drowned himself in the bottom of a bottle for years, remember? It's half the reason we got along so well when the shit hit the fan. We dealt in the same way."

Moving across the room and sitting across from me on the bed, leaning forward and resting on his hands, he stares me down. Never once blinking or wavering in any way. Proving just how serious he is about what he's asking me to do.

"What I mean about you being older is that there is going to be things you'll remember now that will stand out more to you than they did back then. You were eight when she was taken, you weren't focused on anything but that fact for a very long time. You're twenty-five now. Insignificant things back then will make a lot more sense."

Before I can come up with an argument for him, despising what he's asking of me, he continues.

"I know what I'm asking and by now, you have to know that I wouldn't ask you to do it if I really didn't think it would help. You know something, Luke. I can feel it. Something about that day before the babysitter came and we took off to play hockey that is the key to everything."

"How can you be so sure?"

"I don't know. I just know that other than heading home and hounding the police department and your parents, I'm running out of options."

"What about your dad? Isn't he still an option?"

"Yeah, and I'll go to Rachel for his files first thing, but not before I've exhausted the very real option sitting across from me."

Shit. I'm going to give him what he wants. Anything that can get us closer to getting answers about what happened to

Sunday that night, I'll do, even at the risk of throwing myself back under the guilt bus.

I would die for that little girl.

Except she's not that little anymore.

Big, little, it doesn't even matter. She's still my sunshine and I'll do whatever I can for her.

"Fine. I'll do it." I huff out, the weight of the agreement and what it means making it damn near impossible to breathe. Can I really just lay back, close my eyes and go back to that night without any repercussions? It was hard as hell to come through it the first time it happened. I don't think I'll make it through this time.

For the first time in years I'm fucking scared.

"What do you need me to do?"

"Just sit back and try to relax. Imagine something or someone that calms you and focus on it to keep yourself grounded. Once you've done that, bring up that day in your mind and I'll take it from there."

"You really think you should be the one doing this? Don't they have people certified to do this kind of thing?"

"They do," he agrees. "But they don't know you the way I do. If you get in too deep, I can pull you out better than they can."

"If you say so."

"Luke, if you think that doing this is gonna send you someplace you can't come back from, just say the word. I'll get what I need some other way."

"Nah, man. I want to do this for her. I mean, I don't think you're gonna get much, but if even something small can help Sunday, give her peace of mind or get her back to us sooner, I want to do it."

"Seventeen years and you never stopped being her protector."

"Nope. And until a certain jackass I know admits how he really feels, I won't stop either."

Catching the roll of his eyes before he quickly turns away and makes his way back across the room to his bed, I swallow the unease over what I'm about to do and laugh.

It's way too damn easy to push his buttons lately. If I wasn't so hell bent on seeing this thing through to the end, I might spend a little more time ribbing him. Getting a rise out of Cole before was such a pain. This side of him is one I can definitely have some fun with.

"What do you want me to do?"

"Get as comfortable as you can." He says, motioning to the bed and frowning. "And get ready for a horrible walk down memory lane."

Cole

For a second there after I suggested using him to get answers that I needed, I'd been expecting him to walk out of the room.

Aside from his parents, I lived through the change in him. The self-loathing and hatred that became such a huge part of him because he blamed himself for what happened. Believing that if he had just stayed in that night like he was originally going to until the guys down the road coaxed him out, she'd still be here.

We might have been puny little kids back then, but when it came to Sunday, Luke would have turned into a beast to protect her. Feral and animalistic. He would have gladly laid down his eight year old life to guarantee hers.

Him agreeing to do this for me, it's him doing that same thing all over again, only now, as a twenty-five year old. A lot older, smarter and now, because we found her and we know that at least one person out there knew about her existence for years, a whole lot angrier.

This is him protecting her. Giving her back what he still believes deep down he made her lose when she was taken away.

I haven't gotten to interact much with Reagan's family since I've been here, but if the way Jake was with her is any indication, I'm starting to think that at least in terms of having a good life, she wasn't lacking at all. She didn't leave a good home and end up in one like I grew up in. She was loved, cared for and turned out pretty damn good.

Which means that even if Luke had been to blame for what happened, he can't blame himself for her life turning out badly.

She'd gotten lucky. If there's any luck to be had in this situation.

Pulling the chair over from the corner of the room and getting comfortable, I lean back as I watch Luke's chest rise and fall with his relaxed and even breathing. Doing exactly as I said and closing his eyes, transporting himself away from this motel and back to the night when everything turned bad.

"When you got up that morning, was it earlier or later than normal?"

"Everything was right on schedule. Mom was really good about making sure that no one overslept. I wanted to. I was grumpy that morning. I snapped at her when she told me I had to get up, but I did it anyway."

"After you were downstairs, was your dad there? Did you all have breakfast together like you did on the days I stayed over?"

"Nah. He was in a rush that morning. He poured some coffee into a travel mug, kissed her and promised that he would see her for their date after work and he was out the door. It was just her and I until we met up with you outside later."

I have a hard time recalling a lot of what happened when I was younger, only the really big moments sticking out, but with him bringing it up, I can kind of recall that morning myself. Kissing Rachel goodbye before calling out to my dad and telling him I'd see him after school so we could work on my science project together.

"Was there anything different about the street that morning? People that didn't seem like they belonged, or maybe a car that wasn't normally there?"

Luke's eyes remained closed but his head starts shifting around, almost as if he's right back in the moment and searching the street and houses surrounding his. Something that I'm sure the detectives made him go over when they spoke with him during the first interview, but something I'm also sure might not have yielded the proper results.

If there's one thing I've learned working alongside of my dad, is that minute details like a random car out of place or someone walking down the street with their head down and no contact being made, will get ignored and pushed out in favor of something more pleasant or meaningful.

This is why I wanted to do this with him. He might be able to see now what he couldn't then.

"Yeah, there's a lady across the street walking a dog, and another one with a stroller but she looks familiar. Holy shit." He hisses under his breath. "There's a Trans Am parked about twenty feet down from the house. Black with tinted windows."

"If it's that far away how do you know what kind of car it is?"

"I would know a Trans Am anywhere, bro. You know that was my first car when I finally got my license."

"Any distinct markings or decals that might make it stand out in a crowd?"

It's a long shot, the car most likely having been detailed and changed after the kidnapping in order to remain under the radar, but I've got to try. This is a detail that I don't ever remember my dad bringing up with me and one I definitely know Luke never mentioned when we were kids.

This is the break I was looking for.

"Flames on the passenger side door. Red and orange. At least I think they're flames."

"That's good, Luke. That's real good. I can work with that. What's happening now?"

His face distorts, the wide eyed look at realizing the Trans Am's location now changing as his eyes are again closed and something else is taking place in the movie that's playing out in his head, causing him to frown.

"The lady with the dog is walking over to the car. She's bending over and talking to whoever is on the inside. They know each other. Cole…"

"Okay, Luke." I prepare to stop him. I've got more than enough to work with, but before I get him to pull away completely, I have to have one final question answered. "Can you make out a plate number?"

"Negatory. It's too far away. I can see a plate but it's fuzzy."

"But if you saw the car again you would recognize it?"

Popping his eyes open and sitting up on the bed, he nods before his eyes deviate away from mine and across the room, the furrow of his brow a dead giveaway to him being deep in thought.

"I would know that car anywhere, but what are the odds that it's the only one that had that same decal from that time period? I mean flaming out your car was big back then. Everyone was doing it."

"That's true, but not everyone was doing it to a Trans Am. That's a pretty specific detail. I've got one more question before we drop this."

"Go ahead. Can't be any worse than everything I've already remembered."

"What year would you say the car is?"

"If I had to guess, I'd say a '95. Why?"

"Because that narrows down my search. If I can weed out a lot of the newer models, I can get answers faster."

Moving to the end of the bed and tapping his fingers against the blanket beside him, I study him, wondering what he's thinking about. What this new information he came up with is doing to him and right when I get ready to ask, his fist smashes down hard on the bed.

"I know where I've seen that fucking car before!"

He's definitely got my attention now.

"Where?"

"Cole, you've seen it too."

"What are you talking about?"

"The other night when Bethany drove me back to the Carter's to meet up with you. Their garage was open."

Going back in time and pushing past the memory of her body pressed closely to mine while we danced together, the sweet scent of her perfume intoxicating me while the slow movement of her hips against my groin turned the heat up inside me, I force myself to remember how the night ended.

Not the kiss. Not the way her fingers felt gripping onto my hair as she pressed me back against the car and kissed the breath completely out of me or even what happened next at the beach. No, I thought about the garage and closing my eyes, doing what I had just gotten Luke to do minutes before, the image comes up easily in my mind.

Luke was right. The car he's describing, it's an exact replica of the one that was sitting in Reagan's garage.

Holy shit.

Her family was in on it.

Reagan

"Okay, I wasn't gonna say anything, but what's the deal with Mister Hottie? He's been on the phone since he got here."

That's a good question. I've been wondering the same thing myself. Especially since the first thing he asked me when I answered the door not twenty minutes ago was who owned the car in the garage and then immediately jumped on the phone.

"Mister Hottie?" Luke asks, the sound of voice making me jump as he settles in directly behind us.

"Your buddy over there." Bethany explains and Luke laughs.

"Please tell me that it wasn't you that came up with that." He asks me, to which I just shake my head and point to Beth. There's no way in hell I'm taking the fall for that nickname. No matter how accurate it may be.

"Oh please. Reagan would never come right out and say that about someone. She just thinks it and I do it for her."

"Do I have a name?" he pushes and I roll my eyes. He does have a name, but given the way Bethany's cheeks are flushing a whole new shade of pink, something tells me that she's not going to admit it.

So much for her confidence. Looks like she's just like everyone else whenever she gets around Luke. Maybe we're not all that different after all.

"Yeah, Bethy. What did we call Luke?"

"Nothing. We called him nothing, and really, that's not what's important right now."

Right. She's redirecting, which means any minute we're gonna be back on Cole again.

"Who is he talking to?" I ask Luke.

"Cops."

"About the runaway husband?"

"Something like that, yeah."

Great. He's as tight lipped as his friend. Looks like any information I thought I might have been able to score from Luke when Cole was otherwise occupied is going to be a bust.

How the hell am I supposed to learn more about this guy when he doesn't want to tell me and he's got his friend programmed to have his back and keep his mouth shut too?

Getting him to admit that he wanted to be with me was more of a struggle than it should have been, and apparently now, so is this.

Just what the hell am I getting myself into?

"Sorry about that. Work wasn't supposed to be a part of this today, but it couldn't be helped."

Polite as ever. Lips raised in a megawatt smile, one he's used more than once that always seems to have the same response. I can already feel my knees buckling.

"Duty calls, brother. The girls get it, don't ya?"

"Yeah of course. So what's the plan?" Bethany asks, answering the question that I'd been wondering about when Cole called to make plans earlier.

"Honestly, I was hoping you two would have some ideas. Luke and I are out of our element here. What do you guys do for fun?"

"Don't look at me." I say when both guys turn in my direction.

Cole's spent enough time around me to know that when it comes to knowing what's hot in Alliston, I don't have the faintest idea. So why is he leveling me those probing eyes of his? It's already established that I don't get out nearly enough.

"During the day, there's really fuck all to do here. Small town and all." Bethany, sensing my discomfort, explains.

Thank God.

"How about at night?"

"I'd suggest another club, but judging from the way it went down the last time we went, that might not be the safest idea."

She means me. It doesn't take a brain surgeon to figure out with the way she's wiggling her eyes back and forth between me and Cole that she's teasing us about our dance floor seduction.

If only she knew everything that happened after it, maybe she wouldn't be so quick to tease us.

"Well, since we're here already, what do you say to us taking you to lunch?" Cole asks, giving me a lopsided half smile when I meet his gaze.

"Fish gotta swim, birds like Reagan gotta eat." Bethany jokes and when I react and dig my elbow into her side makes Luke laugh. "Ouch! What the hell was that for? It's not like I lied!"

"I am *not* a bird."

Looking up and meeting Luke's eyes before they start to move down from my face, taking me in, I blush before turning away and focusing my attention on Cole. It's already awkward enough with how charged up I am since they got here. Having Luke taking my personal inventory only makes it worse.

"Lunch sounds great. Just let me grab my purse and we can go." I respond softly to Cole before pulling Bethany by the hand and dragging her toward the stairs.

I don't care if she wants to stay where she is. I need a minute away from the eyes that despite not feeling anything other than awkward while it's happening, I still feel can see right through me.

Focusing on getting as far away from the penetrating stare as I can get, I don't even realize I've been holding my breath the entire time until I finally push my way through my bedroom door and release the world's largest sigh. One that alerts Bethany to just how affected I am by what happened downstairs.

"What was that all about?"

"Did you see the way Luke was looking at me after you called me a bird?"

"Yeah, I did actually. I think he agrees with me."

"Bethy, doesn't this feel weird to you at all?"

Laughing, she throws herself all the way back onto my bed, looking up the ceiling and the glow in the dark star décor I asked my parents for a few years before. Humming to herself before finally lifting herself back up to a sitting position and giving me what I'm after.

"Do I think it's weird that these two guys randomly show up at Jake's show and they seem to want to spend as much time as possible with us? Sure, but Reagan, you have seen them right? I'm willing to overlook a little weirdness."

A little weirdness is an understatement, which is funny since in the few times I've been with Cole, I feel anything but weird.

"You don't think it's strange that he wants to be with me?"

"No. I've been telling you for years that you sell yourself short. You're beautiful, Reagan. Inside and out. It's not really a stretch that a guy like Cole would see that, get to experience it firsthand and want more."

"What about Luke?"

"What about him?"

"You two have gotten pretty close."

"Maybe, but that's because I don't operate on the same setting as you do. I like things that move fast. I'm not a fan of the chase. I want what I want and I'm determined to take it."

"And I'm not like that?"

Sliding across the bed and reaching for the book on my nightstand, she picks it up, studying the cover before turning it toward me with a smile.

"This is a prime example of why you're not like me."

"Because I like to read?"

"No. Because you like to read books called *'Stolen Breaths'*. Which if I had to guess, is probably a magical love story between two virtual strangers that find something in one another that they can't live without."

She got the magical part right.

"My point is…I go into every situation with no strings. I don't get attached to anyone. I have fun and I move on. I'm good with that. You want this," she pauses reading over the back and tapping it when she finds what she's after. "Cooper Hudson guy. You want the guy that steals your breath. The one that can see straight into your soul and would stop the world if it meant you would be happy. You want the sweet, sometimes overpowering, and for damn sure life altering love story."

Shit. She knows me too well.

Cooper was pretty great.

The only problem with her logic being, Cooper is a fictional character and Cole is very real.

"Basically, Luke gives me what I want and if you stop overthinking it, I'm pretty sure Cole could be your Cooper."

Is she right? Is all of this one step forward and two steps back stuff I'm doing preventing me from having something that has the potential to be great?

Am I sabotaging my own happiness?

The answer to that question comes a little too easily.

I am, because deep down I don't think I deserve it.

"This would be so much easier if we could just pull guys out of books."

"Rea, you're weird." She laughs, tossing the book back on the table before hopping back off the bed and heading toward the door.

"Now, no offense, but there are two extremely good looking men downstairs that want to take us out to eat and you know me. I won't ever turn down a free meal."

"Only because it's fuel for what you're gonna end up doing after." I shoot back, raising and wiggling my eyebrow at her when she rolls her eyes.

"Well, I won't be able to work off the calories at all if we spend the rest of the day hiding away up here. So put your big girl panties on and let's go."

Cole

"What did you find out?" Luke asks once the girls are safely up the stairs and completely out of earshot.

"The car is in James Carter's name but hasn't always been his. It was previously owned by Martin Francis."

As suspected with the way his face contorts, Luke's clued in pretty quick to just who that is and what it means for how deep this runs.

"Hallie's dad?"

"One and the same."

"Holy shit, Cole. I always thought there was something a little strange about that guy when we were kids, but I just thought it was because he was built like a machine and was covered in tattoos."

This is where things get trickier. Samuel, for as long as I can remember, was on the Grayson kidnapping. Compiling information, hunting down clues and interviewing pretty much every person that lived on our street at the time. He had extensive notes on Martin Francis and a lot of it is because just like Luke, he had his own suspicions about the man.

I remember it. I remember the conversations he had with Rachel when he didn't think I was around where he voiced his belief that Martin was in some way involved. Well, no. It wasn't exactly like that. It was his suspicions about Hallie playing a bigger part that took him down Martin's road.

When John picked up the case when my old man died, he'd gone back to Martin again. I saw notes about it in his journals, but there was no mention in any of them that said he had concrete proof the man was involved.

Not like now.

"Until I can dig into this more later, we have to work on the assumption that the entire Carter family is involved and that at the very least, James and Martin have a personal relationship."

"Cole," Luke says, raking a hand over his face and sighing. "I know you don't want to hear this right now, but we need to tell Reagan. Warn her."

He's right. I don't want to hear it. We're standing in her house for Christ's sakes. We're supposed to be two random guys from Toronto that are into the girls about to make their way downstairs any minute and want to spend as much time with them as possible before heading back to the city. Nothing more.

I'm not dropping this on her.

Luke is right though. I need to warn her about her father, but not this second. Not when she's already skittish.

Did he completely forget the way she bolted up the stairs a few minutes ago?

"You saw the way she took off when you looked at her. How do you think she's gonna react when we say 'hey, your dad might have hired someone to kidnap you seventeen years ago.' It's not the right time."

"With the way this is going, there's never gonna be a good time!"

Before I can tell him to lower his voice, I hear the click of shoes on the floor and turning toward it, find myself face to face with the confused yet inquisitive face of the very girl we're talking about.

"Never gonna be a good time for what?"

Switching gears quickly, Luke grins before grabbing Bethany's hand and pulling her into him. The only sound Reagan's surprised gasp and the sound coming from them as he kisses her.

I've gotta hand it to him, he definitely knows how to change the subject.

"Mmmm." Luke moans once they've separated and he's affectively silenced any questions she might have had. "I stand corrected. Looks like I found a good time after all. Now, who's ready for lunch?"

Chapter Sixteen

Cole

I've never been much of a reader, but ever since I came across that journal entry from John, any spare minute I'm not with Sunday, growing closer and despite my better judgment, falling for her, I'm in this room reading.

This new room. The one I called the owners of the B & B and moved into once I found out they had the space free. The one I chose because I just had to keep myself close possible to Reagan.

If I wasn't going to warn her about James right away, then I was gonna make damn sure she was protected until I did.

I'm sitting in a space that by all rights deserves to be on a show room floor more than lived in, reading dusty journals from a guy in an attempt to learn the truth that for whatever reason, everyone seemed to want to keep me in the dark about. Learning what made the man I once called uncle tick. What drove him to do the things he did. Why I'm the one that's putting the pieces of Sunday's life back together instead of him or even before he passed, my dad.

This latest journal entry giving me a lot more insight into just how deep my part in all of this was then any that have come before it.

Words, thoughts and plans all laid out in his messy scrawl, but with two common themes, as a lot of the entries around this time are. Sunday Grayson and me.

March, 2009
Cole asked about Sunday again today. I didn't have the heart to tell him I found her. That the baby girl he remembers, the one he can't seem to escape even twelve years later, is alive and well living what looks like a peaceful life.

Samuel and I have spoken a great deal about the boy and his feelings for the Grayson girl. A lot of them going back to the way

Cole was when Samuel found him and how it manifests itself now in Cole's insistence that he help locate the sister of his best friend.

So much shit has gone down over the years that I can barely recall what I did during most of my high school years, let alone remember asking about Sunday. In fact, I don't think I can remember a time other than when she first got taken that I talked to either of them about her at all.

Luke and I had conversations about it. I can remember those easily. I wanted to do whatever I could to help him cope at the time, but that was just me trying to be a good friend. I never took it farther than that.

According to John though, I did and more than just the one time. With the way he describes me, and what he believed was going on with me at the time, it's almost like he saw what's happening between Sunday and me before it happened.

I never took the old guy for a psychic, but there can be no denial that when it comes to the way I feel about Sunday, he definitely nailed it years in advance.

There's a determination in the boys eyes. One I don't see often enough outside of my profession. He has done nothing but deny the legacy that he will one day be a part of, yet when he stands before me asking about the Grayson girl, he bleeds this life through and through.

There is a connection between Sunday and Cole. I can't put my finger on what it is they share or why the boy seems so determined to help when it pertains to her, but it's there and with each passing day, despite Samuel's previous excuses for it, it grows stronger.

He is connected to the Grayson's. His friendship with Lucas seeming to be the logical reason, but after much observation, is about more. One only needs to look at Samuel's original observations from the time the girl was brought home to see the truth. He connected to her the moment he held her and his heart from that day forward has been unable to let go. Not of the case and not of her.

A lot of what he's written at first glance reads like total bullshit. Sure, I had a soft spot for Sunday, but past that, there's no real indication that his final words on the page are accurate at all. If my heart really couldn't let go, then why was I able to move on with my life?

I had issues, sure, but everyone does. I still went on to meet and be with Miranda, put the truth about my birth parents in the past and live a productive, albeit lonely life.

Sunday wasn't a part of any of that.

Except she was and the tattoo on my chest is proof.

Dates that no one but Luke and I will understand, but ones since around the time of this journal entry, I've carried in the exact place the tattoo covers.

My heart.

I might think John is touting some serious bullshit, believing there was more going on with me and Sunday then there was because I've blocked it out and can't seem to remember it, but the tattoo, the way I am now and even taking this job to begin with, it's all proof he wasn't wrong.

There *is* something there. There always has been.

But if I thought any of his early words were crazy, it's the final words on the page that chill me to the bone.

June 21, 2014
Cole needs to be the one to bring Sunday home. Knowing what I do now about myself, it's time to put the wheels in motion so in the end that goal can be achieved.

It all makes sense now.

Why he kept the truth hidden. Why he seemed to be so interested in me whenever we were around each other. Why I always caught him staring. It's because he was sizing me up.

Making sure I would be strong enough to do what he wanted me to do.

The reason I'm here with her. The damn stipulation in his will that made everything happening now possible. It was all a damn setup.

He held onto this secret because he thought I needed to be the one to bring her home. That because of this imaginary connection he conjured up in his mind, no one else would do.

John made Sunday and I pawns, knowing that once I became invested I wouldn't be able to walk away.

Just how fucking deep does this go?

I need answers. I need to remember, but most of all, right now, I need her.

I need Sunday.

Reagan

"Jimmy, there's something you're not telling me."
"Don't be absurd, Katherine. You know as much as I do."
"Then what are all the phone calls in the middle of the night about? For the past week you've been holed up in your office, whispering over the line. I would assume you're having an affair, but you would have to stop working long enough to do that."

This is the second time I've woken up to them arguing like this. The first time, I caught the tail end of a conversation where my mom brought up a lawyer and my heart dropped into my stomach. I love my parents and for as far back as I can remember, they've had a solid marriage. If my mom was bringing lawyers into it, it could only mean one thing.

Divorce.

Something I don't want to happen. My parents can't split up. Not when their marriage is what I use as the basis for what I want in my own life.

"Well, I'm glad you have that much faith in me."
"I know we said we would wait to tell her, but if you sneaking out of bed in the middle of the night to take calls has anything at all to do with Reagan, then maybe it's time we rethink our original plan. Maybe we should tell her the truth now."
"Reagan has finally gotten herself to a place where she's happy again. The last thing we need to do is drop this on her,

Katherine. I do agree that we need to tell her the truth soon, but not right now."

The clicking of my mom's heels against the hardwood floor is a giant indicator that much like she always does when she's upset, she's pacing. What she doesn't get because she has no idea that I'm even hearing this right now, is that I need her to stop pacing and start talking because I want to know what it is they're hiding from me.

What truth do they need to tell me? But more importantly, when did this family start keeping secrets in the first place?

"I can't believe I agreed to this."
"Now is not the time to be having second thoughts. You agreed this was the best course of action from the moment I brought her home. You can't go back now."
"Jimmy—"
"No, Katherine. When the IVF didn't work and we made a plan to adopt, you were with me one hundred percent. I need you to continue doing that. I know you have questions and when I have more answers, I'll give you what you want. Right now, you need to trust me. You cannot tell Reagan about any of this."

After a few manic seconds of my head swimming with the knowledge of what I've learned from my father's tightly controlled words, he speaks again and puts the final nail in the coffin of what deep down, I should have known all along.

"Promise me, Katherine. Promise me that you won't tell Reagan about the adoption."

All of the times I stood side by side with my mom taking in all of our differences and having people mention how different Jake and I looked. Always wondering why, but never once going there in my mind. Never believing what my dad just admitted.

I'm adopted.

They've been keeping this from me my entire life. Downplaying it every time I asked them why I didn't look or act like them. Making me believe that the reason we weren't similar was because I was just different. Lying right to my face.

Eighteen years of lies.

Breathe Reagan.

This is big but it's not the end of the world.

If only I could believe that.

It might not be the end of the world, but this secret changes everything. Everything I believed to be true about me and my life. The crap they sold me that I bought into easily is completely shattered.

How long were they planning on keeping this a secret? If I don't call them on it now, are they planning to hold onto it forever? Does Jake know? Is he adopted too or did they get the child they so desperately wanted after they brought home the discarded one?

Discarded. Not wanted. Thrown away. It sounds wrong. It *is* wrong. I've always been wanted. James and Katherine made sure of it.

Oh God. I'm calling them by their first names.

What the hell is wrong with me?

I'm adopted. So what? I'm the child that someone else didn't want that James and Katherine brought into their home and loved unconditionally. Nothing has changed. I'm still Reagan Carter.

Except I'm not.

I don't know who I am anymore.

I just know that whoever I am, I can't be here anymore.

I need to get out, but in order to do that I need to somehow get down the stairs without them realizing it. I need to be able to distance myself from their private fight without completely breaking down. Which right now, is all I really want to do.

Grabbing my phone off the bed and bringing up the messaging window, I write out a quick text and hit send. Praying that what he said after our lunch date with Beth and Luke a few days ago, is still true and that he'll be my escape from here.

I need you. Do you think you can pick me up?

Tapping my foot on the floor nervously, my eyes never once leaving the screen as I'm completely frozen in place waiting for the response, I jump out of my skin when it begins vibrating in

my hand. The shock of everything I've just heard obviously taking enough of a toll to where my nerves are completely shot.
I'm on my way.
He did mean it. He's going to be there for me.

*****

"Thank you for today."
Ever since we all went to lunch together, Cole has been a completely different person. Sitting around the picnic table outside the mall talking and laughing, engaged completely in everything going on around us. The complete opposite of the way I'm used to him being.
From the moment I met him, he's always been closed off. Not giving much in the getting to know you department. Too spooked to get close. That wasn't the case at all today.
So when he leaned in during a moment where Luke and Beth weren't vying for our attention and told me he was moving into the bed and breakfast close to my house, wondering at the same time if I would mind moving some of the work related stuff he'd brought with him over, I'd jumped at the chance.
Sitting in the car now, hours after we'd moved everything over, the urge to thank him hits.
"There's nothing to thank me for, Reagan."
"But there is."
"Okay. Well in that case, you're welcome."
"I like Luke and I like the way he is with Bethany, but..." I feel the need to explain, my words falling flat and fading off the more I start overthinking the way they sound.
"But the way they are is too much sometimes?"
Who knew Cole was a mind reader?
"Exactly."
"The truth is, Luke hasn't acted like that in a long time. Usually the only way to get him to do it before was to ply him with booze. Having him acting like a total idiot with her and doing it sober was a nice change."
"Did you want to keep hanging out with them?"
"No. Don't misunderstand. This is definitely more my speed. I was just commenting on how nice it was to see my friend actually look happy again."

Before I can react, make sense of the warmth that now floods my chest, he leans in until the scruff of the hairs on his face scratch my cheek and he halts my breath altogether.

"I'm right where I want to be."

Flushing, thankful in the moment for the darkness so he won't be able to see the way he's affecting me, I turn my focus to my front door and the brother I can make out standing on the front step waiting.

"I should probably go in."

"Yeah…"

"When you get back to your room and you're settled, will you call me?"

"You worried something is going to happen to me between here and the thirty second drive to my room?"

Laughing awkwardly, not wanting to let on that even though there's nothing to fear, my irrational brain did in fact go there, I blow it off. "No, actually. I just didn't want to go to bed without hearing your voice."

It's been happening a lot lately. This need I have to keep him close. Like if I don't then I'm going to wake up one day and he's going to be gone. Vanished into thin air. I can't explain why, but it's there and I can't be bothered trying to force it down anymore.

Cole is important to me.

"You won't, but Reagan, if this is ever about more than just hearing my voice before you fall asleep; if you need me in any way, call me."

"Really?"

"I'd like it if you did. Especially now that I'm so close. You don't have to be alone anymore. If you need me, I'll be there."

*****

Allowing my body to collapse down on the bed, I close my eyes, blocking out my house, my parents who aren't actually my parents, and my life altogether. Focusing my mind on one thing only.

He's coming.

Cole will help me forget.

Chapter Seventeen

Cole

I can't remember a time where I ever felt needed.

My mom didn't need me. All she needed was her endless string of men and her drugs. That's what completed her. Hell, Sam didn't need me, though he damn sure went out of his way to make me feel like he did. He needed someone and I filled the void, but I don't know beyond all doubt that he actually needed *me*.

Miranda made a good show of needing me, but the reality is, I was the one that needed her. I was on the fast track to nowhere. My attitude was shit, I was pissed all the time, and I was taking it out on anyone stupid enough to give me even the slightest bit of attention.

I was displaced when I met her. I had just gotten through a bullshit attempt to find my birth parents, my dad no longer a mystery thanks to all the time I spent shadowing Samuel and picking up the tools of his trade. Coming face to face with my mom again and seeing her just as messed up as she'd been when I was a kid.

To say I was a fucking mess would be the understatement of the century.

In a way I think that's what Miranda latched onto. The broken parts. My weaknesses. My neediness.

She was the more outgoing one. As strange as it sounds, she was the stronger one of what would later become us. She did everything in the *right way*. She just knew the right buttons to push to get the screwed up guy with a chip on his shoulder to open up.

And to this day I don't know how she did it.

I mean, it wasn't like she was anything special. She didn't stand out in a crowd or anything. None of us did. We were just two random people who happened to take a chance on each other and for a while it worked.

But she didn't need me. No one has ever needed me. Which is why it's so fucking weird staring at this text from Reagan and believing that she does.

How can she need me when all I've given her since I walked into her life is unease and confusion?

Getting close to her and taking things to such a deep level that walking away from her in the end is going to be fucking impossible, then in the next breath pulling away because right when it's about to get too real, the truth slams to the surface and reminds me why this is a bad idea.

Yet she didn't text Beth or her brother. She texted me.

Reagan—*Sunday*—needs me.

The same way I need her.

Lifting my head at the sound of the tap against the window, I see her and catching the smile that lights up her face, I return it with one of my own.

Does she know I was thinking about her? Is that why she's standing here beaming?

Leaning over and flipping the lock on the door, I move back and wait as she slips her body down into the seat. A move she's done countless times over the last week since I showed up, yet one I never tire of watching.

As wrong as it is feeling this way about her, connecting the girl she became with the one she used to be, I like the way it feels when she's going anywhere with me.

I could definitely get used to a life made up of small moments like this with Reagan.

Shit.

Sunday.

"Thanks for coming. I wasn't sure if I was pulling you away from something important."

You are what's important. "It's okay. If you hadn't texted first I would have."

"Really? Why?"

Because I haven't been able to get you out of my head. Your kiss haunts me. Lingering long after it's gone and infiltrating my dreams. Making me ache for something I can never have.

God, I hate my brain. I'm sick of the tug of war between it and my heart. The struggle is way too real.

"I didn't like the way we left things last night."

Well, no. That's not right. The way the night ended was fine, it was what happened before I brought her home that I'm having trouble dealing with now.

How after getting passes from the owners of the B&B, we'd gone to the movies and spent the majority of it wrapped up in the taste and feel of each other before she suggested going to the beach. Wanting a do over after what happened the last time we were there. Agreeing and going there, having one thing lead to another until we were getting closer than we'd ever been.

One innocent kiss turning into something else until we lowered our inhibitions, along with various pieces of clothing and were walking a desire precipice half naked in the sand. A single step away from completely falling into the abyss. Ready and willing to take things to the next level until at the last second, I pulled away.

A moment I'd much rather forget about entirely, but because of the way I feel about her, my heart won't let me let go of.

Not until I've made it right.

"I told you last night. It's okay. Really."

"No, it's definitely not okay. What happened between us, Reagan...I didn't want it to end."

This confuses her. I can tell by the way her nose scrunches. What I'm admitting is the complete opposite of the way I acted then. I can't blame her for being confused. I'm confusing the fuck out of myself.

"Then why did you pull away?" she asks and I can't help thinking it's a damn good question. I wouldn't mind hearing the answer my head blurts out to this, since it's damn sure not gonna let my heart win.

"I haven't been with anyone that way since...well, since the two people I told you about and Miranda. I wasn't planning on ever getting that close with someone again. Sex complicates things. What I've had with you has been the most uncomplicated my life has ever been. I wanted to enjoy it a little longer."

"She really hurt you, didn't she?"

Ugh. I hate questions like this because I'm being honest just as much as I'm lying to her with any answer I give. Yes Miranda screwed with me when she walked out, but she hasn't been lately. It's over and done with. Not wanting to fuck my best friend's sister is what's screwing with me these days.

Even if I know it wouldn't be fucking with Reagan.

"No." I answer honestly, screwing every other response I could give. I'm already keeping shit from her, I won't add more lies to it. "That honor goes to the girl wearing the sundress that's currently taking up space in my passenger seat."

"I'm screwing with you?"

"Not in the way we're discussing, but yes, Reagan. You're most definitely screwing with me."

This seems to give her pause. She goes so quiet, I can't even make out the sound of her breathing. *Shit.* Figures the one time I attempt to be honest it backfires.

I'm not set up for this shit. There's a reason why I stayed away from relationships. This shit is too damn complicated.

"Maybe this wasn't such a good idea." She finally speaks and I resist the urge deep inside to fight her. Let her walk. It's safer because with what I came here to do, it won't be long before I'm breaking her in a whole different way.

"I'm sorry I bothered you."

Her hand goes for the door and I notice the smile from earlier has been stripped. An aching sadness I'm all too familiar with beginning to take up residence on what is the most beautiful face I've ever seen.

This is wrong. She's not supposed to look like this. Look like me.

She was meant to be the sunlight. Not the storm.

"Wait." I call out, moving quickly as she keeps moving, stopping her as she's about to push the door open by resting my hand on hers.

I'm going to hell for this, but I can't let her leave. This is one hell of a mess I've gotten myself into, but I can't stop.

It's hard and it's going to get messy, but I want it that way because I want her.

I want Sunday.

"Please stay. You said you needed me."

"That was before I knew I was screwing with you."

"I didn't mean it like that. Can you forget I said it? It was thoughts that never should have come out of my mouth. Especially when they come out wrong."

I'm damn near babbling now with my attempt to make her stay, but I don't care. I need to make her understand that what I blurted out wasn't true. That I want her here. That I'm always going to want her with me.

That I need her.

"Cole, I can read into a lot. I spend days with all different types of people, learning them inside and out so I can serve them better, but I can't do that with you. You confuse me."

"Feeling's mutual sweetheart."

"What's that supposed to mean?"

"It means..." I pause, sucking in a nervous breath, just as afraid of my own reaction to what I'm about to say next as I am hers. "I need you too."

Reagan

"I'm adopted."

Before I've even processed Cole's admission, one that judging by the pause he took before admitting it, wasn't the easiest thing in the world for him to get out, I'm blurting out the reason I texted him.

This is not going the way I wanted it to at all. I knew we were going to have to talk about what happened between us when we'd gone out, but I was hoping it would have been tabled until later.

So much for that idea.

Years I dated Walker and not once after our first time together did I feel comfortable enough to do it again. It was always the wrong time, awkward, or I just wasn't feeling it. A couple of weeks around Cole and the burn that just being in close proximity to him creates, along with the aching need to touch, taste and give in, and it's all I seem to think about. All I want.

I've never felt this kind of connection with anyone before, but I do know that I want to act on it. I want Cole to be the one I give everything to. Body, mind, heart and soul. He's deserving.

Watching as he swallows and coughs, literally choking on my admission, I frown. If I thought my own surprise was bad, seeing it happen with someone else is far worse.

He's as blown away as I am.

"Excuse me? You're what?"

"I heard them talking. It's why I texted you. I needed to get out. I needed to be with someone that doesn't know me."

Now he's the one frowning. *What the hell?* It's not like anything I said was a lie. He knows about me, but he hasn't been around long enough to know everything.

"I need you to go back to the beginning and start over. You heard who talking?"

"My parents were arguing. My mom wants to tell me the truth. She told my dad I'm old enough and deserve to know, but he's against it. He wants her to keep her mouth shut. He was the one that admitted it. He said I was adopted."

"Rea—" he starts but after a few seconds seems to think better of as his lips tighten and he focuses his attention on the windshield and the cars passing by outside.

Damnit. I made things worse.

"I'm sorry. I shouldn't have blurted it out like that. It wasn't even what we were talking about."

Wasting no time after hearing my apology, he turns back and his body begins to shift, this time moving across the enclosed space instead of further away the way I'd been expecting. Not stopping until his hand is resting on my shoulder and he's squeezing.

The closest thing to a sympathetic embrace as I'm going to get inside the car.

What I really want but can't work up the nerve to say is that I want him to say screw it to consoling me the way I know he wants to. Acting instead on what we shared last night and finishing what he started. What is guaranteed to make me forget. I want him to grab me and pull me across until I'm staring into those piercing blue eyes while straddling his lap. I want his hands on my face, forcing my lips down on his until all I can taste, touch and smell is him. I want him to kiss me like he's never kissed another person in the world before. Lowering the seat back and letting me control his pleasure by using my mouth on him. My tongue sucking, licking and tasting every inch of the body he's holding back from me now, until he loses control and is filling my mouth with his pleasure.

Blocking out everything that came before it.

Making me forget.

Cole doesn't play by those rules, though. He keeps the raw urge I know he has to strip me and screw me under wraps. Only for him. So if I want him to take me and make me forget, I'm going to have to be the one to start it.

Shifting my body in the seat, his hand falling off my shoulder with how quickly I move, I pull his face to mine and before either one of us can second guess the moment, pull away, or worse—change our minds, I nip his lips with my teeth before pressing them down hard, a guttural moan somewhere deep inside unleashing on contact, forcing him to react as he grips my body, his arms not stopping until they're tightly wrapped around me, bringing me as close to him as the car allows.

Feeding me with his need.

"Fuck, Reagan." He growls against my lips before kissing the space where my lips begin and trailing a path backwards across my cheek and down my jaw until he reaches my neck.

Heart pumping, my breathing now erratic yet catching at the same time, I allow my hands the pleasure of feeling his hair, gripping it tighter as he begins to suck so hard on my neck I can feel him marking me. Branding me, inside *and* out.

"Cole," I sigh, his name falling in a breathless plea. "I want you."

He pauses his movement, the suction from his lips lessening as the blood pools back to the surface of my skin and I immediately fear the worst. That just like last night when we were about to give into what this is between us, he's going to put the brakes on.

Pulling back but still keeping his body close, he finds my face and with a soft run of his fingers across my jaw, he presses his lips to mine again, with just as much need as before.

"Don't do that. Don't overthink it. I want you too, Reagan. Just not in my car. You fucking deserve better than that." With another brush of his hand across my face, his eyes moving over me, searching mine for understanding, he leans in again but instead of kissing me, he speaks. "Come home with me?"

"Home?"

"Shit. The B&B." He curses. "That makes it sound even worse."

"It's okay. Take me home, Cole."

Starting the car, he puts it in gear but before pulling away from the curb, turns back again, this time, all traces of desire wiped and a need for assurance in its place.

"Are you sure about this, Rea?"

He's never called me Rea before and this being the first time someone's said it after what I just learned, I expect to hate it. But there's something about the way it comes out sounding like Ray that makes me want to come up with a way to get him to do it again.

"Do you always question a girl this much when she wants to get in your pants?"

Chuckling, he shakes his head before reaching over and slipping his hand over mine.

"No, but this isn't just about letting a girl in my pants and I think the girl agrees."

"If it's not about getting in your pants, then what is it about?" I bite my lip nervously and ask, wanting to hear his response almost as much as I want to hide the fact that he's right and it is about more than just sex.

"Letting the girl into my soul."

If I didn't already know I wanted to be with him, his answer would have sealed it. I was right before. Cole is the right person.

I'm falling again….and this time, I don't want to stop it.

Chapter Eighteen

Cole

I've never been more thankful I came across that bed and breakfast last week than I am right now.

This, with Reagan, it's not something I want to take back to the Super 8. Not when with the type of girl she is; innocent and sweet, demands that she deserves better.

She deserves everything. And with what I know is going to take place once I get up the nerve to step through the door, this is my way of giving it to her.

I'm venturing into unfamiliar territory. The last time I connected with someone so long ago, I'm not even sure I have it in me to be able to give Reagan the perfect moment she deserves.

For months before the implosion things with Miranda were distant. Disconnected. There was no intimacy whatsoever. I felt the loss, but at the time hadn't realized just how deep it went in order to hit pause in an attempt to stop what was happening.

I'd given up on her the same way she did me.

The attraction and then the feelings that developed over the last two weeks with Reagan, is a direct threat to the relative calm I'd managed to achieve after that.

This gravitational pull I have toward her though; the overwhelming need to feel her because of the way with just a touch she seems to penetrate the darkest and roughest parts of me, I can't fight it. I've tried and failed with every attempt I've made. Just a few days in her presence and I believe in something again. She made me believe that if I just give in and allow myself to fall, she'll keep me safe. Not just the twenty-four year old me, but the scared little boy that came before him.

The boy that on a cold day eighteen years ago, came face to face with purity and been forever altered by it.

Reagan, she's bringing back the best parts of who I was and the man I became, wrapping me up in her warm cocoon and giving me a glimpse of what loving her will feel like.

What it already feels like.

I don't care that I've only been around her a couple of weeks. That logically this shouldn't be possible. None of that matters because it's happening. What I feel when I'm with her, it goes back a lot farther than a couple weeks and there's not a damn thing I can do to stop it now that it's been ignited.

"Cole," she gasps after I've unlocked the door and motioned for her to step inside. "It's beautiful!"

Her surprise at the way the room looks knocks me a little off guard. Given that she was the one to tell me about this place, I assumed she'd been here before. As her eyes dance from one end of the room to the other, lingering on some parts, like the bed and the mirror situated on the wall behind it, it's easy to see that this is a first for her.

A first for both of us.

With all the motels I've managed to find myself in over the years, both on cases with my dad and then ones I pursued alone, I've never found myself in one quite like this.

I knew when I stepped in it for the first time that it was perfect, and the wonder in her eyes now as she takes it all in, solidifies it for me.

Stepping forward and applying just the right amount of pressure to her shoulders with my hands, I massage her, unable to mask the arousal that I experience with the shudder that escapes as my hands seems to hit the right spot and her body leans back into mine, giving me full access to her neck. Seizing the opportunity by leaning forward and letting my lips brush against her skin, I begin leaving a trail of kisses from her ear all the way down while using my hands to slip her jacket off her shoulders and down her arms until I'm gripping it tight and she's completely freed.

Pulling back and feeling her eyes on me as I step over to the two footstools at the end of the bed as I place her jacket across them, I look up and seeing myself in the mirror, turn back and motion for her to follow.

"Come closer. I want to show you something."

Doing as I ask and stepping toward me, I revel in the feel of her skin when she reaches me again, pulling her closer into me as I adjust our position, lifting my hand and pointing to what it is I want her to see when we're perfectly aligned.

What I know she's already seen, but what I'm now going to explain the importance of.

"Why do you want to show me the mirror?"

Slipping my arms down until they're resting comfortably on her hips, I shift her body until she's facing it head on and I'm standing behind her. Both of us in the heat of the moment, what I know is simmering on the surface just waiting to be released, on display for the other to see.

"I want you to see what I see when I look at you. I want you to see the person looking back and know that with all of the people I've met, interacted with or just been around, there has never been one as breathtakingly beautiful as you."

"Cole…"

My words are a lot for her to digest, I knew they would be even before I said them, but I have never meant anything more. She's beautiful and here in this moment, where it's just the two of us, I want to erase everything she's ever believed or been told about herself and worship her like the goddess she is.

"I know earlier with how driven we were by the need to have each other that you think that's why I brought you here, but it's not. As good as it felt giving into the need to have you, I want this to be about more than just fulfilling a need or screwing it out of our systems, Reagan. I brought you here, and more specifically, wanted you in front of the mirror, because I want you to be able to see the affect you have on me. How even now, standing fully clothed in front of a mirror with your body resting so damn delicately against mine, you've possessed me. How you own me so completely."

Her body quivering as I speak. Quiet whimpers escaping with the path my hands take across her lower back. Each stroke of my fingers slow and deliberate, feeding into her desire almost as much as my own. Her body molding itself into mine when I reach her thigh and sliding her dress up, begin my descent lower. Each breath she takes and the way it catches in her throat based on every touch of my hand to her now exposed skin adding fuel to my already smoldering fire.

Her body's natural response to me exhilarating. Intoxicating. Addicting.

The soft way she says my name giving me everything I could ever want. Imagining the way it would sound hearing my name in connection with her pleasure paling in comparison to the way it actually sounds as I put pressure behind my hand and begin to rub. The soft moan escaping a dead giveaway that she's as alive in the moment as I am.

Captivated by the vision of us in the mirror, I smile when her eyes begin to roll back before closing. Her body betraying her and giving into me more as she begins to sink deeper the more my fingers brush across her sex. Leaning in until I can feel my lips brushing around her earlobe, I push her further, running my tongue down and bringing it into my mouth and sucking softly when I hear another moan escape.

"Open your eyes, beautiful." I whisper. "I want you to see how fucking gorgeous you are when you surrender."

It takes a few seconds, but she does as I ask. Her eyes fluttering open slowly and raising to meet the reflection of mine as they're trained on her every move in the mirror.

Leaning back, her hips and the circular motion of her ass as it presses into my groin further awakening the beast inside and making it ache to be freed, I capture the skin of her neck with my teeth. Overcome by the taste of her, a low involuntary growl releases as I run my tongue over the light pink bite marks on her skin. Marks that now that I've seen them, begin to suck on in order to soothe. The last thing I want her feeling during our time together, even the slightest bit of pain.

Slipping my hand away from her heat, my eyes trained on the mirror as I watch her unravelling in my arms I lift it until it's resting above the straps of her dress. The last thing I see in her desire ridden eyes being pain of any kind.

Hooking my fingers into the straps of her dress, I slip them down slowly over her shoulders, pressing my lips to every piece of her exposed skin, inhaling her soft powdery scent as I go. Drowning in it. Letting it drive me and fuel me as it shoots the temperature in my body up until I'm ready to combust.

Sliding it down over her waist and mesmerized as it hits the floor, I chance one final look at us in the mirror as her eyes finally give in and shut again. The softest smile playing across her lips as her chest moves up and down slowly.

She's fucking beautiful.

As she steps out from its place around her ankles, I slip a hand down around her legs and with the one across her shoulders, lift her into my arms. Stepping to the bed and laying her down gently, unable to look away as she slides herself slowly back toward the head of the bed, the playful raise to her lips as she watches me watching her tempting me.

Climbing up and making quick work of the distance between us, I capture her lips with mine, biting into her bottom lip. Her sharp intake of breath all the opening I need to slip my tongue into her mouth, actively searching for hers. Feasting on her when our tongues finally collide and we deepen the kiss. Our breathing ragged and heavy with need and the craving we have for more.

Moving away from her lips and kissing down her jawline, I devour her neck with my lips making sure to leave no area untouched as I continue to move down to her chest. Pausing briefly once my mouth grazes the outline of her bra, I nuzzle the area with my nose before placing delicate kisses over each cup as her body begins to move. Taking control of the moment, smiling when she notices my eyes trained on her every move, she slides the straps down and over her arms. Guiding my hand with hers until I can feel my fingers brushing against the clasp on her back, she surrenders the control as I release her breasts by closing her eyes and parting her legs as she lays back on the bed.

A move that to most wouldn't mean anything, but for me witnessing it with Reagan, threatens to break me. Her shedding all of her earlier fears and inhibitions. Stripping away layers of clothing while at the same time opening herself up and showing me the parts she's kept hidden for so long alters me. The level of trust it takes to be able to do this, humbling me. Making me feel unworthy and like the luckiest son of a bitch in the world in the exact same breath.

"Reagan..."

"Hmm?" she murmurs as I lean in and place my lips to the plump softness of her breast, surrendering to the need to taste her and grazing the hardened bud of her nipple with my teeth before beginning to suck.

As her body begins to shift and her exhales become heavier with each brush of my tongue against her nipple, I shift my weight pulling away long enough to repeat the same motion on the other side. So guided by the sound of her breath catching and holding as my tongue teases, flicking over the now hardened nub, I don't realize how lost I am until I hear my own moan of pleasure escape.

Her body with a mind of its own responds to the sound, tightening her legs around me and moving. A silent needy plea for more heard loud and clear as I lean back and start working on my belt.

Moving back off the bed once I've freed myself of the belt and popping the button of my jeans as I pull the zipper down, I slide my legs out, kicking them to the floor before finding her and moving in again. Her smoldering eyes and the way she's biting her lip in anticipation all the motivation I need to strip my shirt off and climb onto the bed again.

Releasing the tight grip on the package in my hand and bringing it to my mouth, I use my teeth to create an opening and slip the condom out. Preparing to put it on, I'm halted when she softly speaks my name and find myself held captive even more when my eyes lift and she's smiling at me seductively. Her intent clear.

Crawling across the bed to me, she slides her hand down until it's wrapped around my cock. Gripping me the way she did the night of our forbidden dance and holding me hostage with her gaze as my pulse quickens with an anticipation of my own. The need for her to move her hand, stroke me and feel what she does to me threatening to burn me from the inside out. A need she gives into as she begins moving her hand, stroking me up and own slowly, pausing just long enough between strokes to pinch the tip of the latex condom before placing it between her lips. In a move so quick and unexpected I have to react she bends over and slips her mouth down over me, teasing and tasting as she rolls the condom down my length, draining away all thought until I'm the one falling victim to the feel of her mouth on me.

"Rea…" I bark though the intense heat flooding me as her mouth reaches the base of my cock and her tongue brushes lightly over my balls.

"Why do you call me that?" she asks as she brings her body back up and our lips connect again. Her eyes dancing when she finally pulls away and begins running her fingers softly over my cheek and down across my jawline. Committing the feel of me to memory the same way I've been doing with her for days.

"You're my ray. You warm me."

"Is that what you were trying to tell me before?"

There was so much I wanted to say to her before, but telling her she's my ray of sunshine, as true as it is, definitely wasn't what I was going for. No. What I wanted to say earlier was something a whole lot bigger than that.

I wanted to tell her I was hers. That I'm always going to be hers for as long as she wants and will let me. But more than all of that, I wanted her to know that I'm falling in love with her.

Words that until the reason I'm really here is out in the open and resolved I can never utter. Not even now when I know that they would mean everything to her.

"Yes."

Slipping her hand down from my face, she runs it across my shoulder and down my arm until her fingers are lightly brushing against mine. Opening my hand at her touch, she slips it into hers and brings it up to rest against her chest. Against her beating heart.

"This is what you do to me, Cole. Do you feel it? How out of sync it seems? It's because every day since I looked across the bank and saw you in that chair, everything has been out of sync. I can't control it. Not the way I want you when I'm awake and asleep or the way I miss you when we're not together. I'm powerless against what's happening here."

"W-which is??" I stammer, choking on the magnitude of her words. What they really mean and the only answer to my question there can be.

"Falling in love with you."

I knew we were connected. That something from our time together as kids had somehow managed to stick around and give us a bond neither one of us understands. But this is more than some silly connection we made as kids. This is what two people who complete each other look like.

I just can't believe that it took getting into a situation like this for me to finally be able to see it for what it is.

This is what love looks like. Not what came before I ended up in her life. This. What's happening between us now and what I hope never has to end.

"You're wrong, Reagan."

Eyes wide, surprise filling them before they start to glisten, I press down where my hand rests on her heart and when her eyes lower to where she feels the pressure, I bring my hand around her back and lay her down, positioning myself silently above her, knowing she's confused but needing to show her the truth.

Words won't do here.

Gripping my cock in my hands and taking control of the desire that drives me after her earlier show with the condom, I stoke it slowly. Once, then twice, slowing down and driving the need higher as I do it a third time as she parts her legs for me again. With her eyes focused on me as I move in closer, my arousal presses into the heat created from her own, my cock twitching with need when she bites her lip.

Guiding myself slowly, I stroke her core, teasing her while readying her for what's about to happen. What does happen when just like before, my name falls from her lips in a plea for everything I've got and am desperate to give.

For all of me.

Lifting my body down slowly, I press my lips to hers capturing them. Capturing her as I guide myself inside her. Moving slowly, conscious of how tight she is, the hold her thighs have grows as she grips me harder, pulling me in until I'm pushing myself in as deep as I can go, my body shuddering and a moan escaping as she takes all of me.

"I'm right, Cole." She purrs, but before I can ask what she's right about, she shifts her body in an attempt to adjust to my size and I freeze.

As good as it feels to experience this moment and every emotion that comes with it, I can't shake the fear I have over moving. Doing something that will inevitably hurt her.

Reagan has only ever been with one person before. She's admitted that it's been a long time. The last thing I want to do is lose myself in the intensity of the moment, how fucking sexy she sounds, or how good she feels gripped around me this way and hurt her.

Raising herself up, she reaches out and brushes my cheek with her fingertips almost as if she knows what I'm feeling. The soft touch of her hand on my skin bringing me out of my fear and back into the moment.

Back to her.

"Cole, it's okay. I want this with you."

With no idea what her words really mean, how they're soothing the fear while at the same time healing the damage inflicted the last time I was in a position like this, I lean forward and kiss her. Infusing her with every ounce of feeling I have to give. And as our bodies begin to move in unison, our hands holding on to each other as if our lives depended on it, I finally release the hold on the words I know she deserves to hear.

"You've already made me fall."

Reagan

When you've spent years hearing people say they love you, after a while it begins to lose its impact. It's why people look for ways show their love. They don't want it to ever lose its meaning.

I always thought that if I ever got to the point where I was ready to hear those words again, they wouldn't affect me. I wouldn't buy into them and give my heart away as easily as I did in the past.

I was wrong. So wrong.

Cole whispering as we made love, telling me that he's already fallen affected me in a way I've never experienced before.

My chest tightened as my breath stilled, my heart swelling before melting as we moved together in unison, our bodies no longer acting like two separate beings, but one fluid one. My hold on him becoming tighter as I gave myself over to him and the pleasure that was ignited the more we moved together. And when after a few minutes he surrendered to his release, his own grip around my body tightening as he repeatedly called out my name, it happened.

The parts of me that I thought were lost, were found.

I was found.

Cole found me. Making me see what he said he wanted to when we got here. What it looks like when we've both been possessed by something greater than the sum of our individual parts.

I felt his devotion as he put my pleasure above his own, each touch and taste making me feel like the most beautiful woman in the world. The *only woman* in the world. A feeling that until now I'd never experienced.

He knelt at the altar that is what we feel for each other and he worshipped me. Letting me experience all he had to give, both of us sated before we'd finally passed out in each other's arms.

Cole Alexander *loved* me.

Loves me, if the tender way he's running his lips across my shoulder now is any indication. So lost was I in what we've shared over the last couple of hours that I didn't even hear him stir.

"Hi…" I smile when I turn toward his touch and am met with his heavy lidded and sated half smile.

"Hey yourself."

"How long have you been awake?"

"A minute or two, but you looked so peaceful I didn't want to say anything and ruin it."

I believe him. He would have stayed wrapped up like this forever if it meant I rested peacefully and wasn't disturbed.

"Rea, how are you feeling? Any pain?"

After we'd made love the first time, cradling me in his arms for a moment or two while he kissed me like he needed it to breathe, he'd slipped from the bed and across to the bathroom. After listening to the water running for a few minutes and wondering what was going on, he'd made his way out and over to the bed, taking the warm wash cloth in his hands and wiping me down gently.

A tenderness I've never known before present as he took care of me. Asking if I was experiencing any pain and not understanding when I smiled and explained that the type of pain I had was the pleasurable kind.

Something he still hasn't wrapped his head around if he's still worried about how I'm feeling.

"It's tender, but it's supposed to be. You can stop worrying." I admonish, playfully slapping at his chest. My hand landing right over the tattoo residing over his heart. One I noticed while we made love, but was too caught up in the moment to ask about.

Chucking softly, he kisses my head before bringing his arm tighter around and pulling me closer.

"If you say so. It's just been a long time since I've had to think about someone else. I tried to control myself, but there were a few times where I couldn't. Where I went faster than I should have. Deeper even."

If I didn't already know I was falling for him, his concern for me would do it. I was there in the moment with him and the way he sees it isn't the way I remember it at all. Yes, there were times where he moved faster and pushed himself deeper, but for everything he was willing to give, I gave back just as much.

What we shared was perfect. Just like I knew it would be before we even started.

"It was perfect, Cole. You were perfect. Please don't overthink it. I do that enough for the both of us."

Brushing my lips against his nose and feeling his lips raise against my chin as he smiles, my heart threatens to explode from the overwhelming feeling of rightness that floods though it.

This…us. The way we are. It's right. It's the way it should be.

It's what happiness is supposed to look like.

"No regrets?"

What an odd question. I think as I burrow my body into his, making sure as he leans back that my head rests securely over his heart.

"Why? Do you have them?"

"No." he answers quickly. "Wait, that's not exactly true."

Swallowing the nervous lump threatening to rise, I lift my head back up until we're face to face again, studying him in the few seconds I have before he speaks again. Cole is closed off again. Whatever he may be thinking or feeling isn't evident in his eyes, which is the complete opposite of the way he was before. Every emotion and thought was on display. He held nothing back.

What changed?

"I regret not finding you sooner, Reagan."

Wow. Bethany was right. I do overthink everything to death.

Placing a soft kiss to his lips, I move down and do the same to his chest. As my lips lightly brush over the tattoo, I give into my curiosity and ask him about it.

"What do these numbers mean?"

"They're dates."

"I figured that, but can I ask what they're for?"

"You can ask me anything you want. This one," he says, pointing to the first date. "Is the day Sam rescued me from the group home."

31.08.1997

"And this one?" I ask, running my index finger over the second date that takes place almost a year later.

03.14.1998

"The day the adoption officially went through. The day he gave me my fresh start and I became Cole Alexander."

"What was your name before?"

"Cole Montgomery." He answers easily, the signs of distress I expect to see bringing up his past nonexistent.

"How many people know?"

"Four besides the woman that gave it to me. Samuel and his wife Rachel, Luke and now you."

Not wanting to let on how honored I am that he trusted me with it, I move on to the next question. The only other one I'm curious about after hearing what the first two dates mean.

"Why were you in a group home?"

My body so attuned to his frequency I don't miss the flinch and how his body, which up until that point had been relaxed and open, tenses as he looks out and away from me before answering. Whatever he's lived through obviously so painful he has to separate himself from everything in order to talk about it.

"My mom was an addict. Heroin, MDMA, Cocaine, it didn't matter. She was up for anything. Add that to the abuse and her need to have a different guy in her bed every night, and I'm sure you can see why I ended up where I did."

That explains a whole lot more about the way he is than I was expecting. If his mom was an addict, it meant he probably spent the majority of his time raising himself and her. Where I grew up in a house that was ripe with people giving affection and making sure we were taken care of and loved, his was the complete opposite. .

It's no wonder he has such a hard time admitting how he feels. Until Samuel he probably didn't even know what love was.

"When did you meet Samuel?"

"He came for a meeting at the end of April that year, but then every week like clockwork after he would come back and visit with me until he finally asked if I wanted to live with him in August."

"What's the third date?"

"Luke's sister. It's the day she was born."

06.10.1997

I know basic details about Luke's sister. That she was taken and never heard from again, but for him to mark his body with her birthday means she meant something more to him than just a sister to his best friend.

"Were you close to her?"

"Pretty close, yeah. There was just something about the way she lit up a room when she smiled and how determined she was to do everything that Luke and I could that bonded me to her. We—Luke and me—we lived for Sunday."

The softness in his eyes and the slight lift to his face, along with the faint trace of a smile that appears as he's talking is a nice look for him. What happened to the little girl is tragic, but having this kind of effect on him for the short period of time they were in each other's lives, is a beautiful thing to see.

He has a soft spot for Sunday. The same way that he does with me if the way he traces a heart on my arm with his fingers before meeting my eyes and blushing means anything.

This is definitely a side of this man I enjoy being a part of.

"What's the last date?"

"The day she was taken."

"Why would you tattoo that? I get all of the others, but losing her had to be hard. Why would you want the reminder?"

"When I went in to get the tattoo done, I thought a lot like you are now. I was just going to get those three dates because they were the most important ones in my life up until that point. Sitting in that chair while the needle outlined them though, I changed my mind."

"What do you mean?"

"The reason the day she disappeared is here is because I made a decision that day to never give up until I found her. For a little while after I had it done, I didn't keep my word. My dad had just died and I wasn't in my right mind. Sunday became the last thing I thought about. But having it here, it meant that even if I put it out of my mind for a little while, I would never forget entirely. I would find her again."

My heart is breaking for him because from the first night we met, I know she hasn't been found. So all this time he's walked around with that date on his chest, reminding him of the fact that he hasn't been able to find the little girl he seems to have cared so much about.

Reminding him that he failed.

"Are you still looking for her?"

Meeting my eyes, he reaches out and brushes my cheek with his hand before kissing me softly. With his lips just resting against mine, his eyes completely shut in a moment so unexpectedly tender, he answers.

"No."

"Why not?"

"I found you."

Me? What do I have to do with Sunday?

"I don't understand."

Pulling himself up in the bed and bringing me with him, cradling me into his side, he places a soft kiss to the top of my hair as he explains.

"The way Sunday used to affect me. The way her smile moved me and made me want to give her the world if she wanted it. Taking all of those jaded and cold parts of me and melting them under the power of her rays. I found it all again in you, Reagan. You gave the light back."

"Cole..."

I don't know what to say or what the right thing to say even is. All I've ever done from the moment I met him was be myself, but with the way he's explaining it, in doing that, I'd somehow given him back what he thought he would never find again.

How is that even possible?

Can one person really exert that much control over a situation and affect someone's life in such a monumental way? It happens in books all the time, but that's because it's fiction. The author can create any type of world they want, especially when it's one better than the one they presently live in. The hero can be forever altered by the heroine and he can speak and alter her perception as well.

But this is real life, and I'm just a girl that works in a bank by day and spends her nights in solitude reading by night. I can barely change my own life, so how is it possible I affect his in the way he's describing?

"You know, until I ended up here, I'd lost track of the way I felt about her in the years since she went missing. I think it bothered me so much that I somehow blocked it out. My uncle would write about the way I was five years ago and it was like he was talking about someone else. That's how good a job I did at burying it. Burying my feelings. Reagan, I know this sounds as crazy as it all did the first night, but you're the reason for everything."

"The reason I remember again. The reason I remember her. It all comes back to you."

If I thought making love to him was going to open me up and leave me raw and exposed, the way he connects me to the way he felt about Sunday shatters me completely.

I want to know more about her. Why his features soften and his entire way of being changes whenever he talks about her. What it was about that little girl that seems to change him back into who I think he may have been before life twisted everything around and changed him.

I need to meet Sunday Grayson so I can know the real Cole.

"Cole?"

"Yes beautiful?"

"Can I meet Sunday? Will you share her with me?"

"Rea," he pauses, his arm as its resting around me tightening again as he sighs heavily. "Why?"

"I want to meet your first love."

Chapter Nineteen

Cole

This is not the way things were supposed to go.

Reagan Carter should have been hands off. My hands, head and heart never should have touched her. She was nothing more than a job.

Even more than that, I never supposed to open up to her like this, treading dangerously close to the edge of telling her everything.

It doesn't matter that I've managed to put together a solid timeline of what I think happened. That I have enough to be able to give her a full and accurate picture of what took place seventeen years ago.

One she can't deny.

None of it matters because despite the promise I made to Luke, where I would tell her everything once I had facts to back me up, I can't do it. I can't be that guy. The one I knew I was going to be when I first saw her in the bank and confirmed what DNA had already told me.

After what I allowed to happen between us, I can't be the person that blows her world apart.

And it's all because of what I just told her. I stopped looking for Sunday because I found her in Reagan. Literally and figuratively. Even if this girl wasn't the one my family has spent years looking for, giving myself to her the way I did tonight, opening myself up and admitting more than I have in years, I would feel the same way.

Sunday is what connects us, but Reagan is what makes me stay.

Can I really add even more shit to this ever brewing pot and give her what she asked for? Can I slip across the room to the box that holds Sunday's life in it and show it to her?

It's not even seeing the younger version of herself in pictures that I'm worried about. She was so young at the time, I don't think she'll be able to put two and two together. It's everything else that's in there. Statements from her parents when they were at their weakest. Tortured words taken after she'd been lost to them and all they were left with was the memory of her. Ones from the day she went missing and every year the case remained active afterwards. Updates until everyone gave up and washed their hands of it.

Can Reagan really see all of that and come away from it the same innocent girl she was when I met her?

I've been a part of this case from the beginning and I know for a fact that I didn't walk away from it unscathed. I can't imagine someone as sweet as her, so pure of heart and caring, being able to do it.

It's not the request she made that's making me drag my feet, though. I know that. Deep down, I could pull that file and hand it over easily. It's what she believes Sunday to be to me that stops me cold.

My first love.

In my experience, love, at least the way she makes it sound, can't happen at the age I was when Sunday was taken. I have no doubt she altered the course of my life in a lot of ways, and I don't even doubt that in some way I did love her, but to call her my first love…I just don't know.

Can a seven year old really love someone? Is he even capable of knowing in his heart what love really is?

For most, maybe, but for me, only knowing pain until Sam found me, it's not possible.

Reagan is wrong. Sunday wasn't my first love.

But she may just be my last.

There's something about looking down and seeing the softness in her eyes, along with the glow of her sated skin and the warmth I can feel all the way through with the way she's nestled into me that forces my hand.

It's time to tell her the truth. Even if it means I have to give her up.

Just not tonight.

I can't be here, experiencing what we just did together, knowing what it took for her to even be here at all, and drop this bomb.

Reagan deserves this moment. This night. I won't be the one that takes it from her. Not when I'm the bastard she trusted enough to give it to her in the first place.

"Cole?"

Successfully brought back by the sweet sound of her voice, I give her the answer she's after.

"If you want to meet Sunday, I need to get out of bed."

Lifting herself up and starting to pull back, giving me the space I need to be able to slip from the bed, I stop her before she can make it too far. If I'm really going to do this, I want to make her as comfortable as possible. Taking what we just shared and expanding on it.

This very well might be the last time I see her. I want to make the memory of it, the one that will last long after I check out of this room and head back to the city, as perfect as possible.

As perfect as she is.

"I'm gonna get the file, but there's something I've got to do first."

"Like what?"

Kissing her nose softly, smiling in an effort to soothe any concerns she might have about what I'm about to do, I release my hold and slip from the bed.

Crossing the room and ignoring the siren like call that seems to emanate from the box as I pass it, I head into the bathroom and shut the door securely behind me. Grabbing the travel sized bottle of bubble bath off the counter beside the sink, I move to the tub and turn the warm water on, letting the flow hit my hand and adjusting it until it's the right temperature.

Squeezing a few drops into the water and laying the bottle on the edge, I watch as the water hits and turns it into bubbles. Leaning back against the door to wait as it fills, I suck in sharply at the cold feel of it against my back. The temperature triggering me until I'm closing my eyes in an effort to fight off the memories I know are coming. Bringing with them a dizziness so crippling it has me gripping onto the door with my hands in an effort to stay upright. Pictures of the last time I was in a situation like this one flooding my mind until it's impossible to fight against and I'm moved from the room with Reagan, right back in the hell that was the last day with my mom.

*****

I can't wait until she gets home and sees what I've done. She's going to be so happy.

Fumbling with the knob for the hot water, I turn it off once I'm sure the water is raised to where she likes it to be. Grabbing the book on the edge of the counter and checking to make sure the bookmark is still where it was, I place it on the shelf beside the tub just in time to hear the sound of the front door open.

Slipping out of the bathroom, making sure to keep the door slightly ajar, I'm heading straight for her when she calls out.

Used to seeing her worn features, when I come around the corner and catch sight of her in the doorway, I'm prepared for what I'm going to find.

What my surprise is going to fix.

Only that's not what I'm met with when I finally make it to her and throw my arms around her stomach. She's not tired and worn. She's pissed and before I can react, move away and distance myself from what I know is coming, I feel the grip of her hand on the back of my shirt as she pulls me off my feet and completely away from her.

"You little shit! Where is it?"

Where is what? I don't understand?

"W-where I-is w-what?" I stutter, scared and needing to know what she's talking about so I can make it better.

Slitting her eyes, she grins at me before releasing her hold and letting me fall to the floor like a limp rag doll. Before I can ask why she's smiling, she gives me her answer as I feel the harsh rubber of the front of her shoe as she slams it into my stomach.

Curling into a ball, I force the tears I can feel building away. Afraid of what she'll do if she catches me crying, I ask her again what she's talking about.

"The money you worthless piece of shit! I know you took it!"

None of this makes any sense. I don't know about any money. She doesn't tell me things like that. Why does she think I took it?

"I don't know, Mommy!" I cry a little louder. Secure in my answer until her foot makes contact again. Instead of reacting and giving her a reason to beat on me for acting like a little girl the way she has before, I curl myself tighter into a ball, all the while thinking about the surprise waiting for her in the bathroom.

A bubble bath. One that was supposed to prevent this from happening again.

Mommy is always happiest when she's soaking in the water and relaxing with her stories. She even smiles at me, which hasn't happened in so long I've almost forgotten what it looks like.

I want her to do it again. She's so beautiful when she smiles.

"Cole, where the fuck did you put it?"

"I don't know, Mommy. I swear I don't know."

Flinching in expectation of what I'm sure is about to be another hit, I hold my breath until my head is swimming, but nothing comes.

Opening my eyes and unclenching my body, I chance a look out around my legs. Looking up when I can't make out any sign of her, I shift and attempt to sit up, swallowing down the sharp pain that comes from the places she's marked with her boot and coming face to face with her steely blue eyes.

The ones that tell me without a word being spoken that the bath I ran for her is never going to be seen.

I'm never going to see her face when she finally does step into the bathroom because she's done with me.

It's worse than all the other times. Bending over and gripping me tight again, she yanks me up from my place on the floor, her nostrils flaring and her eyes wild, the obvious drug high overriding her senses and making her believe in something that's just not true. Smiling at me again, this one even more sadistic and angry than the first, she throws me and closing my eyes, I ready for the impact that comes seconds later as my body collides with the wall.

"You're nothing but a useless stain." She screams through the haze in my head. "How I ever thought I could love a little piece of shit like you is beyond me. You're nothing but a lying sack of shit. The worst decision I ever made in my life."

"T-that's not true, Mommy. I know it's not. You don't mean it."

Proving me wrong, she falls to her knees and bringing her hands around my neck she begins to squeeze. Each passing second the hold she has getting tighter until all I can see is spots in front of my eyes.

I should have known this was how it would be. It doesn't matter what I do to please her.

Pulling one last burst of energy from somewhere deep inside, I attempt to shake her hands off me. Break the hold so I can get my bearings again, repeating the one thing I hope will change this. Fix it. My voice coming out as choppy and broken up as I am.

Words that should be able to fix anything.
"I love you…"

*****

Allowing my body to give in to the effect of the memory, I slide down the door until my ass hits the floor. Shaking under the force of what I can see so clearly happening after that.

The spots I'd been experiencing as she tried to choke the life from me turning more solid. Feeling the sharp sting of what I learned later was a knife being plunged into me and everything going black a few minutes later when I passed out. Assuming I was dead until I was woken a few hours later in the hospital by a nurse. Her soothing voice repeatedly telling me I was safe even when I tried to crawl my broken body away from her.

Wondering where my mom was. Why she wasn't there with me and how I even ended up there to begin with. All questions that took hours to come, but when they finally did, I understood.

Someone heard what was happening inside the apartment and called the police. They heard me calling out and had gotten me out of there before the unthinkable happened.

Before I died.

"Cole?" Reagan knocks against the door lightly.

Shit. How long was I out? Is the water even good anymore? Did I screw the entire thing up?

Is she going to be mad when she sees?

"Cole, please say something. You've been in there for a while and I'm worried."

Forcing my body up off the floor and shaking myself in an attempt to scrub the memory, I plaster a smile I don't feel on my face and open the door right as she's about to knock again.

"Please don't be mad." I beg when our eyes finally meet and despite her not knowing what just happened in here, her eyes are filled with understanding.

"Never. Not ever." She whispers, stepping forward and wrapping me up in her embrace. The sound of something being dropped to the floor the only sound I hear as my shaking legs finally buckle, giving in and causing us to fall to the floor. Her grip tightening around me when our knees hit the floor, pulling me in closer as the dam breaks and I fall apart.

"Don't leave me." I beg her. "I know I fucked up, but Sunday, please don't leave."

When her body tenses I realize what I've done. The slip I made.

Damn you, Mom. Why did you have to haunt me now?

I need to do damage control. I need to tell her what happened. Make her see that the reason I just called her Sunday isn't because it's who she is, but because my mind is so twisted that I'm not thinking clearly.

She can't know. Not yet.

Not tonight.

"Reagan, I'm sorry."

"This is my fault, Cole. I should be the one apologizing."

Say what? "No, this is on me. I wanted to do something nice for you and it got all twisted."

"Yes it did and it's because I asked about the tattoo. I twisted things."

"No Rea, you didn't. You made it beautiful. You're beautiful. I'm the one that's ugly. I did this. I always do this."

If she wants to argue the point she doesn't show it. Her hands now running soothingly over my back as she presses her face to mine. The sound of her breathing mixed with the feel of it against my skin settling the nightmares buried inside.

"I made her a bath. She loved relaxing with a book in a bubble bath. We'd had a rough week and I wanted to be the one to make it better. I was so excited for her to get home and see it. I believed it was going to turn everything around."

Lifting her head and looking over to the bath waiting a few feet away, the very bath I wanted to run for her in order to soothe what I know had to be soreness after we'd made love, she sighs. The sound of it as it falls making the ache in my chest worse.

This isn't how I wanted it to be. She isn't supposed to be hurt by what I wanted to do.

Mom was right. I really am useless.

So much for giving her a perfect night. I've turned it into a nightmare.

"She's not here, Cole, and what you did for me," she turns to me smiling. "I love it."

"You do?"

"I really do. No one's ever done this for me before."

Of course not. She dated a prick who didn't care about anything but where he was going to shove his dick next. But at least he didn't attempt it and then fall apart on her before she could enjoy it.

That's all on me.

"She never saw what I did, Reagan. I blacked out after she started choking me. She was arrested. She never got to see…"

"She didn't deserve to see. She didn't deserve you."

"Just like I don't deserve you."

Shaking her head in disagreement, she lays her hand out between us and when I look at her, search her eyes in an effort to understand, she nods to my hand and waits as I slip it across and into hers.

"Come on," she says, using her strength to pull us to our feet. "She doesn't get to ruin this. What you've done here, it's just for us."

With a soft brush of her lips against my cheek, she steps away and lowers her hands to the strap of the robe she's wearing. Slipping her fingers through the knot, she untangles it. Lifting it up and over her shoulders she lets it slip down her arms until it's falling to the floor. Giving herself to me the same way she'd done when we made love.

Not giving my body the chance to react to how beautiful she looks standing in front of me naked, she takes me by the hand and guides me to the tub. Motioning once we've stopped for me to get in.

But a request I don't think I can fill when I take in the actual size of the tub.

"Rea, there's not enough room for the both of us."

"There will be, you'll see. Just get in."

Doing what she asks, I get in and lower my body down into the now lukewarm water. Turning once I'm situated comfortably and holding out my hand for her to take.

Stepping into the bath she lowers her body down, moving back until her back is pressed against my chest, laughing softly when the water splashes around us. Expecting her to stretch her legs out now that she's in, she takes me by surprise when she bends her legs and brings them to her chest. Moving back again until she's nestled tight against my back.

"See," she says with a sweep of her hand around us. "I told you there'd be enough room."

"Except you're crunched up while I'm completely stretched out. That's not what I was going for."

"Correct me if I'm wrong, but wanting the both of us in the bath wasn't your idea. It was mine. So hush it. I'm right where I want to be."

"Okay, you win." I concede laughing. The knotted up feeling from a few minutes ago beginning to crumble. Reagan laying it to waste.

"You are so perfect." I murmur softly against her ear. Nuzzling my face in her neck when I hear the soft sigh that escapes.

"I could say the same thing about you. Everything about tonight has been perfect. I'm almost afraid to leave the room because I don't want the way this feels to end."

A sentiment we both share.

"So, I might have done something while I was waiting for you to come back out."

Leaning into her, burying my face in her hair and giving myself over to the feel of having her this close, I don't register what she means right away. "Hmm?"

"I went through your boxes."

The peaceful feeling of calm that surrounded me from the time she entered the bathroom begins to fade as the reality of what she's saying kicks in.

No, no, no. This can't be happening right now. Not when everything was finally starting to even back out.

"You did?"

"Yeah. I hope that's alright. I just wanted to help by grabbing what I could find with her name on it."

"What did you grab?" I nervously ask, amazed I can even get the question out at all with the way my throat seems to constrict.

"Just a file. I didn't disturb anything else. I figured that when I opened it and saw the picture of her that was paper clipped to the front page, I'd found what you were going to grab."

"Where is it?"

Lifting her hand out of the water and pointing behind us to the ceramic container, I see it. Twisting around enough to reach it, I grab it and bring it down over the both of us. Careful as I slide myself back down into the bath not to let the water that continues to splash around us get on it and breathing easier once I see what file it is she came across.

"You saw her before you knocked on the door?"

"Yes. I'm sorry, Cole. I was curious."

"It's fine. You have nothing to be sorry for." I reply even though it's not the way I feel. Nothing about how close she is to this right now is fine. I'm right back on the damn tightrope again. It won't be long now before I slip and fall off completely.

"I noticed something kind of interesting when I picked it up. The photo from the hospital the day she was born, she had blue eyes, but when I looked at another one of her a couple of months later, they were brown."

"Yeah."

"How does that even happen? It's cool and all, but isn't that weird? Don't you keep the eye color you're born with?"

"Not always. It's basic biology. Didn't you learn about it in school?"

"I might have." She shrugs. "I don't remember."

"Sunday's eyes changed and eventually matched her dad's."

"Does her mom have blue eyes?"

"She does. They're the same color as Luke's."

"That's awesome. I don't think mine ever did that."

"What do you mean you don't think? Don't you have baby pictures?"

When she shakes her head, my heart aches for her. Of course she doesn't. Sunday wasn't with the Carter's when she was a baby.

"I always wondered why. I asked about it a bunch. My mom and dad just said that before we moved into the house we're in now, we lived in a little apartment and there had been a fire. One that destroyed a lot of the stuff from when I was young. With what I overheard today, though, I'm starting to think it was a lie."

Right. The reason she reached out and we're even here right now. She's overheard them talking about her being adopted.

"Would it be such a bad thing if you were?"

"No. Bethany's adopted and she's amazing. If anything it made her life better. I just don't understand why they couldn't tell me. Why did I have to hear about it the way I did?"

The answer I want to give her and the truth are two different things. With the part I now know her father played in everything that happened, his reasons for keeping silent are obvious. What her mom gets out of it though, when I can't find one tangible piece of evidence that says she knew or was a part of it, I don't get.

"They love you and my guess is, they didn't want that to change. I mean, if they told you the truth, do you really think you'd look at them the same way?"

"No. Things definitely would have changed. I wouldn't have stopped loving them or walked away, they're my parents, but it would have made things awkward for a while."

"Then it's easy to see why they didn't tell you. If I were them, knowing the way you are, I wouldn't have wanted to risk losing that either."

"I would rather tell the truth and risk losing something, giving the person the option of forgiveness than continue to lie to them until there's no choice anymore."

"Yeah..."

We need to get off this topic. I'm already guilty as hell. Having Reagan pretty much tell me what's going to happen when the truth is exposed is only making the knife twist in worse.

Tomorrow. I need to tell her everything tomorrow and hope like hell that when I do that option of forgiveness she's talking about is still there.

I don't know what I'm going to do, especially now, if it's too late.

If I lose her after only having found her again, I don't think I'll survive it.

I'll be as dead to the world as she believes Sunday is.

Chapter Twenty

Reagan

Sneaking through the door and making a beeline for the stairs, I manage to get halfway up before the loud clearing of a throat stops me cold.

"Reagan."

Ugh. Great.

Out of the three people that live here, why couldn't it have been Jake or my mom that caught me? Why did it have to be him?

I might be an adult, but living at home means living by their rules, and I know for a fact I'm breaking a pretty big one right now. Staying out all night and trying to sneak upstairs at seven in the morning isn't going to go over well.

"Dad," I start when I've turned around to face him. "I can explain."

"You can explain what exactly? Why you didn't pick up your phone all night? Or maybe you care to share who it was that kept you out all night."

This is what I'm talking about. I'm almost nineteen but here he is treating me like I'm twelve. It's only going to get worse. There's nothing I can say, no excuse I can give that's going to satisfy him. You can't satisfy James Carter. He's got his mind made up before even giving you the chance to explain.

"I didn't answer any of your calls or texts because I left my phone in Cole's car. By the time I realized it was there and went to get it, the battery was dead."

Crossing his arms across his chest means he doesn't believe me. Of course he doesn't. The only truth he knows is his own.

"So you did spend the night with him."

"Yes."

I'm not going to lie or treat Cole like he's some dirty secret. What we shared last night was something more than sex. He opened up to me. Gave me everything he had and trusted me with the broken parts. It was beautiful.

As pissed as my dad is right now, he's not going to take that away from me.

"What do you even know about this guy, Reagan? Up until a couple of weeks ago he didn't even exist."

"I know enough."

"That's what all naïve little girls say." He quips snidely and I resist the urge to slap him.

He can be a protective asshole all he wants, but I don't have to sit here and take him treating me like dirt on top of it. Especially with what I know he's been holding back from me.

The man that doesn't like secrets sure is keeping a pretty big one.

"I am not being naïve, and you know nothing about my relationship with Cole, so don't stand there pretending you do. He makes me feel good. I like the way I am when I'm with him. Something that after what I went through with Walker, I would have thought would make you happy. Not turn you into even more of a judgmental prick."

"Watch your mouth, young lady! You may be an adult but I am still your father!"

"No actually, you're not."

Oh no.

Why did I have to go and do that? Now I'm never going to be able to escape upstairs. There's no way he's going to let that go.

"Excuse me?"

Since I've already stepped in it, there's no sense trying to backtrack now. I knew I was going to have to confront what I overheard sooner or later.

Sooner it is.

"You heard me, James. You are *not* my father."

"That's not funny."

"I wasn't trying to be. I've never been more serious."

Watching as he tries to maintain his composure, not wanting to appear affected at all by what I've said, I wait him out. When he doesn't say anything or follow it up with a question after a few minutes of silence, I finish it.

"I know about the adoption, Dad. I heard you and Mom arguing. I know everything."

Turning and running up the stairs, I ignore the sound of his voice demanding I come back so we can talk things out. Blocking everything out and keeping it together until I'm safely tucked away inside my room with the door locked tight.

Let him stew on that for a while. Maybe if he's focused on me knowing the truth, he'll leave Cole alone.

Swallowing down the empty feeling that's building because of what happened with my dad, I step away from the door and make my way over to my bed, taking the room in as I go.

So much has happened here. For as far back as I can remember, this room has always been my sanctuary. My safe place to fall.

It's where I experienced my first kiss. The nervous jitters before my first date. Making out with Walker when we should have been studying. All night gossip sessions with Bethany. My bed the place that housed the tears that I cried every time someone hurt me. The carpet to this day altered by the imprint of my boots from all the pacing, dancing and jumping around that took place through every stage of my life.

Everything about the little girl I was and the woman I became is wrapped up in this room. In this house. With a family that for the longest time I thought I was the luckiest girl in the world to have, but that now doesn't even seem real.

Looks like Cole and I aren't so different after all.

Sitting on the edge of the bed and looking around, I'm unsure what to do now. I could sleep considering I hadn't gotten much the night before, but with the fight with my dad looming, I know I won't be able to. Which leaves either the computer or laying back on the bed with the one thing I own that has never let me down. My books.

Resting my eyes on the bedside table, I'm reminded of the conversation with Bethany the last time we were here. The same book lying face down calling to me like an old friend. Begging me to give in and pick up where I left off. To let it work its magic and help me forget.

Fingering the paperback in my hand, I open it to where I left off and moving back on the bed, settle in to read. Getting more than a few paragraphs in and like always, falling under the spell that being taken to another world creates, I almost miss the light tap on my door.

"Go away, Dad!"

"It's not Dad."

Jake. The only person in the house that I'm not totally pissed at right now.

Thank God.

Putting the book down and making my way to the door, I flip the lock and open it before heading back over to my spot. Bringing my knees up and resting my face against them while he makes his way in, making sure to flip the lock on my door again once he's in.

"What happened?"

"I spent the night with Cole. Before you give me the speech, I already heard it from Dad when I got in. I'm not in the mood."

"No speech, Rea."

"So if you're not here to gloat over me breaking the house rules, what do you want?"

"Well, at first I was waiting for you to come home so I could ask if you'd come watch me play tonight. We signed with Divinity and they're doing this All Ages music thing tonight that I don't really wanna do without you. But since I heard you and Dad going at it, now I just wanna know if you're okay."

"Since when do you care how I'm feeling?"

"Since I was old enough to understand what feelings were. So uh, twelve years?"

"Yeah right."

"Rea, I know I act like an ass sometimes, but it's kind of in the job description. You're my big sister. I'm supposed to annoy the shit out of you. So can you stop riding me and tell me if you're okay?"

"I'm fine."

"That's believable."

"What do you want me to say Jake? How much of that shit with Dad did you catch?"

"Enough to know you're adopted."

This is another reason I feel so twisted up inside. Why I'm so mad at my parents.

Jake is their son. Their biological, carried for nine months and loved, baby. Something that for eighteen years I thought I was, but now I know I'm not.

I'm just the stray they took in while they waited for him to come along. And despite not wanting it to happen, I can't help looking at him differently because of it.

He's not the annoying little brother I love more than life anymore. I don't even know what he is, but that's because right now I'm not even sure who I am.

"Yeah, I am."

"Reagan, there's something you need to know."

No way. I don't like the sound of this. With the way he's averting his eyes I just know he's going to tell me something that's going to make all of this worse.

"What?"

"I knew about it. I overheard them a couple of years ago. I didn't understand what I was hearing at the time, but they were arguing. Mom was pissed, Dad was trying to get her to settle down and well, I was outside the office door hearing every word of it."

Oh God. Jake knew.

"Why didn't you tell me?"

"I didn't want it to change anything."

"It wouldn't have, Jakey."

"Look at how you are right now. I can see the tears starting, Rea. Things have already changed and it's just going to keep doing it now that you know the truth."

No. That's not right. As upset as I am with my parents for keeping it from me, it doesn't change the fact that I love them and we'll get through it.

They're still my family. It changes nothing.

"Looks like you were wrong a few weeks ago. I'm not the golden child. You are. I'm just the stray dog they brought home from the pound."

"Reagan! Jesus Christ!" he curses before shaking his head and beginning to pace the floor in front of me.

"What?" I yell back, pausing him mid pace. "Are you really gonna stand there and tell me different? How many times have they told us the story of how we came to be, Jake? How they tried for years to have a kid. I've heard it so many times I can recite it word for word. For years they made me think I was their little miracle, but the reality is, you were. It's always been you. I'm just what they settled for when they gave up trying to have you."

"That's fucking bullshit."

"No, Jake. It's fact."

"Why aren't you more like your sister? You need to focus more on school and less on frivolous pursuits that will get you nowhere. Reagan didn't drop out of school so neither will you. Reagan this…Reagan that." He quotes our parents as he crosses the room and sits down beside me. "It's always been about you. So how you can think that you were what they settled for is messed up."

"Jake…"

"What?"

"I don't want to feel like this. I just had the best night of my life and instead of riding the wave of it, I'm sitting here empty inside. It was like everything drained away when I walked through the door and I can't get it back."

"I'm sure Dad going off on you had something to do with that."

"Maybe, but it's more than that. I liked the way things were. Why couldn't they stay that way?"

"I don't know, Rea. You're not the only one screwed up over this. Their lie is messing with me too."

I can't believe I entertained the thought of Jake not being my brother. He's still the same kid he's always been. He's still my Jakey.

The only one I'll ever have.

"I love you, Jake."

"Not nearly as much as I love you, Rea."

"I don't know what to do…"

"Are you open for suggestions?"

"Sure."

"Text Cole and Beth. Tell them to meet you at the bar and come out with me. I know it's not your scene, but it's better than sticking around here and risking another round with Dad. Come out, dance a bit and take your mind off all of this."

Can I really do that? Go support Jake and pretend like nothing's wrong?

"Please come, Rea. I don't think I'll be able to do it without you."

"Now I know that's bullshit." I joke.

If there's anyone that can get on that stage tonight and blow the world away with his music, it's my brother.

My brother.

I like the sound of that way too much.

"Fine. Its bullshit, but it doesn't change the fact that I want you there."

"Then there is where I'll be."

I just hope I don't live to regret it.

James

This is worse than I feared.

I knew arguing with Katherine when both kids were home and at risk of overhearing was trouble and now, with the cold steel of her eyes as she admitted knowing the truth playing on an endless loop in my head, my worst fears are confirmed.

Reagan knows the truth.

It won't be long, especially after the look I caught when she admitted spending the night with Cole, until she learns the rest.

What the hell is Martin's problem? Why is this even an issue? He was contracted to get rid of the Alexander kid. Not disappear into the night, giving Cole the opportunity to grow even closer to Reagan.

Closer to Sunday.

Looks like the old adage is true after all.

When you want something done right, you have got to do it yourself.

Despite not wanting to get my hands dirty, it appears as though I am going to have to because I cannot let this nonsense between Cole and Reagan continue. I have to keep her clueless about who she really is.

Even if it means breaking her heart.

"What the hell was that about?" Katherine demands angrily, the clicking of her heels across the floor making the migraine that appeared after my altercation with Reagan even worse.

"It was nothing."

"When are you going to realize that using that line on me is pointless because I can see through it? Clearly what happened between you and Reagan was a whole lot more than nothing."

Squeezing the bridge of my nose tightly, I sigh. Fine. If she wants to know what happened so damn bad, I'll tell her.

"She knows the truth about the adoption. She claims she overheard us talking about it, but I think there is more to it."

"What more could there be? We *have* been arguing about it lately and we're not exactly the quietest people when our emotions get involved."

"The man she's been seeing. This Cole character. I know him, Katherine. His entire family is made up of private investigators. I think that in their time together, he may have been the one to enlighten her."

"Not this again!" she exclaims and I step toward her, slapping my hand down hard over her mouth. Gripping tighter when she protests by trying to pull away.

"Keep your voice down!" I seethe, holding my hand in place until her body begins to relax. "Now do you think we can talk about this like two rational adults?"

"You claim to want to talk like two rational people, yet you're standing here acting completely irrational."

"How so?"

"She heard us, Jimmy. This has nothing to do with the boy she's been seeing and even if it did, who cares? She deserves to know the truth and honestly, from what I've been able to tell from her behavior over the last couple of weeks, this *Cole character* is making her happy. Reagan is acting like herself again."

Katherine is wrong. Cole Alexander is nothing but a virus that needs to be eradicated. A virus that I'm going to enjoy bringing to his knees before taking him out completely.

"Regardless of whatever changes you think you see in her, it changes nothing. She knows the truth and now we need to get our stories straight."

"Do you even hear yourself? What story do we need to get straight? The one where we tried every available option at the time and were unable to conceive? Or the one where we jumped at the chance to adopt her when her parents wanted to give her up?"

Damnit. This has shaken me more than I expected it to. She's right. There is no story we need to get straight. What she believes happened, her truth, it's already straight. It's only up to me to sort the fiction and lies from the reality.

"I'm sorry. You're right. What happened with our daughter has left me rattled. I never expected to be in this position."

"So what you said before is true."

"Yes. I never wanted her to know. There was no reason she needed to. She's had the best life she could possibly have. Telling her that when she was a baby, she was no longer wanted by her parents so we adopted her would hurt her. I want to spare her that pain."

"While you go behind her back and background check every person that comes into her life. Vetting them out in private before letting them get close to her. A method that failed with Walker, I might add. How is that better than telling her the truth?"

"Walker knew the score. He knew I meant business and if it weren't for our daughter and the feelings that remained long after they broke up, that son of a bitch would have been dealt with properly. In the only way a snake like him deserves."

Take a step back James.

Just because you've had your hand in some pretty shady dealings over the years, been a co-conspirator and accomplice to murder, Katherine doesn't know that. She still believes her husband to be the fine upstanding man she fell in love with years ago.

The way it has to remain.

"There's something else going on here. Something you're not telling me. I've tried ignoring it, but I can't anymore. Whatever it is, it's bad and it's changing you."

"You're being absurd. You know as much as I do. We share everything."

"Maybe before that was true, but these last few months, I've been seeing a different side of you. One that scares me. You're not acting like the man I married anymore. I only wish I knew why."

She may not want to admit it, but deep down she knows the answer. She knows why I'm not the same man she fell in love with. It's all around her now. This house, the money that we have to be able to live comfortably, it's all because of who I became.

Katherine may not like the person I've turned into and the way our marriage has changed, but she cannot deny that it's been beneficial.

"We tried to do things the right way, remember? We were living in a house that on a good day we were thankful didn't collapse around us. In debt over bad investments and the damn in vitro treatments. We may have had love, but that was all we had. In order to have all that we do now, things had to change. I had to change. Katherine, as much as I know you hate it, I have no regrets."

Gasping, she takes a step back. Her eyes wide and alert, the fear she spoke of not a few minutes ago now completely on display. What I've said and how I reacted to the way things are, it's too much for her to take.

She hates the monster standing in front of her.

And as I step forward in an effort to soothe her, try and fix the damage that my brutal honesty has caused, she speaks again, proving with words what her body language has already told me loud and clear.

"You don't regret it now, but maybe you will when it causes you to lose everything you love. I'm sorry, Jimmy. I am. I tried to ignore all of this, but I can't anymore. Not when whenever we're alone like this, I don't know what version of you I'm going to get. I love you, but I can't do this anymore. I'm leaving."

Chapter Twenty-One

Cole

Are you busy?
If by busy she means staring at my phone every other second while trying to ignore everything I've tried to piece together about her then yeah. I'm swamped.

But since I know she doesn't because she's still as clueless as she was the day I met her, that's not how I answer back.

No. Just lying in bed wishing a certain girl didn't leave this morning. That she was here with me still.

Don't do that. She texts back, but before I can question what she means, another text bubble pops up on the screen.
Don't say sweet things and make me regret leaving even more than I already do.

Telling her the truth and having her call it sweet is cute. Different. Definitely not the way I'm used to having my words received.

But teasing you is fun.

The response is instantaneous. She's not messing around.
Cole!

Reagan! I text back, unable to control the grin that takes over my face knowing what me goading her like this is going to cause.

You're incorrigible.
And you love it.

Pausing right before I hit send, I stare at the word on the screen. The one that despite knowing I feel it, still seems so foreign. Love hasn't been a part of my vocabulary in years. I never thought it would be again. Not even in the innocent way it is here. I'm not even sure with what I feel for Reagan that I can even call what I shared with Miranda love anymore. It was something, but nowhere near what I share with the girl in text with me now.

Hitting send, I start typing out another message. One designed to fit with what I should have said to begin with.

Now stop using big words to try and confuse me and tell me why you asked if I was busy.

It takes a few minutes but when it comes through, I'm smiling again. What she wants I can definitely give her.

Jake is playing a show tonight and wants me to be there. I was thinking if you weren't busy, you might want to meet us there. I want to see you.

There's a tugging in my chest as I read over her words. I'm reacting because I feel the same. Nothing's been right since she left. If I'd given into what my heart wanted this morning, she never would have. I would have kept her here with me forever.

I want to see you too, so I'm in. It's taking every ounce of restraint I have not to come over there now.

I wish you would.

Now this wasn't the response I expected.

I figured when she said she wanted to go home, she needed a break from everything. What we'd experienced the night before, especially what happened toward the end of the night had to be a lot to take in. She was going to need time to adjust. Decompress from the emotional toll it took on the both of us. If the roles were reversed, I know I would.

This response speaks to there being more going on than her missing me.

Are you okay? Did something happen? Do you want me to come get you?

I promised her that if she ever needed me, I would be there and I meant it. Even before I slept with her I meant it. The need to do it is just stronger now. If she doesn't want to be there then she's coming with me.

I'm never going to let her go through anything alone.

Physically I'm fine. Not so sure about the rest of it. You don't need to come get me. Jake's here. I just...

Jesus Christ. I know what those dots mean and it's not that she's thinking of what to say next. Its girl speak for 'I'm about to drop something heavy on you'. Whatever she's about to say next, I know I'm not going to like. It's going to make me wanna bail on meeting up with Luke so I can go to her.

I told my dad that I know about the adoption. Things are tense here which is why Jake suggested all of us go out.

Yeah, I was right. I wanna leave.

Are you sure you don't want me to come get you?
No Cole. I'll just see you tonight.

Looking up from the phone at the sound of the rough knock, I toss the phone down and head over and answer it. Coming face to face with Luke once it's open, I usher him in before heading back to the bed and picking up the phone again. Waving him off when he attempts to start a conversation.

Reagan, I'm not gonna push, but if you need me to pick you and Jake up, just say the word.

I will. He says thank you.

He doesn't need to thank me.

No, but he did and now so am I. Thank you, Cole. I'll see you soon. Love you. XOXO

Here's my shot. She's not standing directly in front of me with those eyes that display every emotion she's feeling. I can tell her how I feel. Utter those three little words that I haven't said in forever, but that I would mean more than anything.

So if that's true, why does it feel like my chest is about to seize up just thinking about it? If this is so meant to be, why does the thought of admitting it scare the living shit out of me?

"Cole, man, are you alright? You look like you're gonna be sick."

Three little words typed on a screen. I can do this. Two if I just copy what she did.

Two of the scariest words in the English language.

"Fuck!"

"I wasn't going to say anything, but with the way the bed looks this morning, I'm thinking that's exactly what you were up to last night." Luke jokes. Ignoring him and giving the phone my full attention, I close my eyes and focus on what I want to say. Determined to get it out once and for all.

Love you more, Rea. So much more.

Slamming my thumb down hard on the send button, I toss the phone back on the bed before blowing my way past my best friend and straight into the bathroom.

Making my way over to the sink and lifting my hand to the tap, I try to control the shaking in my hand in order to turn the water on. Water I'm going to need to force out the panic attack I can feel coming now that I've finally said the words.

Meant the words.

Reagan isn't my mom. She won't turn them around on me. Reagan isn't like the rest. She won't break me. She'll be the one putting me back together.

Leaning my face over the sink and letting the outpouring of water soak my face, the cold doing what I need it to and freeing me from the seize of panic. Turning the tap off and staring at my reflection in the mirror, I'm reminded of what's waiting for me on the other side of the door.

Luke.

I've gotta fill him in and hope like hell he doesn't want to go off half-cocked and do something stupid.

Something that in the end could cost us the very thing we're here for.

Sunday.

Luke

I've heard and seen some sick shit in my life but this has to be the worst.

It's so fucked up I'm not even sure it's real.

There's no way this kind of sick shit happens in real life. That a man desperate to have a child would resort to something like this. Let alone killing his own partners' kid in the process.

"You sure about this?"

"Yeah. It took a bit of digging around and a whole lot of Frank Greavey ass kissing, but it's all true."

"So the guy that I remember in the Trans Am really was Hallie's dad, and he was hired by Carter to take her?"

"And make John's death appear to be of natural causes, yeah."

"Who does that shit? Throws his daughter under the bus for money?"

"Martin Francis does. But with all the background I've been able to pull up on him, I'm not surprised that's the road he took. He's a person of interest in a lot more cases than just this one, Luke. Hallie knew about it. She wasn't as innocent as the cops and press made her out to be. This wasn't a 'in the wrong place at the wrong time' thing. She wanted to do this."

"Is there enough there to nail them?"

"Not yet, but there is enough to be able to give Sunday the complete picture. I've also given everything I have to the police. I know they gave up on her, which is the whole reason your parents wanted me to find her, but I'm hoping with everything I've managed to come up with, they'll do something now."

"Cole," I say, taking a breath in an attempt to soothe the uncontrollable anger beginning to flow through me. "I want Carter's head."

"I know you do, but if we want to do this right, guarantee he pays, then we need to go on the way we have been."

"That's fucking bullshit! There was nothing *right* about him stealing a little girl away from the family that loved her! Why the fuck should he get to walk around free?"

He can't argue this. I know Cole better than anyone. He's calm on the surface, but I know for a fact that he's as pissed off as I am. If anyone wants to make that sadistic asshole pay as much as I do, it's the man in the room with me now. Especially with everything I've been able to put together since I got here.

It's pretty obvious that things between him and Sunday finally spilled over and have been taken to the next level.

"He shouldn't, but from a legal standpoint, what we've got here isn't enough. So it's the hand we're being dealt."

"No, it's the hand you're dealt. I don't have to do anything."

"Luke, I agreed to let you stick around and be a part of this, but only if you agreed to follow my lead. You can't go off half-cocked. You do that and you'll just end up making everything worse."

"Fine!" I snap, knowing he's being the voice of reason, but still hating him for it. "I won't fucking do anything until you say so, but what do we do in the meantime?"

"Meet the girls at the bar and watch Jake's band play."

"So basically you want me to pretend like I don't know her father is a sick son of a bitch that needs to be put down for taking her from us?"

"Pretty much." He scowls. "But not for long, I swear. I'm ending this tonight."

"Why tonight?"

"Luke…"

"If you tell me you slept with my sister and then try to blow it off or say it's something that can't happen again, I will fucking end you."

Stepping back and throwing himself down on the bed, he lets out a shaky breath before raking a hand through his hair. The truth of what happened here the night before crystal clear in his eyes when he finally meets my gaze.

"Did you sleep with her?"

"Yeah."

"Do you love her?"

It's shaky once he deviates from the silent shaking of his head to speaking it aloud, but there can be no denying it. Cole is in love with Sunday.

"Yeah, Luke. I do."

Fuck. This complicates shit. The truth hasn't even come out yet and I can already feel my loyalty being tested. When the shit finally does go down, it's not just my sister that's gonna be affected. It's my brother too. Cole and I might not be related by blood but he's as close to a brother as I'm ever going to have. A brother that it took years to get back and with one bombshell, I'm in danger of losing.

"What are you going to do?"

"Whatever she needs me to. Stay away. Give her space. Be her punching bag or hold her and never let go. Anything she needs, I'll do."

"And if she wants nothing to do with you?"

"I knew going into this that her turning away from the truth was a possibility. I knew I would be the person that singlehandedly destroyed her life. If she wants me gone, I'll go, but it won't change my feelings."

"Cole—"

"No, Luke. I blurred the line between personal and business. I did that. I knew it was wrong getting close to her knowing what I did, but I was a selfish prick who did it anyway. I fell for her. I have to live with that for the rest of my life. If I have to do it alone, so be it. If it means that you hate me for hurting her, I'll own that shit too."

We haven't even done it yet and he's already preparing for everyone to walk away. He's still the same stupid ass kid he was before. Always waiting for the other shoe to drop. Honestly believing that he doesn't deserve to be happy.

He's wrong, but I don't know how to make him see it. How do you combat six years of torture and lies? Especially when those six have stuck around and messed with the other eighteen?

I can't do it. He's got to be the one to beat this, but if he loses Sunday the way he assumes he will tonight, it's something I don't ever see happening.

It's not his temper, the whiskey, or the memories from his past that's going to kill Cole in the end.

It's going to be love.

Chapter Twenty-Two

Reagan

Jake was right. This is exactly what I needed.

Him on stage where he belongs living his dream, and me sitting at a table a few feet away surrounded by the people I care about most.

Luke is a new addition, but with his connection to Cole and what seems to be happening between him and Bethany, along with everything I know now about his sister, it's only natural that he's a part of it. I've opened my heart to him the same way I have to Cole and there's no going back.

"Holy shit! Did you just see that?" Bethany yells and following her hand as she points toward the stage, I see what she's getting at.

Just like the first time his band played here, there's girls all over the stage area desperate for a bit of attention from my brother and his band mates. The only difference between the past and this time is that there's a girl up there that just stripped her top off.

Which wouldn't be all that shocking if she was wearing a bra when she did it. Looks like someone forgot how to dress themselves.

For a small dive bar, you wouldn't expect much in the way of security, but no sooner does she put her Tata's on display then Angel and Davis are in her face and using her now discarded shirt to cover her as they remove her from the stage.

Turning away from the spectacle, I bury my face into Cole's shoulder as the laughter rips through me.

"Looks like Alliston knows how to party." Luke jokes, which only serves to send another wave of laughter through me.

"S-s-stop." I plead as Cole begins rubbing my back. "I'm gonna pee my pants!"

"Please don't." Luke begs, holding his hands up in prayer. "While I'm sure Cole has no problem imagining you sans clothes and soaking wet, it's not real high on my list of sights to see."

"Gross!" Bethany calls out before leaning over and whacking him in the back of the head.

"Bethy, I'm gonna kill you." I threaten when I'm finally able to catch my breath.

Needing a break before I make good on the earlier threat of pissing my pants, I slip off the comfort of Cole's lap and make my way around the table to where Bethany is waiting with a grin on her face.

"On that note, I do believe we need the little girl's room." Bethany announces, following my lead and getting up out of her seat.

"Yeah and because I want to spare Luke anymore trauma, we'll be right back." I continue the farce, returning Cole's understanding smile with one of my own before leaning across the table and pressing my lips softly to his. "I'll be right back."

"You better be." He mouths before turning his attention back to Jake as the band launches into another song.

"Thanks for that, Sunshine!" Luke calls out when we've turned and started making our way toward the washroom. "Shacking up with Cole was traumatizing enough!"

Laughing as I hear the slap that follows, I release Bethany's hand when we get the door and push my way through.

"I was going to ask if you really peed your pants, but since I can see they're clean, you wanna tell me why we're here?"

"There's something I wanted to talk to you about, but I couldn't do it with the guys around."

"Like what?"

"I did it." I waste no time blurting out.

"That's great. I'm happy for you. What exactly did you do again?"

"Cole."

"You did what with Cole?"

"I slept with him. Spent the night. Made love. Whatever you wanna call it. Bethy, I did it."

"Pretty sure it takes two to tango there Rea, so you both did it, but seriously?"

When I nod she squeals before stepping forward and throwing her arms around me in a hug.

"How do you feel? Was it okay? Did he take care of you?" she begins firing off questions when she's stepped back and given me the space needed to breathe again.

"I feel great. It was better than okay and he took such good care of me, Bethy. It was completely different than the way it was with Walker."

"That's because Walker was a dick with a person attached. He didn't care about anybody but himself."

She's not gonna get an argument from me on that. It may have taken being with someone who genuinely cared about me in order to see it, but now that I have, there's no going back. Walker was a first class asshole.

"So what does this mean?"

"What does what mean?"

"You and Cole. What does it mean? You're the one that said he was in town on business, right? What happens when he's done and has to go back to the city?"

We haven't exactly talked about it, but with how close we've gotten over the last couple of weeks, it's a pretty safe bet that when he does go back, we're going to be giving the long distance thing a try.

That's what I want anyway.

"We're gonna suffer through the distance."

"Which would be a lot easier to do if Jake didn't wreck your car."

She's right, but Cole driving and my car not going to be in the shop forever means that it won't be long before we're both racking up the miles.

"We'll figure it out."

"Looks like you found your Cooper after all."

"Yeah." I smile. Warmed by the idea that Cole is to me what Cooper was to Lily.

Looks like real life can mirror books after all.

"So what about you and Luke?" I change the subject. "I gotta figure he's gonna be heading back soon too. Any idea what the two of you are going to do?"

"Oh no you don't! Whatever is going on between me and Grayson is nothing like you and Cole. A love story we're not."

"Ahh, that's right. You're not into that kind of thing and you're just *having fun* with each other."

"Exactly."

"So have you slept with him?"

"As a matter of fact, Ms. Too Nosy for Her Own Good, we have."

"And?" I prod her, laughing when she flips me off. "Come on, Bethy! Inquiring minds wanna know!"

"Annnd…it was nice."

"Just nice?"

When her cheeks flush I know I've got her. Not much can make Bethany react the way she is now. Knowing Luke has made enough of an impression on her to make it happen only solidifies what I thought was happening over the last couple of weeks. Bethany is into him and for the first time since we became friend's years ago, she's in it for something more than just fun. Even if she's not ready to admit it.

"It was hot, okay? He knows what he's doing, but I swear to god, Reagan. That's all you're getting!"

"I can handle hot."

"Good. Now can we please get back out there before they start thinking one of us tripped, fell in and we've drowned in a sea of our own piss?"

"Really? All the things you could have come up with and that's what you chose? You're as gross as Luke."

"Well, we have screwed around. Maybe some of him seeped into me."

Gagging as she laughs, I head for the door, not even bothering to hold it open for her, only a few steps away from plugging my ears so I don't have to hear whatever other dirty thoughts she's looking to share.

Pausing to wait for her to follow me out, my eyes fall on the table where we left the guys. Both of them talking with their hands in what looks like pretty heated discussion. Taking a view seconds to admire the view, my suspicions about their conversation are confirmed when Cole throws himself back in his seat and scowls.

Holding Bethany back when she makes her way out of the washroom and pointing to the table, I press my finger to my lips before taking the first couple of steps toward the table. Picking up on their voices the closer I get, but not their attention.

"How much longer, Cole?"

"A couple of hours. I told you I'm gonna do it. You were right from the start. This has gone on long enough."

"I know you want to protect her, man. Do things differently than you normally would because of how deeply involved you are. But I don't know how much longer I can sit here pretending I'm just your best friend from the city and not her brother."

What is he talking about? Is this about Bethany?

Knowing she's adopted and how tight lipped she's been about her birth family, especially about any brothers and sisters she might have, could this be about her?

Looking back to my friend and seeing the frown on her face, I can see she's gone to the same place I have. The only place she really can go considering I haven't filled her in about me.

There's something that doesn't add up though. If Luke's talking about Bethany, why is he saying that Cole is deeply involved? The only time they've spent together since they got here has been with me.

Unless there's something they're not telling me.

Leaning in, praying that I'm just overthinking what we've heard, I tap her on the shoulder and when she looks my way, spill my guts.

"Have you spent any time with Cole when I'm not around?"

"No. Why?"

Oh no.

No, no, no.

This is not happening.

The only person here tonight that Cole is deeply involved with besides the person that he's talking to is me.

This is about me.

They've been holding something back. Cole's been lying to me.

I'm going to be sick.

"We talked about this earlier. I know what I have to do and I'm gonna do it. She deserves to know the truth. I just wish that I didn't have to be the one doing it."

"Better you than some cop that doesn't give a shit about her."

"Maybe." Cole shrugs. "Doesn't change how I feel though."

"I know, man. I get it. I don't know how much longer I can go on pretending that Reagan isn't Sunday."

Oh my God.

Chapter Twenty-Three

Cole

She gasps first, alerting me to the fact that she's no longer safely tucked away in the bathroom with Bethany the way we assumed when we started talking, but it's the way her legs seem to give out on her that I'm focused on most.

Gripping onto the chair to the right and steadying herself, I'm up and out of my own seat in record time. Crossing the floor to get to her until Bethany steps in and blocks me.

"I should have known." She mutters before turning and focusing all of her attention on Reagan. Doing nothing to mask the anger seething behind her normally friendly eyes. A look she, like Reagan, has every right to have.

"Sunday…" Luke says as he pushes his chair back and makes quick work of the distance separating him from his sister. A move that once he gets close enough, I reach out to stop. If I can't be anywhere near her then neither can he. With everything I'm sure she caught from our conversation, she's going to need time.

"Don't call me that!" Reagan cries and the break in her voice hits me like a sledgehammer to the chest. Knocking the wind out of me and making the need to reach out and take her into my arms even more powerful than it was before. "It's not true. You're lying! Playing some kind of game. I don't know why, but I'm not your sister. I'm not Sunday!"

She has no idea how badly I want to tell her she's right. That she's not Sunday Grayson and I wasn't sent here to bring her back to her real family. That this is all some sick joke Luke and I thought up on one of the nights we were getting blasted off our ass.

But I can't do that. I can't lie. I won't. Not anymore.

Before I can piece together the right combination of words that will somehow turn this entire situation around, Luke beats me to the punch.

"It's true. Everything you heard is true. Reagan, you *are* Sunday. You're my sister."

"Luke…" I interject in an attempt to stop him, but he just shakes his head and pushes his way toward her again.

"No, Cole. I played it your way for weeks. I'm not doing it anymore. She needs to know."

"Yeah she does, but not like this."

"Because your way has worked out real well so far, huh?"

I hate how right he is. How badly I screwed up. Choosing to get to know her and become a part of her life the way I had been, it was selfish.

My time with her was healing me, but not doing a damn bit of good for the case.

And now for the second time in my life because of my selfishness, I'm about to lose the only thing that matters.

Shaking her arm free of Bethany's hold, she runs away. Four words falling in the air around us before she makes her exit through the door and disappears out of sight.

"I can't be here."

Every instinct I have is screaming to go after her, but I can't move. I'm stuck. My feet are nailed like lead to the floor and with Luke and Bethany not moving, it appears I'm not the only one too blown away to react.

"We need to go after her. Explain everything and fix this." Luke says.

"You're not getting anywhere near her! You hear me, you lying piece of crap! Never again! I can't believe I was so stupid! I knew there was something off about the two of you."

"Beth—"

"Oh no you don't, asshole!" She snaps before stepping closer and shoving her hands into my chest. "You don't get to call me that *ever* again! Only people I trust get to call me Beth and you're sure as hell not someone I trust. Not anymore."

"I deserve that."

"You deserve a whole lot more than that, Cole." She seethes. "If that's even your real name. I mean, really? Is there anything about you that's real or was all of this like Reagan said and just some sick joke?"

Before I can answer, her eyes go wide, almost as if in the midst of her anger she's realized something that despite how much we know, neither Luke or I are privy to.

"I can't believe I didn't think of this sooner. Did Walker do this? Pay you guys to mess with her? It sure smells like something he'd do."

Walker is the last person I would do something for, and with the way I react every time his name is brought up, I assumed Bethany would know that. I loathe that prick more than they do.

I just don't have the first clue how to make her see that with as angry as she is right now. I do know that even if she's right about everything being fake, the one thing that isn't is how I feel about her best friend. If those were fake, my heart wouldn't be breaking into splintered pieces along with hers right now. I'd feel nothing. And despite wishing I could shut the shit off, all I can do right now is feel.

"What I feel for Reagan is real."

"And his real name *is* Cole, I *am* Luke, and the girl that we just let run from here alone in the middle of the night *is* my sister. My Sunday."

Ignoring Luke and his attempt at getting through to Beth, I put my focus back on the one part of what he said that I do care about. Her running from here alone.

"Where would she go?" I turn and ask Bethany, doing my best to ignore the hard stance of her body and the death glare in her eyes in an effort to get answers.

Even if Reagan says she never wants to see me again, there's no way I'm leaving her out there on her own.

All of this might have started as a job, but its way more than that now. I want to make sure she's alright and it's got nothing to do with bringing her back to Orangeville to Allison and Xander. Reagan Carter got to me. Burrowed her way under my skin and into my heart and it won't be able to rest until I know she's okay.

"I know you're pissed off and want to kill us. Maybe even a little confused and want answers, but right now, I need to know where she would go. When she's gotten upset in the past, what does she do?"

"She goes home." Bethany admits, the edge to her voice proof that she hates giving me even that little sliver of information.

Nodding as way of thanks, I turn to Luke and lay out what's going to happen now.

"I'm going to the house."

"You're not going alone."

"Yeah, I am." I argue. "You're right. My way of doing things is what got us into this mess. I let my feelings override the job and everything blew up in my face. She's running scared. We dropped a huge bomb and if we both show up, it's only going to make her run harder and faster. That can't happen."

"Then what the hell am I supposed to do? I'm not sitting around on my ass. I've done that long enough."

"Stay here with Bethany. Fill her in, even if it's the last thing she wants to hear. I'm pretty sure after tonight, Reagan is gonna need her best friend."

It's not a lot, but some part of what I've said seems to get through to Beth because she moves and this time, it's toward my now vacated chair instead of my face. She may want to kill me, but her love for Reagan is keeping her from doing it.

"There's only one problem with this little plan of yours." She offers up after she's made herself comfortable at the table.

"What's that?"

"Jake."

"What about him?"

"He's gonna come off the stage any minute and he'll wanna know where she is."

"Then Luke can explain it to him too."

"Because that worked out so well the last time they got within a foot of each other." She snips sarcastically. "If what the two of you were talking about when we came out of the ladies room is right, then she has two brothers that take their roles as such seriously. You need to head that off."

"She's got a point, Cole. It also wouldn't be a bad idea to have someone with access to her house on our side."

"Oh, he won't be on your side." Bethany laughs. "He's going to want to beat your asses the same way I do."

She's not saying anything I don't already know. I noticed it the first time I saw them together. Jake, despite not being biologically related, cares about her more than he lets on and when he learns the truth, it's not going to go over well.

He's going to want to make me hurt the same way I did his sister and I'm going to be deserving of it all. Going into this, I thought about Reagan and her reaction to the news. Wanting to soften the blow for her when I finally told her the truth. I didn't think about everyone else that would be affected.

I'm not only destroying one life here. I'm destroying two.

Hearing the final chords of the song play out and Jake's voice through the mic as he ends his set just the way Bethany predicted, I move and reach him just as he jumps down from the stage. When he sees me, his face lights up before moving behind me to where I know he's expecting to see Reagan waiting and frowning when she's nowhere to be found.

"Where's Reagan?" He asks and forcing myself to pick my eyes up from the floor, determined to do what Bethany said and tell him the truth, I motion with my head toward the door.

"That's what I wanted to talk to you about. You think we can talk outside?"

"Why can't we talk here?"

Because when I tell you who I really am and what I'm doing here, you're going to want to kill me and I can't have it happening in front of a room full of people that came here to see your band play. I can't ruin you the same way I ruined Reagan.

It all sounds so damn noble in my head, but no amount of trying to do the right thing now is going to make up for the wreckage I've caused tonight. The wreckage I've been causing for weeks by holding back.

"Figured after being up under the lights, you might want some air."

Laughing easily, not realizing what's really going on, he starts walking toward the door. Heading around the side of the bar once we're out and jumping up on the old picnic table hidden away in the corner. Proving himself to be more than just a pretty face when he levels me with a knowing look and forces my hand.

"What's really going on, Cole?"

It's time to tell Jake the truth.

To blow another life apart and hope that when the dust settles, his concern and love for Reagan will override his urge to kill and he'll help me get close to her again.

Even if it's the last thing I deserve.

Luke

"Start from the beginning and don't even think of leaving anything out."

You know, if this was any other situation we found ourselves in, I'd probably be complimenting her on her ball busting ability, but since it's not and I'm not looking to be shredded to bits tonight, that's not at all what's gonna happen.

If she wants the truth than it's about damn time I gave it to her.

"You remember me talking about my sister."

"Yeah, I do, and if I remember right, you said you lost her. Always talking about her in the past tense. You made it seem like she died."

"Well, considering that up until a couple of weeks ago, I hadn't seen her for seventeen years, she was dead. I'd written off ever finding her. I figured she was as dead as the babysitter that took her."

"Okay stop and rewind back a bit. I need to know what the story is with the babysitter."

"This isn't a movie, Beth."

Growling, she slugs me in the arm. "What did I say about that?"

"Fine. *Bethany*. This isn't a movie, and despite what you called it earlier, it's not a sick joke either."

"Then tell me what it really is."

"Sunday was taken by our babysitter when she was eighteen months old. Not long after she was taken, the babysitter was found dead. No sign of Sunday."

"Okay, and why do you think Reagan is Sunday?"

"DNA match mostly, but if you've spent the amount of time around my family that Cole has, it's the more subtle things. The way she smiles, the way she displays emotion, and some other unconscious mannerisms. She also looks a lot like my dad."

Releasing a heavy breath, she runs her fingers though her hair and I know that the small bit of information I've given her, no matter how pissed off she is at me, she believes.

Bethany is starting to realize that Reagan is Sunday.

"It's true."

"Yeah, it's true. Leaving DNA out of it, I knew it the first time we came here. She looked in my eyes and it was like my dad was staring back at me."

"There's something I don't get. If she's Sunday and you've finally found her after all this time, why the cloak and dagger? Why lie about it?"

"You'd have to ask Cole that, sweetheart. I was ready to dump the truth on her the first night."

"Cole's not here, Luke. So I'm asking you. Why would he want to keep it quiet?"

How do I explain what I've been watching happen? What it had taken Cole weeks to admit? That he did it because he has feelings for my sister and didn't want to be the one to tear her entire world apart.

"Something tells me that you already know the answer to that."

"He cares about her, doesn't he?"

Nodding, but not feeling like it's a good enough response, I attempt to explain.

"I've known Cole a long time. He's the closest thing I have to a brother and when I tell you that he doesn't care about much other than family, I mean it. Caring about Sunday is an understatement."

"What would you call it?"

"I call it him being in love with her and hating himself for it."

"Why would he hate himself for loving?"

"Because it's how Cole operates. We're like brothers and he's in love with my sister. Not to mention everything he's holding back from said sister because of his need to do right by her. He thinks that feeling something for her is going to make things worse than they already are."

"So he's afraid of losing you?"

"Something like that. Cole is a wicked smart guy, but sometimes when it comes to things that are staring him right in the face, he's an idiot."

"How do you feel about Cole and Reag—Sunday?" Bethany asks, proving just how quickly she's caught up to the whole Sunday thing by changing the name at the last second.

"Well, up until you both came out of the bathroom and overheard us, I felt pretty good about it. He told me earlier that he was tired of holding back and was going to tell her the truth tonight. That he loved her. Which, if you know him the way I do, you'd know wasn't an easy thing for him to admit. Even knowing it was my sister, I was happy for him, especially after the year he's had. Now though, I'm not sure how I feel."

"She'll forgive him."

Not much can take me by surprise, but the way Bethany has been with me since Cole left us alone, does. How perceptive she is even though like Reagan, she's learning everything the hard way.

I know Cole better than I know myself, so I know why Bethany is saying what she is, but even with me and Cole, there's still a bit of doubt under the surface. With Bethany, there's none of that. She really believes Reagan will forgive him.

"How can you be so sure?"

"It's the person she is. No matter how hurt she is by someone, she somehow sees past it and forgives easily. I can't say she'll ever forget it, but she will forgive. Honestly, I'm a little jealous of her ability to do that."

"I take it you've seen her do it before?"

"Remember Walker? For months there were signs he was cheating. Really obvious ones. The one time she confronted him about it and he spun some lie about the girl coming onto him, she forgave it. His excuses were total bullshit, but she just heard him out and forgave. She'll do the same with Cole too even if right now I don't think he deserves it."

"What about me?"

"What about you?"

"Do you think she'll forgive me?"

"Depends."

"On?"

"Whether you've done something you need to be forgiven for."

Hasn't she been listening to anything I've said? What kind of question is that?

"I lied to her, same as Cole did. I think that's something I need to be forgiven for. At least it does if I ever plan on having a relationship with my sister again."

"You're right. You did keep the truth from her, but she didn't spend weeks falling in love with you, did she? That's all on him. You didn't lead her on. You didn't create a lie and continue living it despite knowing the damage it would cause when the truth came out. Honestly, the only thing you're really guilty of is not having the balls to go against your friend and tell her the truth."

"Leave my lack of balls out of this." I joke and when she laughs, I start believing that maybe this situation might be able to be saved after all.

"You were close, weren't you?" Bethany quietly asks once the laughter subsides.

"Close with who? Cole?"

"Sunday."

As hard as it is to talk about the way things were before she was taken, I owe Bethany the truth. No matter how painful it might be.

"Sunday was the heartbeat of our family. She was *my* heartbeat. I lived for my little sister. I gave up looking for her. Stopped believing she was alive. Pushing hope away and not giving faith the time of day because if Sunday wasn't alive, I didn't want to be alive either. Close doesn't even begin to describe how things were with me and Sunday."

"I'm sorry." She says and when I look up and meet her eyes, the softness I see there speaks to her sincerity. She really means it. "I know this can't be easy to talk about, but I've got one more question. You think you can hang in there long enough to answer it?"

"Shoot."

"Now that you know she's alive and for the most part, happy, do you think you can do it again?"

"Depends what you're getting at."

"Live for her."

"Why would you ask me that?"

"Simple. When everything comes out, and I don't mean the shortened version you've given me, but everything, she's going to need something I can't give her."

"And what does that have to do with me?"

"Because the same way I think Reagan is going to need Jake, Sunday is going to need someone too. You."

Chapter Twenty-Four

Reagan

Pushing my way through the front door and taking the steps two at a time until the bathroom is in my sights, I fly through and slam the door so hard behind me I can feel the walls shaking.

Standing in front of the mirror—a move I've done more times than I count without even realizing the significance of it—I lean in close, placing my hand on my cheek, keeping it still for a few seconds before gathering up the courage to move it. My fingers brushing against all the parts of my face that until a few minutes ago I always found so insignificant. Starting at my cheeks before lifting up to my forehead and dipping down over my eyes. Running over my nose and around to the other side and back down to my chin.

Eyes the color of chocolate, fair skin tone and raised cheekbones. Full pink lips. Features that I've lived with for the last eighteen years and believed made me unique, but now know the difference of.

Luke's confession, the one he didn't realize I had been close enough to hear rings in my ears, combining with the shattered look on Cole's face the moment he realized I heard it all. Who he really is, who I really am and just why I look so different from the rest of my family.

Putting to bed years of unspoken and unanswered questions, while at the same time, bringing to life real feelings of loss and betrayal.

Cole hadn't come to Alliston to find a runaway husband and happened upon me at the bank. I'd been the target all the long.

"I know you want to protect her, man. Do things differently than you normally would because of how deeply involved you are. But I don't know how much longer I can sit here pretending I'm just your best friend from the city and not her brother."

"We talked about this earlier. I know what I have to do and I'm gonna do it. She deserves to know the truth. I just wish that I didn't have to be the one doing it."

"I know, man. I get it. I don't know how much longer I can go on pretending that Reagan isn't Sunday."

I'm Sunday Grayson.

Even thinking that twists me up inside. There's something so foreign about it, but at the same time, something equally as right.

Didn't I spent a straight year when I was ten harassing my parents at every turn? Determined after a talk with the girls at school that I had secretly been adopted and wanting them to just admit it so that things could make sense again?

An adoption that a few days ago I learned the truth about. And then just like in the movies, the picture changed, becoming more distorted until it wasn't just an adoption I was dealing with, but a kidnapping.

From my dark brown hair and eyes to my short stature when compared to the height that both of my parents commanded, it was there staring me in the face. Instead of delving deeper into it, I just bought into what my parents wanted me to.

I was unique. Different. One of a kind.

They were right. I was all of those things.

But only because I was her.

I'm gonna be sick.

Moving from my position in front of the mirror, I drop to my knees in front of the toilet, flipping up the seat and bending over. My head spinning so rapidly it doesn't take long for the fries and burger I'd had at the bar to come up and spill out in a violent tidal wave. The tears now streaming down my face both burning and soothing my broken heart and tortured soul.

I don't know what hurts more. Finding out that I'm not who I thought I was or that I trusted Cole with my heart only to find out that just like Walker before him, it was all a game.

I'm drowning in both, but it seems like the hole that's opened in my heart over Cole's betrayal is by far the worst.

After Walker and his cheating, I didn't think I could ever feel as low again, but I'm being shown otherwise.

It's happening again.

"I hate you!" I scream into the empty room, waiting for the tightness around my heart to alleviate with the admission, but falling further into despair when it only seems to hurt worse.

It won't stop hurting because I'm lying.

I'm in pain. Absolute heart wrenching agony, but I still care. I'm still feeling the high that I experienced every time we were together, and just below the surface is the soft hand of forgiveness that despite wanting to do the opposite, I'm powerless to fight against.

The last thing I feel right now is hate. I'm angry, broken and torn in two, but I don't hate because even though it would be easier if I could, I can't hate anyone. Least of all the first person to make me feel since Walker put my heart through a meat grinder and laughed as it turned to mush.

Letting the cold tile have its way with the pain now drilling a hole through my head, I lower my body to the floor. Enjoying the cool brush of it against my skin as it begins to permeate its way through and soothe the damage the night had done.

But in my world, a moment of pleasurable relief is just that. A fleeting moment in time. No sooner do I begin to drift and fade, the heavy knock comes and just as quickly as my heart and mind had been able to find peace, it's gone.

The same way Reagan Carter is gone.

"Rea!" Jake's voice filters through. "I know you're in there. Open the door!"

Of all the people I expected to knock, the last person I'd been expecting was Jake. Especially since the last time I'd seen him, he'd been preoccupied with singing and the blondes practically humping the stage.

Cole, Luke, hell, even Bethany knocking on the door since they were all there and a part of this made more sense than my little brother.

It doesn't even matter. I don't want to see any of them. Including Jake.

I can't. It will hurt even worse than it already does and right now it feels so heavy that I'm sure it's got the power to kill me.

"Go away, Jake!"

"Not happening, Rea. Open the damn door!"

"No!"

Banging again, his sigh so loud that even from my place against the tub, I can hear him as if he's standing directly in front of me, he yells through again.

"He told me what happened. Please let me in."

The slow drawn out way he says please threatens to be my undoing. It's the one thing that despite all of the years and experience I've had hearing it, I can't turn away from. It's not often Jake says please which means he meant what he said. He knows what happened and despite knowing what it means in the end, he still wants to help.

He still wants to be my little brother.

Lifting myself up until I'm on my knees again, I begin to stand but freeze in place the second I hear the softer tone of the voice with him. A voice so familiar that even through my pain it makes my stomach do somersaults.

"She needs to open the door in the next two minutes or I won't be responsible for ripping it off its hinges."

Cole.

He's here.

Cole

"You know, I think I always knew." Jake admits after pounding on the bathroom door again heeds the same response as the last twenty times he's done it, and he leans back against the wall exasperated.

"Don't do that to yourself, man. You couldn't have known about this. It just feels that way right now because it's how your mind is dealing with it."

"So you're saying that I'm feeling this way because I'm in shock?"

"Pretty much."

It's strange. When I told Jake everything back at the bar, aside from the moment he jumped off the table and lunged at me like a wild animal, he's been surprisingly calm. Almost like in a way, he understood it. Accepted it. Knew it.

Only that can't be right because no one knew just how deep this ran. Hell, I didn't even know Sunday was alive until a couple of weeks ago. There's no way, even with everything I know Reagan overheard with her parents that Jake knew about it.

"I need to get in there." I repeat motioning to the door again. "I'm afraid of what she'll do if we leave her alone too long."

"Do you even know her at all?" He snaps. "Reagan's been through a lot of shit lately and all she does is hide away in her room. She's not gonna hurt herself."

"That's not what I meant."

"Then why don't you clue me in on what you did mean because from where I sit, you're worried she's gonna take a blade to her wrists or something. Which is total bullshit, by the way."

How do I even begin to explain that I'm not afraid of her harming herself, but scared that she'll turn into me? That everything she learned tonight will twist in her mind until she's flooded with memories she can't control.

"I just don't want her to be alone, Jake."

Stepping back off the wall and heading to the door again, he pounds on it. The force of the hit so heavy that for a split second I can see it shaking from the impact. What I said earlier might not even have to happen at the rate he's going. He'll take the door off its hinges all on his own.

"Reagan, I know what happened. Cole told me everything. Please come out so we can talk about it." When no response comes, either through the sound of her sobbing the way we heard when we first got here or the sound of her giving in and opening the door, he pleads with her again. "Please, Rea. Open the door."

"Where's her room?"

"Why?"

"I think you'll have a better shot of getting her out if I'm not nearby."

"Reagan's room is her sanctuary, Cole. Even if I got her to come out, it would be the first place she'd go. Surrounded by her books and shit is where she feels safest. If you're in there, it's only going to make her run and then we'd never get her back out."

"We can always get her out, Jake. There's a reason tools were invented."

"No shit. Look, maybe you should just take off. Head back to the bar and pick up your friend. Head back to the motel or something. I can take care of this."

That's not happening. No way in hell am I leaving. I know full well that when she does come out of that room, she's going to be guided by my betrayal and she's going to kick me out of her life. I've accepted it. But until that happens, I'm not going anywhere.

I made her a promise. Even if I was holding something pretty significant back from her when I made it, I won't leave her. I'll always be there, even when it's the last thing she wants.

"I'm not going anywhere until Reagan tells me to."

Turning toward the door, this time using my own fists and beating on it, I repeat what I just told Jake, loud enough so I know beyond a shadow of a doubt she's heard me.

"You hear me, Reagan? If you want me out of your life then you're gonna have to come out of the bathroom and kick me out of it!"

I don't know how I know that it's going to work, but I do. The need inside of her to see me gone, the hatred I know is building with each passing second she spends locked away, is going to be what gets her out.

"I'm going to her room. When she opens the door, which we both know she will, bring her there please. I need to explain this."

"And if she doesn't want to hear anything you've got to say?"

"That's not an option. I'm going to give her all the facts the way I should have from the start and then I'll do whatever she wants me to after that."

"So you'll leave if that's what she wants?"

As much as it pains me to say it. Admit that I will walk away from the first thing in years that's truly made me feel, I have to do it.

"Yes. I'll leave."

Turning, I head toward the bedroom but before I can slip myself completely inside, Jake calls out and stops me in my tracks.

"Was any of it real?"

I know what he means. The feelings. The way we were together. The look in my eyes and the feeling of total completion inside me whenever she was around. He wants to know if I was playing a role.

"I lied about why I was really here and kept the truth from her for weeks. Those are my crimes and I'm willing to pay for them, but Jake, everything else and I do mean everything, was as real as the urge inside you right now to beat the shit out of me for hurting her."

Nodding in acceptance of my answer, I turn and make my way into her room as I hear him step forward again and bang on the door. After a few seconds of his feet pacing back and forth across the carpeted floor, muttering curses under his breath, what I'm after happens as I hear the click of the door.

From my place in her room, it's not loud, but the sound is clear. I was right. Her need to get me out of her life for good has brought her out.

"Rea…"

"Where is he?" she demands, the shaking in her voice betraying the tough stance she's taking. "I know he's still here. Where is he, Jake?"

"Your room."

"Okay."

Taking in the sound of her boots across the floor, I lower myself to the bed and prepare myself for what's about to happen. What I wish I could go back in time and take back so that it didn't have to go down this way.

I prepare to lose Sunday again.

Chapter Twenty-Five

Reagan

Why is this so hard?
Walking into my room and seeing him sitting on my bed with the pained expression on his face, I wasn't supposed to react. Yet here I am wanting to forget the last couple of hours even happened and bury myself in his arms.

The arms that not twenty-four hours ago had made me feel safe. Protected. But more than that, loved. Adored. Like I was the only woman on the planet.

"Rea—"

"No." I quickly cut him off. I can't stand to hear him call me Rea, especially knowing why he does it. "Don't call me that."

"What do you suggest I call you then?"

"Sunday." I state matter of fact. "That's who I am, right?"

"Yes, but under the circumstances it doesn't feel right."

Now he's worried about what feels right? Where was this concern for what was right when he spent the last two weeks making me believe in a lie?

"It's a little too late for what's right, don't you think, Cole? If that's even your name."

"If it would make you feel better, would you like to see my license? Birth certificate? I can prove that I'm exactly who I said I was."

"Bullshit."

"Reagan please. I don't wanna argue with you."

What exactly is it that he's expecting from me? It's been a little over twenty-four hours since I was dealt the blow of being adopted. Fast forward a couple of hours and there's yet another blast going off that blows that one completely out of the water. I'm a whole lot more than adopted.

Kidnapped. Stolen. Someone else.

Almost eighteen years I've spent living a life that wasn't meant to be mine in the first place.

No. No. No. Shit. I knew looking up was a mistake.

He can't look at me like that. I'm not strong enough to combat the agony I see laid out all over his face. The way his body is starting to sag only adding to it. I need more time before he does this.

"Stop it!" I cry, desperate for him to go back to the way he looked when I first walked in. Where the need to go to him was there but wasn't quite this strong. When I could still fight the impulse.

"Stop what? Hating myself for what I've caused? Wanting to kick my own ass for not telling you the truth that first day at the bank? What Reagan? Tell me what you want me to stop and I'll do it!"

I don't get it. He lied to me. Hid the truth for weeks, making me believe he was in town for a reason separate from the one I now know about and even with all of that, I can't seem to hate him.

I want to hate him so bad it hurts, but every time I even so much as think about it, I feel like I'm going to throw up.

"Stop lying to me."

"I'm not lying. Not anymore. I never should have done it to begin with. It was selfish. So fucking selfish, but I couldn't tell you the truth. Not when I knew what it would do. I had to protect you. As stupid as that sounds now after…" he pauses, scrubbing a hand over his face as it twists, obviously as sick of his own words as I am. "Well, after everything."

"The only thing you had to protect me from was you."

I'm cold. So freaking cold, but I know it's not the temperate causing it. It's the betrayal. The pain. The ache inside my heart that's making it almost impossible to breathe. It's stealing the warmth from my body until all that's gonna be left is a frozen block of ice.

"You're right."

"That's the first thing you've said tonight that I actually believe."

Flinching from the impact of my words, he releases yet another heavy breath before raking his hand through his hair and standing.

Not knowing what he's planning on doing next, I immediately start backing up. Not trusting myself to get close to him. Afraid of what my heart will do. How it will betray me and give in. Something that can't happen.

I can never let Cole close again.

"I know you don't want me here. If I were you, I wouldn't want me here either, but I can't leave. Not until you know the truth. I wouldn't be able to live with myself if I walked away before giving you that."

"The truth." I laugh. "What a joke."

"Reagan..."

"Stop it! I'm not Reagan, remember? Reagan Carter is dead. No. It's even worse than that. She never existed! So you either call me by the name you know me as or get out."

"I can't do that. I'm sorry. Ask me for anything else and I'll do it, but not that."

Of course he can't give me that. Sunday is his precious light. She made his life worth living for the year and a half she was in it. She got the best parts of him. Why would he call me by her name? It would tarnish the memory he has of her.

Except that's not true.

I'm his light.
I had the very best parts of Cole.
I'm the one that made his life worth living.
I'm Sunday.

"Can't or won't?"

"Both."

"Then I'm done. Get out."

When he doesn't make a move to go, accept what I've asked and leave the way I want him to something inside me snaps. Crossing the room and readying my fist at my side, I stop just shy of our bodies touching and I hit him as hard as I can. Just the way Jake taught me after I learned about Walker.

With no reaction coming, I do it again and just like before, he takes it. The only indication he's even feeling the abuse is the way his body stiffens with each blow I land.

Maybe he wasn't lying after all. He really will take whatever I have to give. He really does hate himself that much.

"Get it out, Reagan. Hit me until you can't anymore. I can take it. I *want* to take it. I deserve it."

Something about the way his voice sounds when he says he deserves it reminds me of the night before. The memory of his mother and the beatings he took at her hand. His voice sounds so different. So broken down and weak because just like last night, he's gone back in time and become that six year old boy again.

And I've become his mother.

Pulling back quickly, I turn my back on him. Bending over as I feel the bile growing in my throat. Desperate for its release, I slide a finger down as deep as I can go until it's spilling out. Making my carpet as stained, dirty and gross as I feel after what I've done.

"I'm sorry." I choke through the tears that with a mind of their own have started to fall. "I didn't think."

"It's okay. I told you to do it." He responds softly, bridging the distance between us until he's the one with his hands soothingly rubbing my back. My body betraying me despite the turmoil it feels and enjoying the way it feels.

"No it's not. It doesn't matter what you did. I never should have put my hands on you."

"I deserve a lot worse than a few punches to the chest, Reagan. What my mom did when I was a kid might have been wrong and maybe I didn't deserve it, but what I did to you? That I did deserve."

I hate that he's right. That there's even a small part of me that believes beating on him is the right thing. Violence has never been the answer before. There's no reason it should start now. My entire world crashing down around me or not.

"You need to go."

"Not before I've told you everything."

"Cole, I can't do this right now. It's too much."

Stepping back once I attempt to right myself, he makes his way back over to my bed and lowers himself down onto it before speaking again.

"If I give you what you want and go back to the B&B, when you've had time to process what you learned tonight, are you going to come find me so I can tell you the rest?"

He's asking a question that judging by the way his lips hover in a straight line, he already knows the answer to. I won't go looking for him after tonight. I'll never go looking for him again.

"No."

"Then you need to hear me out now. I'll give you what I know you need most right now, but not before I put an end to what made this happen in the first place. Not until I tell you the whole truth."

Do I want to hear this? Can I take hearing it? It's pretty obvious that by the truth, he means something other than what I already know from overhearing him and Luke. Which means this is about the kidnapping.

How I ended up as Reagan.

"Fine."

"You might want to sit down. It's a lot to take in and at the bar, you almost passed out when you overheard us. I don't want that happening again."

"I'm not a porcelain doll, Cole. I can handle whatever you've got to say."

There it is. My backbone. What for so long after my heart was put through the grinder seemed to fly the coop and go on a permanent vacation. Looks like I'm not as weak as I thought.

At least I'm not until Cole speaks again and blows it all to shit.

"He did it, Reagan. James…Your dad. He was the one that kidnapped you."

Cole

My childhood may have been shit, but there's one thing I do know from being around others whose weren't like mine.

There's a special bond that a father has with his daughter. No one can determine with any degree of accuracy what makes this happen, but there's just something that occurs when a father sees his daughter for the first time. And it doesn't end at that first meeting. It lasts forever. Long after one or both have passed.

It's a bond that has the power to span the test of time.

So being here while everything is so raw and telling Reagan that her dad orchestrated all of this, brought about what's happening now, well, it cuts deep. I'm attacking that bond even if James was the one that started it first by preventing the bond she should have had with Xander.

James Carter doesn't deserve to have that bond. I'm not much better with what I've done, but this piece of work is worse than I could ever dream to be.

Reagan does deserve it though and I hate being the one that strips it away from her. I've already taken enough.

"I'm sorry, but I don't think I heard you right. I could have sworn you just said that my dad kidnapped me."

This is what I'm talking about. Her calm response. The way she needs me to repeat myself because she can't wrap her mind around what I've just said. This is the bond. She could think her dad is the biggest jackoff on the planet and she still wouldn't believe what I'm telling her now.

Which is exactly why I came armed with proof. Something physical that no bond in the world can deny, even if it still tries.

Leaning over the bed and grabbing the journal and the file resting underneath, I turn back and hold it out between us.

"I figured you would say that, so here. It's all there in black and white."

The journal doesn't seem to faze her as she takes them from my hands, but when she brings the file to rest on top, it seems to hit a nerve as her eyes go wide.

"This is the file we went through together last night."

"No it's not. That one only held basic information on Sunday Grayson. It's the decoy file my uncle set up so that if and when someone caught on to him investigating, they wouldn't know just how deeply he was involved."

"So what's in this one?"

"Every piece of evidence he ever uncovered, the police handed over after the case went cold for the hundredth time, or he got from my dad and other various sources."

When Luke and I had been scouring over the journals, there were two we hadn't gotten around to. Two books that after going through like ten other ones, we were too damn worn out to actually read. We overlooked it and in doing so, overlooked everything John really wanted us to see.

Like the location for the real file. The one Reagan now has a death grip on.

"In addition to planning and executing Sunday's kidnapping, he also hired someone to make my Uncle John's death look like an accident. He found out he was sick awhile back, so they used that to their advantage and ran him off the road. That's in the file too. Along with the book you're holding if you're having trouble believing me."

I know I severed the trust she had in me when I kept the truth from her, but I meant what I said earlier. I will tell her everything now and I will make sure that everything I do say is backed up.

"But isn't this just your uncle's word against my dad's?"

"Yes, but only because the people involved covered their tracks well. If you look in the file, you'll see that there's other evidence there that gives them away. It's only a matter of time before it comes back to him."

Stepping toward me but refusing to meet my eyes, she sits down on the bed and laying the journal down on the blanket, does exactly what I told her she should. Opening the file to study it.

After minutes pass with no movement or sound, she finally looks up as a sigh escapes and motions with her hand for me to sit.

"I have a question."

"You can ask me anything."

"I figured that, but I'm more worried about getting a straight answer."

I deserve that, I know I do, but damn. It still fucking hurts like hell.

"No more secrets."

"Hallie Francis. Her death. There's something about it that doesn't make sense."

"What about it?"

"We talked before about the fact that she was murdered. The file backs that up, but how do you know that my father had something to do with it?"

"The day Sunday went missing, there's an eyewitness report of a Trans-Am being parked just down the street. It was there for hours. Almost as if it was doing surveillance. The same Trans-Am was also reported being in the vicinity of where they found Hallie's body."

She reaches the conclusion Luke and I came to a lot faster than I expect her to, though I feel like kicking myself for doubting how quickly she could catch on. She's smart. I already knew that, so doubting her at all is just another way I'm letting her down.

Another reason for me to hate myself.

"Let me guess. It was black with flames down the sides."

Reaching across the space, our hands brushing and her jumping back as I slip the file out of her hand, my intent innocent but another stabbing pain where my heart resides appearing when she reacts, I flip a few more pages and hand it back over. This time making sure as I do to release it before we can connect.

As much as I want to pull her to me and hold her as all of this sinks in, I can't do it. I need to stick to the promise I made.

Give her the facts and leave.

No deviations.

"That's the eye witness report from the day Hallie was found. The police put out an APB on it right away, but the owner at the time sold it to someone else and well, they drove it out of Orangeville."

"Parking it in a garage for years and letting it gather dust."

This is new. With the bitter way she says it, she knows a whole lot more about that car then I originally thought.

"He never drove it after?"

"I don't know for sure, but as far back as I can remember, no. Up until six months ago the thing was buried under a tarp."

"What happened six months ago?" I ask, needing to know what it was about that time that would make him take it out of hiding.

"Jake got his licence. Dad said he was gonna do some work on it and when it was done it would be his car."

"Holy shit." I mutter under my breath once I put the pieces together.

This had nothing to do with Jake getting the car because he'd gotten his licence. This was James shifting the car onto someone else. Believing after John he was in the clear, but wanting the car out of his name so that just in case it ever did come up, his name wasn't attached.

Sure, his son's name being on the title would trace easily back to him if it was ever looked into, but with as confident as James must have been at the time, I don't think it even occurred to him. He'd make his first mistake.

"He couldn't have pinned it on Jake, Cole." She reads my mind easily. "He wouldn't do that. He adores Jake."

"You're probably right, but you can't deny it's suspicious."

"No, I can't. I can't deny any of this." Finally raising her eyes to mine, the toughness she's been projecting stops and I see the cracks shining through. "He really did this, didn't he? He really planned this out and kidnapped me."

"I'm sorry, Reagan. I know it doesn't make any of this better, but yeah. He did it. He's been doing it for years."

"Your uncle..." she cries softly. "I'm sorry, Cole."

This is where things are going to get harder because there's more. It's there in John's journal and it's something that if she's going to know it all, she has to read.

John wasn't his only attempt.

He'd tried the same thing with Samuel too. Only the old man kicked it before they could succeed.

"Rea," I test out the shortened name, only able to settle the roughness inside when she doesn't protest. "He tried to do the same to Sam."

In a move so unexpected I'm in no way ready for it when it happens, her arms fly around my neck and her body presses into mine. The contact giving her what she needs as I hear her begin to quietly sob into my shirt.

"I—I...I don't even know w-what to say."

"You don't need to say anything." I tell her, taking advantage of the moment and selfishly running my hand soothingly over her hair.

I mean every word. She doesn't have to say anything. That's not what this was about. I didn't tell her about Sam because I wanted sympathy or wanted to make her hurt. I told her because it all ties together.

But really, she also needed to know because it won't be long before they come for me.

It's not a coincidence, both my dad and uncle having attempts made on their lives when they started digging around the Carters. So it stands to reason that it's only a matter of time before it comes knocking at my door. The only difference between me and them being, I'm going to be ready.

If they want to take me then I'm gonna make damn sure I take both of them down with me. Along with anyone else who may have helped put this in motion seventeen years ago.

Pulling out of my arms and leaning back on the bed as she wipes at her eyes, she looks at me again. Those damn eyes of hers just as open and telling as the day I met her. Every emotion on display and breaking my heart all over again.

"Are you in danger?"

"Not right now."

"But you will be."

"Only if he knows I'm here."

Her head lowers and my heart drops into my stomach. *Shit.* I had a feeling she told her parents about me, but I'd been hoping that she'd left my last name out of it. There's no way they would have figured I was part of this if that wasn't talked about.

"How long has he known?"

"A few days maybe. A week tops. That's it, I swear. Cole…"

Her concern for me is overriding her anger. She feels bad and because I hate not being able to look her in the eye, I feel bad. She shouldn't be going through any of this. None of us should.

I never should have left Toronto.

"Reagan look at me." When she doesn't make a move to raise her head, I try again, this time, my request bordering on begging. "Please look at me."

Only when I reach out and touch her chin in an effort to bring her eyes to my level does she move and when she does, I see the pool of tears just waiting to spill out. Dealing with them first, I wipe them away before pressing my forehead to hers, making sure as I do that I keep my eyes open and trained solely on hers.

"You have nothing to be sorry for, but more than that, nothing to worry about. I'm going to be fine. I'm going to get in my car and go home. You have everything you need now. It's up to you what you do with it. If you don't ever want to explore this, then he never has to know that you know. It can be *our* secret."

Despite wiping her eyes, tears still manage to find their way out anyway and each one that falls makes the need to kiss them away stronger. The need to take away her fear and pain impossible to fight.

"I love you so god damned much I can't see straight. I've never felt like this before, Rea. I have no idea what to do with it, but I do know that in order to make this right—to prove to you that everything I felt and still feel is as true as the day I admitted it to you, I need to leave."

"So many wrong turns I took with you, sweetheart. So many times I could have told the truth and I held back because I wanted to do it the right way. I messed everything up and am completely at fault for the hate you feel in your heart. Hate you never should have had to feel at all. In order to do my part to make that better, I have to prove to you that I mean it."

"Cole…"

"I love you, Sunday."

Not giving her a chance to respond to what I've finally given her after fighting it every step of the way, I press my lips to hers. Lingering as I commit the feel of them to memory, knowing this is going to be the last time I'll ever feel them.

Pulling away, I stand, but instead of heading for the door, I move in again. This time bending over her and pressing my lips to her hair as I breathe her in.

"Goodbye Sunshine."

Making quick work of the distance from her bed to the door, feeling the tidal wave of it hit me before I'd even made the decision to kiss her head that final time, I book it through just in time for the first one to fall.

Something eighteen years in the making, but now that they're here, I know I won't be able to stop.

Tears.

Chapter Twenty-Six

Cole

Twelve.

The amount of bars I've passed on my way from Alliston back to Orangeville. Twelve different dives that it took every bit of self-control I had not to pull off and head into.

Twelve different places that I could have drowned my sorrows in, each shot effectively erasing Reagan Carter forever.

Erasing Sunday.

I didn't do it though. I didn't give in. I couldn't.

Because I can't erase her.

I couldn't do it for all those years after she went missing and I can't do it now. Especially not after what we shared.

Doing what she wanted and leaving her house was probably the hardest thing I've ever had to do. I'd done it, but only because I'd already screwed everything up enough. Made too many mistakes. I wasn't about to compound it by adding more.

I would give her what she wanted.

The only solace; what made walking down that driveway easier, being that I didn't leave her empty handed. When I kissed her that final time and felt the soft warmth of her breath across my face before I turned away, I'd left her with the part of me she affected the most.

My heart.

Making sure with each mile travelled, the empty shell that rests just under my skin dries out even more so that by the time I pull into the Grayson's driveway and turn the key in the ignition, I can say there is nothing left.

All that was left of the feelings I'd stupidly thought I could have stretched out over the highways and streets between their house and Reagan.

It had to be this way. What I have to do now demands it. I need to remain impartial as I sit Allison and Xander down and tell them everything I've learned about their little girl.

They can never know what I let happen. Luke may have been cool, but that's because our friendship spanned years. Decades even. The same can't be said for his parents. Even with the respect I have for them, I can't say we were all that close. There's no way they would be understanding of my feelings for their daughter. Especially after I spent the last two weeks keeping it and her a secret.

Leaning back against my car, I watch the house. The low light of the television filtering through the curtains proving that even though the hour is later than I expected it to be, they're awake.

Now all I need to do is psyche myself up for the moment they answer the door, sweep me in and I tell them I found Sunday alive and well. Then in the same breath tell them that she knows the truth and had chosen to stay where she is.

Breaking their hearts for a second time.

With Sunday being of age—an adult in the eyes of the law and government—in the end it would be up to her what she did with everything she learned tonight. The decision on what happened to James was different of course, his fate coming down another way, but I don't much care what happens to him at this point. It's only her I can see.

If she wanted to stay where she was and live her life, at the end of the day, it was her decision to make. Though I wouldn't ignore the small sliver of hope I had that after the news settled in, she'd choose differently.

She would come home to Xander and Allison.

Come home to me.

Time really has changed things. I wasn't much for wishful thinking before, never having anything to have real hope and faith in or about, but just like she'd touched the darkness in my soul, it seemed she was doing the same with my way of thinking too.

Despite believing deep down that she wouldn't change her mind, there was still a part of me that held out hope she would.

Forgoing the use of the bell and lifting my hand to the door, I rap my knuckles against it once, pausing for a moment before doing it again and taking a step back to wait out the response inside. Using the few seconds I have to shake away the memory of Sunday in order to bring forward what I can control.

Facts. Truth.

The job.

A job I never asked for, but one I know I'll never be able to entirely walk away from, even after this case is closed.

The light turns on first alerting to me to their presence and after a few seconds, the door swings open and the worn face of Xander Grayson comes into view. Light spilling out over him. Light that in the dark of night reminds me a lot of the woman I just walked away from.

Damnit, Cole. Keep it together. You're here to tell them everything you know and put the wheels in motion to make the people that took Sunday pay. That's it. You've done your job. Once you do this, you can wash your hands of it for good.

Not the pep talk I was going for, but valid despite my dislike of it.

Once I do this, I will have done what John wanted me to do when he put the damn stipulation in the will. I'll be able to move on with my life.

What's left of it anyway.

"Cole?"

"Hey Xander. I'm sorry for stopping by so late, but this couldn't wait."

There we go. Much better. Getting right to business.

Maybe I've got a little PI in me after all.

Xander's eyes lift when he processes what I've said and he steps back from the door, ushering me in with his hand and waiting patiently as I step through. My focus going straight to the family room where I can make out Allison resting peacefully on the sofa.

I was hoping to be able to say all of this to both of them. The ache in my chest that I can't seem to rid myself of making me want to get this over in one shot, but with as comfortable as she looks even from where I stand, it looks like I'm not going to be able to pull away as quickly as I thought.

"Is this about Sunday?" Xander whispers, pointing to the kitchen before motioning to where his wife rests and presses two fingers to his lips. "Did you find something?"

"I found a whole lot more than something, Xander." I admit when we've finally made our way in. Waiting until we're both seated at the table before dropping the rest. "I found her."

Blinking rapidly, taken back by what I've admitted, he releases a heavy breath. Rubbing a hand over his face as his foot begins tapping on the floor, he exhales again. Only acknowledging that I'm there with him when he's managed to get control of himself.

"She's...She's alive?"

"Yes. The girl that you were tipped off about, it's Sunday."

Pounding his hand down hard on the table, I'm jolted back in my seat, flinching from the sound of the chair hitting the floor as he stands and shoves it back. Watching as he stalks over to the kitchen counter, curses under his breath before doing the same to it.

There's no set way to act when something like this happens. Sure, I'd come in here expecting him to be happier about the news, but after seventeen years of trying to hold onto a hope that was slowly dying, I can't imagine there being much happiness left. Xander was probably a lot like Luke. Assuming she was dead. He just hadn't been able to say it out loud.

He couldn't make it real.

"Where did you find her?"

"Alliston. She's been there the entire time, other than the week or so that it took them to get out of town."

"Wait," he says, catching on to what I've just admitted and turning back to face me. "Are you saying that she was here for almost a week afterwards?"

Twisted up by the sadness staring back at me in his eyes and not trusting my voice to answer without giving away my own feelings on the subject, I just nod.

"Oh my God."

My sentiments exactly. When I learned James had kept her closed off from the rest of the community so she wouldn't be detected until he could move them out, I'd been just as surprised.

Guilt ridden.

She had been under my nose for a week. A full week where if I had just tried a little harder, I could have found her. Instead, I failed and she'd been slipped out of town.

Never to be seen again.

Until two weeks ago.

"If you know that much then you must know who took her." He surmises and catching me avert my eyes, he tries again. "You know who it is, don't you?"

"I do."

"Cole, you need to tell me."

I fully plan on doing that, but not before making sure he can keep his cool. With as insistent as he'd been slipping us into the kitchen and away from Allison, it's obvious that he wants to keep her away from it. What I'm about to tell him, it's not going to be easy to hear. It could risk that distance he wants.

"Are you sure you're ready to hear it?"

"I've been ready to know the truth for seventeen years, Cole."

"What about Allison? Do you want to wake her?"

"No. It's been a rough night for her. I came home to find her in the attic with some of Sunday's old things. I finally got her calm enough to rest. She needs the peace that sleep is providing right now."

"You better sit. It's a lot." I keep on task, not wanting to get into how agonizing it is hearing about the trauma his wife is still suffering with.

"Just tell me everything you know. I'll decide what I need."

The edge to his voice and the rigidness of his body tell a different story, but I'm not here to argue. If Xander believes he can stand for this; that the information I'm about lay on him isn't going to bring him to his knees, I've got to accept it even if I know better.

I could barely handle everything I learned and I wasn't her father.

Wanting to ease him into this as easily as I can, not ready to just name drop and let his mind run wild, I choose a different route.

"How well did you know the Francis family?"

"Not well. We'd talked in passing a few times, which is how we knew about Hallie's babysitting business, but we weren't close." He pauses, his mind going exactly where I knew it would when I asked the question. "Are you saying that Martin and Marie had something to do with this? Hallie? Did that son of a bitch take my daughter? Has he been holding her all this time?"

"Martin was involved, yes. So was Hallie. I think deep down you knew that there was more going on with her at the time than the police and the media spoke about. She was a whole lot more than a pawn."

"You didn't answer my question." He says, leaning over the chair he'd just vacated and leveling me with his steeled gaze. "Does Martin have her?"

"No."

"What aren't you telling me?"

"Xander, look. I know you said you want to hear it all, but you're already on edge just hearing that Martin Francis was a part of this. What are you going to do when I finally do tell you who she's with?"

"You're right. Of course you are. I'm sorry. I know this isn't on you, Cole. You're just the messenger. I asked for this. I just never thought we'd be having this conversation. I gave into what Allison wanted and called you in because I wanted her to have closure. I thought this conversation would be you telling us she was gone. I don't know what to do with this."

I was right. He did let her go. At least in his head. I don't think that ever would have happened in his heart. Even if I'd come back here and told them both that she was dead, he would have still held on in his heart.

The way a real parent is supposed to.

"Martin was hired to steal her, which is where Hallie came in. Her babysitting Sunday that night and Luke out playing hockey with me and the guys gave them the opening they were after."

"Gave who? Who put this in motion?"

"James Carter."

Shit! It happens so fast that he's on the floor before I've even had the chance to react and try to prevent it. Damnit. I knew that this was going to happen. I just thought I'd have a bit more time before it did. I had no idea that just hearing his name would cripple Xander.

Hitting the floor, hearing the crack of my knees as they hit the hard tiles, I grab a hold of the now broken man and using as much strength as I've got, lift him from the ground until I've got his body slumped down into the seat.

His eyes wide but vacant. His mouth frozen in a straight line. The only sign that he's still here with me the slight movement of his chest as he breathes.

"Xander, what is it? Why did you react to Carter?"

"I handled a case for him. I was just starting out back then, you remember. He'd come asking for my help. Countless hours we spent together after that. I knew everything about him by the time our business concluded. Right down to the struggles his wife was going through and the toll it was taking on their marriage. His need to give her a child. I knew it all. He was…well, he was my friend."

Well, this is news I didn't have. When I found out about the plan and how it was executed, I just as well assumed that it was a stranger abduction. Not that Xander and James had been close.

This just got a whole lot messier. There's no way with what he's just admitted that he's not going to blame himself. The same way that Luke and I have been doing for years. All of us doing it for different reasons, but with the same end result.

I can't let it change him the way it did us. Especially now that we know she's alive.

"What were you hired to do for him?"

"Nothing too deep. Just standard business contracts as he was taking a new approach to what had been up until that point a losing venture. He wanted a complete makeover and hired me to help."

"Did he ever spend time here?"

"A few times, yes. We had dinners together."

It's a hard question for me to ask, but in order for me to get an even clearer picture of what happened here that night, I need to know. Even if it's going to make me want to go against the right way to do things and end James Carter myself.

"Did you ever notice him or Katherine interacting with Sunday in what might have been considered an inappropriate way?"

"Katherine held her a few times and spoke to her, telling her she was beautiful, but a lot of the women in the neighborhood did that when they were over and caught sight of her. She had that effect on people, but it was never a cause for concern."

Don't I know it? She still has that effect.

"What about James?"

"He interacted with Lucas. While the women gushed over Sunday, we'd head out and screw around tossing the football or shooting baskets. He never once interacted with her."

Interesting.

Maybe Luke being out playing hockey with us that night saved his life.

"You're telling me that James has had Sunday all this time? That he hired Martin and Hallie to steal her so that he could raise her as his own?"

"Yes. They did eventually have a child of their own, but it was a couple of years after they grabbed Sunday."

"There's another child involved here?"

"Yes, and before you ask, he's biologically theirs. I checked. Twice."

"Well, at least there's one good thing to come out of this. I would hate to think he had done this to another family."

"Katherine went through years of in vitro fertilization. At one point they even looked into a surrogate when they realized the problem did not originate with James. But just like you talked about before, they couldn't afford it. Adoption wouldn't have been possible either, for the same reasons."

"So they decided to steal someone else's child? Who the fuck does that?"

Xander Grayson has only ever been one way. Calm. The voice of reason. Never showing anger or rage. The complete opposite of the life I was born into. It's part of the reason I liked spending so much time here in the beginning. Sam and him were a lot alike. I think this might be the first time I've ever heard the man curse. That's how regulated he was.

"Sick people do it, Xander. People with no conscience."

"Did you get close enough to meet her?"

We got a whole lot closer than that.

"I did." I offer up, wanting to say more but the chokehold that the memory of us being closer than Xander knows about preventing me from giving up anything else. It's still too raw.

"How is she, Cole? Is she happy? Well taken care of? Was she loved? Appreciated? Did she grow up to be as beautiful as her mother?"

His questions are like knives straight into my soul. The pleading sound as he asks each one threatening to undo me. Making me want to tell him everything. The entire truth.

That she's beautiful. Even more than her mother. That she's amazing and that she owns me and she always will.

That I'm in love with her.

"She looks like both of you, Xander. Her eyes are expressive like yours. You wouldn't need her to tell you how she was feeling, it's all just there in the way she looks at you. She's got Luke's smile. His attitude too. She's…well. She's…"

"She's what, son?"

"She's perfect."

Turning away before he can react to my loose lipped choice of words, I follow the same path he used before, making my way over to the counter and leaning over it. I wasn't supposed to show emotion here. I was supposed to deliver the facts and then take the steps needed to bring everyone to justice.

Not let Xander find out I'm in love with his daughter.

Remembering the one thing I have in my possession that I can use to show him what my words can't get out, I slip my hand into my back pocket and pull out my phone. Scrolling through the photo album until I come across what I'm after.

Pictures I took of her. A few she didn't know about and some others she did and even posed for. My favorite being the one she wanted to take of the two of us. Her tongue sticking out and licking up my face while I tried not to laugh as I snapped the shot.

Closing my eyes and taking a deep breath, attempting to settle the ache inside, I walk back to the table and hand over my phone.

"Xander," I start once he's taken it from my hands and sees what's waiting for him. "Meet your daughter."

His finger starts moving across the screen before I get the chance to stop him. Picture after picture flying by on the screen that I don't move quick enough to stop. Pausing on the final one, he taps the screen a couple of times before placing it down on the table and twisting himself around in the chair to face me.

"How long, Cole?"

Looking down to where the phone rests, confronted with the photo he's paused on, I physically feel my heart freeze before it drops.

A picture of us at our most intimate. Taken the night before at the B&B after we'd made love for a third time.

Shit.

I knew it was there so why didn't I move it? I hadn't scrolled that deep into the pictures when I pulled out my phone but that's no excuse. I should have known better.

Looks like the secrets out.

"How long what?" I ask, even though I already know where he's going to go. And when he speaks again, he doesn't disappoint.

"How long have you been in love with my daughter?"

Luke

"Lucas Devlin Grayson! Get your ass in here."

Seven years I've been living away from my parents and with three words he's put the fear of god into me.

When I got Cole's message about heading back to Orangeville to see my parents, I'd dropped Bethany off and hightailed it after him. Judging from the tone of my dad's voice now though, I wasn't fast enough.

Cole must have told him everything.

Making my way into the kitchen and seeing the stress lines across my old man's forehead and the dejected look on Cole's, I've got my answer.

He knows.

"Now that you're both here, maybe you can explain why this wasn't your first stop the second you found out Sunday was alive."

"Xander..." Cole speaks up first, throwing me a look that tells me to keep my mouth shut. "I didn't bring it to you right away for the same reason that I didn't tell her the truth the second I realized it. I needed to make sure I knew everything first."

"That's bullshit, Cole!" My dad yells and again, the same way we did as kids when he would raise his voice, we both flinch from the sound. "We should have been your first call!"

Not caring what Cole wants me to do and unable to let his shit fly, I step toward my dad, staring him down before letting him have it. I don't care how messed up he is over what he learned. He's not going to treat the person that found her like garbage.

"No, Dad. The way Cole did it was right. Not all of it, sure, but that first day. I should have been his first call."

"How do you figure?"

"Luke—" Cole attempts to interrupt and I just shove my hand in the air to silence him.

"Cole, no offense, but this is between me and him. You did what they hired you to do. They know the truth now. You don't owe them anything else."

He shakes his head and I slam my fist down on the table before turning and staring down my dad. "He called me first because he knew that I had given up. That I believed she was dead. He needed to right that wrong. You never gave up! Years and years of me having to listen to you spouting your hope. Your belief that she would eventually come home and all of my arguments to the contrary. Cole called me because I needed it more than you!"

"You're wrong, Lucas. Your mother didn't give up, but I did."

Well, I wasn't expecting that. He had been so adamant in the past that she was out there somewhere. Hearing that even for a second he thought she was dead, settles a small bit of the growing rage I have over his attitude.

It still doesn't explain why he's shooting accusations around, though. Shouldn't he be asking when he can see her?

"None of this shit matters. She's alive. We didn't call you because we needed time to piece everything together, but also get to know her again. I needed to be around her, Dad. Be pissed all you want at me, but don't you dare take it out on Cole."

"That's where you're wrong, Lucas. The both of you not coming to us and explaining what you found I can overlook. What I can't seem to wrap my mind around, much less accept, is what you did with the time you were given."

What the hell is that supposed to mean? What we did?

Spend time with her? Get to know her?

What the hell did we do wrong?

Able to sense my confusion, Cole doesn't waste any time explaining.

"He's talking about me. He knows about us, Luke. He knows it all."

For the second time since I got back here and was beckoned into the kitchen, I'm surprised. With the way everything went down earlier tonight and the way he sounded on the phone before he took off back here, I would have assumed Cole was going to keep what happened between him and Sunday a secret.

Not blurt the whole thing out to my dad.

"You told him? Why?"

Motioning with his hand to his phone on the table, I move over and grab it. Unlocking the screen and coming face to face with the exact reason Cole had opened up when I knew it was the last thing he wanted to do.

Tossing the phone back down, still not getting why my dad is so pissed, I shake my head and lay out a truth that's been hidden for too damn long inside my head.

"What the hell does this shit even matter?" I point to the phone. "Cole's been like this with Sunday since she was born. It was always the two of them when she wasn't with me. You should be fucking happy that it happened with him."

"Calm down, Luke." Cole calls from across the room and I just shake my head.

"No, Cole. They lived this shit just like I did. They could see the way you were with her back then. They knew the kind of guy you were and the way being here with her seemed to change you for the better. They should be happy that it's you in those pictures and not Walker."

This seems to get my dad's attention. Apparently Cole hadn't gotten around to explaining just how much he knows about Sunday's life since she was taken.

"Who's Walker and what does he have to do with Sunday?"

"Nothing. He was just Sunday's ex. You know, the one who repeatedly cheated on her, broke her heart and turned her into a quiet little hermit with everyone but her brother Jake and her best friend Bethany."

God. I am so fucking over this shit.

"Instead of being so damn offended by Cole and Sunday being in love, can we maybe focus on what we're going to do now? How we're going to deal with the sadistic bastard that took her in the first place?"

Ever since I found out that James Carter had been the one to take my sister and keep her hidden away all these years, all I see is red. Blood red. Vision after sick vision of the various ways I want to see him dead in front of me.

By my hand.

That is what we need to be focusing on. Not what went down between my sister and Cole.

"He's got a point, Xander. I know you're not happy about the way that we handled the situation, especially how I chose to do things, but it's too late to go back and change it now. We need to take what we know to the police and start setting things right."

Two against one is never going to be good odds, especially when the two people are me and Cole. It's been this way for years. No matter what the issue is, we're always going to fall on the same side because we understand each other. Our lives may have been totally different until we became friends at seven, but deep down where it counts we've always been the same. It's what makes us such good friends.

Judging from the way my dad's body falls into the chair and the fire in his eyes is replaced again with the calm we're used to, it looks like we might be able to get back on track after all.

We got through.

"You're right. Whatever my feelings are about what's already happened, it's what happens next that has to take precedence.

So," he says, looking from me to Cole and back again. "What do we do now?"

"Well, you two need to wake Allison and fill her in. While you do that, I'm going to head over and talk with local PD. Give them everything I've got and get them to coordinate with the OPP in Alliston. I want a few plain clothed officers on Reagan's place now that she knows the truth."

It's not lost on me that he can't call her Sunday. That even after the truth came out he can't bring himself to turn them into the same person.

Which considering what I'm sure is going to happen next, might be a good thing. If he can keep himself separated from his feelings for my sister we might be able to keep everyone safe.

"Are you going to head back and keep an eye on her after you're done in town?"

"No. She told me to stay away and I'm gonna keep my word. I won't hurt her any more than I already have. When you're done here maybe you can head back up."

"Cole, I know you wanna prove yourself, but I really think you should be the one up there. When this comes out and we both know it's going to, she's safer with you. You've been trained to deal with stuff like this."

"I may have tactical training, but I've never been trained to deal with something like this, Luke."

He doesn't need to explain. Whenever he's gotten himself into a situation in the past, it's always been easily handled because he wasn't personally involved. He could keep the job separate from everything else. When it comes to Sunday that's impossible. His feelings and the job, they're all wrapped up tight together.

They always will be too if the dejected look in his eye is any indication.

Cole is going to love Sunday forever. Even if he has to do it from afar.

"Lucas, leave your mother to me. Do what Cole said and head back to Alliston. I'd prefer there was someone that knows and cares about her looking out for her."

Nodding to my dad and turning to Cole, watching as he finally lowers his head toward the table and the screen that is still staring back at him, I feel my heart breaking the same way as his.

I can feel his pain. Agonizing waves of it.

Remorse, guilt, hate. Loathing and aching sadness. It's all there, one wave after another until the once solid and secure stance of my best friend completely shatters in two and he shows his weakness.

My dad hasn't caught it yet, but I have. The solitary tear that runs down his face before he drags his hand across his face. Washing it away just as another one falls.

Cole is crying.

Chapter Twenty-Seven

Luke

This has been painful to watch.

It's been three days since Reagan learned the truth about who she really is. Three days since she kicked Cole—and me—out of her life. Closing the door and completely locking us out.

It's also been three days of constant surveillance on Cole's orders. He's still refusing to go against his promise and be the one here watching over her. So it's been up to me and the Orangeville PD. Well them in conjunction with the OPP.

All of us watching and waiting for James Carter to slip up. Just one little mistake so that like Cole said, we can get him the right way.

My right way and his differ exponentially. It's taken everything in me not to beat down the door and do to him what he did to our family when he took Sunday from us. Make the piece of shit pay.

With the way the nightmares have come back, guilt over my not being there forefront in my mind again along with the twisted road my thoughts have taken, I'm starting to think Cole's not alone in the way he's been since the truth came out.

Dark. Jaded. Broken.

Distant.

Two weeks of having Sunday back in our lives only to have her ripped away again.

We knew this is how things would go down, but I think both of us hoped for a different outcome. That when she did learn the truth, after taking a few days to herself, she'd come around.

It's only been three days so that could still happen, but something tells me it won't. At least not without a little interference. This is a stain you can't just wipe away. Especially with what we know Sunday had already been through before we found her again.

Trust was already an issue for her and we'd taken what she gave us and threw it back in her face with our omissions.

What Bethany said was beautiful but untrue. Sunday didn't need me. Why would she? Every time she looked at me now, she would see Cole. What they shared, what he hid, and all of the pain associated with it.

Sometimes, like now, when thinking about all of this becomes too much to bear, especially with my proximity to her, I think it would have been better if we'd just gone on believing she was dead.

That when Hallie died, Sunday died too.

We were starting to heal from what happened. We had all moved on with our lives, even though they were never the same. We could have gone on like that forever and no one would be feeling what they are now.

Sunday destroyed. Cole vacant. My parents feeling the loss of their girl all over again and me, well, me a combination of all of the others.

But this isn't about me. I hate that I have to resort to watching over my sister from a distance, but I can handle it. I will do what I have to in order to be close to her.

It's Cole I'm worried about.

With each day that passes without word from her, he's been on this precipice. He's in danger of falling off the edge and when that happens there won't be anything that brings him back.

To an outsider the way he is sounds overdramatic, but that's because outsiders don't know the real Cole. I'm not talking about the guy he was when he first landed in Orangeville with Sam. The cold and jaded guy comes easy for him. He's venturing that way again now with everything he's been dealt. That's not the real Cole though. It's the one he became a few months after he moved in I'm talking about. The one who cared too much. The kid so desperate for love and acceptance that he sought out ways to get it because he didn't have any experience with it.

Until he met us.

Met Sunday.

The moment it all changed. When I saw the emptiness he presented with at seven begin to dissipate like the rain does when it runs off into the sewer drain during a thunderstorm, is one that even after all these years, I've never forgotten.

A memory that with all the time that's passed I've never shared with another soul.

Not even Cole.

Back then every day seemed to always start off the same, and this one was no different. I'd gotten up after some prodding from my mom, annoyed that even on a weekend she wouldn't let me sleep in and I'd made my way downstairs with just enough time to inhale my cereal before his familiar knock came across our front door.

And just like every other day, the second Cole stepped in the door, he gravitated toward Sunday. Sure, he'd grunted out a good morning to me and my mom, but his attention immediately went to her.

The same way it did when he landed in Alliston two weeks ago.

*****

"Cole, would you mind taking Sunday out of the high chair and bringing her into the family room?"

"Sure, Mrs. Grayson."

Turning away from us she focuses her attention on the empty bowl in front of me, sighing before walking over and plucking it off the table.

"Lucas Grayson! How many times have I told you to put your dishes in the sink when you're done?"

Scowling when Cole starts laughing, I bring my hand behind my back so my mom can't see and give him the middle finger.

"Sorry, Mom."

I'm not really sorry. I just finished. She could have waited a couple more minutes before nagging me about it. It's not like I'd gone upstairs and left it there for her to deal with. I'm still sitting here.

Turning her back and bending over the sink, I hear the sloshing of the water as she washes the dishes. Looking over to Cole as he pulls Sunday out of her chair, the grin on his face matching the toothless one she's wearing, I follow him out until they're both settled in the family room.

"Five minutes and we'll go." I tell him and he just nods, not once taking his eyes off of Sunday.

The way he is with her should make me happy. The focus and attention he gives her whenever he's over here being what I wanted him to do when I first told him about her, but that's not how I feel at all. There's this gross feeling in my chest when I see them together that makes me want to punch him.

It's the same feeling I had when my parents told me they were having a baby nine months before she came.

Jealousy.

She's my sister. Not his. It should be my job to make her smile like that. Not his. Best friend or not, I'm really starting to hate this. I hate that she seems to like him more.

Growling under my breath at being brushed off, I turn and head up the stairs to get dressed. After doing it in record time, determined not leave him alone with her for too long, I barrel down the stairs. Pausing on the third step from the bottom when I hear Cole talking. His voice is quiet, but the pitch fluctuating the way it is means that whatever he's saying, he's happy about.

Sunday has that effect. Doesn't matter how pissed off we get. Whenever we're around her it disappears. It's like she won't have it any other way.

"You like stories, don't you Ray? I think I've got one you're really going to like."

When she gurgle responds, the vibration of the sound mixing with the drool I know is pooling in her mouth and he laughs, my stomach twists.

She really does like him best.

Settling in on the stairs and leaning my head against the wall, breathing in and out through my nose in an attempt to tame the green eyed monster, I listen.

Hearing her giggle again, right before it's drowned out the sound of him laughing, I edge even closer to the wall and wait for what's about to come.

I'd much rather be heading back in there and cutting him off. Ripping on him for the way he sounds like a little girl when he talks to her, but I can't do it. My body doesn't wanna cooperate. No matter how pissed I am that he seems to have a way with my *sister that I don't, there's something about this moment that I can't interrupt. I need to know how it's gonna play out.*

"Once upon a time," He starts. "In a land far away from the one we're living in now, there was a boy. An angry, hurt and scared little boy that didn't know his place in the world. He was alive, he knew that much, but most days he questioned why and how. Darkness seemed to follow him no matter where he went, but it was always the worst at night. When he stepped through the doors of his dilapidated castle, hatred descended on him. The boy's mother angry yet again over something she believed he had done."

Jesus. He's telling her a horror story.

"Day after day the boy took what came from the anger, but he secretly ached for something more. Something better. A light to break through the darkness. A light that would see what he was living with and do everything in its power to save him. Cover him in its protective rays, the force of them so strong that not even the pure evil of his mother could break through."

"Years he wished for this one thing. Never giving up on the belief that it was going to happen. The ache of each day that passed with no change building as time went on, until one day, he couldn't wish anymore. Couldn't hope. The darkness; the overwhelming hatred that seemed to dominate his life had finally won. Beaten, broken, and with his body filled with nothing but anguish, he finally accepted that he would never find the light and gave into the darkness."

I don't know what to do here. Part of me is curious to see where he's going to take this, but the other part of me wants to end it. Sunday is just a baby. She doesn't know how screwed up life can be. She shouldn't have to know. Especially not how screwed up Cole's life was before Sam.

She's small. Innocent. She needs to be protected from this. Not subjected to it.

What the hell is Cole thinking?

"That's how it remained, at least it did until the man with the nice smile stepped into his life at the end of summer the year he turned seven. Bringing with him a small ray of hope. It wasn't much. The boy was still so untrusting that he couldn't believe that *it was real, but the more time that passed, the less he could deny it was there."*

"*For months he took in this new world he was a part of. Waiting for the other shoe to drop. For the cracks now present in his layers of darkness to completely shade over again. It never came.*"

"*Then, at the beginning of October, the cracks that the man created turned into gaping holes, the light streaming through so forcefully that the boy was in no way prepared for what was happening. His fear of never seeing the light turned into fear of what the light would actually mean now that it was there.*"

Holy shit!

I know what he's doing. Samuel was the crack that appeared in his darkness. That's easy to see considering when Cole came to live with him. It's the gaping holes of light that get me most.

The ones that were caused by the person in the room right now listening to the story.

Cole's admitting that Sunday changed him.

Whoa.

"*A fiery ball of energy on a path that he would soon see ended with him. A ray of light so pure that it did what nothing he had tried before had been able to. Reaching deep inside to the heart he thought was close to beating its final beat and bringing it back to life. Making him believe again.*"

"*Hope, love and happiness. All of the things that after that last fight with the darkness he had given up on filled him again. Replacing the cold with constant bursts of warmth until the boy; the one so broken and worn down by the way his life had been was healed.*"

"*And you want to know the best part of the story, Sunday? The ball of energy that wouldn't take no for an answer, it wasn't just a light. It was a real life person. The **best** person.*"

Here it is. I just know he's going to say it. Admit once and for all that she's his light. She's the force strong enough to break through the darkness.

Sunday is Cole's salvation.

"*The ray of light that saved the little boy was you, Sunday Grayson. Just you.*"

*****

He changed after he told her that story. For a couple of years after sharing that moment with her, he was the person I believe he was meant to be.

The one that had she not gone missing he would have continued to be. The guy that the darkness he spoke of in his story had come back around and stolen.

For as much as I blamed myself after it all happened, I knew deep down it was worse for him. Where I was driven under by the guilt of not being there when she needed me most, he was driven by the knowledge that he had brought the darkness into our lives.

Exposing Sunday to his life that day with his story made this happen. I knew it as easily as I knew my own pain from back then. His inability to let go and the years he spent haunted, even when he managed to bury it during his time with Miranda, it was because of this.

The reason for his tattoo, hounding Sam and John when we were kids, and then every move he took when he finally found her again. Why he couldn't tell her the truth right away. It's all based on around his guilt.

His belief that he was the reason this happened in the first place.

Eighteen years I've held onto that memory. It's been eighteen years too long.

It's time to confront it. Only Cole's not the one that needs to hear it. It would only throw him deeper into his despair.

No. If I want to change things, make them right, there's someone that needs to hear it more. The one resting less than fifty feet away from me now.

Sunday needs to hear this and I'm not leaving until she does. Cole's life depends on it.

Reagan

After spending days avoiding anything to do with the outside world, the doorbell happening to go off right when I'm making my way past it to the stairs, takes me by surprise.

With Bethany and Jake being the only two people I've said more than two words to since everything came out days ago, I know it's not them. Bethany would have called first and Jake has never rang the doorbell anywhere in his life, least of all a place he has the key for.

Part of me thinks that maybe it's my mom. That she's ringing the bell because when she took off almost a week ago, she'd left her keys behind. It's that thought that has me swinging it open and coming face to face with the second last person I want to see. The first being this guy's best friend.

Luke Grayson.

Yay.

"I know you want nothing to do with me, but there's some things you need to hear. Things that might change everything."

Fat chance of that happening. Nothing Luke can say will change anything here. I won't flip a switch and forget. I spent years letting people hurt me, blocking out what they were doing and moving through life oblivious. I'm not doing it anymore.

I'm better than that.

Stronger.

"Not interested. I heard everything Cole said. That's the end of it."

"No it's not, and deep down I think you know that. It's the reason you haven't left the house in days." He announces and when I raise my eyebrow he laughs. "Sweetheart, it can't really surprise you that I know that."

He's wrong. It does. Cole swore he would leave me alone. Back himself out of my life and give me the space I needed to come to terms with everything he told me. What I know after three days of phone calls, digging around in my dad's office when he wasn't home and the loss of my mom when she took a suitcase and walked out, is true.

Everything he told me was the truth even if it all started with a lie.

A fact that has the binds around my heart, the ones determined to push him away, weakening because it means it wasn't all lies.

Meaning the two weeks with him that seemed to shower everything in the most vibrant of color, wasn't all wrong. Fabricated. Based on lies.

No, wait. That last ones still true.

"Actually it does, but you know what? It doesn't even matter. I'm still not interested in anything you have to say."

"Even if it's about Cole?"

Keep it together, Reagan. Do not let him see you react.

"Especially that."

"He's a wreck, sweetheart. He hasn't slept in days, he won't leave his apartment and he's missing meetings."

Meetings?

"What does him not showing up for work meetings have to do with me?"

"Not work, Sunday. Counselling. You know, the ones people go to when they need help getting control of an addiction."

Remembering our time together, things he said, places we went together and his reactions while we were there, I know what Luke means now. Cole made no secret of his disdain for drinking which means the meetings he's getting at have to do with that. What happened with us throwing him right off the wagon of sobriety.

"That's not my fault."

"No, you're right. It's not. None of it is. But you are the one holding all the cards. The only one with the power to make things better."

"And if I said I didn't want to do that?"

"Well, I'd call bullshit because you'd be lying."

How can he just see through me so easily? Know the truth when half the time I'm not even sure what my feelings are anymore. It's been three days of nonstop rollercoaster emotions. One minute I'm sick to my stomach with tears streaming from my eyes with no apparent cause, and the next in a blind rage over having been duped by yet another guy. Then there's the disappointment knowing that the life I thought was mine was all a mirage and anger knowing that the man I grew up adoring was the cause of it.

If I can't get the way I'm feeling under control and understand it, how is it that Luke can?

Meeting his eyes for the first time since the truth came out, I see the similarities between us. The bone structure of his face, while masculine, is crafted from the same design as my own.

"You're wrong."

"I'm right, actually, but I'm not here to argue with you. There's some things I need you to hear. Things I think will change things, but even if they don't, you still deserve to know. You have almost eighteen years of memories as Reagan. It's time you had some of Sunday too. I want to be the one to give them to you."

"And if I don't want them?"

"Then you just listen to what I have to say and push it away, like you've been doing with everyone around you for the last three days. I'm only here to fill you in. Not tell you what to do with it once I'm done."

"Aren't you a gentleman?"

"Your sarcasm is cute, sweetheart, but I never claimed to be a gentleman. I never claimed to be anything."

"Except my brother."

"Yeah, but in my defense, that one's a fact. Even if I wanted to change it, I couldn't. It is what it is."

I can't believe this guy. Cole told me how close he was to his sister. Luke even told me himself in the way he would talk about her before. I heard the pain in his voice the day I overheard them talking. How can he stand here acting like he doesn't give a shit about anything to do with me when I've experienced different?

The worst part of this though, is the way I'm affected by his attitude. How it feels like someone just shoved a knife in my chest and twisted it hard. The need I have to scream at him. Tell him to stop lying to me and just admit how he really feels.

The same way I wanted to do with Cole.

Still want to do with Cole.

No matter how far I push him away, I still want him to fight. I want it to be him in the patrol cars I see passing by every few hours. Him in the car just down the street. I want him to be the one standing here fighting tooth and nail to keep us together. Proving that what we had wasn't this fleeting moment in time, but something that like the way he felt about Sunday, would withstand anything.

Which makes me just as stupid as people thought I was after Walker.

"Now who's the one lying?"

Unable to stop the satisfaction that comes from seeing him look away, I laugh. The sound vibrating off my chest but no part of me believing in it. It's not real. As smug as I am that I managed to catch him in a lie, there's nothing humorous about it.

"Seems we have something in common after all. We're both lying to protect ourselves from more hurt."

"You're wrong."

Huffing out a breath and dragging a hand down over his face, he steps forward and before I can react, his arms are around me and I'm being pulled into an embrace.

"You can fight me all you want, Sunday, but it won't change a damn thing. You were my god damned sunshine then and you're my god damned sunshine now."

When he pulls back, I'm frozen in place with my mouth agape. Completely in shock and to do anything to snap myself out of it.

Taking a breath and tapping his foot against the ground, his eyes averted away from mine and focused toward the ceiling, he waits for me to count a total of ten breaths before brushing past me and heading straight for the stairs.

"What the hell do you think you're doing?"

Pausing in his climb and turning back, he smiles.

"I've waited a long time for this."

"For what? I ask, taking a few tentative steps toward where he's waiting.

"For the chance to tell you a story. So come on, sunshine. It's story time."

Not knowing whether I should do as he asks and follow him, or make a break for the family room where I can call and have him forcibly removed, I don't move from my position at the bottom of the stairs.

I can't explain it, but there's something about the way he sounds that makes me want to follow him. I'm curious. I need to know why those few simple words seem to spark something in me that until now has been dormant. Why it's them that give me the first feeling in three days that wasn't completely drenched in pain.

"The voice in your head that's telling you to trust me, you should listen to it."

"Oh yeah? Why's that?"

"Because you always have in the past."

The past.

I trusted Luke in the past.

Fed up with fighting, I put a stop to the tug of war going on between my head and my heart and head for the stairs. A move that when I finally pause beside him and he smiles, serves as another reminder of who I really am.

"You always did like story time." He shares before turning and heading up the stairs and down the hall into my room. And once I follow him in and shut the door behind us, he does exactly what he said he would.

He tells me a story.

Chapter Twenty-Eight

Cole

Things have been too quiet.

With everything out in the open, I would have expected to hear of some blowback by now. Too many people had been filled in on the truth for there not to be some kind of reaction.

It didn't even have to come from Reagan. It could be Jake losing his cool and spilling his guts, or Bethany and Reagan talking and someone overhearing them. Katherine finally learning the truth and forcing James' hand. Anything really. Not this, though.

Not dead silence.

In a situation like this, with how many people are involved and what it would mean for all of them once the truth comes out, this level of quiet would have people letting their guards down.

The typical mindset being, if nothing's happened yet then it won't.

But that's not my mindset.

I know different. This is *too* quiet. Which makes me think there's something else in play here. A plan of some kind that in the end is going to cause a lot more shit than what me keeping the truth to myself for two weeks had done.

I'm on alert all the time. Every movement, shadow and sound around me only serving to put me more on edge.

Something is coming. I can feel it. I just wish I knew what it was.

Truth is, this is probably just all in my head. A delusion based on the fact that I haven't been eating or sleeping right for days. Five of them. Five long days that I've spent apart from the woman I love.

One night in some random bed and breakfast was all it took to change the way I've been living my life for years. A few hours and every method of coping I'd utilized in the past when the memories were too much, does nothing but fail now.

Making love to her had connected more than our bodies. It connected our souls. It's what makes sleeping without her beside me agonizing and closing my eyes even worse. She's everywhere.

I smell her powdery soft scent every morning as I make coffee, our first meeting so real in my head that it feels like I'm living it all over again. I'm even seeing me telling her the truth over hot chocolate and having her accept it.

Accept me.

It's only when the phone rings or Luke sends me a text updating me on things in Orangeville and Alliston that reality is slammed back into me and the vision that I thought was real, turns out to be a day dream. What my mind wants to hold onto so desperately because the alternative is threatening to kill me.

I hear her sweet soothing voice and feel her breath warm across my face every time I step into a room, but no room more than the bathroom. The place and situation different but the memories not caring. Working together with all of my senses and making me relive every part of what happened that night at the B&B. The way she reached into my soul easily and started to mend the broken tethers left behind from my time with my mom.

I'm told a person can live up to seven days consecutively with no sleep before their body simply gives out and they die. Of course, it's not as simple as that. You're dealing with the nausea first, then the spots that appear and seem to grow bigger with each day you don't sleep. Next stop is the delusions. Hearing and seeing things that aren't there. The finally the paranoia that comes from it all gripping you with its mighty hand, bringing you to your knees and making you its bitch.

This is day five in my seven day stretch and as much as I want—no—need to protect Reagan; make sure nothing happens to her or takes her away from me in an even more permanent, end of the line sense, I'm ready for the end to come.

Faced with a life without the people I love most, closing my eyes and never opening them again seems like the better alternative.

I struggled for years to find my purpose. The path I was meant to take. I found that in her. I didn't know it when I was a kid, but there's no denying it now. John wanting me to be the one to bring her home wasn't him playing a game. It was him knowing what I was too damn afraid to admit.

Bringing Sunday home was my purpose.

Xander and Allison knowing where she is now and taking legal action to get her back means I've succeeded.

I've brought her home.

But if I've accomplished what I was meant to do, why does it hurt so much? Why is the pain so debilitating I know it's only a matter of time before I give into the urge to press a shot glass to my lips in order to drown and forget?

The answer is so simple even a baby could figure it out.

It's because she wasn't supposed to just come home to her family.

My purpose can't be fulfilled because she was supposed to come home to me.

Reagan

"Are you sure about this, Rea?" Bethany asks as I pull another drawer open and yank out the clothes inside. Turning and tossing it into the suitcases I have lined across my bed before heading back to do it again.

"I'm sure. Luke was right. I'm not living. I'm just hiding away."

"And you think going to Orangeville is the answer?"

"Honestly? No. I'm not sure what the answer is. All I know for sure is that if I don't go there and meet them, I'm going to regret it."

"What about Cole?"

Cole's complicated.

After spending the day barricaded away with Luke and listening as he brought his memories to life for me in such vivid detail that I felt as though I really had been a part of them, he'd gotten to what he remembers of my time with Cole. What he called the real story time.

Cole wasn't lying about my effect on his life. The impact I made. Which only seems to validate the struggle I've been living with since I made him leave. How easy it was to focus on the ways he breathed new life into me and not the horrible secret he kept from me.

I really was his light in a world full of dark.

A world that when I pushed him away, I sent him straight back into.

"I love him, Bethy."

"I know you do. That's not what I meant. What I'm wondering is, what are your plans after you meet the Grayson's? Will you go to Toronto?"

"No."

"Why not?"

Dumping another load of clothes into the second suitcase, I turn back to the dresser, doing my best to hide the smile that's rising on my face. The first natural one I've had in days.

"Reagan Marie Carter, I see that smirk on your face. You sly bitch! You're up to something."

Focusing on packing, not giving her an inch as I move back and forth on auto-pilot, she finally reaches out and pulls me to a dead stop.

"I know it wasn't easy admitting you love him with the way I've been hating on him lately, but you did. Which makes me think that in addition to finally admitting what you've spent days trying to deny, you've also planned out what you're going to do with it. What I don't get is why you won't tell me."

Pushing one of the suitcases back on the bed and sitting, I take a deep breath and give her what she's after. What I've been keeping to myself in an effort to keep him safe. Ever since Cole showed me proof of the attacks made on his family, ones that my father put in motion, anything that has to do with a location, I've kept to myself. The last thing I want is for my dad to get wind of where Cole is and decide to finish what he started with Sam and John.

"I'm not going to the city because he's not there."

"Where is he?"

With a quick glance to the door and finding it empty, I fill her in. "He's working out of the Grayson's."

Thanks for that, Luke.

"So you going back there isn't about them. You're going back for him."

"Yes and no, Beth. It's complicated."

"I'm pretty sure it's easy, Rea. It's only you that makes it complicated. Now stop stalling and tell me what your plans are."

I want to tell her everything, I just don't know where to start.

It took two days after the visit with Luke for me to finally admit to myself what I think I'd known all along. That I belonged with Cole.

It wasn't an easy decision. I thought about it every waking moment and even struggled with it in my dreams. I yelled, screamed and cried over it, but no matter how much time I spent weighing pro's and con's, it always came back to the same thing.

He saved my life.

It started the first night in the bar and then again over hot chocolate, culminating in the night at the beach when we finally broke down the barriers between us and admitted what up until that point we'd been denying ourselves. The happiness that comes from letting go of the pain of the past and allowing yourself to fall in deep with the one that's meant for you.

Cole's journey to that place was harder than my own. It only became undeniable to him when we made love. It was then that the final wall crumbled and he gave himself over to his feeling for me. Letting me see every flawed layer he spent a lifetime burying.

He also admitted the truth to me that night, even if I didn't know it at the time and it had taken me days apart from him to finally see.

The reason he stopped looking for Sunday was because he found her. I'd given him the light back, just like he saved me from drowning.

Even though it didn't last nearly as long as it should have, I'd fought the demons and darkness of his past and won.

"Did you know he tried telling me the truth the first night? I misread the way he was acting and almost walked away from him for it."

"Yeah, but you said he was being evasive and it reminded you of Walker."

"It did, but that's just because Walker was all I knew. I think it would have happened the same way with anyone that came in after him. Cole just happened to be the unlucky one who got to deal with it."

"Or maybe he's the lucky one because he didn't let you go."

"He tried again after we left the club and went to the beach." I quickly change the subject, not wanting to let on how wrong she is. Because while he didn't let me go, the same couldn't be said for me.

"Then," I continue. "The night at the B&B he actually did tell me the truth. He said the reason he stopped looking for Sunday is because he found me. Me, Beth. He found Sunday."

"Does that mean what I think it means?"

"Yeah, Bethy. What I said to Cole that night is true. Reagan's dead."

"So is this where you tell me I've gotta call you Sunday?"

"I'd never tell you that, Bethy. It's just a name."

I can tell by the way it takes a few minutes for her eyes to settle and her mouth to pop back into place that she wasn't expecting this from me when she offered to come over and help me pack. With the dark cloud that seemed to be following me around for days, I can only assume she expected more of the same.

"So what happens now?"

"I put these suitcases in the car, drive over to the B&B to see my mom and then head out."

"What about your dad?"

"What about him?"

"Now I know you're playing. You know as well as I do that he's not going to let you leave town."

"He doesn't get a say anymore. He lost the right to that when he put his selfish need for a child above everything else."

"Rea…"

Standing from the bed and heading back over to where my books lay waiting to be packed, turning my back on the concern I hear dripping in Bethany's voice, I lay them in between two shirts to keep them protected and say the only thing I have left now that everything else has come out.

"He kidnapped me, Bethy. He plotted, planned and executed this twisted and sick plan all because he didn't have the patience to keep fighting."

"Reagan stop."

"No. It needs to be said. I've spent five days keeping it buried. The only way I can go there and see this through with the Grayson's and attempt to move on, is to admit what that selfish son of a bitch did to all of us."

It's only when I turn toward the sound of a throat clearing behind me and come face to face with the cold and angry eyes of my dad that I finally understand what she was doing.

Warning me.

"Oh, Sunday. Why couldn't you leave well enough alone?"

Chapter Twenty-Nine

James

I knew it was only a matter of time before she learned the truth.

A truth I never wanted her to know, but it appears Cole Alexander just had to make sure came to light.

I'd made a mistake pulling the plug on my plans for Alexander after witnessing the change in Reagan over the last few days. A change that only a breakup or a separation could cause.

After learning that Cole had indeed left town and returned to whatever hole he crawled out of back in the city.

It appears as though Alexander is not as gone as I first thought.

"Leave well enough alone? Really *James*?" she replies, her nose turning up and her tone snide as she uses my given name. "You're the one that should have left well enough alone."

She has no idea what she's talking about. She's just in shock over what she's learned. If she knew the struggle that got us to the point where I felt the need to do the things I did, she would see I'm not as bad as she thinks.

She would see that everything I've done has been for her.

"How long did you think you could hide it?"

I ready myself as she steps forward, preparing for her now clenched fists to make contact, but at the last second she swerves around me until she's shutting the door. As I ready myself to answer her question, she makes her way around again. This time anger driving her as she pokes her fingers hard into my chest and her voice raises as she questions me again.

"Why did you do it? Why me?"

Lying has become as familiar to me as breathing these days, so standing here now, I want to continue spinning the web I created the day I took her, but something stops me.

For years we've been the only family she has ever known. Despite not always seeing eye to eye, I know for a fact she's loyal to us. To me.

It may be wrong, but it's that love and utter devotion she has that I need to exploit and use to my advantage.

"I'm afraid the answer to that question is not easily explained."

"Try me."

Looking from her to the now shaking form of the Thompson girl and seeing the fear in her eyes. I decide to humor her. .

Slipping around my daughter, gripping her by the wrist when she weakly attempts to stop me from coming any closer and using my height and strength to get her to back down, I make my way over to the bed and lean into it once seated.

"The struggle to have a child and the toll it took on us finally reached a point seventeen years ago where the lines of right and wrong became blurred. And then, by some miracle, you were there. I just knew the first time I saw you that you were meant for me."

She doesn't even attempt to hide her feelings from me, the disgust present in her eyes at the same moment her nose scrunches up and a scowl crosses her face. I've made something that was so pure at the time sound perverse. Wrong.

"Our savings was depleted and we'd just gotten word that the latest IVF attempt had been a fail and there you were. A gift sent from Heaven. The answer to our prayers. The daughter we so desperately wanted but could never seem to have."

"How do you do that?" Reagan snaps, turning away from me and beginning to pace. "How do you make something so wrong sound acceptable?"

"It wasn't wrong."

Shock evident as her eyes widen, she begins shaking her head feverously. "You're so damn twisted you actually believe that. You really can't see that what you did is wrong!"

"It has nothing to do with being twisted, Reagan. I don't see the choice I made back then as wrong, nor what I put in motion being as evil as you seem to think it is because for me, the motivation behind it was pure."

Stepping toward me, her face a tortured mask of all the hatred, anger and disgust she feels, she leans over the bed until her face brushes against mine. Her next words showing her hand and telling me just how much she knows.

"And what was John Alexander, Dad?" She sneers. "Was what you and your sick buddies did to him motivated by that same purity too?"

"No. John got what he deserved."

If she is aware of John, it means Cole has given her everything. It won't be long now before she is throwing Hallie Francis and the attempt on Sam in my face as well.

"Why, Dad? Because he found out what you did and wanted to make it right?"

"Sending you back into that home will never be what's right, but because I'm growing rather tired of this lame attempt you're making to appear stronger than you are, I'll agree."

Sucking in a breath and lowering her eyes to her hand as she begins flexing it out and in, she turns toward her friend. Taking her by the hand, she begins backing toward the door and the escape she assumes I'm going to let her make.

"You're sick." She hisses across the room to me. Her belief in her words radiating from her now fiery irises. "And I'm done with this. I've heard enough."

I give her two steps and half of a third before I'm up and off the bed, capitalizing on her mistake as I grab her from behind. One arm gripping her tight across her midsection while the other resting across her throat, applying just enough pressure to drain her completely of the fight she may have been about to mount.

Hating that things had reached this point, having hoped when I overheard her today that I would have been able to make her see sense, I mull over my next move.

Bethany is a wild card. She never factored in to any of the visions I had over the years of what would happen when this day came. How I planned on dealing with things when it did. Neither had Cole, since the Intel I had been given when John was our primary target was incomplete.

But wild cards don't have to mean problems. They can—with the frozen look of fear in the young girls eyes now—be beneficial to the end game.

Tightening the hold around Reagan's neck, I release the hand around her midsection and slip it into my pants, quickly pulling out the pocket knife. A weaker tool, but one that serves its purpose as I bring it around to both girl's now widened eyes. Slamming the reality of the situation home.

I would never intentionally hurt Reagan. After Katherine, she's the reason I live and breathe, but I am not against using the fear of the unknown to my advantage.

This isn't about Reagan anymore. It's about rectifying my mistake.

If I have to take my own baby girl hostage in order to make it happen, so be it.

"Let her go." Bethany demands. The words falling flat when she sees just how close I've pressed the blade into my daughter's side.

"You know I can't do that, Ms. Thompson. She knows too much."

Feeling Reagan's struggle, hearing the gurgling sound that escapes as I lay the full weight of my arm across her larynx, I give into the pleasure of the sound and laugh.

"The more you fight, the worse this will be." I whisper before turning my attention back to the friend to gauge what her next move will be.

Is she worried about Reagan enough to comply with any demands I give her? Or is this the moment she attempts to be a badass and comes at me in what will be a failed attempt to be a hero?

Which road will Bethany Thompson take?

While I wait for her to decide, her eyes flickering back and forth between me and her friend, I decide to give them both a sneak peek at what's going to happen now.

"It should have ended with John. When Martin ran him off the road and made his death look like an accident, it was supposed to be the end. This was never supposed to be a problem again. I hadn't banked on Samuel's need to bring in strays off the street becoming a problem later down the line, but it appears as though he is. And you know what they say about problems right? They need to be eradicated. Taken out. Wiped off the face of the earth."

This gets the young girls attention. Where she had been debating whether to be the hero and save her friend, she backs down. Her body sagging back against the door and her voice when she speaks again, weak and submissive.

Just the way I like them to be.

"What do you want?"

"Well, young lady, it's quite simple. I want Cole Alexander."

"He left. Went back to the city. If you want him so bad, let Reagan go and get him yourself. You don't need me for that."

"That's where you're wrong. You're the only one besides this little angel right here," I pause, running the hand with the blade over Reagan's now exposed side. "That's capable of getting me what I want."

"How am I supposed to get him?"

This girl. I expected more from her. She might act ditzy the majority of the time, but underneath the façade, she's smart. She should already know how this is going to go and what her part in it is.

"Bethany, you are aware of the way he feels about our Reagan. What lengths he will go to in order to keep her safe. We're going to use that to our advantage."

"No!" she yells, lunging forward and falling to the floor as I shift myself and Reagan at the last second. *Stupid girl.* I knew she would attempt this. It was there buried under her fear waiting for the right moment to strike. A moment she has to realize now, will never come.

Once she's down, I use the opportunity presented and laughing as she attempts to right herself, sit up in what I know will be another attempt to break her friend free, I slam my boot down into her spine. Her cry of pain echoing through the room as she crumples and pulls into herself. Not content that one hit will be enough, I do it again. This time to the back of her leg, where her knee meets her calf, causing another mangled cry to escape into the quiet room. Reagan looking on horrified, with tears beginning to fall when I follow that hit up with another. The moan that escapes through the Thompson girl's mouth flooding me with a rush of pleasure like I've never known.

Maybe Martin and I are not so different after all.

"Now before you forget who it is you're dealing with and try to do something stupid, let me make something clear. If you try to be the hero, Reagan will get hurt."

As much as it's going to pain me to do this, they need to know I mean business. So when Bethany begins to move again, her breath releasing in low hisses as she fights through the pain in order to meet my eyes, I prove it to them.

Angling the blade at just the right position, falling back on my knowledge of the human body to be sure it's done correctly, I pull back and shove the blade deep into Reagan's side. Gripping her tightly as she screams through the pain and her legs begin to give out.

As Bethany reacts, the sound of her guttural no bouncing off the walls of the enclosed space, I'm sated. A last minute rash decision turned out to be the best one after all.

There's no doubt with what she's seen me do now that she'll go against me again. Let alone attempt to be the hero.

"Now that I've got your attention, we can really get this party started."

"W—what do you want f—from me?"

"I told you dear. I want Cole, and you, when you're managed to collect yourself off of the floor, are going to bring him to me."

"And if he doesn't come?"

"There is no if, Bethany. Cole is like the rest of his pathetic family. A protector. He *will* come. You my dear, just need to get him there."

"Get him where?"

"Back where it all began of course. Monora Park." When the name doesn't seem to spark any response, I realize that for everything Cole told Reagan, she must not have passed it on. Bethany has no idea what the place is, much less what it means. "He'll know the place."

"Don't hurt her." She pleads as she begins to crawl across the floor, dragging herself over until her hand is gripping onto my boot. "I'll do whatever you want, just don't hurt her."

"Well, Bethany, that all depends on you. Reagan's fate is in your hands now."

Reagan

Stirred awake by the feel of a hand across my face, I shift and lean into it. Craving the softness of the touch until the voice attached to it speaks and my stomach twists so tight, I feel like I'm going to be sick.

"Rise and shine, princess."

Twisting in the seat, opening my eyes and being met with darkness, I realize quickly what's going on. He's got me blindfolded. Shifting my hands and attempting to lift them to my face to make certain I'm right, I cry out as the friction caused from moving makes my wrists burn.

Make that tied up and blindfolded.

"Screw you!"

"You don't mean that, angel. You're just upset right now."

Oh my God. He really has lost his mind.

The anger I expect with me fighting so hard against him isn't there. He's acting like it's just another day. Like what happened in my room, the hazy pictures I can make out in my head of Bethany on the floor and the knife plunging into my side, didn't even happen.

That I didn't just learn how sick he really is.

"You're delusional!" I cry out, shoving my body forward until just like I hoped, I connect with his. Hard too judging by the string of expletives that follow before I feel the sting of his arm across my face in retaliation.

"I didn't want to do that, Reagan. You left me no choice. I won't have you speaking to me that way."

Feeling his breath hot on my face, I ready myself to do it again. Use every bit of strength I have to smash my head into his. Maybe gain the upper hand by hitting him so hard I knock him out. It's only when I shift forward and I'm thrown back forcefully that I realize where we are. Better yet, I realize we're not alone.

James is in the backseat with me while someone else is driving. Someone who obviously has no control over the ten tons of steel he's been tasked with driving as we hit another bump and I fall back again. This time, my head smashing off the door causing me to sink into the seat as a moan of pain escapes.

"Are you trying to get us killed?" he snaps at the driver and that's when I hear it. The voice of the other person in the car. It's one I recognize.

"I'm sorry, James, but you're the one that chose this location. The road is shit out here."

It's my dad's lawyer, Cameron.

Oh God. How deep does this go? How many more people are involved? Better yet, where is he taking me? It's there, I can feel it just on the edge of memory but I can't get close enough to remember.

"Well next time be more fucking careful! We're carrying precious cargo."

He means me. I'm gonna be sick.

"What makes you think he'll show?" Cam continues, reminding me again of what else happened in my room before the blood loss from the stab wound made me pass out. Cole. This is all about Cole. "Or if he does show, that he'll come alone?"

"Because Alexander is a lot of things, but stupid is not one of them. He wouldn't risk anything happening to his little whore."

"Fuck you!" I scream, my anger boiling over when they mention Cole. I could care less about him calling me a whore. I'm not gonna listen to them talk shit about Cole when they don't know the first thing about him.

Even if what he said is right.

Cole is definitely not stupid and even though things ended pretty badly with us, I know he wouldn't risk something worse happening. He'd hurt himself a million times over before he'd let that happen.

Feeding into the familiar ache that comes whenever I think of him and the way we left things, I close my eyes and attempt to fight back the tears. It's only when I hear James laugh that I know I've failed.

"Would you look at that? Even after lying to you for weeks you still care about him."

"Shut up. You know nothing about us."

"Such strong words from such a weak girl."

"You didn't seem to have a problem with this weak girl when you kidnapped her."

This seems to silence him. Good. I'm already torn up enough, the wound on my side making the pain even more real. If I have to sit here and listen to him drone on like a crazy person, I'm going to really lose it. I can't afford to do that. I need to stay strong.

It's the only way I'm going to make it out of this alive.

"I'm sorry, angel. Please forgive me. You're right. I didn't mean what I said before."

I'm sure he doesn't. Too bad I don't care.

"Why are you doing this?" I ask and the sound of his sadistic laughter filling the car sends chills down my spine.

"You're asking the wrong person. I didn't do anything but defend what's mine. You want the answer to why this is happening, ask Cole."

"And how am I supposed to that? I seethe, my nostrils burning from the anger flaring up in them over this nonsensical responses. "You've got me blindfolded, bound and god knows what else in a car with you and your pervert lawyer. He's nowhere in sight."

"All will be revealed soon. Daddy promises."

Swallowing the urge to puke that comes when he calls himself daddy, I close my eyes. Knowing what he's getting at, but willingly putting myself into the darkness in order to distance myself from it. What he's going to reveal, it's in what he told Bethany to do before he whisked me way.

He's going to kill Cole.

Unable to fight the nausea I'm experiencing or the fear ripping through me in waves, I roll over in the seat and just like the night I found out who I really am, I empty my stomach. The sound of my gagging successfully blocking out not only James's sick laughter, but Cam's too.

"That's right, Reagan. Get it all out. Purge the junkie from your system."

No way. He might have me at his mercy right now, but there's no way I'm gonna listen to him call Cole a junkie.

Screw him. He doesn't know shit.

"Go. To. Hell."

"Try as I might, I could never wrap my head around why instead of settling down with a nice young man, you chose instead to continuously give your heart over to the undeserving."

Considering how undeserving this bastard was of the love I gave him the last seventeen years, my answer to his quandary is clear.

"I guess I wanted to be with a man like my father."

Take that asshole.

Sensing him over me, I flinch at the exact moment his fist comes down hard across my face. Crying out before burying my face into the leather seat in an attempt to protect myself from whatever it is he's planning to do to me next.

"That useless piece of shit Samuel Alexander took pity on is NOTHING like me!" he roars and I start laughing.

The pain in my side, the insanity of this moment and the fear, finally coming together and making me lose it until I'm falling as deep into the crazy as him.

Maybe we're related after all.

"No, he's not." I choke out my agreement. "He's better."

With another hit to the back of my head, I bite down on my tongue, holding back the sob that wants to escape, not wanting anything to ruin what I'm going to say next. The only thing left to say that matters.

He's going to know how I feel.

"He's the exception."

Chapter Thirty

Cole

"You need to see this." Luke says, shoving his way through the door and throwing his phone down on the desk.

Ever since I came back to Orangeville, I've been holed up in Xander's office when I'm not at the police station going over every piece of evidence I have. All in an attempt to keep myself as busy as possible so that even if I wanted to, I wouldn't have time to think about her.

Miss her.

Need her.

I thought that after five days, I wouldn't see her everywhere I turned, but that's not the case because she's everywhere here the same way she was when I went home to Toronto.

And with what I'm staring at now, it looks like she's always going to be.

A text from Bethany. News to me considering that in all the time we've spent together over the last few days he's never mentioned keeping in contact with her. I was under the impression she hated him as much as she did me.

"When was it sent?"

"A little over an hour ago."

He knows.

Two words that to anyone else could be in reference to just about anyone, but ones I know mean one person specifically.

James Carter.

He's aware that Reagan knows the truth, which means it won't be long until he comes for me.

"Is she safe?"

Luke's shrug sends chills down my spine. If James knows and Luke hasn't heard anything back from Bethany, it means nothing good.

Reagan is in danger.

Something that when I walked away she never should have been. James has the issue with me and my family. Not her. I thought giving her what she wanted and leaving, but keeping eyes on her every possible way I could was the right move. I was wrong.

I never should have left.

I failed her again.

"You call local PD?"

"Yeah and apparently the two guys out front aren't responding. This isn't good, Cole."

No shit.

When Reagan was oblivious, she was safe, but if he's aware now of just how much she knows, that safe bubble has burst and there's no telling what he'll do.

James may have done all of this in the beginning out of some deep seeded need to make his wife happy. To have the family he wanted so badly. Twisted for sure, but relatively understandable. Desperate men and all that. But things are different now. It's become about a whole lot more than his need for a child.

It's about control. Reagan knowing the truth means he doesn't have it anymore and he's going to do whatever it takes to get it back. Including hurting the woman I love.

Screw giving her what she wants. I never should have done that. I should have stayed at that damn bed and breakfast and fought for her. Maybe if I had, this wouldn't be happening.

That ends now. It's time to do what I should have from the start.

Bring her home.

Picking up the phone and tossing it to him as I push the chair back and stand, I motion to the door. "Let's go."

Grabbing my jacket off the sofa and throwing it on, slipping my own cell into my back pocket and holstering my gun to my pants, I head out with Luke hot on my heels, taking the stairs two at a time until I'm standing at the door, about to swing it open and head through until the doorbell going off stops me in my tracks.

Figuring he'll know what this is about, I turn to Luke. "You expecting someone?"

"Nope. You?"

"No. I was going to head into town later to look into a couple of things, but that was the extent of my day until you came in."

"Well we can't just leave them standing out there." Luke offers up as the ring of the bell goes off again. Pushing down the sliver of fear that builds, imagining that we're too late and when I hope the door, it's going to be James Carter or worse—Martin Francis waiting on the other side, I push the handle down and pull the door back.

Registering Luke's curse first, I turn and catch the look of absolute shock on his face before turning to face what made him react. When I do, coming face to face with the last person I ever expected to see standing on this doorstep.

Bethany.

And judging by the way she's leaning into the doorframe, one hand on her back and her eyes shut, one that just lived through hell.

"Beth?" I ask dumbfounded as Luke slips his way past me and pulls her into his arms.

"He's got her, Cole. James knows everything and he took her."

"Took her where?"

"Monora Park. He said you'd know it. Said he was taking her back to where it started."

"Shit." Luke curses and I second the sentiment. It isn't exactly where everything started, but it is where this turned from kidnapping to murder.

If he's bringing Sunday there, it can only mean one thing. He's going to re-create what happened to Hallie.

"What is it?" Bethany asks, slipping herself out of Luke's embrace and attempting to stand on her own. The sharp hiss that escapes before she collapses into the doorway again enough incentive for Luke to pull her back into him.

More concerned in the moment with her wellbeing, I ignore her question and switch gears. "What did he do to you, Beth?"

"Got his point across. Loud *and* clear."

"Specifics please."

"I tried to get her away from him. I failed and he kicked the shit out of me."

"Where?" Luke demands, sizing her up. "Besides your back, where did that son of a bitch touch you?"

"Back of my legs and then the parting shot to my face before he took off with her." Bethany gives up easily before repeating her earlier question. "Why did you guys freak when I told you the place?"

"Do you remember the babysitter Luke told you about? It's where they found her body."

Sucking in sharply, she comes to the same conclusion I made a few minutes before.

"No!" she yells. "He said he wasn't going to hurt her! Said that if I got you there she would be fine. The knife was supposed to be the end of it!"

Say what? What knife?

"Wait." I say, needing a minute to wrap my mind around what I'm hearing as a string of expletives falls from Luke before he slams his fist down hard on the doorframe. "He stabbed her?"

"Yes." She chokes up. "I never should have tried to get her away. I made him do that."

"No, you didn't. James being a sick son of a bitch is what caused that." Luke attempts to soothe, tightening his hold around her and not letting go despite her weak attempt to keep distance.

"Bethany, go inside with Luke. Give him your phone and let him call Jake. If he's willing to hurt Reagan, there's no telling what he'll do to him."

"What are you going to do?" Luke asks once he's got her inside.

"Give him what he wants."

"He already has that. He's got Sunday."

"No he doesn't." Beth interrupts. "He doesn't have what he wants."

"What do you mean?"

"Luke, he's using Sunday as bait."

"Exactly." I agree, knowing exactly what it is that he wants. The one remaining link to the whole kidnapping. The one with all the answers. And to save her life, I'm going to give it to him.

He's not thinking clearly enough to realize that there are more people than just Sunday, Beth and I that know, which means what he thinks will end this and leave him scot free is me.

"Cole, you're not fucking going there alone."

"Yeah, Luke, I am. This begins and ends with me. He's not targeting you. I'm gonna keep it that way."

"Over my dead body."

"Stay here with Beth and do what I said. Make sure Jake and Katherine are safe. Get them here if you have to. I've got the rest."

I can tell he wants to argue, but he can't. He knows how imperative it is that everyone in Reagan's life is safe.

The rest is my fight.

One I don't plan to lose.

Luke

I can't just sit around here doing nothing.

No fucking way.

It had taken a whole ten minutes to get Jake on the phone after Cole booked it out of the house. Another thirty minutes after that for him to get his mom on the phone and make sure she was tucked away somewhere safe, securing a ride from one of the guys in his band in order to get here where we'd be able to keep an eye on him.

Another ten minutes to get my dad on the phone and explain what was going on so that he could get home.

Cole can say this is his fight all he wants, but I'm not letting him go through this alone. This isn't about him anymore. It's about my sister. James has my Sunday and I won't rest until I get her back. Even if it means getting hurt in the process. Or worse.

It's evident with the way he only wants Cole that he won't be expecting me. It gives me an advantage. With both Cole and James unaware that when my dad steps through the door I'm going to be making a play straight for them, I might be able to pull this off without anyone getting hurt.

I might even be able to make up for what happened when she was taken, and this time be there for her the way I should have been then.

Save her.

"Where the fuck is he?" I pace back and forth in the doorway, feeling Bethany's eyes on me but not giving into them. Not wanting to see the look of fear that's been in her eyes since she landed on the doorstep.

Fear that despite this supposedly being about fun between us, guts me to see.

"You said he was working when you called, right? Maybe it was harder than he thought it would be to get away."

"Maybe. I just wish he'd hurry the fuck up already."

Silence falls again as I continue to look from my watch to the door and back. Nearly climbing out of my skin with need to do something more than stand around looking like a chump.

"Luke, maybe you should listen to Cole."

"Not happening, Beth. That's my sister out there."

"I know that, but you didn't see James. The look in his eyes or the sick twisted smile he had when he stabbed her."

"I don't need to see it. He killed someone that was paid to help him. He tried to take out Cole's dad and actually succeeded with John. I know how sick he is. Which is exactly why I need to be there. Why Cole shouldn't be going in alone."

"Why? So you can end up dead?"

"Who said anything about dying?"

Leveling me with the coldest look I've ever seen, her eyes as dead as her body is numb, she sighs. "What do you think is going to happen when you go after her? It's either you or him."

"And you're betting on Carter?"

"No, you dumb oaf! I'm not betting on anyone. I just don't want to see anyone else get hurt."

"No one will."

Lowering her head and breaking eye contact, she stares intently at the carpet, silence beginning to descend on us again until a quiet sniffle breaks it.

"Beth…"

"No, Luke. Nothing you say right now is going to make this feeling go away."

Crossing the room and sitting beside her on the sofa, I pull her into my arms and physically attempt to soothe her worries. Starting with her head, I comb through her hair with my fingers, using my other hand when after a few strokes she responds by sinking deeper into me, to rub lightly over her back.

Surrounded by the sound of her even breathing as she finally gives in to me and relaxes, I think about everything she said that brought us to this point. .

She's right. What I'm going to do could go all kind of sideways and end up getting someone other than Carter hurt. I know that. I'm just choosing to believe that it won't go down that way. I refuse to let this play out any other way than the way it ends in my head.

Me pulling the trigger and killing the man that took my sister away.

Freeing us from the nightmare once and for all.

"Please just trust me, Beth."

"I do trust you. If there's anyone that bring her back, it's you two superheroes. I just…I just can't believe this is really happening."

"That makes two of us, sweetheart."

Hearing the turn of the key in the front door, signaling my dad's arrival, I place a soft kiss to the top of her head and pull away. Crossing the room and meeting him at the door before he's even made it all the way in.

"Okay, I'm here." He says once he's stripped off his jacket and thrown it on the hook. "You wanna tell me what's going on?"

"It's happening again, Dad. James…he took Sunday. He's using her as bait to get to Cole and the idiot took it. He's gone off after them."

The thought of Cole out there alone with no backup rips me to shreds. Since the second that guy ended up here, it's been the two of us against the world. I'm not even sure I'd be standing here if it wasn't for him. He might have been stuck on his own destructive path for years, but he never gave up until he saved me from mine. I hate not being able to do the same for him.

"Of course he did." My dad calmly says, like everything I've just told him is something he hears every day.

Did he miss the part where I said his daughter was taken again?

"What's that supposed to mean?"

"Lucas, you know how we feel about Cole. How we believe he's as much a Grayson as you are. But him running off half-cocked and giving James what he wants under the guise of getting Sunday back, it's been a long time coming."

"I don't follow." I admit, genuinely confused. What does my dad know about this that I don't? What am I not seeing?

"He's been looking for a way out for a long time, son. This is his shot. He's going there to finish this. He's so quick to react because while I believe he fully plans on getting Sunday out safe, he's not expecting the same for himself."

"No way. Cole wouldn't do that. He's been through a lot over the last few years, but he wouldn't go on a suicide mission."

I refuse to believe the reason he wanted me to stay here is so I wouldn't be a witness to what he hopes is the end. Cole is a lot of things, but using this as a way to take his own life? No way.

"Maybe I'm wrong then. You know him better than anyone."

"Yeah, Dad, I do."

Needing to make him understand why I think he's wrong and only having one example I can use to do it, I don't hesitate dropping it on him.

He's going to have to accept it sooner or later.

"You didn't see the together, Dad. She brought him back to life. Remember the way he was when we were kids? He was like that again. The one that cares about shit and fights for what he loves. She did that."

"Lucas..."

"I've heard you out, now you need to hear me. I know the two of them being together is hard for you to accept because of the age difference and how close we all were as kids. I get it, I do. But it's the way things are. Cole is in love with her. Sacrificing himself like this and giving into James, it's because of that. He kept the truth from her, selfishly fell for her instead of staying true to the job and this is his way of fixing that. Making it right."

"How is throwing himself into a lion's den making things right?"

There's no way with as rigid as his stance is on this, I can make him see the truth. Standing here trying is just wasting time we no longer have.

It's only when I hear the padded footsteps across the floor, the wood creaking under the weight as Bethany makes her way to my side and slips her hand through mine, that she puts an end to it.

"Because it's going to be what saves her. The same way he believes she saved him seventeen years ago."

Chapter Thirty-One

Cole

Monora Park Ravine.
I should have known this is where James would want it all to end.
He's wrong about it being where everything started, but with everything that's happened here since Hallie's body was found, and as twisted as he is in the head, there really is no better place.
What had once been a sea of green is now multiple housing developments, some of which are in the process of being added onto or rebuilt. It's the perfect place for whatever he had planned. It would give him the privacy he needed to keep her hidden, or the attention he'd be after if things didn't go according to his plan.
Attention I definitely don't want.
I'm going to make sure this is where it ends.
I *won't* lose Sunday again.
Pulling off the road and into the development, I slow to a crawl before putting the car in park about halfway in. With a quick glance out the windshield and seeing no sign of James or anyone else around, I grab my phone and read over the text from Reagan's phone that came through while I was driving. Committing the location to memory—the third house from the end—I slip it back into my back pocket, leaning over into the backseat and grabbing my gun and holster before getting out and slamming the door behind me. Hooking it to my waist before bending over to check the second piece buried on the inside of my boot.
I'm not used to carrying this much, but I'm not taking any chances. Not knowing what I'm walking into means having to be prepared for everything.

Coming up on the house and doing a sweep down the space that separates it from the one next door, I cross the lawn, making sure when I get to the front windows I duck down low enough to pass by undetected. Crawling across the grass until I've made it all the way around to the other side, I do the same until I'm content in the knowledge that at least from the outside, no one is waiting to ambush me.

Heading up porch steps, I pause at the door. Taking a deep breath and readying myself for whatever's waiting on the other side, I try the handle. Slipping myself in quickly and quietly when I find it unlocked.

Stepping deeper into the house and being met by a winding staircase, I head toward it. A loud bang from the end of the hall taking me by surprise and pulling my attention away at the exact moment my foot lands on the first step.

As the first bang is followed up by another, the second sounding more like a crash than a bang, I pull myself back completely and turn toward the hall.

Whatever might be upstairs is gonna have to wait. If there's even the slightest chance the crash I heard has something to do with Sunday, nothing else matters.

Making quick work of the distance, glancing behind me repeatedly as I continue moving, I pause once I get to the sliding doors.

Leaning against the wood and honing in on the voices I can make out, it only takes a minute or two to figure out what's taking place on the other side.

"Come on, Sugar. It's time to pay up."

Gripping onto the handle and closing my eyes I focus on my breathing as I try to simmer the rage that's building, I pull the door back slowly until I can make out two bodies on the far side of the room.

One bound to a chair with her eyes shut tight with tape over her mouth and rope binding her legs. The second a taller man, bald and slightly overweight. Familiar.

"Yeah, that's it. Keep fighting me sweetheart. Your resistance turns me on."

As his hands begin to move over her chest, lingering on the buttons of her sweater and laughing as she tries unsuccessfully to fight him off, I hear the sound of the first button popping and something inside of me snaps.

"Back off!" I growl, the sound of which gives me what I'm after as the pervert pauses and turns to face me.

I was right. I do know him.

It's been a long time since we've crossed paths, the last time being when I was a kid, but I'd recognize him and the scar running down the right hand side of his face anywhere.

Martin Francis.

With his hands all over my girl.

The sick look in his eye as he directs his perverse smile my way solidifying what I already knew he was attempting to do.

Something that now that I'm here, is going to happen over my dead body. He's not going to lay another hand on Sunday.

Pulling my gun and aiming it straight at his heart, flexing my finger on the trigger as he laughs, his cold eyes and sadistic smile never once breaks from mine as he pops another two buttons on the sweater.

"You're the same scared little brat you were seventeen years ago, Alexander. You don't have it in you to pull the trigger." He taunts me, his belief giving him the confidence needed to tear his eyes away from mine and focus his attention back on her.

Crossing the room and taking in the sight of her bare skin as the sweater slides down over her shoulder, my tightly wound bubble of control snaps.

Something about any part of her body being visible to this pervert more than I can stand. Driven by the adrenaline coursing through my veins, I move in quickly, threatening him as I rush him.

"Maybe you didn't hear me the first time, asshole. I said back the fuck off."

Shoving the full weight of my body into his, he loses his footing and we both land on the floor hard, the full weight of my body landing on his successfully knocking the breath out of him. Keeping the pressure on, I chance a look at Sunday. The fear I'm met with in her widened eyes gutting me from the inside out.

Wanting to help her but needing to deal with Martin more, I turn my attention back on him, pressing the barrel of the gun to his head. "Don't even think about moving. I won't hesitate to blow your brains out all over the floor."

Forcing myself back to my feet, gun trained on him, I get my point across as I slam him in the chest and back with repeated blows from my boot. The same moves Carter used on Bethany. His groans filling the room as he curls into himself and I do it again. One after the other until there's no sound coming from him at all.

With Martin down, I turn my attention back to Sunday and see her shaking her head manically back and forth, rocking the chair under the force of her trying to free herself.

"Baby, I'm gonna remove the tape so you can tell me what you're fighting so hard to say." I explain as I peel back a corner of the tape, gripping it securely and counting down from 3 in an effort to ready her for the burn that's going to come.

Yanking across when I finally get to one, she gasps as I free her and immediately begins sucking in all the air she can manage. After a minute or two letting her catch her breath and keeping my eyes locked on the still unmoving Martin, she finally speaks.

"It's a setup, Cole."

"Say what?"

"He wanted you to find me with Martin. He planned it all."

"What do you mean he planned it?"

"When he knocked Bethany out and took me, I heard him talking to someone. His lawyer, Cameron. Cole, he told Cam that Martin was a loose end and instead of doing it himself, he was going to get the job done another way."

"Get me to take him out." I surmise.

"Yes. He's sick, Cole. The things he said. What he wants to do to you…"

Leaning in and pressing my forehead to hers, immediately warmed by the small act of closeness, I let my heart enjoy her. Being here, knowing she's okay and that it's almost over settles the sense of drowning I've been experiencing since learning she was in danger.

As long as she's okay I can get through whatever that son of a bitch has planned next.

"I don't want you to think about that. Nothing's going to happen to me."

"Cole..."

Turning as she says my name, sensing the same movement that she must have seen, I shift away, shielding her eyes with my free hand as I aim the gun at Martins head. Without hesitation after seeing the perverse glint in his eye as he attempted to get to his feet, I pull the trigger.

As the bullet makes contact with his skull, Sunday screams. Ignoring the guilt that's building...the need inside of me to comfort her—shield her from what I've just done, I focus instead on the ties binding her.

Pulling my hand away, I slip around the chair, gripping the rope in my hands as I struggle to untie them. The knots obviously made by an amateur, but the sheer amount of them frustrating the hell out of me.

"Only a couple more, sweetheart. Then I'll do the ones on your legs so we can get out of here."

"You shot him."

"I didn't have a choice."

"I know, I just..." she pauses, wiping either sweat or tears away from her face. "Thank you."

"Thank me when we get out of here. Until then, I'm just doing what needs to be done. What anyone would do."

Crouching down until I'm level with her legs, I repeat what I'd done to free her hands. Allowing myself to breathe again when she's finally able to move them freely.

"Can you walk?"

"I think so, yeah."

When she stands from the chair, her legs wobbling and causing her to stumble, I reach out and pull her to me, noticing the blood stains across the left side of her abdomen as her body rests into mine.

"He really did it."

"Did what?" She asks as I slowly move us forward. Her pale coloring all the proof I need that she's not as okay as I first thought.

If what she said when I finally freed her mouth of the tape is true and this is all a trap, I've got to move quickly. This is going to turn all sorts of bad if I don't.

For her and for me.

"Stabbed you. Bethany told us he did it."

"He did a lot more than that." She spits out angrily, a fire rising in her eyes that based on how worn out she is, seems off. With as weak as she is right now, her eyes are saying something different.

She's livid and now that I've seen it, I'm going to use it to my advantage.

"What do you mean he did more than that?"

"He pistol whipped me in the car on the way here. Used his hands to beat on me before rubbing my face in my own vomit because I wouldn't give him what he wanted. Then he told Martin to break me."

"He was going to let Martin rape you."

"Yes." She shakily responds, tightening her grip on me as we finally reach the door.

"The only touch you're ever going to feel from now on is one filled with love, Sunday. I swear to you. No one will ever get close enough to hurt you that way again."

As her head pulls back and her eyes find mine, she smiles and returning it with one of my own, the first genuine one I've been able to have since I told her truth, gives me the fuel I need to finish this.

Get her out of this house and back where she belongs.

Hearing the crack before my head explodes in a sea of spots, I make out the sound of Sunday screaming before my entire body goes weightless and I crumble to the floor. The last thing I see before everything goes black—James pulling her to him as the shiny glint from a blade comes to rest across her neck.

I failed.

*****

"Cole Alexander." The confident voice smoothly calls as the clicking of shoes brings me out of the black. Gripping the back of my head and sucking in a breath as another wave of pain slices its way through me, I attempt to shake the cobwebs away and focus. "I'd say it's nice to finally meet you, but we both know that would be a lie."

Forcing my eyes open, I take him in.

Standing about ten feet back with a smug as shit grin on his face, he holds a knife to Sunday's throat. His belief that he's won with me incapacitated clearly evident even from this distance.

Forcing my body to lean forward in an effort to reach my gun, two things become clear. The first not concerning me nearly as much as it should once I realize my holster is empty. That not only does he have Sunday right where he wants her, but now he's also got my primary weapon at his disposal.

My hands being bound behind my back I can work with. Having to worry about not breaking free fast enough and James using the gun on her, that's where the panic sets in.

She warned me this was a setup, but blinded by my need to reassure her that everything was going to be okay, I'd lowered my guard. Lost my focus.

Put her at risk. *Again*. Proving what I've known from the start.

She was safer away from me. I'm only going to get her hurt…or with the way it looks from the position of his hand pressing the blade into her neck, something even worse.

Keeping my gaze trained on him and limiting my movements as much as I possible, I focus all of strength into freeing my hands from the rope. The friction from just the light tugging I'm doing causing enough pain that I'm having to bite down on the tongue to keep from hollering out and giving myself away.

"You got what you wanted. I'm here." I seethe. "So why don't you let her go?"

Ensuring her compliance, he grips the back of her neck tighter as he runs the blade down the side of her face. Laughing when not only do I struggle to rise in order to save her and end up falling back as my legs collapse around me, but Sunday begins to shake.

"It appears as though my earlier assumptions about you were wrong, Mr. Alexander."

I'm sure they were. I can't imagine anything he thought about me being right. Sick fuck that he is.

"What assumption would that be?" I humor him, hoping that by engaging him I can somehow figure out a way to gain the upper hand despite the lengths he went to after clocking me over the head to keep me grounded and immobile. Using the silence as I wait for his answer to again attack the binds that hold me and feeling them start to loosen just as he begins to speak.

The loosening of the binds what I hope is only the first of many mistakes he's made.

"I assumed when my daughter mentioned you being here that you were here with the sole purpose of bringing her back to the Grayson's. That anything you may have felt for her was as phony and fabricated as the business you're in. It appears as though rapist-junkie spawn is capable of feeling after all."

I'll show him rapist-junkie spawn. He's way too cocky for his own good, especially when I've still got the element of surprise on my side. Or I would if now that my hands are freed, I could get my damn leg to move in order to reach it.

My secondary firearm. The one James has no knowledge of. My only defense if I ever got into a situation I couldn't get myself out of.

Like the one I find myself in now.

My body really needs to cooperate.

"I never figured you for a name caller, James. With everything you've already done, that seems to be a bit beneath you."

Appealing to the cockiness, complimenting him on what he's managed to pull off thus far is the right move. The way his body responds by standing a little straighter, I can tell I'm getting through to him. At least it is until after a few minutes, he calls me on my bullshit. Seeing right through me and going straight for the one thing he knows will make me react.

Sunday.

"You're right, but it appears as though Reagan and I have more in common than I first thought. Because much the way you believe name calling is beneath me, you are also beneath her. Isn't that right, angel?"

Opening her eyes but not giving him the satisfaction of seeing her react, she steels her expression, her gaze becoming dull and lifeless and I can't help smiling, impressed.

"Screw you."

Pressing the blade deeper into her skin, I notice the addition of red that appears as he pierces her skin. Crying out, she fights against him. Struggling to break free before he can do it again. My inability to get to her in order to help as I watch it unfold making me feel like I'm the one that's been cut wide open.

"I've got to hand it to you, Cole. You managed to do quite the number on our little Reagan. She told me the funniest thing on the way here."

"Before or after you beat the shit out of her?" I seethe, keeping my eyes trained on his as I push through the pain of whatever he'd done to my leg and shift it across the floor toward me.

The need to see him pay for what he's done the only thing keeping me going as my head begins to swim.

"Before actually. It appears as though she believes that all of her boyfriends in the past never worked out because they were too much like me. The reason that I didn't like them, went out of my way to ensure they didn't last long, was because I couldn't accept her dating someone like myself. Isn't that funny?"

There's nothing funny about any of this, but I'm not going to give him the satisfaction of letting him know that. There is something buried in what he's said though that I do want to know about.

"You sabotaged her relationships?"

"Of course. You can't tell me you're surprised by that. When she didn't heed my warnings, I had to show her."

Sunday, obviously hearing this for the first time the same as I am, gasps, causing James to laugh even louder.

"It can't be that much of a surprise, angel." He leans into her and repeats. "Did you really think I was going to let you continue to see them when I knew in the end all they would do is break your heart?"

"You tempted Walker with girls, didn't you?" I interrupt before she has the chance to respond. "You knew he'd done it in the past and used it to your advantage."

"I'm impressed, Cole. You catch on quickly. It's a shame that things have to go down this way. If the situation were different, I might have been able to use someone like you."

Over my dead fucking body that's happening. He's no better than the string of assholes my mom used to bring home when I was a kid.

No better than the junkie he keeps reminding me I am.

"Never. Gonna. Happen."

"I'm aware. Your time with Samuel ruined you in that regard. You have a sense of morality and rightness that is not made for the world I live in. You feel too much for your own good. I'm sure you can see by now that those feelings can put you in danger, or worse, get you killed."

Feelings will get me killed.

There's no way I'm missing the point he's trying to make with that. Especially with his earlier statement. I also can't ignore the fact that he's right. Caring about Sunday and wanting to do right by her instead of preparing her weeks ago and possibly preventing what's happening now had been my downfall.

Even if he doesn't use my own gun on me, the feelings will kill me.

"What I can't for the life of me understand is what possessed you to follow in their footsteps. You saw how the job drained the life out of them. Why would you willingly sign up for that?"

As much as I hate to admit that anything that comes out of his mouth is a good question, I can't deny this is. For years I thought the reason I absorbed myself so deeply into what Sam and John were doing was because of my injury. When it stripped away my chance at playing in the NHL, I figured this would be my fall back option. Then, as time went on, I started to think it was because of Sunday. My need to find her, even if it was finding her body and bringing her home to rest, drove me into a business that when I was still riding the hockey wave, I'd balked against. The truth is, it wasn't either of those things. It's because of what he just said.

"Sam and John were worn down because the job effected every aspect of their life. Their relationships with their spouses, their family, everything. I didn't have anyone I cared about, so doing this job made sense. I could do it well because I wouldn't let anything stand in the way of getting the job done."

"Yet here we are." James laughs. "Looks like you're not the robot you appear to be after all."

No, I'm not. Reagan—Sunday, they both showed me that. They made it okay to love again and despite the mess it put us in, I don't regret it happening. I might not do it all over again the same way, but I would definitely do it over again if it meant feeling the way I did for those weeks with her.

The way I still feel now as I watch her starting to succumb to her injuries.

"Guess not. Looks like I've got you to thank for that."

This gets him.

He doesn't understand what I would be thanking him for and with the look of surprise all over Sunday's face, she doesn't either. When he shifts his eyes away from mine, I use it to my advantage, slipping my hand deep inside my boot and gripping the pistol tight before pulling it and shifting my body to conceal it from view.

"If it weren't for you slipping up." I continue pushing him. "Not keeping better tabs on your daughter and sealing the hospital record when she got hurt a couple of months ago, I never would have found her. So thank you, James. You gave Sunday back to me."

"Reagan! Her name is Reagan!"

Rage getting the better of him and losing control, the blade begins to shake in his hand. But before I can push myself to my feet, aim the weapon and fire, Sunday acts first. Bringing her elbow hard and deep into his side, knocking the knife to the ground and diving out of his grasp as he attempts to adjust himself to regain control.

Time slows to a crawl as his hand slams down over her wrist and her survival instincts kick in. At the same moment as he attempts to drag her back, she spins, leveling him with a hit to the chest before bringing her leg up and kneeing him in the groin. A hit so forceful it knocks him completely off balance and his body falls to the ground with a resounding thud.

Turning toward me and pausing as our eyes meet, she exhales heavily before erasing the distance between us. In the time it takes me to blink on her knees in front of me, breath ragged and eyes filled with dread. Tears beginning to pool in the corners of her eyes when she runs her gaze over me and sees the damage that's been done.

Tears I know are also brought on by the adrenaline. The fight or flight response that kicked in when she took the biggest risk of her life in order to break free.

"If I had known that calling me Sunday would set him off, I would have done it in the car." She rambles as she surveys the damage to my head with her hands. Moving down and doing the same thing with my leg when she's gotten what she's after. The tears that had been filling her eyes finally falling as the trauma of the last few hours slams straight into her.

"He hurt you." She swipes at her eyes. "He hurt you and I couldn't stop him."

Understanding what's happening, but needing her to focus so we can use what little time we have while James is incapacitated on the floor to get out of here, I take her hand in mine and squeeze tight until her eyes lift and meet mine again.

"We need to get out of here. He's not going to stay down for long and he's got my gun." *Well, one of them.* "But I need your help with something first."

"What?"

"My knee. He dislocated it. I need to use your body weight to pop it back into place so I can move."

Laughing under her breath, she proceeds to do as I ask and forces all of her body weight down onto my leg, filling me in as she does to what I said that was funny.

"If I didn't know you were serious, I would have slapped you for that weight comment. Didn't anyone ever tell you not to comment on a girl's size?"

"Beaten, kidnapped, almost raped and held at knifepoint and you're worried about your weight? Really?"

Slipping her hand into mine and squeezing, I cry out as the pain shoots straight through me as she resets my knee, taking the support she's given in me in holding my hand and giving it back to her by squeezing back hard, seething through my teeth as I wait for the pain to subside.

"I'm a weird chick, what can I say?" she jokes softly as she pulls me up to our feet.

Slipping her arm through and around my back, carrying as much of my weight as she can handle, she starts guiding us toward the entrance to the room. Managing to make it a couple of steps out before I feel the strain and collapse back against the wall, releasing her, I hear the sound of movement in the room.

Like a switch flipping inside my head, my body prepares for what's to come, going rigid as I pull myself off the wall and wait.

"Cole, what are you doing? We have to go!"

"Not yet."

Pulling on my arm in an attempt to wake me up and get me out of there, I shrug her off. I know I should listen. Cross what's left of the distance between where we are and what will be our freedom, but I can't.

I can't walk away until it's over.

"Cole, please!" She pleads, her voice filled with the same worry and fear I'd seen in her eyes when I freed her from the restraints what now feels like forever ago.

I don't want to scare her or be the reason she sounds like this, but I also know I can't walk away.

I can't play by the rules anymore.

Not wanting to waste a second, hyper aware of every sound and movement, I pull her into my body, lifting her head until her frightened eyes are locked on my determined ones.

As much as I need to stay here and finish this, I also need to keep her safe, which means she's got to go. Now. I've got to make her leave me behind and run.

"You need to go."

"I'm not going anywhere without you."

"I'll be right behind you."

"I don't believe you."

Frustrated despite knowing she wasn't going to be that easy to convince, I try again, this time using her feelings for me to my advantage and hoping she takes the bait.

Cupping her cheek with my hand, I press my lips to hers forcefully, hopefully making my intentions very clear. This is not going to be the last time I kiss her. Taste and touch her. This is a promise to do it again.

Once this is over and James Carter is out of our lives for good, I'm going to do this again and this time never stop.

"Please, Sunday. Go. I need to end this, but I can't do that until I know for certain that you're safe."

"I don't want to leave you. He's already hurt you once."

"The first shot was free. I let my guard down. That won't happen again. He won't get a second chance to finish the job."

"You don't know that." She whispers softly, her mouth so close to mine I can feel the warmth of her breath as it falls across my face. A warmth that even now fills me with need.

"I do know that. If you're safe, I'll be able to put my focus into ending this the way I should have weeks ago. I can finish this, Sunday. I *need* to finish this. I can't let him hurt you again."

"You don't need to save me anymore, Cole. You already did it. You've been doing it for weeks."

Shaken by her words and the honesty I can sense in them, but not agreeing, only able to see the danger I brought into her life, I shake my head in denial.

The only person saved here was me.

A debt I can never repay, but one that in ending this—ending James—once and for all, will at least start the ball rolling until I can.

This is the one thing I can give her.

Her life back.

Pressing my lips to her forehead one last time, I pull back. Pushing her away as forcefully as I can without actually hurting her, motioning to the door before turning back to where I know James is lying in wait.

Hearing her strangled cry as my back turns, I force myself not to react by looking back the way my heart wants me to. Waiting her out, I'm given what I need when after a few seconds I hear the sound of her feet moving quickly across the floor. The sound growing quieter as she does what I need her to.

I hate that it has to be this way, but what happens next, she can't be here for.

There's only two ways this is going to go and while I had tried focusing on the positive one when she was with me, I know deep down that I'd be just as okay with the other option.

About to step into the room, I'm thrown back started at the sound of wood breaking and the loud voices that follow it. One voice in particular standing out above the rest in its familiarity.

Luke.

Turning toward my best friend, the anger at him being here and ignoring what I said about letting me do this alone getting the better of me, I call out but before I can even register what's happening, I hear it.

The firing of a weapon. Followed up by the sound of Luke's muffled cry.

"Gun!"

Luke

When I get to the ravine and see all the houses lined in rows, I start to lose hope. Finding Cole and Sunday in the sea of houses was going to be like finding a needle in a haystack.

It's only when I make my way down to the four houses on the end that are in various stages of rebuilding that I find it easier to breathe.

After clearing two of the four and pulling on the door of the third only to have Sunday's body run straight into mine, not only was I able to breathe, but I started to believe I made it in time.

Holding her close and brushing away the string of tears flowing from her eyes with the promise that everything would be okay before handing her off to the officers on scene, I felt even better.

Though injured and in need of medical care, she was alive. She was standing on her own two feet and more than that, she looked relieved to see me. A look that with everything that's happened, was one I never thought I'd see again.

Once I'm positive that she's out safe, I move deeper into the house, my attention locking on the sliding doors leading into the living room as a body makes its way out around it. Gun in hand, he lifts it and aims it straight at Cole. Calling out and attempting to warn him at the last second, I was too late as the gun fired before the words even formed. Cole reacting a second too late as the bullet hit its mark, slamming straight into him and dropping his 210 pound body to the floor like it was nothing.

As a sea of officers, guns drawn and ready, flood in around me, making quick work of the distance as they combat Carter's shots with ones of their own as they attempt to cage him in, I drop to my knees in front of Cole.

Blood soaking the back of his shirt, I lean over and place my hands to his neck searching for a pulse, one that after a few minutes I find, but one that's fading fast.

"Don't even think about it, asshole." I threaten as he groans weakly. "I'm not losing you like this. Not in an ugly ass model home."

Careful not to move him too much, but needing to know the severity of the wound, I lift him up off the ground slowly, pulling him into my arms as my eyes fall on the bullet shaped tear, identical to the one in his back protruding through the front of his shirt.

"Sun-day." Cole moans quietly and I grip him even tighter to me. Apparently getting shot isn't enough to stop him from worrying about her.

"She's safe. You did it man. She's outside with the paramedics. You saved her."

Repeating her name again, his eyes start to flutter closed as he gives into the blood loss and needing him to fight, needing him to keep going despite how desperate I know he has to be to pass out from the pain, I shake him lightly before telling him the way it has to be.

"Don't close your eyes. Stay awake, Cole. Keep fighting."

"Lu-ke." He coughs, blood spraying out of his mouth and landing across the front of my shirt.

"I'm right here, buddy. I'm not leaving. I'm gonna take you to Sunday. You just gotta stay awake. Fight it and stay with me. Stay for her."

Not having much experience with gunshot wounds to understand whether its good or bad that the bullet doesn't seem to be stuck in his chest, but knowing deep in my gut that the spray of blood that he just spit up can't be good, I grab onto him as tightly as I can manage, careful of every step I take and start dragging him back toward the front door where I know paramedics are waiting.

Where Sunday is waiting.

Hollering out over the noise of the takedown and getting the attention of the officers standing guard on the lawn, they rush toward me, two paramedics' quick on their heels with medical bags in hand. As their knees hit the ground, I register their pleas for me to let him go and give them the space they need to work on him and after a few seconds, release the hold and let him go.

Collapsing against the door, I watch them work, unable to look away until they've lifted and settled his body on the stretcher. Slowly getting to my feet once I see him being loaded into the ambulance, I stalk back into the house just in time to see the takedown of James Carter playing out at the end of the hall. Looking from the floor coated in Cole's blood, sickened by just how much of it there is, I look up again just in time to come face to face with James. Who despite having his plan thwarted, is still grinning like he won. His smile turning into a full on laugh the second he lowers his eyes to the floor and sees Cole's blood.

Like something out of a movie, time seems to slow as Carter's eyes find mine. A silent recognition passing between us, even though up until now, we'd never actually crossed paths. Flashing me the same grin as he'd been sporting when he shot Cole, he quickly shifts, shoving the full force of his body weight into the officer to his right, knocking him off and balance and laughing as he falls. Predicting his next move before he has a chance to do it and catching sight of the fun attached to the other officer's side as James rushes toward him, I seize the opportunity, pulling it out and aiming it straight at my target.

My mind overrun with images of Sunday and then holding Cole while he lay bleeding in my arms, I pull back on the trigger and put an end to this once and for all.

As the bullet leaves the gun, all sound evaporates. The impact it makes with his skull as he drops to the floor the only thing I can hear as chaos reigns down around me. Two officers grabbing me and pulling me through the door, the force of their hold keeping me grounded and contained, while the realization of what I've done sets in and a calm washes over me.

It's over.
I did it.
James Carter is dead.

Chapter Thirty-Two

Reagan

What the hell is taking them so long?

It's been hours since they brought Cole in. Hours since I stood and watched the ER doors slam harsh and cruel behind the doctors as they took him in and attempted to save his life

Someone must know something by now. Something to settle the panic that's gripped me since I saw them taking him from the house.

Frustration reaching a fever pitch after pacing the corridor for what feels like the hundredth time since I got here, I feel three sets of eyes on me again. Silently studying me from their place a little ways down, the way they have been since they got here a little over an hour ago. The only three people in the world that might understand what it's like to be me right now. But three people that also happen to be family.

Stressed and worried enough and not needing the pressure of the Grayson's being here on top of it, I throw my body down into the seat, exasperated. Exhaling deeply before bringing my hands up and covering my face in an effort to block them all out.

An effort that fails as my legs begin to shake, the force of which is so bad that my boots are actually clicking loudly off the floor. Before I can attempt to right myself, will my body to cooperate and stop the incessant shaking, I hear the sound of the chair beside me being pulled out and as fast as I can take a breath, I'm being pulled into the person's warmth.

A person I see when I pull my hands away is someone else I almost lost today.

Luke.

Crashing into him when Cole pushed me away, demanding I leave and get to safety so I wouldn't be around for what would end up happening. His desperate need to keep me safe equal parts sweet and frightening, but definitely impossible to fight against. Having him hand me off to the officers who then whisked me into the ambulance, treating the wound on my side and attempting to make me as comfortable as possible while my mind screamed at me to go back inside.

To be with Cole.

I'd pushed him away when I found out the truth, and even though it had only taken a few days to come to terms with it, it still looked like I'd given up on what we shared.

Like I'd given up on him.

He didn't have to save me. It wasn't his job anymore. He found me the way he was hired to, which meant his job was done. Yet he still showed up. Fought for me. Killed a man who if he hadn't shown up when he did—when James wanted him to—would have hurt me in a way I may never have been able to come back from.

Cole saved my life, and all I wanted while stuck in the back of that ambulance was to run back inside and do the same for him.

Something I never got the chance to do.

He has to be okay.

I need to tell him I understand now. That I get it and my feelings haven't changed. Tell him I love him.

"He can't...he can't die, Luke."

I can tell with the way he tightens his hold that he agrees, but his lack of argument or fight speaks volumes to the uncertainty of the moment. He can't say that Cole is going to be okay because he was there when it happened and with the amount of blood he was covered in, he knows it can go either way.

Knowing I'm his sister and caring about me the way I know he does and he still won't lie to me. Even knowing it would bring me peace if he did.

"Cole is a lot of things, but most of all he's a fighter. He's strong. If there's anyone that can come back from this, it's him."

"I wasn't there. I left him."

Pausing his hand mid-stroke over my back, he pulls back, using his hand and lifting my chin up until our faces are level.

"Why would you say that?"

"Because it's true, Luke. I should have ignored what he said. Stayed with him. I could have stopped this from happening if I did."

"That's the stupidest shit I've ever heard, Sunday. Fuck. Is that really what you think?"

"He wouldn't have shot Cole if I was there. He wanted to get to Cole through me sure, but he wouldn't take the risk that I would get hurt."

As sick as my dad was, deep down I know he cared about me. It just wasn't the type of love you expect between a father and daughter. His was more twisted.

"That's bullshit. James Carter was sick in the head. The second he found out you knew the truth, he stopped being your dad. He became your captor."

I don't believe that, but with as tense as Luke already is, I don't want to make it worse by arguing it with him. It doesn't matter anyway with the way it all played out in the end. Whatever James was or wasn't no longer matters because he's dead.

When I got to the hospital with Cole and overheard two officers talking about what happened at the house and remembering the way Luke was brought out, I'd put two and two together. I knew Luke had been the one to end it.

What I didn't expect was how numb to it all I was.

James has been the only father I've ever known. Seventeen years of taking care of me, making sure I wanted for nothing and guiding me to be the person I am now. I should feel his loss. The loss of my dad. But I felt and still feel nothing. I'm completely numb to it. The only feeling I'm able to summon when I think about him anger at what he did to Cole.

And what he will have taken from me if Cole dies.

"You wanna know the truth?" Luke says, interrupting and pulling me away from the sinking feeling in my chest over my last thought.

"Truth about what?"

"Cole."

"Sure."

"I knew the day he called and said he had something he needed to tell me that you'd done it."

"Done what?"

"Gotten under his skin. I knew then that we'd end up here. Maybe not right where we are now, but in a world where the two of you ended up together."

"How could you have known that?"

I'm genuinely confused by his words. From what I remember of the first day, all we'd had was coffee together before Luke showed up in town. So, something he wasn't even around for. How he could have known what would happen off of that is beyond me.

"There was something in his voice. You've heard him talk when it's all of us together. You know how he normally sounds. His tone is rough because he's guarded. He's way too damn serious and getting the dude to crack a smile? Yeah right. Unlikely. During that call, he wasn't like that at all."

"What was he like?"

"Surprisingly calm. Which wouldn't have been enough to tip me off, but when he would mention you, his voice would change. It was like we were eight all over again and he was giving Mrs. Jackson her missing dog back. Same damn thing."

"So he's been bringing the missing home since he was eight? It didn't start with me?"

Laughing despite the seriousness of the situation, caught up in the memories I know he has to be reliving in his mind, he shakes his head.

"Hell no. He started that shit at seven, but he did it for a couple of years. Tracker in training. Come to think of it," he pauses, his eyes lighting up as he remembers. "It started right around the time he met you. I guess that's how I knew how this thing with you guys would play out."

"Why you accepted it so easily."

"Yeah. I liked the way he was when he was with you. Whether you're Sunday or Reagan, he was at his best when the two of you were around each other. If you'd lived through what he was like before the way I have, it would make perfect sense why I accept it."

"Accept? As in still do? Even after I pushed him away?"

"Considering why you did it, yeah. You had a nuclear blast blow your entire world apart. I'd say he got off light. You could have easily killed him for what he did."

My heart seizing when he speaks of killing, Luke catches himself almost immediately and tries to make it right, not realizing the damage has already been done. The ease of the moment; how light it felt distancing ourselves from reality and talking about something as simple as my feelings for Cole, is gone.

"It was supposed to be a joke. I didn't think."

"It's okay, Luke. It was fun while it lasted, but it's not like we can escape what's happening forever."

"No, you're right." He concedes easily, his eyes breaking away from mine for a second as he looks down the hall, studying the emergency room door that I've been doing my best not to glare a hole through and turning back to face me again, a smile playing on his face. "We might not be able to escape it forever, but we can escape for a little bit."

"What do you mean?"

"I don't know about you, but sitting here and waiting like this is driving me fucking crazy. I need a break."

I can't leave him. Someone could come out any minute and if I'm not here when they do, I'll never be able to live with myself. I'm finding it hard enough to do that as it is.

"You can go, but I need to stay here. I need to be here when he comes out."

"Can I make a suggestion?"

"Sure."

"We talk to my parents. Let them know we're gonna head out for some air and get them to call me if anything happens while we're gone."

"Why do you want me with you so bad?"

Raising his eyebrow in surprise, my question obviously taking off guard, he brings his hand up and slaps it off the back of my head, smirking at me when my knee jerk reaction is to punch him back.

"Looks like you got a bit of Grayson in you after all." He jokes. "But I'm kind of disappointed you even had to ask that question."

"Will you just answer it?"

"Oh, I don't know Sunday. Maybe it's because you're my sister and out of all the people in the world I could choose to be around, there's no one better than you. All this messed up shit I'm feeling and being worried fucking sick about Cole, you get it."

Stunned by his words almost as much as I'm touched by them, but not wanting to make the moment any more emotional than it already has been, I reach up and slap my hand hard into his back, grinning at him the second the scowl appears.

"Well, *brother*. Why didn't you just say that in the first place? Let's get some air."

Luke

Of all the things I imagined doing with my sister when we found her—if we ever did, pushing her in a swing wasn't one of them.

I gave up on being able to take my sister to the park about five years after we lost her. The years turning me older, and if she was out there somewhere, her getting older to, it just seemed like something we'd never get to experience together. But here we are.

And it's even better than the last time we were like this.

"Are you ever going to swing yourself?" I call out when she heads back toward me, my hand again shoving forward into her back as she soars upward into the air, her laughter so loud it spills out for miles as she comes back for another push.

"No!" she yells back, gripping onto the chains tighter and keeping her legs bent and unmoving. Purposely making it so I can't slip down into the swing beside her and join in on her fun.

"That's such a Reagan answer!" I joke and am struck frozen momentarily as I wait for her reaction. For the second time today saying something before thinking it all the way through and wanting to kick myself for it.

"Only because it's one that doesn't agree with you!" she calls back, easing and erasing my earlier fears as she follows it up with another laugh.

Stepping inward, ready to push her again as she makes her way back, I'm taken off guard when instead of hitting her back, it's the hard sole of her shoe smacking hard against my hand. A move that I see when I look up and she's laughing, she did intentionally.

"Now that," she laughs. "Was a Reagan answer!"

"You're gonna pay for that!"

Grabbing her feet when she makes her way back again and using as much force as I can, I slow her down and letting her go, watch as she lowers her feet straight to the dirt below, dragging it along until she's bringing herself to a full stop.

"Bring it, Grayson." She goads me. "Just remember. Payback is a bitch!"

"You're forgetting one thing." I tease as I step toward her, answering her call for me to bring it until I'm hovering over her from behind, chuckling when she sucks in a sharp breath.

"And what's that?"

"I've got a year and a half's worth of tricks that I'm willing to bet still work."

Not giving her a chance to react, my hands grip her waist and I start tickling. My earlier statement coming to life in front of me as she starts giggling, unable to catch her breath as I continue the assault. Sliding off the swing and down to the ground in an effort to end the cycle of torture, she curls into a ball right when I pull my hands back.

"And the winner is..."

"A cheating asshole!" she calls out.

"Otherwise known as Luke." I joke, enjoying the way she looks as she tries to hide her own laughter, but her body betraying her as the shaking gives her away.

Reaching over and holding out my hand as our laughter dies down and pulling her to her feet, I'm hit with a blast from the past.

All those weeks in Alliston that I spent time around her, getting to see her smile and I never noticed it until now. Her smile, while aged because of the years that have passed since the last time I've seen it...it's the exact same toothy one she used to flash me and Cole when she was a baby.

Damnit Cole...Wake the fuck up already. You're missing it.

"So this is a lot better than the last time we did this, huh?" she asks once she's successfully wiped all the dirt off the knees of her jeans and I freeze.

What the hell did she just say?

We talked a lot on the way over here, but it was about trivial stuff. Her asking about Bethany and telling me random stuff about Jake and the deal he signed with the record label. Nothing that could possibly set us off. Definitely not memories that only I could have of the past. There's no way she could have known about any of that.

Taking a chance and meeting her eyes, seeing my own shock reflected back at me through hers, I get her to repeat herself. Needing to know that even though she's as surprised as I am, I didn't imagine it.

"Say what?"

"We did this before, right? At night?"

My heat picking up speed, sweat beginning to form as the shock settles itself in for an extended stay, and further blown away by the fact that she remembers the time of day we did this, I curse under my breath and she groans. Completely misreading my reason for swearing as she works herself up and rapidly starts talking.

"I thought I was remembering something we did together. I'm sorry. I didn't mean to do that. It must have been something I did with my mom." Her groan turning into a full on scowl as she mentions Katherine Carter.

"No, wait." I stop her. Shaking off the shock and finally accepting the moment for what it is. She's remembering Sunday. "You were right. We did do this before. You woke up one night crying so I snuck you out with me to meet up with Cole and a couple of the other guys. I didn't want you to wake mom and dad. Turns out when the guys saw that I had a baby with me, they all bailed pretty quickly. Well, everyone but Cole. You probably already know that though. You owned him even then."

"You mean, it really happened? What I remember?"

"Yeah, but is it just being at a park at night that you remember or was there more?"

Holding up a hand, she makes her way over to the baby swings and lifting one up in her hands before turning back to me, she motions toward it and smiles.

"You put me in one of these and pushed me."

I don't know what to say. The need to go pick her up off the ground and swing her around celebrating this is strong inside me, but the lead that my boots are now apparently made of prevents me from doing anything at all.

I'm stuck. This is too much to take in right now. Especially after everything we've already managed to live through today.

What was supposed to be a quiet walk in the park to clear our heads has officially turned into more.

She's really back.

Sunday's home.

Chapter Thirty-Three

Sunday

When I was on the swing, there was this moment as my body lifted into the air where I experienced a feeling of déjà vu. A fuzzy picture sitting on the outskirts of my memory that made me think I'd done this before.

At first, I chalked it up to all of the times my parents had taken me and Jake to the park when we were little, but as I continued swinging and joking around with Luke, it kept chipping away at me.

It's only when he tickled me that the fogginess of the picture cleared and I saw the younger version of myself being pushed in a swing in the middle of the night. The only light around for miles the street lamps that lined the parking lot adjacent.

When Luke swore under his breath though, I thought for sure I was imagining things. That it wasn't a memory of my life before, but something I'd made up in order to accept the truth easier.

Knowing differently now with Luke filling in some of the blanks that the picture wouldn't have been able to tell me, it's shifted everything between us.

Moving from his spot by the swings, mumbling something about wanting to get back to check on Cole, he makes his way over like before and slips his hand in mine. Only this time his grip is different. Tighter. Like he's afraid if he doesn't hold me tight enough, I'll slip away.

He'll lose me again.

Walking back up the hill and across the parking lot, I pull us to a stop before we can head back in.

"What is it?"

"Did I ever call you Wuke?"

Laughing at how ridiculous I sound, I shake my head and look away embarrassed. Not wanting any part of the response I know is coming with how awkward I felt even saying it.

"Welcome home, Sunday." He says softly, surprising me as he steps forward and brings me into his arms. "Your Wuke missed you so damn much."

Another fuzzy memory now pieced successfully together and proven to be real. One that seems to make him react even more than the last one did. Pulling back, he swipes a hand over his eye, leaving a streak line behind. A streak I know can only be caused by two things. One of which—sweat—not a factor since his face is otherwise clear.

Which means what he wiped away was a tear. One he didn't want me to see, but one that even if the streak line wasn't visible I would still be able to feel. Both of us held captive by the emotion of the moment.

"I missed you too."

Looking down to our hands, the ones that even with us stopped are still connected, watching as he does the same, I take a deep breath and finish my thought. "I'm sorry I was gone for so long."

Tightening his grip on my hand, his eyes never once drifting away from them, he wipes at his face with his free hand again before breaking away and shaking himself. "Alright, that's enough of that. Let's go harass someone for information."

Smiling softly and nodding my acceptance when his eyes finally meet mine again, I let him guide us through the door, keeping pace until we're making our way around the corner and down into the familiar corridor again. Luke's parents—*our* parents sitting right where we left them.

Beginning to slow down the closer we get to them, Luke feeling the change slows to a crawl again. This time his body being the one to stop first.

"It's okay. They know this is a lot. They're not going to push you."

"And if I want to be pushed?"

"Say what?"

"I know that there's a lot we need to talk about when we're not so worried about Cole, but Luke, I think I want to meet them. Do you think that would be alright?"

"You're kidding right?" He asks, laughing when I reach out and punch him in the arm. "Sunday, they would love that."

"Yeah?"

"They've been waiting for years to meet you, so yeah, I'm sure. Come on."

Walking again, we make quick work of the distance and before I've had the chance to settle my racing heartbeat, suck in a breath and prepare for what's about to happen once we get to them, we've stopped again and Luke's tapping our dad on the shoulder.

"Dad, there's someone here that wants to meet you."

As the man I now know to be Xander Grayson turns and acknowledges Luke, his eyes move over to me and just like that, I'm hit with a replica of my own staring back at me. Eyes that grow softer the longer he stares.

"Allie…" he calls softly to his wife—my mom—without so much as a twitch giving away that he's about to break contact. Her small frame coming into my line of vision almost the second after he's spoken her name.

"Mom…Dad, meet Sunday." Luke introduces us and I squeeze his hand in silent thanks before breaking the connection and stepping toward them, my nerves so tightly wound my legs shake and begin to wobble.

"Hello, Sunday." Xander greets and whether it's the relief I hear in his voice, the softness of it or the fact that with everything I've been remembering today, I feel close to him, I rush toward him.

And just like I hoped, his arms are there to catch me as he pulls me into the warmest embrace I've ever experienced.

"Oh, honey." He whispers as I tighten my hold on him. "Welcome home."

*****

Despite the way I embraced Xander, when the time came for me to do the same with Allison, I couldn't and it wasn't hard to see why. It's because of what James told me when he first threw me into the car in Alliston.

Katherine didn't know about any of it.

Not only was he lying to me all this time, but he'd done the same to her too. She believed in her heart that I had been her miracle. Never knowing that in order to have it, she'd had to take it from someone else.

Knowing this makes it harder to detach from her. She's still my mom. So as much as I wanted to be able to hug Allison the same way I did with Xander, I couldn't and I'm not sure when I'll be able to.

But by some stroke of luck, I was saved from having to explain all of this to them as the doctor chose that moment to make his way out and update us on Cole.

"Cole is very lucky. A few inches to the left and it would have been lodged in his spine. It was the fibers from his clothing we found in the wound that caused us the most concern. Having since extracted them in effort to minimize the risk of complications in the future, we're confident he'll make a full recovery."

Remembering the way he looked when he was pulled from the house and lifted into the ambulance, the blood stains soaking through both the front and back of his shirt and the size of the hole the bullet made when it impacted, my stomach recoils.

Turning away and bumping straight into Luke's chest, he holds me still, rubbing my back as the doctor continues to update our parents.

"They're gonna put him in recovery." Luke says. "They want to monitor him overnight, but as long as everything stays the same, we can spring him in a couple days. He's gonna be just fine."

Releasing the world's biggest sigh of relief, the sound so loud that it's even got the doctor and our parents pausing to acknowledge it, Luke laughs before apologizing and asking the only other question I want an answer to.

"When can we see him?"

"As soon as he's settled in recovery." The doctor answers easily. "But only one of you at a time and immediate family only."

Immediate family.
Something I'm not.

"Oh no you don't, sunshine." He admonishes once he's seen my frown. "There is no one more immediate than you when it comes to Cole. You're going in there the second we're allowed in."

It's what I want and it warms my heart, Luke being so adamant that it's what's gonna happen, but he's wrong. I might be important to Cole, but so is he. If there's anyone that should go in first, it's him.

I'm not even sure after everything that's happened Cole is going to want to see me.

"No, Luke. You should go in first."

"Is this more of that blaming yourself shit? Because we've been over it already. What happened to Cole is on James."

Shaking my head, willing the tears I can feel building to stay down a little longer, I swallow them, open my mouth to talk and the flood gates burst.

Saliva rising to the surface and choking me up before I can even get the words out, the force of it making my eyes water and laying waste to my big plan to explain this calmly.

"What is it? Shit. What did I say?" Running his hand over my eyes and wiping away the waterworks, he searches my eyes frantically for answers. "Tell me what the hell I said."

"You didn't say anything. I wanted to explain why you should go first but I got all choked up."

"Well, take a deep breath and try again. But before you tell me, just remember that whatever it is, it's not going to change anything. I'm not gonna freak out or go off on you. You can tell me anything."

The way Luke is right now is the way I imagine Jake being in a few years. The calm to my storm. The way all brothers should be with their sisters.

"I'm scared he's not going to want to see me, Luke. That despite everything he said before he tried to get me out of the house, he's going to hate me for putting him through this."

"Oh, sunshine," he sighs. "The only person Cole is capable of hating is anyone that threatens his family and himself. He could never hate you. Not when I'm pretty sure he's been wired to love you."

"Wired to love me?"

"I know it makes no sense right now, but I'm asking you to trust me on this. That guy in there. The one that went and got himself shot because he was more worried about you and me than he was himself, he could never hate you."

"Does he know what a good friend he has?"

"Pretty sure he does, yeah, but if you ever wanna smack him up the side of the head and remind him, I won't stop you. Sis over bro and all that." He jokes and I allow myself the peace that comes from laughing, joining him until water of a whole different kind is falling from my eyes.

When we've both gotten control of ourselves, Luke takes my hand and just as I'm about to ask what he's doing, he lifts our hands and points down to the hall where recovery is waiting.

"Let's go. We'll be closer to him there."

When I don't move and resist the hold he has on my hand, he swears under his breath.

"That shit's getting real old. You wanna tell me why you're stalling now?"

"Why do you call me Sunshine? Cole did it once too."

"He didn't tell you?"

"No. Considering he only did it the one time and you've done it a bunch, I figure it's not his story to tell."

"Your name technically means the day of the week, right? Well, for me, it's always meant more. You know the way the sun feels when it first hits your face?"

"Yeah. What about it?"

"That's the feeling I had whenever I was around you. So, *sunshine*," he grins. "Now that I've answered the question, can we go see our boy now?"

"Lead the way."

Cole

She's here.

I might not be coherent enough to make out where she is in the room, all the narcotics they've plied me with making it damn near impossible to keep my eyes open, but I don't need to see her to know.

Sunday is a trigger for all my senses. She has been from the day I met her.

Her soft powdery scent tickling my nose is what signals her arrival first, the soft sound of her feet as they move across the floor, hopefully toward me, all I'm able to focus on next. As the anticipation of her touch builds to a fever pitch, I force my eyes to stay closed, needing to know what she'll do next when she's working under the belief that I'm resting.

Unprepared for the electrical surge I experience as her fingers lightly brush across the top of mine, I bite back the urge to give myself away. The pleasurable feel of her touch almost too powerful to fight against.

"I was going to wait a little longer before I came in, but Luke wouldn't let me. He said that if I didn't come in on my own, he'd drag me in kicking and screaming."

Captivated by the soft laugh that escapes, my lips begin to twitch and it's not long before I can feel my face lifting from the smile I can't stop.

"Why didn't you tell me my brother was an asshole?"

The almost musical lilt of her voice, a result of the subdued happiness present as she refers to Luke as her brother, breaks me. Blowing my plan to play possum out of the water until she's not the only person in the room that's laughing, though the differences between them are astounding.

My rough and hoarse to her soft and gentle.

Our laughter the perfect reflection of us, both together and apart.

Darkness and light. Broken and Unbroken. Cold and unfeeling to warm and expressive.

The perfect balance in an imperfect situation.

"It's funnier when people come to that conclusion on their own."

It's not exactly what I want to say, but works in the moment as it gives me what I've been after since I heard the door open and she was the person that came through. A sight for a pair of extremely tired and sore eyes that now that she's closer, I want to experience again.

Up close and personal this time.

"Hey." She smiles when I've succeeded in opening my eyes enough to be able to take all of her in. Including the stains left behind on her face from the obvious tears she's shed.

Ignoring the twist in my gut that's born from the hatred I feel at being the reason she looks like this, I smile back.

"Hey beautiful."

"How are you feeling?"

"Better now." I answer easily, leaving no room for doubt in what I mean as I look at the way our fingers are tangled together. "You have no idea how good it is to see you."

"I'm pretty sure that should be my line, Cole."

"Nope. Sorry. This one's all mine. You're going to have to get your own."

Meant to be a joke, I'm not at all prepared for the water that pools in the corners of her eyes and falls.

"Sunday—"

"I was so afraid, Cole." She sniffles softly as she tightens her grip on my hand. "When they brought you out of the house and I saw all the blood…"

"Sunday, look at me." When she doesn't lift her head or shift at all, I try again. "Please look at me. It's the only way this is going to work."

Talking becoming increasingly easier the more I do it, the rasp in my voice finally beginning to clear, I wait her out. My eyes growing heavier the longer she takes, until it's taking everything I've got in me to fight against them closing completely.

I can't give into what I know my body needs until she knows. Until she sees that even though I know she was scared, I'm here and okay.

"Why won't you look at me?"

With the way she's been keeping her head lowered, her eyes locked on our hands even when I'm pleading with her, I don't expect her to answer this question either. So when she does, this time meeting my eyes at the same time, I'm stunned.

"Because I didn't—don't want to see what I know is gonna be there when I do."

"And what exactly are you expecting to see?"

"Anger…but disgust more."

What the hell?

"Alright, well you're looking at me now, sweetheart. What do you see?"

I know what I see. What I've always seen and want to share with her, but this isn't about me. It's about her. Making her see that she's wrong.

So damn wrong.

"I don't see those things."

"That doesn't answer my question."

Shit. It's happening again. My damn head is swimming, I'm starting to feel dizzy and my damn eyes are closing. That can't happen.

"Worry, alright? I can see that you're worried about me."

Not what I was going for, but since it's not a lie and I am worried, I'll work with it. "What else?"

"Exhaustion." She says, biting her lip nervously, nailing another one that I won't even bother denying.

"I'm dead on my feet, sweetheart. So you're right, but try again. This time, really look. It's the same thing that's reflecting back at me in yours."

"Love." She answers with no sign of hesitation. "I see love."

"Exactly. What you came in here expecting to find, you're not going to get. Ever. I don't care if you hate my guts and want to run me down with your car after everything that happened with us. It will never change the way I look at you. The way I feel. So please don't ever be afraid to look at me."

"I thought I lost you."

"Never."

Even though I know she's going to have her doubts, I've never meant something more than what I've just said. She will never lose me. Where we stand may be up in the air. We might never be able to move past it or go back to the way things were before, but that isn't. I belong to her. She owns me and she always will.

"When I kissed you back at the house, it was a message. I'm just not sure who the message was for anymore."

"I don't understand."

"That was not going to be the last time I kissed you. Held you. It wasn't a goodbye. I thought that I did it for you so you'd know how I felt...how I still feel, but I'm starting to think I did it for me. I needed to remind myself of what, or rather—who, I was fighting for."

"Sunday, I was fighting to come back to you and even though I'm losing the fight now, my damn eyes wanting to betray me and shut down, I'm still fighting."

"But this is my fault!" she cries. "If I didn't make you leave that night, none of this would have happened. You should hate me. Why don't you hate me?"

Hate has always been the emotion I summon easiest. A child of rape, I was born with a deep intense hatred for the father I never knew. It didn't mix well with the naïve love that even after being beaten down physically and emotionally by my mom I still had for her. Once the love faded though, it was open season. I hated her with a passion.

The truth is, I've been hating everything for so long that until Sunday, I thought it was the only thing I'd ever be able to feel. It was my destiny despite all of the people that tried to get me to see differently.

Sunday changed all of that, which is why I don't understand this.

How could I hate the person that taught me how to love?

I can't.

"I need you to listen to me, okay? Even if you don't agree with anything I'm saying, don't say anything. Just listen."

Sighing her compliance, she slips her hand out of mine and starts to turn, her eyes falling on the chair in the corner, but before she can step away, I reach out and stop her.

"Don't."

"Why not?"

"Because I want you here with me." I admit with a smile as I run my hand over the blanket. "I need to have you close."

I know I've touched her with the way her eyes soften and she's having to wipe away tears as they slip out, but it's not enough for her to give into me. At least not yet.

"You weren't the only one that was worried. That still is."

"There's not enough room."

"Wasn't that my line before?"

This gets me what I'm after as her lips betray her and lift into a smile at the memory of the last time we'd had to make due in a smaller space. Stepping closer, she lifts herself onto the bed and curls her body into mine. Aware with every move she makes of where not to put pressure, the sweetest sigh escaping when her head finally rests on my shoulder.

"I'm ready to listen."

"What happened today would have happened regardless. From the second Xander and Allison hired me to find you, it was the road we were destined to go down. It's only the severity of it that no one could predict. When I told you everything, I sped up the clock. I knew it was only a matter of time before James reacted. I just didn't realize he was going to go after you to get to me. I thought leaving the way I did would have kept you safe. If there's anyone at fault for what happened, Sunday, it's me."

"It's not your fault."

"I wasn't finished." I laugh when she interrupts and chancing the pain I know is going to come when I do it, lean over and kiss the top of her head. "Can't even give the wounded guy a break, huh?"

"Where's the fun in that?" she allows herself to joke back, the sound of the laugh escaping out over my chest and into the air warming my heart. Giving me back something that I never expected to have again.

Hope. Forgiveness.

Love.

"What I was going to say before I was so rudely interrupted," I tease, laughing when she lifts her head and rolls her eyes. "Was, even knowing that at some point we were going to end up where we did, it didn't change anything. I would have walked through fire if it meant keeping you safe."

Up until now, she's been relentless in her pursuit to have me blame her. For the first time the roles reversing between us and her being the one to feed into her fears. Only seeing the bad things, while all I see is that I'm here. Breathing and alive, with the girl of my dreams in my arms. Right where I always want her to be.

All the good that came out of this nightmare.

I'm fighting a battle I'll probably never win, especially with my body threatening a revolt, but I've still got one more card I can play. The only card left in the deck. One that when I use it, I hope gets me the only prize worth having.

A future with her.

"I can't give you what you want. I won't ever be able to give it to you because it's impossible. I will never hate you, not even when you think you deserve it. For the first time in my life, I don't want to be ruled by my anger or the rage and hate that I've based my life around since I was old enough to understand what those things were."

Shifting her body and propping herself up, lifting her eyes to meet mine, she gives me the opening I'm hoping for.

"What is it you want now?"

"A chance."

"To do what?"

"To start again." Pausing and feeling the weight of all the words we'd left unsaid between us beginning to melt away, I follow it up with words I should have told her five days ago, but had been too trapped in my own devastation to admit. "I want to make my first love my last love."

Lifting her hand to her after sucking in a breath, I seize the opportunity before it has the chance to pass me by.

"My only love."

Resting my head on hers, locking our fingers together before pulling her hand away from her face to the bed, I brush my nose against hers softly before finally surrendering to the moment and kissing her with everything I have.

A kiss so intense it makes the rest of the world, where we are and why drift away completely until her voice whispering soft words of her own is all I hear once it ends. Soothed by the sound, I finally give into my body's need to rest. Closing my eyes and beginning to drift off to dream in her arms.

"You like stories, don't you, Cole? Because I think I've got one you're really going to like…"

Epilogue
Six Months Later

Cole

"Are you sure you want to do this?"

We've been here a total of fifteen minutes and this is the tenth time she's asked me that question. You'd think with the answer never changing, she'd have accepted it by now, but apparently it's going to take a few more tries before it sinks in.

It's not even like I just dropped this on her and she's trying to come to terms with it. I brought it to her immediately when it came to me a few weeks ago. I wanted her thoughts, considering it had everything to do with her.

The final decision on whether or not to mark my body with ink again was mine of course, but from the moment I sensed her in my hospital room months ago, life stopped being just about me. It became, especially after she told me her story, about us. I wouldn't do this—couldn't, if she wasn't completely on board.

Getting shot and by some stroke of luck having the bullet miss all vital organs was a wakeup call. My wakeup call.

It was my chance to do something that looking back now, I hadn't been doing for a long time. Living. And with it, a second chance to do right by the girl that played the biggest part in my ability to finally see it.

Sunday made me see that while what happened to me in the past might have been horrible, it didn't have to define every moment that came next.

It didn't need to define me.

"You know I am. So what's this really about? What's going through that beautiful head of yours?"

"I just don't want you thinking that you need to do this because I said I wanted to when you first brought it up. You could just," she pauses biting her lip nervously. "Hold my hand or something."

"This is me doing the *or something*, Sunday. And when he calls us back in a few minutes, you can be damn sure I'll do the hand holding too." I grin motioning with my head toward the tables set up side by side in the room at the end of the hall.

One of the perks of looking into something for the owner in the past and maintaining a relationship once it was done.

"Cole..." she starts up again, the nervousness over what we're about to share attempting to get the better of her, but failing as I pull her body lush to mine. Lifting her hand to my lips and using them to soothe her worries as I kiss across the tips of each one of her fingers.

"I know you don't get it, but I *need* to do this." I explain just as Mark tips us off to his arrival by clearing his throat and smiling once we turn to greet him.

"You two ready?" he asks, pausing when he catches the look of sheer panic on Sunday's face. "Or you still need some time?"

"Give us another minute."

"Sure thing, man. Just holler when you're ready."

Flashing me a look of understanding he turns and heads to the back and shifting my focus back to the nervous girl in my arms, with her back now to me after Mark's interruption, I spin her around to face me.

"I'm sorry." She blurts out the second my hand reaches out to tuck a stray tendril of hair behind her ear. My eyes catching hers and keeping them grounded. Needing for her to see what's reflected back in mine.

The need to do this. Not just for her, but for us.

What it means in the long term. What its means for what's already etched into my chest and what it means for the future.

"You have nothing to be sorry for. You wanted to do this when we talked about it and now that it's here and we're actually going through with it, you're scared. I get it."

"I just don't want you to end up regretting it."

If she wasn't so damn adorable, I'd definitely call her on her craziness.

How I could ever regret walking into that room with her while Mark adds one final date onto the tattoo that rests over my heart, is beyond me. Especially with what the date means.

11.18. 2014

The day that if I had been asked about it seven months ago, I would have said would never come. But one we've been living every day for the last six.

The day she came home to us.

To me.

Add all of that to her wanting to mark the space over her own heart with the day she remembered Sunday and there is no way in hell I could ever regret what's about to happen.

"I regret not finding you sooner. I regret not telling you the truth the second our eyes met in the bank, but nothing else, Sunday. Not one damn thing. I will never regret any moment spent with you. Past, present or future."

Lifting her hand and laying it over my heart, I start moving our hands slowly over and down until we're tracing over the spot under my shirt where my scar resides. The scar that for months afterward made her cry every time she saw it. Pausing our hands over it, she sighs and something about the soft way it falls tells me that I've won.

Something I've said, along with the tender way she touches the only remaining physical evidence of what we both lived through six months ago has broken through the fear and given her peace.

"I love you." She whispers against my lips before wrapping her arms around my neck and kissing me. Awakening feelings that even now, I'm still adjusting to. Emotions that until she was thrown back into my life again, I never thought I was allowed to feel, much less be capable of actually feeling for someone.

"Not nearly as much as I love you."

Words that I mean whole heartedly and that I can tell she wants to argue with me, but won't because of everything we went through to get here.

I don't remember much of what happened, but I do remember waking up and her tear stained face being the first one I saw. Her love for me spilling over in waves as she disobeyed hospital rules and settled into the bed with me. Never letting me go despite how in and out of consciousness I was.

Wrapping me up in her warmth. Giving me everything I needed for my body to relax and begin what would be months of the healing process. Secure in the knowledge that she wasn't leaving.

That neither of us would leave the other again.

"I'm about to let some guy stick a needle in my chest. I'm pretty sure that means I'm the one that loves *you* more." She jokes playfully.

"You keep believing that, sweetheart. I'm okay being the only one that knows the truth."

Pressing my lips to her forehead and silencing any argument she might have coming, I do what Mark said and call out to him, content in the knowledge that now that we've talked, we're ready to get this show on the road.

"You ready, Sunshine?"

"About as ready as you are for the beating you'll get from Luke when he finds out you called me that." She jokes.

Peeking his head out around the door at the exact moment my hand makes contact with her ass and she's cursing my name, he motions for us to come through.

Searching her eyes one last time for any sign she's about to change her mind and finding nothing but her glistening eyes shining back, I slip my hand through hers and start walking.

It's time to put an end to the chapter of my life that has been dominated by pain, failure and loss, and officially start a new one.

A chapter that a seven year old predicted would happen seventeen years before it actually did.

One that begins and ends with the light.

Sunday

Maybe Jake was right.

Being here *is* weird and I'm starting to think maybe even a little morbid.

It's too late to go back, but standing in the cemetery now and seeing the physical result of the decision I made months ago, is a little unbelievable.

Nothing about this entire experience seems real.

When Jake agreed to go with me, citing his own reasons to be here, there'd been a split second where I wanted to talk him out of it. Make him see that in the end, being here would do more harm than good.

Harm that even now, knowing who I really am and accepting it, I couldn't allow Cole to be a part of. Which is why he's over an hour away waiting for me at home and I'd given in to Jake and brought him with me.

The city. Toronto. Another decision made about a week after the shooting, but one that still, months after deciding, still makes the most sense.

As familiar as Alliston was I knew I couldn't go back.

None of us could.

So two days after Cole was released from the hospital, we'd left the horrors of Alliston and Orangeville behind and gone back to his place in the city.

The first step in our agreement to start over. A clean break from the past and a place to begin again as Sunday.

It wasn't just Cole and I that were starting over, though. It was a new beginning for the other two people that had lived through James Carter and his lies. A clean break as it turns out, they were more than ready for.

It had taken a bit of work and a lot of going back and forth between two different realtors, but within a few weeks of speaking to Katherine and Jake, I'd gone back to Alliston to nail a For Sale sign into the ground of our familial home and gotten their signatures on the lease for a new place in Toronto.

A decision that based on all of the opportunities that have been thrown Jake's way with his music since he made the move, was definitely the right one.

"Remind me again why you wanted to do this?"

This—what he still can't wrap his mind around no matter how many times I've explained it, is the two tombstones we're standing in front of in the cemetery. One baring the name of

Jake's father and the other marked with the name of the child he coveted most.

Here lies Reagan Carter. Beloved Daughter. 1997-2014

With Cameron behind bars awaiting trial, and the other two people involved in the kidnapping dead and buried after paying the ultimate price for their crimes, it was the only thing left to do. Cole and Luke had both been cleared of their part in that final showdown and we were all ready to move on from it.

The empty grave symbolizing not only that Reagan died the day her father did, but also serving as a reminder that for almost eighteen years she did exist. She was real and she did mean something to the world.

Even if it was never meant to last.

"I needed to say goodbye. Same as you."

"I get that. I guess what I don't get is why you're doing it now when it's been like this for months."

"I wasn't ready to face it before, Jake. I am now. It's time."

"Are you ever going to tell me what it says?" he asks, motioning to where my hand rests over my heart and the bandage now covering it. The tattoo somehow infusing me with the strength I need to get through what's going to happen next.

"When you do what you came here to do."

Kneeling down in between both graves, his gaze lingers on the first stone, before he shakes his head and looks away. Obviously as stunned by everything that happened today as he was when he learned the extent of it months ago. Looking to me and molding his hands into the shape of a heart, he presses it against his own heart. Silently holding it in place for a few minutes before turning to face the last remaining remnant of the sister he loved.

Laying the bouquet on the empty plot and leaning in as close to the stone marker bearing her name as he can get, he begins to speak and that's when I turn and step away. What he's here to do—saying goodbye to Reagan—needs to be done in private.

Hearing the crunch of his feet across the grass a few minutes later, I don't make a move to turn around and face him until he nestles his head on my shoulder and tells me he's done.

"Are you sure? When you mentioned coming here earlier, I assumed it was to say goodbye to him." I motion to James's stone. The one that while he had stared at it for a while after we got here, he never once said goodbye to.

"I've got nothing to say to him, Sunday. I thought I did, but when we got here and I saw the grave, whatever I did want to say didn't seem so important anymore."

This has been the hardest part of all of this. The losses that not only Jake suffered when the truth came out, but Katherine as well.

Ones that no matter how badly I want to, I can never erase the pain from.

"Don't you have someplace to be?" Jake interrupts and I nod, remembering the plans I made with Cole before I dropped him off at the Grayson's.

"If by plans you mean Cole being up to something and keeping it a secret from me."

It hasn't happened a lot in the last six months, but in the rare times it does, I'm reminded of just how lucky I have it. Not only did I end up having two of the best guys in the world as my brothers, but what had started out as them on opposite sides and adversarial, now has them united with a common goal.

My happiness.

Eyebrow raised and a frown on his face, he throws his arm around me and lays the protective brother act on thick. "You need me to kick his ass?"

"As tempting as that is," I laugh softly. "I think he's got more to fear from me doing it than you, little brother. You stick to melting girls panties off with your music. It's what you're good at. A fighter you're not."

"That hurts, Sunday. I could totally beat his ass into the ground if I wanted to."

"Sure you could, Jakey, but since I know you love him almost as much I do, you won't."

"So what do you think he's hiding?"

"If I had to guess, he's planning some kind of surprise for me. Something big enough that he couldn't do it on his own, which is why he had me drop him off at the Grayson's before coming to get you."

"Any idea what it might be?"

"No clue, but with all the classes he's taking lately, I can't see him having a whole lot of time to plan anything. The only reason we got the tattoos earlier was because he planned it weeks in advance."

Cole's classes. Another decision we made as a team that was born from a memory of his time with Samuel.

Police College.

"Before he died," he explained. *"Samuel sat me down and asked me what I would want to do with my life if my dream of playing in the NHL never amounted to anything. At the time, I never gave it a lot of thought. I mean, playing hockey was all I ever say myself doing, but he didn't let up. So after a couple of days thinking about it, I told him I wanted to be a cop. That if I couldn't be on the ice, I wanted to do the only other thing that made sense."*

"Saving the world." I surmised, laughing the second his head starts shaking in disagreement.

"Not likely. More like giving back what Sam gave me."

"Like I said. Saving the world. Face it. The second you took on the Alexander name, you became a hero."

"You're letting your books bleed into reality again."

In most cases, I would agree with him, especially with the way some of the heroes are written in the books I read, but in this case, it's not about fiction. It's about our reality.

Cole, for all of his arguments to the contrary, is a hero.

Mine.

"Think what you want, but I know the truth. You're a hero and I'm living proof."

"Well, when you put it that way..." he concedes, his eyes softening as he leans in and kisses me softly.

"Let's save the world, Officer Alexander."

When he said we were going to start over, he'd meant it. We might have been brought together by a secret, but it was honesty and truth that kept us together and what made this work.

Unable to stop the response to the memory, the same one that always appears whenever I think about just how far we've come over the last six months, I pull out of Jake's arms and turn back toward the graves, ready to face what up until now I couldn't.

Pointing to the grave when he looks back and forth between me and the grave confused, I walk over and running my hand over my name, say the only thing left to say.

"Thank you, Reagan. For everything." Turning back to Jake, I point to the gate and the car that's waiting for us in the distance. "Let's go home, Jakey."

*****

"Can I take it off yet?" I ask when after hearing the sound of the key in the lock, he's lifted me off my feet and carried me into what I know is our apartment.

He's had me wearing this blindfold since I ended up back at my parents' house. Ushering me back out the door and into the car before I'd even gotten the chance to say hello to any of them.

Luke's smirk from the door as Cole slipped the soft material down over my eyes the last thing I saw before he'd settled me in the car and driven here.

"In a minute. We're almost there."

"Almost where?"

"Where we need to be."

"You're taking this whole over the threshold thing a little too far don't you think?" I joke and he laughs.

"Consider it practice for the future."

He's serious and knowing me a little too well, when no response comes, he teases me.

"Cat got your tongue, sweetheart?"

"You wish."

"You might be right about that." He jokes, laughter spilling out as I shove my elbow into his side. "Just for that, I'm calling the whole thing off."

When his threat falls flat, me reacting the opposite way I would have in the past and elbowing him again, he places a soft kiss to my cheek before lowering me to the ground.

"You can take the blindfold off now."

Slipping the blindfold off when my feet hit the floor, I'm met with his playful smile first. Unable to resist the urge, I lean forward and press my lips to his softly before focusing on exactly where we are.

"What are we doing in here?"

Our bedroom. A room that other than the candles I can make out in all four corners of the room and the books I can make out on the bed, looks the same as it did this morning.

"Would you believe me if I said I brought you in here so I could have my way with you?"

"Normally, yes. But since you don't need old books for that, I'm gonna go with no."

"It was worth a shot."

Taking my hand and squeezing, he guides me across the room until we're both seated on the edge of the bed. Slipping his hand out of mine just long enough to reach over and pick up the first of the books, he holds it out for me to take holding it out between us for me to take.

"Cole," I try again. "What is all of this?"

"What does it look like?"

Ugh. Frustrating man. He knows how I am about answering a question with a question. It's also pretty obvious with the way he's smirking at me that he doesn't care.

"It looks like you're trying to seduce me with old books." I joke. "Is that what all of this is about? Are you finally ready to accept that you're not the only man in my life?"

"Hate to break it to you, sweetheart, but I accepted that fact a long time ago. So that's not what this is. Though I do like the seducing you part."

"Well you won't be seducing anything if you don't explain what's going on."

Shifting and moving across the bed until his body is settled in beside mine, he taps the book before slipping his hand down into mine.

"Open it."

Guiding our hands down to the edge and pulling the cover back, I shift the book until it's resting between our laps as I take in what's inside.

Pictures.

What I can see by flipping each page, is a lot of them. Some new and others, judging by the grooves and cracks present, a lot older.

"What is this?"

Turning to the last page and studying the pictures, all of them familiar and taken over the last six months, my need for an answer fades as I realize what I'm looking at.

Me. The greatest hits of Sunday Grayson. Then and now.

"Cole…"

"Before you say anything, take a look at this one." He says, twisting on the bed and grabbing the second album. Handing it over and smiling when I place it between us and open it. "It's the same subject with a twist."

It's easy to see what he means. This one, unlike the one before it, is all photos of us. Our time together. The greatest hits of Cole and Sunday.

"How did you…when did you? I— I don't even knew what to say." I stumble over my feelings at what I'm seeing. The books sitting in our laps and what's contained in them taking my breath away.

The same way he does.

"I've been planning on doing it since I was staying at the B&B." He explains. "I never had any pictures growing up, but with the way things were when I was little, I was thankful for that. Living with the nightmares was more than enough. I didn't want reminders from that time. But Sunday, it wasn't the same for you. You had those memories taken from you."

"That night when you admitted not having any baby pictures, how sad and lost you looked, it hurt me. I promised myself that no matter what happened in the end, whatever paths our lives took when the truth came out, I would give you this. I wanted to be able to give them back to you."

"You wanted to help me remember Sunday."

"Yes, but with the second book, I was selfish. I put that one together because I wanted you to remember me too."

We've talked about this before. The memories I've had of the time before I was taken and how they'll probably be the only ones I have. It was something we both just had to accept. Cole understanding that while the impact I made on his life as a child stayed with him long after I was gone, the same couldn't be said for me.

But in moments like the one we're in now, I really wish I could remember our time together so that I could give him back a small piece of what he's given me.

"There's nothing about any of this that's selfish, Cole. All I see is you wanting to do something nice. Give me something more tangible than just the memories of others. It's beautiful."

"This one," he says, pointing to a picture of what I now know is us. "Is my favorite."

"Why?"

"In order for it to make sense, you have to look at the picture directly to the right of it. The one of us now." Watching as my eyes move across the page and fall to the photo he's talking about, he points to his face. "Look at my eyes in this one. The way they're trained on you and not the camera. No look back to the one from when we were kids."

Doing as he says and studying both shots, it doesn't take long for me to see what he's getting at. While my expression is different in both shots, his isn't. It's exactly the same. His head is angled down, his eyes trained on the person in the photo with him. On me. The faintest trace of a smile on his lips, and nothing but adoration shining in his eyes.

Even at seven he only had eyes for me.

It was always meant to be Cole.

"You see it, don't you?"

"I do."

"I found you." He leans in whispering as memories of our first time flood to the surface inside me. The truth of those words even more meaningful than it was the first time around. "My one and only love."

Sighing as his lips brush lightly across mine, tempted to give into the tenderness of the moment but remembering the other book still lying closed in the middle of the bed, I softly kiss him back before pulling away long enough to grab it.

"If the first two albums are for us to remember, what's this one for?"

"Shit," he curses before standing and sliding his hand around to the back of his jeans, grabbing something from his back pocket and handing it over a few seconds later. "The last book. This is a part of it."

This photo he's holding out to me is different than all of the others. It's a Polaroid shot, and if that wasn't enough to set it apart from the ones that line the other albums, what's in it does.

Earlier at the tattoo parlour when Mark was finished with me, he'd set to work on Cole, but what had only taken a little over a half hour with me, seemed to go on far longer with him.

Now I'm getting to see the reason why.

Not only is there a new date on his chest, but there's something else surrounding it. His entire pectoral muscle covered in it.

A sun, complete with shaded in yellow rays cover all of the dates including the new one and what looks like tiny stars rest at the beginning and end of each date.

Taking what had warmed my heart originally and setting fire to it until I'm melting.

"Why didn't you tell me?"

"Because I thought it was better if I show you." He offers up before taking his place on the bed again. This time pulling me to him and once he's got me resting comfortable, lifting the third book into our lap. "Open it."

Doing as he says, again I'm met with the same two pictures we'd just been looking at in the last album, but with one subtle difference. There's a space between them this time. One that's just big enough to fit the picture in my hand.

Placing it into the spot, the stickiness of the paper locking the picture in position, I try and make sense of what it all means. What Cole could be trying to say having these three pictures resting the way they are.

"Three days after that first picture was taken, I told you a story. Almost eighteen years later when you thought I was sleeping, you told me one."

Wait, what? He heard the story?

"Yes, I heard every word you said." He laughs softly when my sharp intake of breath gives away my surprise.

"The tattoo is a reflection of that. The light in my life and the magic in yours." He explains, reaching out and tracing the stars with his fingers, brushing against mine as I do the same with the yellow rays. Meeting in the middle and curling into each other, the same way our bodies do whenever we're close.

"And when the time comes, there's enough space for two more dates." He continues, tapping the album with his hand as he lifts and turns the page. "Two dates that are the reason this third album exists."

"Two more?" I ask confused. *What more could he want to add?* Everything that means anything to him is already there.

"The day you become Sunday Alexander and the day our first child is born…that is, if we decide we want children." He backtracks, his eyes searching mine for some kind of confirmation that what he's said isn't too much too soon.

Something I'm not going to hesitate giving him.

Leaning over and placing the album down on the bed, I settle back into his arms. Turning my body into his, I lift my hand and running it down the side of his face, am amazed at how easily he gives himself over to me. His eyes closing and his body melting into mine almost the second we connect. Tracing a line down and along his jaw, I pause when I reach the edges of his mouth. As a soft breath falls when his lips part in anticipation, I lean in and press my lips to his. Kissing him as tenderly as he did me minutes before. Confirming with our closeness and the words that follow next, what he should have known all along.

It's always been him.

"Let's make some memories."

The End

Sunday's Story

Once upon a time, there was a girl. One that to the rest of the world appeared to be just like every other girl on the planet, but who deep inside, felt so very different.

This girl you see, didn't look like the rest of her family. From the day she was born, there was nothing about her that looked remotely like them. Her eyes brown to their blue, hair brown to their blonde. Short to their tall, soft to their hard, the differences continued to build until she was completely weighed down by them. Making her feel completely alone. Like a puzzle piece that didn't fit.

And it only got worse the older she became.

Her idea of fun, the things she enjoyed, even the way she thought, there was no one else quite like her, which only made the loneliness grow until she truly felt invisible.

It was in these moments of solitude, where she felt the most alone, in a world not of her making that she experienced her first taste of love.

The person she gave her heart to a character from one of her many stories.

Characters that she wished could be brought to life, but not for all. Just for her.

You see, this girl was a dreamer. She had no idea that magic like that didn't exist in real life. That it could only exist in the fictional realm.

Al she knew was that even if it took the rest of her life, she wouldn't rest until she found it.

Until she found the heart that beat to match her own.

A heart that after stepping out into the world and taking a chance, she finally believed she found. But by the time she realized how wrong it was, her heart had become invested and ultimately in the end, broken.

This girl, for all of the strength she had been told she held inside her was barely hanging on by a thread.

Pulling away from the world and her life, she welcomed invisibility. Losing a little bit more of herself each and every day until one day, months later as she prepared for another day of going through the motions, something changed.

She met him.

A beautifully broken man with piercing blue eyes and secrets he was too afraid to share.

He changed it.

Changed her.

Pulling her back from the edge of loneliness and the darkness of despair.

His heart—when he finally opened himself up and let her see deep inside it—the one she knew beat to match her own.

Bringing to life what she believed from the start.

Magic does exist.

You just need to surrender yourself to love in order to see it.

Remembering Sunday Playlist

The Crow & The Butterfly by *Shinedown*

Here & Now by *Seether*

Bloodstream by *Ed Sheeran*

She Is The Sunlight by *Trading Yesterday*

The Edge Of Tonight by *All Time Low*

Lover Killer by *My Brightest Diamond*

I Just Want You by *Ozzy Osbourne*

Set Me Free by *Griffin Peterson*

Let Her Go by *Passenger*

Pieces by *Sum 41*

Stormy by *Hedley*

With Me *by Sum 41*

The Truth by *Good Charlotte*

Incomplete by *Backstreet Boys*

Unwritten by *Natasha Bedingfield*

You're Beautiful by *James Blunt*

Only One in Color by *Trapt*

I Am Machine by *Three Days Grace*

Remembering Sunday by *All Time Low*

Acknowledgements

No book is ever a solitary effort. Sure, it may take you on a journey, but there are people that touch you along the way that take what starts as just an innocent idea and help turn it into something else. People that become just as big a part of the journey you as the writer take while writing as the characters themselves. That is no more apparent than with **Remembering Sunday**.

Without the people listed below, this book probably would have stayed hidden away in a file on my original laptop (I am now on laptop #3). So this is where I take a time out now that the story is done and thank those people for being such an integral part of my journey.

For believing in Cole and Sunday even when I didn't.

Pamela Sparkman

You were there right from the beginning. When I wasn't sure whether there was a story here at all. Almost a year later and a million conversations in the interim, there was and with your support it's now getting to see the light of day. I thank you, but more than that, Sunday and Cole thank you for believing in us.

Joey Winchester

Embarrassing my best friend in 3…2…1.
Not to get all emotional, but we both know that if it wasn't for you and everything that you do on a daily basis and will continue to do in the future, this wouldn't be happening right now. You've seen me at my best and seen me at my absolute worst and loved and supported me through it all. So as much as it embarrasses you, me thanking you this way, you're gonna have to suck it up buttercup. I will not let a book or day go by

without telling you just how much I appreciate you. I love you JJ and it's an always and forever kind of thing.

Chauna

Though the book was started a year ago, long before we ever met, it's because of you in the here and now that it's finished. You will never truly know how much it means to me, the hours you spent reading what I had before the Cole block of 2015 set in and believing in my ability to finish it. How thankful I am that I had you there in the background, rooting for Cole while also at the same rooting for me. Believing that I could finish this and do so in a memorable way. From the bottom of my heart, thank you for everything you did and everything you will do in the future. I treasure your friendship.

My Beta Readers

You guys are the best. Taking time out of your lives, and busy schedules to read over something I've written and giving me honest feedback (even when it hurts) so that in the end I could put out the best possible book for me and story for these characters. It means the world to me and I would be lost without you.

Bloggers

Before I decided that I wanted to try my hand doing what I've loved doing since I was a child, I had another love. Reading. As a kid, I never understood the power of reviews and what they meant, both to the author and to other readers, but now, having written 16 books and starting a blog of my own, I see it. I get it.

You are the heartbeat of the book world. Without you, we as authors wouldn't be able to do what we do. So from the bottom of my heart (and for all other authors out there), thank you for reaching out, supporting us (even if you end up not liking the books we write) and for doing what you do. I love and appreciate you all.

Anyone Taking a Chance on This Book

Well, anyone taking a chance on anything I write. Thank you. There are so many books, by so many wonderful authors out there in the world and the fact that you stopped on one of mine, bought it no less, it means the absolute world to me. The idea that someone out there can find enjoyment in my words, in the stories that for so long stayed buried in my head, well it's humbling. There will never be enough thank you's given, enough appreciation or enough declarations of love to explain entirely how much you taking the chance and then supporting me truly means. From the bottom of my very humbled heart…thank you all.

About The Author

Melyssa is a mother of four from Toronto, Ontario, Canada.

She's currently working on standalone title **Heroine**, as well as Luke Grayson's story, **Ready When You Are**, as well as the seventh book in the *Count On Me* Series, **Through the Storm**.

When she's not writing, you can find her buried under the covers with her portable DVD player, watching marathons of Supernatural and Veronica Mars. When those aren't available, she can be found curled up in a corner with her e-reader and a plethora of books, falling in love with characters written so well she deems them her book boyfriends and girlfriends. If you want to find her, check Facebook or Twitter (@WinchesterBooks) as she may just have an addiction to both. If those don't work you can always keep up with her progress on her personal site.

Made in United States
North Haven, CT
08 October 2023

42498839R00202